Computational Neuropharmacology

Scrivener Publishing
100 Cummings Center, Suite 541J
Beverly, MA 01915-6106

Publishers at Scrivener
Martin Scrivener (martin@scrivenerpublishing.com)
Phillip Carmical (pcarmical@scrivenerpublishing.com)

Computational Neuropharmacology

Fundamentals and Clinical Aspects

Edited by

Bhupendra Prajapati

Shree S.K. Patel College of Pharmaceutical Education and Research,
Ganpat University, Gujarat, India
Department of Industrial Pharmacy, Faculty of Pharmacy, Silpakorn University,
Nakhon Pathom, Thailand

Alok Tripathi

Era College of Pharmacy, Era University Lucknow, Uttar Pradesh, India

Rishabha Malviya

Department of Pharmacy, School of Medical and Allied Sciences,
Galgotias University, India

and

Lucy Mohapatra

Amity Institute of Pharmacy, Amity University, Noida, Uttar Pradesh, India

Scrivener Publishing

WILEY

This edition first published 2025 by John Wiley & Sons, Inc., 111 River Street, Hoboken, NJ 07030, USA and Scrivener Publishing LLC, 100 Cummings Center, Suite 541J, Beverly, MA 01915, USA

For more information about Scrivener publications please visit www.scrivenerpublishing.com.

Wiley Global Headquarters
111 River Street, Hoboken, NJ 07030, USA

For details of our global editorial offices, customer services, and more information about Wiley products visit us at www.wiley.com.

Library of Congress Cataloging-in-Publication Data

ISBN 978-1-394-24244-3

Front cover image courtesy of Adobe Firefly
Cover design by Russell Richardson

Set in size of 11pt and Minion Pro by Manila Typesetting Company, Makati, Philippines

Printed in the USA

10 9 8 7 6 5 4 3 2 1

Contents

Foreword

The book *Computational Neuropharmacology: Fundamentals and Clinical Aspects* explores the synergies between computational science and neuropharmacology. The human brain, with its complexity and ambiguous functions, has long confounded scientists and clinicians, and traditional pharmacological treatments often fall short in treating neurological and psychiatric disorders. Computational neuropharmacology offers more precise, efficient, and personalized therapeutic options.

The book, edited by Dr. Rishabha Malviya and his esteemed team, is a significant contribution to neuropharmacology, offering a comprehensive guide to understanding how computational methods are revolutionizing approaches to neuroscience and pharmacology. It begins with fundamental principles and progresses to specialized topics, providing insights into computational techniques that are transforming drug development, theoretical neuroscience, clinical applications, and psychiatric treatment approaches.

The book's interdisciplinary approach bridges the gaps between computer science, pharmacology, and neuroscience, providing unique and invaluable information. The integration of molecular docking, machine learning, and in silico modeling demonstrates how these tools collectively enhance the rational design of neuropharmacological agents.

The future of neuropharmacology depends on the seamless integration of computational techniques with traditional methodologies. This book celebrates the remarkable progress made so far and offers a glimpse into the fascinating possibilities that lie ahead. Through this book, researchers can anticipate significant breakthroughs that will enhance their ability to diagnose, treat, and ultimately understand neurological and psychiatric disorders.

This book is more than just an academic publication; it serves as a resource for the future and a source of encouragement. The editors and contributors have developed an excellent resource through this book that will undoubtedly affect the course of neuropharmacology for years to come.

I welcome this opportunity to congratulate Dr. Rishabha Malviya and his team for their remarkable achievement.

Shri Mahendra Patel
Senior General Manager & Head
R&D
Aculife Healthcare Private Ltd (Nirma Ltd)
Ahmedabad, India

Preface

This book is a detailed exploration of the current state and prospects of computational techniques in neuropharmacology. Its objective is to provide a complete overview of the computational tools currently utilized in the field, spanning from molecular modeling to clinical applications. Beginning with a discussion of the fundamental principles of computational neuropharmacology, it continues into specific chapters, including theoretical neuroscience, in silico modeling, and the use of computational approaches in counseling psychology and psychiatry.

The book's interdisciplinary approach integrates neuroscience, pharmacology, and computer science, offering researchers, physicians, and students a comprehensive perspective. Its chapters on drug discovery for specific neurological disorders—such as Parkinson's disease, epilepsy, Alzheimer's disease, autism spectrum disorder, depression, and anxiety—emphasize personalized approaches and innovative methodologies being developed to address these complex medical conditions. Another key chapter explores the integration of molecular docking and machine learning, providing powerful tools for identifying and optimizing medicinal molecules. Additionally, the book includes thorough case studies and practical examples to help readers acquire the knowledge and skills needed to apply these technologies in their own research and clinical practice.

Computational neuropharmacology will continue to play an important role in identifying and developing novel medications for neurological and psychiatric illnesses. The editors thank the contributing authors for their expertise and dedication, as well as the readers for their interest and participation. In brief, *Computational Neuropharmacology: Fundamentals and Clinical Aspects* is more than just a collection of chapters; it reflects the persistent search for a better understanding of the brain and its interactions with pharmaceutical drugs.

The editors are grateful to the reviewers who have contributed to improving the quality of the book through their constructive comments. The editors also thank Martin Scrivener and Scrivener Publishing for their support and publication.

Part 1

FUNDAMENTALS OF COMPUTATIONAL NEUROPHARMACOLOGY

1

Basic Principles of Computational Neuropharmacology: Neuroscience Meeting Pharmacology

Lucy Mohapatra[1], Alok S. Tripathi[2*], Deepak Mishra[1], Alka[1,3] and Sambit Kumar Parida[4]

[1]Amity Institute of Pharmacy, Lucknow, Amity University Uttar Pradesh, Noida, Uttar Pradesh, India
[2]Era College of Pharmacy, Era University, Lucknow, Uttar Pradesh, India
[3]Faculty of Pharmaceutical Sciences, Rama University, Mandhana, Kanpur (Uttar Pradesh), India
[4]Amity Institute of Pharmacy, Amity University Rajasthan, RIICO Kant Kalwar Industrial Area, Jaipur-Delhi Highway (Main Road), Jaipur, Rajasthan, India

Abstract

The graded features of the brain are protruding in the pharmacologic handling of psychiatric ailments, chiefly directing the receptors that outspread mounting to inherent connectivity within various sections of the brain, inter-provincial connectivity and clinical annotations, including electroencephalogram. The identification and depiction of impending pharmacologic targets in neurology and psychiatry is an elementary enigma at the connection linking medicinal chemistry and the field of neurosciences. Breathtaking novel procedures including proteomics and genomics have encouraged brisk advance, prologue copious inquiries as to the operational importance of drug receptor interactions. A cohesive understanding of neuropharmacological negotiators necessitates linking the breach amid their molecular mechanisms and the biophysical determining factors of neural role as it has been observed that numerous psycho and neuronal active medications characteristically slog in nerve cells by distressing several facets of electrophysiological activities. Computational neuropharmacology lays off a key character in

**Corresponding author*: shloksk@gmail.com

Bhupendra Prajapati, Alok Tripathi, Rishabha Malviya and Lucy Mohapatra (eds.) Computational Neuropharmacology: Fundamentals and Clinical Aspects, (3–30) © 2025 Scrivener Publishing LLC

this operable correlation. Vigorous numerical simulations have been established communicating all foremost dynamic rind stuffs under endogenic and exogenic chemical control in the brain. These primarily incorporate voltage-dependent ion channels, GPCRs and neurohumoral transmissions. This chapter describes neuropharmacology from the computational viewpoint and delivers persuasive illustrations of how several compartmental simulations can clarify, elucidate, and forecast the consequence of neuro chemical agonists and antagonists in the brain. It also highlights the recent advancements in the field of computational neuropharmacology to enable understanding the mechanism of action of neuropharmacological intervention on neurones.

Keywords: Neuropharmacology, computational neuroscience, neuroimaging, neurotransmitters, compartmental simulations

Abbreviations

5-HT	5-hydroxytryptamine
Ach	Acetylcholine
ADD	Attention deficit disorder
ADHD	Attention deficit hyperactivity disorder
ALS	Amyotrophic lateral sclerosis
AMPA	α-amino-3-hydroxy-5-methyl-4-isoxazolepropionic acid
CNS	Central nervous system
DA	Dopamine
EEG	Electroencephalography
fMRI	Magnetic resonance imaging
GABA	Gamma-aminobutyric acid.
K^+	Potassium
MEG	Magnetoencephalography
NA	Noradrenaline
Na^{+2}	Sodium
QSAR	Quantitative Structure-Activity Relationships
QS	Quantitative systems pharmacology
REM	Rapid-eye-movement
SBD	Sleep behavior disorder

1.1 Introduction

Computational neuroscience has the potential to leverage our increasing knowledge about the brain to understand and address brain malfunctions in diseases [1]. In recent years, the lines between imaging and computational neuroscience have blurred, with many practitioners in fMRI and electrophysiology adopting computational approaches due to the statistical and computational skills required for analyzing neuroimaging data [2, 3]. Two key areas in computational neuroscience relevant to neuroimaging are brain activity modelling, which seeks to explain understanding, behaviour, and cognition, as well as biophysical models of neural dynamics, which are inspired by biophysics [4]. Brain diseases affect around 16 million adult Americans, leading to cognitive and behavioural impairments such as memory and motor control issues. As life expectancy increases, the prevalence of this neuropathologist also rises, posing potential social and economic challenges [5]. While pharmaceutical interventions are the most often used therapies for neuropsychiatric patients, numerous neurological and mental illnesses entail unique biological processes, localized parts of the nervous system, and different cellular types [6, 7]. Neuropharmacology aims to understand how therapeutic interventions impact the nervous system, but drugs can have dynamic effects on an organism's functional phenotypes and may cause side effects. To enhance patient outcomes for neurological illnesses, it is crucial to integrate biomedical and healthcare data and establish treatment pathways supported by precision medicine [8]. Neuroinformatics, Bioinformatics, and Computational Chemistry perform an important function in this context. A special issue of Modern Neuropharmacology focuses on these topics, presenting research and review papers that range from computational and statistical analyses to clinical investigations [9, 10]. The issue covers various subjects, including computer docking investigations, comparative pharmacological assessments, artificial intelligence, molecular modelling, and structure-based drug development as shown in Figure 1.1 [11]. It includes research on neuro-disorders such as Alzheimer's disease, schizophrenia, Parkinson's disease, depression, epilepsy, dementia, migraine, stroke, Niemann–Pick type C disease, rapid-eye-movement (REM) sleep behavior disorder (SBD), amyotrophic lateral sclerosis (ALS), and Huntington's disease, highlighting the importance of computational techniques in the research and advancement of therapeutics [11, 12].

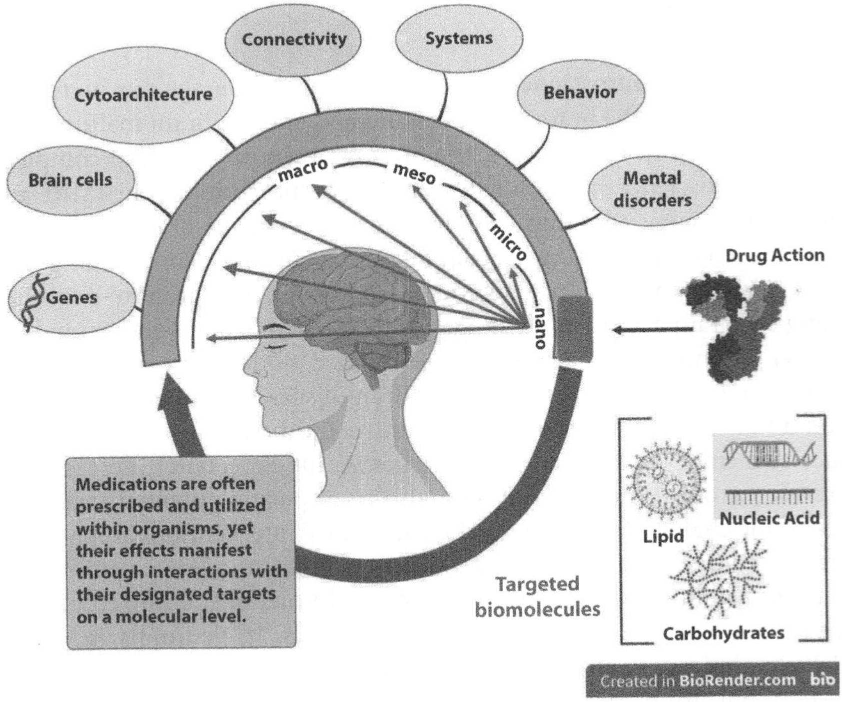

Figure 1.1 Different targeted approach used in computational neuropharmacology. The different aspects of Neuropharmacology are covered such as brain cells, cytoarchitecture, behavior and mental disorder in which targeted biomolecules can show action by using a computational approach.

1.2 Basics of Computational Neuropharmacology

Advancements in computational models have the potential to significantly benefit medicinal chemistry, a field that is already extensively utilizing computational techniques in fundamental neuroscience. As we get a greater comprehension of molecular mechanisms and their interactions with biophysical factors influencing neural activity, the process of discovering and characterizing new pharmaceutical are neurological and psychiatric are going to become more effective and successful in its goals [13]. The future of this functional relationship between computational neuroscience and bioinformatics looks promising. One misconception that needs to be addressed is the belief that computer simulations rely on arbitrary assumptions that cannot be verified by the modeler. Most computational models are built on implicit assumptions that underlie the Scientists everywhere use conceptual frameworks. Developing a novel computer model can be a fascinating scientific endeavor, aiming to explore potential explanations for complex experimental data that might otherwise be challenging to understand [14]. Early versions of these models can generate hypotheses that predict how specific situations, such as drug exposure, might affect individuals, even before such predictions have been experimentally validated. When experimental data becomes available for these scenarios, the model can be validated or refined using the new information, leading to the formulation of new scientific hypotheses that can be rigorously tested [15]. Although constructing a completely predictive model for neural processes remained a challenging assignment due to their intricacy and the continuous discovery of new mechanisms, the process of building these models often hints at the existence of novel mechanisms. Importantly, when a model's predictions are confirmed by experimental results, its applicability for neuropharmacology is enhanced, providing valuable insights for drug development and treatment strategies [16]. Table 1.1. depicts the interactions among drugs and the various neurotransmitters on which they act upon.

Table 1.1 Interactions between drugs and neurotransmitter receptors.

G-Protein family	Brand name of the drug	Region of the brain	Drug action	Condition	References
Acetylcholine (Ach)					
Muscarinic 1/5 (Gq)	Donepezil	Cortex, hippocampus	Breakdown inhibitor	Alzheimer	[17]
Nicotinic*2	HP-184	Hippocampus	Release stimulator (also for 5-HT)	Spinal cord injury	[18]
	Nicotine	Thalamus, striatum, anterior cingulate cortex	Agonist	Smoking cessation	[19]
	Philanthotoxin	Muscles cells, cortex	Noncompetitive antagonist	Neurodegenerative disease, cognitive disorder	[20]

(*Continued*)

Table 1.1 Interactions between drugs and neurotransmitter receptors. (*Continued*)

G-Protein family	Brand name of the drug	Region of the brain	Drug action	Condition	References
Dopamine (DA)					
d1/5 (Gs)	Adrogolide	Ventral tegmental area, limbic forebrain, nucleus acubens	Agonist	Cocaine addiction	[21]
	Bupropion (Zyban, WELLBUTRIN)	Frontal cortex, hippocampus, striatum, thalamus	Reuptake inhibitor	Smoking cessation, depression	[22]
	Haloperidol (Haldol)	Mesocotex, limbic system, nigrostriatal pathway	Antagonist (also for d2)	Antipsychotic	[23]
	Methylphenidate (Ritalin)	Prefrontal cortex, hippocampus, striatum, hypothalamus	Reuptake inhibitor	ADD, ADHD	[24]
	Reserpine	Striatum, frontal cortex	Ruptake inhibitor (also for NE and 5HT)	Antipsychotic	[25]
Gamma-aminobutyric acid (GABA)					
GABAA/C[*3]	Diazepam (Valium, Diastat)	Cortex, telencephalon, hypothalamus	Agonist	Anxiety, epilepsy	[26]
	Methohexital (Brevital Sod)	Thalamus, hippocampus	Agonist	Anestetic	[27]
	Sodium valproate (Depakote)	Occipital, parietal, and frontal cortex, accumbens, substantia nigra	Breakdown inhibitor (also NE, DA, and 5-HT modulator)	Mania, migraine, epilepsy, GEFS+ (generalized epilepsy with febrile seizures plus)	[28]
	Topiramate (Topamax)	Prefrontal and occipital cortex	Agonist (Na+ blocker)	Epilepsy, SMEI (severe myoclonic epilepsy of infancy)	[29]

(*Continued*)

Table 1.1 Interactions between drugs and neurotransmitter receptors. (*Continued*)

G-Protein family	Brand name of the drug	Region of the brain	Drug action	Condition	References
GABAB (Gi)	Baclofen (Lioresal)	Hypothalamus, hippocampus/ DG, frontal cortex	Agonist	Antispastic agent for multiple sclerosis.	[29]
5-hydroxytryptamine (5-HT)					
5-HT1A, B,D (Gi/o)	Eletriptan (Relpax)	Trigeminal sensory nerve-endings, intracranial blood vessels	Agonist	Migraine	[30]
	Fluoxetine (Prozac)	Dentate gyrus	Reuptake inhibitor	Depression, obsessive-compulsive disorder	[31]
	Vilazodone	Cortex, hippocampus	Partial Agonist (SSRI)	Depression [51, 52]	[32]
5-HT2A,B,C (Gq/11)	Quetiapine (Seroquel)	Prefrontal cortex, striatum, limbic system, anterior pituitary gland	Antagonist	Bipolar disorder, schizophrenia	[33]
	TGBA01AD	NA	Reuptake inhibitor (5HT2/1A agonist)	Depression	[34]
5-HT5A (Gi/o)	SGS-518	N/A	Antagonist	Schizophrenia	[35]
5-HT4A, B (Gs)	Benzoate	Frontal cortex, striatum, hippocampus	Antagonist (also for 5HT3)	Neurode-generative diseases	[36]
5-HT6 (Gq/s)	Olanzapine (Zyprexa)	Frontal cortex, hippocampus	Antagonist (also for 5HT2, DA)	Schizophrenia	[37]
5-HT3 [*1]	Mirtazapine (Remeron)	Central amygdala, anterior insula, septum	Antagonist	Depression	[38]

(*Continued*)

Table 1.1 Interactions between drugs and neurotransmitter receptors. (*Continued*)

G-Protein family	Brand name of the drug	Region of the brain	Drug action	Condition	References
Glutamic acid (Glu)					
AMPA*1	S-18986	Frontal cortex, dorsal hippocampus	Modulator	Neurode-generation, cardiovascular ischemia, cognitive disorder	[39]
	Talampanel	Amygdala	Noncompetitive antagonist	Epilepsy, Parkinson	[40]
mGluR1,5 (Gq) mGluR2,3, 4,6,7,8 (Gi/o)	Lamotrigine (Lamictal)	Striatum, thalamus, cortex, hippocampus, amygdala	Release inhibitor (Ca2+ and Ih blocker)	Epilepsy, bipolar disorder.	[41]
	LY354740	Striatum, hippocampus, cortex, cerebellum	Agonist	Anxiety	[18]
	MPEP	Striatum, thalamus, cortex, amygdala	Antagonist (also for AMPA)	Epilepsy, pain, neurode-generative diseases, anxiety	[42]
Kainate*1	LY293558	Hippocampus, lateral and medial habenulae, superior and inferior colli	Antagonist (also for AMPA)	Postoperative pain, anxiety	[43]
NMDA 84] *2	Aptiganel	Hippocampus, cortex	Antagonist	Neurode-generation, Parkinson, brain injury, cardiovascular ischemia	[44]
	Ifenprodil	Hippocampus/ CA3, piriform cortex, amygdala, striatum	Noncompetitive antagonist (Ca2+ blocker)	Schizophrenia	[45]
	TAN-950A	Hippocampus/ CA1	Agonist	Neurodegenera-tive disease [65, 66]	[46]

(*Continued*)

Table 1.1 Interactions between drugs and neurotransmitter receptors. (*Continued*)

G-Protein family	Brand name of the drug	Region of the brain	Drug action	Condition	References
Noradrenaline (NA)					
α1A,B,D (Gq) α2A,B,C(Gs) β1,2,3 (Gs)	Atomoxetine (Strattera)	Caudate putamen, cortico (prefrontal ventral and orbital)-baso thalamic loop	Release inhibitor	ADHD	[47]
	Dextroamphetamine (Dexedrine)	Insular cortex	Reuptake inhibitor (also for DA and 5-HT)	ADHD	[48]
	DOV 21947	Striatum, meso-corticolimbic, cerebrospinal fluid	Reuptake inhibitor (also for DA and 5-HT)	Depression	[49]

All receptors are G-coupled unless noted with* (ionotropic); *1 Monovalent cations; *2 Mono/divalent cations; *3 Chloride.

1.3 Multiple Aspects of Computational Neuropharmacology

Computational psychiatry is a developing field that bridges the gap between psychiatric and computational neuroscience. Computational neuroscience is a discipline of neuroscience that uses theoretical investigation, mathematical representations, and brain abstractions to comprehend the principles underlying the progression, structure, physiology, affective, and cognitive processes of the nervous system. The different aspects of computational neuropharmacology include drug databases, computational fMRI, compartmental simulations of neuronal electrophysiology along with quantitative systems pharmacology and others has been discussed in this section.

1.3.1 Drug Databases

As neuropharmacology expands, medicinal chemists face challenges in gathering comprehensive information on significant molecules within their areas of interest. Given that new drugs target specific brain regions, it is crucial to understand the systems these medications collaborate with each other, as well as their modes of action. Formalizing this information would be extremely beneficial to experimental and theoretical researchers associated with drug discovery, advancement, and interactions [50]. The Neuroscience Database Gateway provides relevant resources, involving datasets of synaptic proteins, GPCRs, and disease-specific neuroinformatic archives and tools. Currently, medicinal chemistry and biomedical literature organize available data primarily by medicine (molecular structure) or disease (pathological state). However, for computational neuropharmacology, a more effective approach would be to group information based on underlying sub-cellular processes [51]. This strategy would allow researchers to explore medication effects at the level of essential chemical components, cell types, and brain regions, enabling the application of new models and simulations with references to quantitative findings in the literature. For instance, the Brain Pharm database focuses on compounds acting on neuronal receptors and signal transduction pathways in both healthy and unhealthy brains, facilitating research on medications for various neurological disorders [52]. This database serves as a source of information that can aid in creating computational neuroscience models relevant to medicinal chemistry. By leveraging this knowledge, researchers can apply it to different categories of dynamical models to demonstrate its potential. The first category involves single-neuron models with compartmentalized electrophysiology and membrane biophysics. The second category comprises computer simulations of enzyme pathways in subcellular compartments of a single neuron. Utilizing this information within these models can offer valuable insights into drug effects and their mechanisms of action, bridging the gap between computational neuroscience and medicinal chemistry as shown in Table 1.2 [53].

1.3.2 Computational fMRI

In this section, we present an approach for describing fMRI data using computational models of brain activity and provide an illustrative example highlighting key concepts. Over the course of the past five years, innovative data illustrating philosophy has emerged, moving away from simply modelling noticed neurological signals about experimental variables (e.g., conventional ANOVA models) and towards explaining the data in terms of quantities that the brain must encode, focused on simplified assumptions about brain function [54]. This paradigm often rests on the idea that the brain seeks to optimize various processes. Some researchers propose that perception aims to maximize the reciprocal information among sensory signals and cognitive representations of their sources or minimize the prediction error [55, 56]. Similarly, for

Table 1.2 Drugs that communicate with ionic channels [18].

Ionic channel	Family	Drug name	Mechanism of action	Region of brain involved	Neurological condition
Potassium	K_{dr} (noninactivating)	Flindokalner	Stimulation	N/A	Cerebrovascular ischemia
	K_{ir} (inward rectifying)	Spironolactone	K+-sparing diuretic	Lateral septum, hippocampus	Andersen-Tawil syndrome
	K_M (muscarinic dependent)	DMP-543	Inhibitor (Ach release enhancer)	Hippocampus	Alzheimer's, epilepsy
	K_{ATP}	BAY-X-9227	Stimulation	CNS neurons	Neurodegenerative disease
Sodium	Nap (slowly inactivating)	Topiramate	Inhibitor (GABA-A agonist)	Hippocampus, hypothalamus	Migraine prevention, epilepsy
	$Na_{v\alpha 1.1/1.9}$	ADCI	Inhibitor (NMDA antagonist)	Hippocampus	Neurodegenerative disease, anticonvulsant
		Lithium	Alteration of nerve Na+ transport	Hypothalamus	Bipolar disorder, mood stabilizer

(*Continued*)

Table 1.2 Drugs that communicate with ionic channels [18]. (*Continued*)

Ionic channel	Family	Drug name	Mechanism of action	Region of brain involved	Neurological condition
Mixed	Hyperpolarization-activated (NA+ , K+)	Lamotrigine	Shifts the inactivation curve	Striatum, thalamus, cortex, hippocampus, amygdala	Epilepsy, bipolar disorder
Acetylcholine	Muscarinic1/5 (Gq)	Donepezil	Breakdown inhibitor	Hippocampus	Alzheimer
	Nicotinic	HP-184	Release stimulator (also for 5-HT)	Hippocampus	Spinal cord injury
		Nicotine	Agonist	Thalamus	Smoking cessation
GABA	GABAB (Gi)	Baclofen	Agonist	Hypothalamus	Antispastic agent for multiple sclerosis
Nor-adrenaline	α1A,B,D (Gq) α2A,B,C (Gs) α1,2,3 (Gs)	DOV 21947	DA and 5-HT reuptake inhibitor	Cerebrospinal fluid	Depression
		Atomoxetine	Release inhibitor	Caudate putamen	ADHD

motor coordination, many cost functions have been proposed, which the brain strives to minimizes throughout activities [57]. A classic example of this kind of technique is the use of encouraging prediction errors in the development of fMRI regressors. The reward prediction error is a significant latent variable in models of optimal control and reinforcement learning, capturing the discrepancy between anticipated and actual rewards [58, 59]. The value function, representing the expected reward, guides the best behavioural strategy once the mistake in reward estimation is minimized. The underlying functional anatomy of value acquisition along with the brain underpinnings of reward in prediction have been widely researched., emphasizing brain regions like the ventral striatum [60, 61]. A crucial aspect of computational fMRI is the ability to compare various computer models using actual fMRI results. This enables researchers to discern differences in the neurophysiological implementations of different computational approaches and their corresponding neural correlates of assumed mental processes [62]. Additionally, refining neuromeric functions, which translate computational variables to observable neural responses, helps establish a precise mapping from computing modifications to representations of neural networks. To demonstrate how to get psychological states and utilized to explain neuronal activity, we provide a small example in the following paragraphs [63].

1.3.3 Compartmental Simulations of Neuronal Electrophysiology

Ion channels present in complex dendritic arborizations enable neurons to combine and generate electrical impulses. The changes in dendritic membrane potential over space and time can be estimated through the solution of partially differential equations. However, the realistic complexity of dendrites, including branching, varying diameters, and multiple non-linear ion channels, makes analytical solutions impractical [64]. To model neurons computationally, researchers use a collection of interconnected, tiny, cylindrical compartments, each representing an isopotential unit with membrane resistance and capacitance, as well as interior resistance or conductivity. These compartments serve as numerical representations of electric circuits. Precise neuronal morphologies are imported into established software environments like NEURON and Genesis from digital reconstructions of actual cells scanned through microscopy [65]. By incorporating membrane and synaptic features, such as the kinetics of voltage- and ligand-gated ionic channels, throughout the arborization of the nerves, researchers can mimic neuronal behavior and match experimental results or test new theories. In this computational framework, researchers can also simulate the impact of medications on neuronal activity by altering neurotransmitter receptors and electric currents. This approach allows for a detailed and flexible exploration of neuronal dynamics and responses to various pharmacological interventions [66].

1.3.4 The Concept of System Thinking Makes an Appearance in Neuropharmacology

It is essential to recognize that most modern medications used in neurology and psychiatry interact with multiple targets, and this is a consequence of their historical origins. The conventional animal screening method for central nervous system (CNS) drug development is no longer a viable or acceptable option for several compelling reasons [67]. Besides the low productivity and moral concerns for animal welfare, there are considerable inter-species discrepancies in gene expression levels of several proteins important for neuropathologies and medication distribution in the brain among rats and humans. These distinctions influence extrapolations from animal models to human brain diseases [68]. Furthermore, the inability of animals to communicate hampers their ability to accurately mimic human neuropsychiatric illnesses, making it challenging to model symptoms such as hallucinations, sadness, or headaches effectively. To overcome these limitations and enhance translation from animal research to the human clinic, the field of systems pharmacology has been growing [69]. Systems pharmacology aims to eliminate barriers by using computer modeling to create multi-target medications that can effectively act on various molecular entities, thereby restoring the disrupted interconnected network caused by specific diseases. This approach seeks to develop drugs from the ground up, designed to target multiple components and pathways, making them more effective in treating complex neurological and psychiatric disorders [70].

1.3.5 Quantitative Systems Pharmacology

The emerging paradigm of quantitative systems pharmacology (QSP) builds upon neurophysiology, pathology, and genomics data used in preclinical animal models but places a strong emphasis on human data. QSP aims to be a valuable translational tool by mathematically modelling actual biological mechanisms in the appropriate medical region, as well as combining preclinical neurophysiology with clinical information on pathology, visualization, and clinical scales [71]. In neuro-psychopharmacology, it is well-established that medications affecting the behavior and functional state of patients with neuro-psychiatric disorders primarily work by altering the effects of synaptic mediator combinations. This aligns well with the systems pharmacology approach, which does not solely focus on one neurotransmitter, as seen in target-oriented drug development programs [72]. Instead, it aims to simultaneously optimize various inhibitory and facilitator effects on multiple targets. While this may seem challenging compared to a single-target approach, a considerable number of computational methodologies, bioinformatics, and cheminformatics resources have been established and are rapidly evolving to support this systemic approach. A consistent systems biology-based drug discovery method for neuropharmacology could

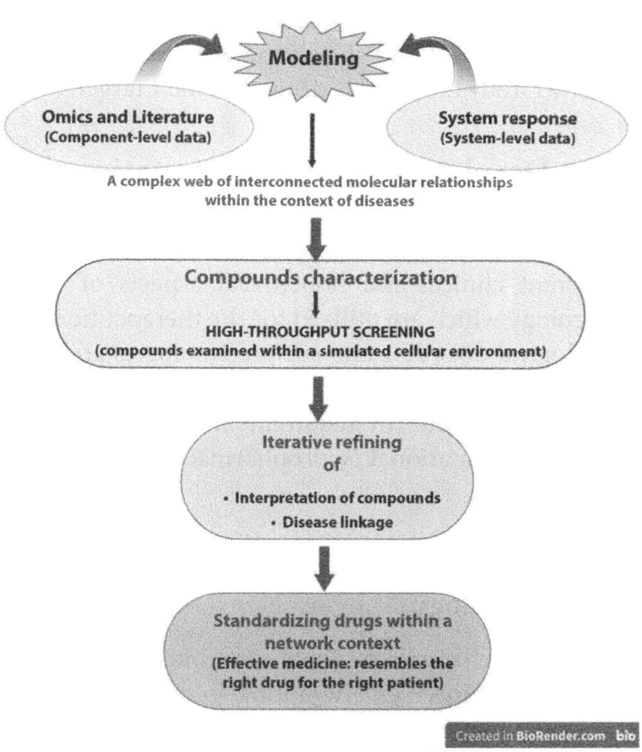

Figure 1.2 Use of computational models for the study of drugs and their targets. The figure depicts the way of selecting the correct drug for the individual patients by using various computational techniques in which the literature and system response is taken into consideration. The compound characterization is done by the help of High throughput screening where the compounds are examined inside a simulated cellular environment where they need to exhibit their action. Then, the iterative refining is done for the compounds as per the disease that it needs to target are linked together. At last, the standardization of drug is achieved where the effective medicines are finalized for the patient as per their requirement and condition.

involve several key phases [73]. The first step is to get complete information of the molecular interaction network underlying the specific illness state. This involves delving into the precise biology of the neuropsychiatric disorder at the cellular and circuit level before selecting appropriate targets and developing tailored therapeutic approaches to treat the condition. Multiple approaches that enable integrated biological techniques, such as transcriptomics, proteomics, and metabolomics, can be used to detect and characterize pathway responses in healthy versus sick brain states. Computational tools are then used to find connections in massive biological databases [74]. By integrating human data, computational models, and cutting-edge biological techniques, quantitative systems of pharmacology offer promising avenues for advancing drug discovery and development in the field of neuropharmacology. Figure 1.2. has shown the way to use the computational models in order to study the various drugs and their target.

1.4 Recent Developments in Computational Neuropharmacology

There are different clinical and nonclinical aspects of Computational Neuropharmacology which are utilized for the therapeutic intervention of different neurological disorders and such technologies are utilized for the future perspective and development of drug regimen for safe and effective use of medicines as per the requirement of the patients. The recent developments in Computational Neuropharmacology has been discussed further in this section.

1.4.1 Compartmental Simulations of Neuronal Electrophysiology

Neurons rely on ion channels to enable the integration and generation of electrical impulses. However, attempting to calculate the intricate and dynamic variation in dendritic membrane potential across space and time purely through solving partial differential equations proves impractical due to the intricate dendritic branching, varying diameters, and the presence of multiple nonlinear ion channels [64]. The estimation of dendritic membrane potential, a pragmatic approach involves conceptualizing the neuron as an assembly of interconnected, minute, uniform cylindrical compartments. In a mathematical framework, these distinct "entities" can be represented by an electrical circuit that comprises both internal characteristics such as longitudinal resistance or conductivity, as well as exterior characteristics such as membrane resistance and capacitance. This depiction takes both the internal and exterior qualities of these units into consideration [65, 66]. This model draws inspiration from specific neuronal structures derived from digitally reconstructed images of actual cells obtained through microscopic scans. These neuronal models are then imported into well-established software platforms tailored for neuron simulation and modeling [18]. The computational model incorporates digitally reconstructed dendritic morphology [75], and integrates key ion currents, including a sodium current, extended rectification potassium currents, A-type potassium currents, and a mixed sodium/potassium current that responds to excessive polarization of the membrane. This comprehensive model has been evaluated against select experimental observations [76].

1.4.2 Quantitative Structure–Activity Relationships (QSAR) for Alzheimer's Disease Treatment

Alzheimer's disease poses a significant a public health problem by impairing cognitive functions in patients [77]. To address this issue, Babita and colleagues have harnessed chemoinformatics techniques, exemplifying their use in uncovering the physicochemical attributes of inhibitors for acetylcholine esterase [78]. This investigation has the potential to lay the foundation for future research endeavours aimed at exploring the efficacy of these compounds in preclinical models [79].

1.4.3 Composite Machine Learning Algorithms for Schizophrenia Treatment

The essential inhibitory neurotransmitter gamma-aminobutyric acid (GABA) holds pivotal importance. Sahila *et al.* used machine learning, computational chemistry, and phytochemistry to identify the chemical characteristics of natural inhibitors aimed at reducing the morbidity and death rates associated with schizophrenia [80]. The pursuit of natural substances as a means of alleviating and managing symptoms offers a pioneering avenue for addressing the intricacies of this neurological affliction.

1.4.4 Discovery and Evaluation of Dual Target Ligands for Parkinson's Disease

Perez-Castillo and associates present a groundbreaking endeavor in structural bioinformatics, employing mathematical methodologies to derive docking scores that elucidate protein-ligand interactions targeting not one, but two distinct protein targets [81–83]. The generation and assessment of these docking scores hold significant importance in the realm of structure-based drug discovery. This innovative and rigorous study is poised to unlock novel pathways for assessing therapeutic targets that could concurrently influence pleiotropic protein entities, thus enhancing the potential for drug repurposing [84, 85].

1.4.5 Utilizing Structure-Based Drug Design for Neurological Disorder Therapies

The amalgamation of extensive molecular class-specific information has emerged as an asset in the development of models and algorithms for forecasting are elucidating intricate mechanisms like 3D domain switching, a hallmark of neurological disorders such as Alzheimer's and Parkinson's diseases [86–88]. Recent endeavors aiming to integrate genetic variances and pharmacological target data have unveiled numerous potential therapeutic links within the domain of neurological disorders [89–91]. A comprehensive overview of contemporary advancements in neurological drug discovery is presented by Aarthy *et al.* This review probes into the realm of neurological ailments, harnessing the power of structural bioinformatics and cheminformatics methodologies [92]. Diverse neurological conditions, including Alzheimer's disease, Niemann–Pick type C ailment, REM-SBD, ALS, epilepsy, dementia, migraine, and stroke, are meticulously examined. The review adeptly intertwines clinical, biochemical, and bioinformatics approaches, thereby furnishing a valuable resource for students, researchers, and clinicians seeking holistic insights into the multifaceted landscape of drug development [92].

1.4.6 Structural Modeling of Voltage-Gated Sodium Ion Channel from *Anopheles gambiae*

Despite sustained global endeavors to combat mosquito-borne infections and the illnesses they propagate, there remains an imperative to identify efficacious vector control strategies [93]. A spectrum of maladies spread via mosquitoes, including malaria, dengue fever, West Nile virus, encephalitis, and Zika fever, manifests neurological repercussions [94, 95]. Gaining insights into the molecular intricacies of vector proteins holds pivotal significance for the creation of repellents, vector-targeting agents, and diverse chemical interventions aimed at curtailing mosquito-borne viral spread [96, 97].

1.5 Limitations of Computational Neuropharmacology

While theory-driven modelling techniques have provided valuable insights into the underlying processes and mechanisms of various illnesses across different research levels, their application has predominantly centralized around clinical issues [98, 99]. Nevertheless, these computational methods come with significant limitations. A prominent drawback is their demand for substantial expertise, often rendering them perplexing and bewildering to nonspecialists [100]. Consequently, there exists a challenge in effectively sharing pertinent information among medical professionals, trial practitioners, experimental researchers, and theoreticians. A potential solution lies in accentuating the utility of computational techniques through their active integration within clinical trials [101, 102]. The convergence of theory-driven and data-driven approaches holds promise, and indeed, the amalgamation of these techniques can yield tangible benefits in practical applications. When the existing knowledge and mechanistic comprehension of a disorder are suitably robust, theory-driven methods offer the capacity to evaluate specific facets relevant to that ailment [103]. This serves to notably curtail data complexity by narrowing down the dataset to pivotal features and parameters. Subsequently, data-driven techniques, often rooted in machine learning, can be applied to this streamlined dataset with heightened dependability and efficiency. This hybrid methodology has been previously deployed in investigating conditions like Huntington's disease [104] and schizophrenia, culminating in noteworthy enhancements in the categorization of behaviour and subtypes within these disorders. Thus, a computational model emerges as a formidable instrument, transcending mere theory, and holds the potential for integration with clinical tools for patient assessment [105]. For the attainment of a fully applicable model, it becomes imperative to harness each computational approach while mitigating their individual constraints and complexities. Indeed, the intricate nature of mental disorders and the existing gap between biological and phenomenological understanding can be gained from a multi-tier strategy that seamlessly integrates diverse modes of simulation and modelling [106]. This entails a top-down, high-level approach focused on phenotype and symptoms, in conjunction with a bottom-up, biologically oriented model at a granular level.

1.6 Conclusion

Neuropharmacology addresses seriously insufficient medical resources, with neuro-psychic illnesses becoming more common and prevalent in an ageing community, even though inadequate medicinal product is available for serious medical conditions such as stroke and most neurodegenerative diseases, especially Alzheimer's dementia. Moreover, current neuro-drugs simply reduce or, at most, eradicate some neurologic/psychiatric symptoms without addressing the underlying causes of the disease, which in the case of most neuro-psychic diseases are frequently unknown. This is primarily due to neuroscience's strong reductionist focus, which has resulted in precise information about the cellular level but only a far poorer grasp of integrative brain functions, which are emergent qualities displayed exclusively at the organismic level. We have shown in this study how computational neuroscience may employ formal models at various levels to give a mechanistic and functional perspective for interpreting psychopathology and its underlying pathophysiology. Different novel models need a high level of competence in a variety of disciplines, including cellular and molecular neuroscience, network neuroscience, cognitive neuroscience, computational neuroscience, psychiatry, psychology, computer science, mathematics, and engineering. This contributes to the vital multidisciplinary topic of computational psychiatry.

References

1. Obermayer, K., Computational Neuroscience and Modeling of Diseases: Do We Need New Paradigms. *Basic Clin. Neurosci.*, 3, 3, 3–4, 2012 Jul 10.
2. Duman, R.S., Neuropharmacology in the next millennium: promise for breakthrough discoveries. *Neuropsychopharmacology*, 20, 2, 97–8, 1999 Feb.
3. Gong, S., Sheng, P., Jin, H., He, H., Qi, E., Chen, W., Dong, Y., Hou, L., Effect of methylphenidate in patients with cancer-related fatigue: a systematic review and meta-analysis. *PLoS One*, 9, 1, e84391, 2014 Jan 8.
4. Green, A.R. and Aronson, J.K., An agenda for UK clinical pharmacology: From basic to clinical neuropharmacology: targetophilia or pharmacodynamics? *Br. J. Clin. Pharmacol.*, 73, 6, 959–67, 2012 Jun.
5. Hill, R.G., Neuropharmacology of the injured spinal cord. *Spinal Cord*, 25, 3, 209–11, 1987 Jun.
6. Margineanu, D.G., Neuropharmacology beyond reductionism–a likely prospect. *Biosystems*, 141, 1–9, 2016 Mar 1.

7. Bielinski, S.J., Olson, J.E., Pathak, J., Weinshilboum, R.M., Wang, L., Lyke, K.J., Ryu, E., Targonski, P.V., Van Norstrand, M.D., Hathcock, M.A., Takahashi, P.Y., Preemptive genotyping for personalized medicine: design of the right drug, right dose, right time—using genomic data to individualize treatment protocol. *Mayo Clin. Proc.*, 89, 1, 25–33, 2014 Jan 1, Elsevier.
8. Overby, C.L., Rasmussen, L.V., Hartzler, A., Connolly, J.J., Peterson, J.F., Hedberg, R.E., Freimuth, R.R., Shirts, B.H., Denny, J.C., Larson, E.B., Chute, C.G., A template for authoring and adapting genomic medicine content in the eMERGE infobutton project. *AMIA Annu. Symp. Proc.*, 2014, 944, 2014, American Medical Informatics Association.
9. Tenenbaum, J.D., Translational bioinformatics: past, present, and future. *Genom. Proteom. Bioinf.*, 14, 1, 31–41, 2016 Feb 1.
10. Polgár, T. and Keseru G, M., Integration of virtual and high throughput screening in lead discovery settings. *Comb. Chem. High Throughput Screen.*, 14, 10, 889–97, 2011 Dec 1.
11. Bajorath, J., Integration of virtual and high-throughput screening. *Nat. Rev. Drug Discov.*, 1, 11, 882–94, 2002 Nov 1.
12. Shameer, K., Nayarisseri, A., Duran, F.X., González-Díaz, H., Improving neuropharmacology using big data, machine learning and computational algorithms. *Curr. Neuropharmacol.*, 15, 8, 1058, 2017 Nov.
13. Migliore, M. and Shepherd, G.M., Emerging rules for the distributions of active dendritic conductances. Nat. Rev. Neurosci., 3, 5, 362–70, 2002 May.
14. Hastings, H.M., Sobel, S.G., Chaterpaul, S., Frank, C., Russell, E., Pekor, J., Chemical amplifiers, in: *Computational Fluid and Solid Mechanics 2003*, pp. 1705–1707, Hempstead-NY, Elsevier Science Ltd, 2003 Jan 1.
15. Hines, M.L. and Carnevale, N.T., The NEURON simulation environment. *Neural Comput.*, 15, 9, 6, 1179–209, 1997 Aug 15.
16. Bower, J.M. and Beeman, D., *The book of GENESIS: exploring realistic neural models with the GEneral NEural SImulation System*, Springer Science & Business Media, Boulder, USA, 2012 Dec 6.
17. Fujiki, M., Kobayashi, H., Uchida, S., Inoue, R., Ishii, K., Neuroprotective effect of donepezil, a nicotinic acetylcholine-receptor activator, on cerebral infarction in rats. *Brain Res.*, 1043, 236–41, 2005.
18. Ferrante, M., Blackwell, K.T., Migliore, M., Ascoli, G.A., Computational models of neuronal biophysics and the characterization of potential neuropharmacological targets. *Curr. Med. Chem.*, 15, 24, 2456–71, 2008 Oct 1.
19. Levin, E.D., McClernon, F.J., Rezvani, A.H., Nicotinic effects on cognitive function: behavioral characterization, pharmacological specification, and anatomic localization. *Psychopharmacology*, 184, 523–39, 2006 Feb.
20. Anis, N., Sherby, S., Goodnow, R., Niwa, M., Konno, K., Kallimopoulos, T., Bukownik, R., Nakanishi, K., Usherwood, P., Eldefrawi, A., Structure-activity relationships of philanthotoxin analogs and polyamines on

N-methyl-D-aspartate and nicotinic acetylcholine receptors. *J. Pharmacol. Exp. Ther.*, 254, 3, 764–73, 1990 Sep 1.
21. Foster, J.D., Cervinski, M.A., Gorentla, B.K., Vaughan, R.A., Regulation of the dopamine transporter by phosphorylation, in: *Neurotransmitter Transporters*, pp. 197–214, 2006.
22. Vann, R.E., Rosecrans, J.A., James, J.R., Philibin, S.D., Robinson, S.E., Neurochemical and behavioral effects of bupropion and mecamylamine in the presence of nicotine. *Brain Res.*, 1117, 1, 18–24, 2006 Oct 30.
23. Juckel, G., Schlagenhauf, F., Koslowski, M., Wüstenberg, T., Villringer, A., Knutson, B., Wrase, J., Heinz, A., Dysfunction of ventral striatal reward prediction in schizophrenia. *Neuroimage*, 29, 2, 409–16, 2006 Jan 15.
24. Reneman, L., De Bruin, K., Lavalaye, J., Gunning, W.B., Booij, J., Addition of a 5-HT receptor agonist to methylphenidate potentiates the reduction of [123I] FP-CIT binding to dopamine transporters in rat frontal cortex and hippocampus. *Synapse*, 39, 3, 193–200, 2001 Mar 1.
25. O'Leary, O.F., Bechtholt, A.J., Crowley, J.J., Hill, T.E., Page, M.E., Lucki, I., Depletion of serotonin and catecholamines block the acute behavioral response to different classes of antidepressant drugs in the mouse tail suspension test. *Psychopharmacology*, 192, 357–71, 2007 Jun.
26. Vargas, M.L., Abella, C., Hernandez, J., Diazepam increases the hypothalamic-pituitary-adrenocortical (HPA) axis activity by a cyclic AMP-dependent mechanism. *Br. J. Pharmacol.*, 133, 8, 1355–61, 2001 Aug.
27. Zhang, L., Zhang, Y., Wennberg, R., Multiple actions of methohexital on hippocampal CA1 and cortical neurons of rat brain slices. *J. Pharmacol. Exp. Ther.*, 286, 3, 1177–82, 1998 Sep 1.
28. Jansen, J.F., Aldenkamp, A.P., Majoie, H.M., Reijs, R.P., de Krom, M.C., Hofman, P.A., Kooi, M.E., Nicolay, K., Backes, W.H., Functional MRI reveals declined prefrontal cortex activation in patients with epilepsy on topiramate therapy. *Epilepsy Behav.*, 9, 1, 181–5, 2006 Aug 1.
29. Obrietan, K. and van den Pol, A., GABAB receptor-mediated regulation of glutamate-activated calcium transients in hypothalamic and cortical neuron development. *J. Neurophysiol.*, 82, 1, 94–102, 1999 Jul 1.
30. Sandrini, G., Perrotta, A., Nappi, G., Eletriptan: a review and new perspectives. *Expert Rev. Neurother.*, 6, 10, 1413–21, 2006 Oct 1.
31. Gourion, D., Perrin, E., Quintin, P., Fluoxetine: an update of its use in major depressive disorder in adults. *L'encephale*, 30, 4, 392–9, 2004 Jul 1.
32. Hughes, Z.A., Starr, K.R., Langmead, C.J., Hill, M., Bartoszyk, G.D., Hagan, J.J., Middlemiss, D.N., Dawson, L.A., Neurochemical evaluation of the novel 5-HT1A receptor partial agonist/serotonin reuptake inhibitor, vilazodone. *Eur. J. Pharmacol.*, 510, 1–2, 49–57, 2005 Mar 7.
33. Gefvert, O., Lundberg, T., Wieselgren, M., Bergström, M., Långström, B., Wiesel, F.A., Lindström, L., D2 and 5HT2A receptor occupancy

of different doses of quetiapine in schizophrenia: a PET study. *Eur. Neuropsychopharmacol.*, 11, 2, 105–10, 2001 Apr 1.

34. Szarfman, A., Tonning, J.M., Levine, J.G., Doraiswamy, P.M., Atypical antipsychotics and pituitary tumors: a pharmacovigilance study. *Pharmacother. J. Hum. Pharmacol. Drug Ther.*, 26, 6, 748–58, 2006 Jun.
35. Gaster, L.M. and King, F.D., Serotonin 5-HT3 and 5-HT4 receptor antagonists. *Med. Res. Rev.*, 17, 2, 163–214, 1997 Mar.
35. Yan, Z., Regulation of GABAergic inhibition by serotonin signaling in prefrontal cortex: molecular mechanisms and functional implications. *Mol. Neurobiol.*, 26, 203–16, 2002 Oct.
37. Montgomery, S.A., Baldwin, D.S., Blier, P., Fineberg, N.A., Kasper, S., Lader, M., Lam, R.W., Lépine, J.P., Möller, H.J., Nutt, D.J., Rouillon, F., Which antidepressants have demonstrated superior efficacy? A review of the evidence. *Int. Clin. Psychopharmacol.*, 22, 6, 323–9, 2007 Nov 1.
38. Carrard, A., Elsayed, M., Margineanu, M., Boury-Jamot, B., Fragnière, L., Meylan, E.M., Petit, J.M., Fiumelli, H., Magistretti, P.J., Martin, J.L., Peripheral administration of lactate produces antidepressant-like effects. *Mol. Psychiatry*, 23, 2, 392–9, 2018 Feb.
39. Lee, H.J., Pogatzki-Zahn, E.M., Brennan, T.J., The effect of the AMPA/kainate receptor antagonist LY293558 in a rat model of postoperative pain. *J. Pain*, 7, 10, 768–77, 2006 Oct 1.
40. Hoyte, L., Barber, P.A., Buchan, A.M., Hill, M.D., The rise and fall of NMDA antagonists for ischemic stroke. *Curr. Mol. Med.*, 4, 2, 131–6, 2004 Mar 1.
41. Poolos, N.P., Migliore, M., Johnston, D., Pharmacological upregulation of h-channels reduces the excitability of pyramidal neuron dendrites. *Nat. Neurosci.*, 5, 8, 767–74, 2002 Aug 1.
42. Slassi, A., Isaac, M., Edwards, L., Minidis, A., Wensbo, D., Mattsson, J., Nilsson, K., Raboisson, P., McLeod, D., Stormann, T.M., Hammerland, L.G., Recent advances in non-competitive mGlu5 receptor antagonists and their potential therapeutic applications. *Curr. Top. Med. Chem.*, 5, 9, 897–911, 2005 Aug 1.
43. Jones, N., O'Neill, M.J., Tricklebank, M., Libri, V., Williams, S.C., Examining the neural targets of the AMPA receptor potentiator LY404187 in the rat brain using pharmacological magnetic resonance imaging. *Psychopharmacology*, 180, 4, 743–51, 2005 Aug.
44. Kroppenstedt, S.N., Schneider, G.H., Thomale, U.W., Unterberg, A.W., Neuroprotective Properties of Aptiganel HCL (Cerestat©) following Controlled Cortical Impact Injury, in: *Intracranial Pressure and Neuromonitoring in Brain Injury: Proceedings of the Tenth International ICP Symposium, Williamsburg, Virginia, May 25–29, 1997*, pp. 114–116, Springer Vienna, Berlin, Germany, 1998.
45. Berger, M.L. and Rebernik, P., Zinc and Ifenprodil Allosterically Inhibit Two Separate Polyamine-Sensitive Sites atN-Methyl-d-Aspartate Receptor Complex. *J. Pharmacol. Exp. Ther.*, 289, 3, 1584–91, 1999 Jun 1.

46. Tamura, N., Iwama, T., Itoh, K., Synthesis and glutamate-agonistic acitivity of (S)-2-amino-3-(2, 5-dihydro-5-oxo-3-isoxazolyl)-propanoic acid derivatives. *Chem. Pharm. Bull.*, 40, 2, 381–6, 1992 Feb 25.
47. Pattij, T. and Vanderschuren, L.J., The neuropharmacology of impulsive behaviour. *Trends Pharmacol. Sci.*, 29, 4, 192–9, 2008 Apr 1.
48. Schwarz, A.J., Gozzi, A., Reese, T., Bifone, A., *In vivo* mapping of functional connectivity in neurotransmitter systems using pharmacological MRI. *Neuroimage*, 34, 4, 1627–36, 2007 Feb 15.
49. Skolnick, P. and Basile, A.S., Triple reuptake inhibitors ("broad spectrum" antidepressants). *CNS Neurol. Disord. Drug Targets (Formerly Curr. Drug Targets-CNS Neurol. Disorders)*, 6, 2, 141–9, 2007 Apr 1.
50. Gardner, D. and Shepherd, G.M., A gateway to the future of neuroinformatics. *Neuroinformatics*, 2, 271–4, 2004 Sep.
51. Zhang, W., Zhang, Y., Zheng, H., Zhang, C., Xiong, W., Olyarchuk, J.G., Walker, M., Xu, W., Zhao, M., Zhao, S., Zhou, Z., SynDB: a Synapse protein DataBase based on synapse ontology. *Nucleic Acids Res.*, 35, suppl_1, D737–41, 2007 Jan 1.
52. Elefsinioti, A.L., Bagos, P.G., Spyropoulos, I.C., Hamodrakas, S.J., A database for G proteins and their interaction with GPCRs. *BMC Bioinf.*, 5, 1, 1–8, 2004 Dec.
53. Lam, H.Y., Marenco, L., Clark, T., Gao, Y., Kinoshita, J., Shepherd, G., Miller, P., Wu, E., Wong, G.T., Liu, N., Crasto, C., AlzPharm: integration of neurodegeneration data using RDF. *BMC Bioinf.*, 8, 1–2, 2007 May.
54. Haruno, M. and Kawato, M., Heterarchical reinforcement-learning model for integration of multiple cortico-striatal loops: fMRI examination in stimulus-action-reward association learning. *Neural Netw.*, 19, 8, 1242–54, 2006 Oct 1.
55. Mumford, D., On the computational architecture of the neocortex: II The role of cortico-cortical loops. *Biol. Cybern.*, 66, 3, 241–51, 1992 Jan.
56. Rao, R.P. and Ballard, D.H., Predictive coding in the visual cortex: a functional interpretation of some extra-classical receptive-field effects. *Nat. Neurosci.*, 2, 1, 79–87, 1999 Jan.
57. Murray, S.O., Kersten, D., Olshausen, B.A., Schrater, P., Woods, D.L., Shape perception reduces activity in human primary visual cortex. *Proc. Natl. Acad. Sci.*, 99, 23, 15164–9, 2002 Nov 12.
58. Todorov, E. and Jordan, M.I., Smoothness maximization along a predefined path accurately predicts the speed profiles of complex arm movements. *J. Neurophysiol.*, 80, 2, 696–714, 1998 Aug 1.
59. Kakade, S. and Dayan, P., Dopamine: generalization and bonuses. *Neural Netw.*, 15, 4-6, 549–59, 2002 Jun 1.
60. Shadmehr, R. and Krakauer, J.W., A computational neuroanatomy for motor control. *Exp. Brain Res.*, 185, 359–81, 2008 Mar.
61. Daw, N.D. and Doya, K., The computational neurobiology of learning and reward. *Curr. Opin. Neurobiol.*, 16, 2, 199–204, 2006 Apr 1.

62. O'Doherty, J.P., Hampton, A., Kim, H., Model-based fMRI and its application to reward learning and decision making. *Ann. N. Y. Acad. Sci.*, 1104, 1, 35–53, 2007 May.
63. Rangel, A., Camerer, C., Montague, P.R., A framework for studying the neurobiology of value-based decision making. *Nat. Rev. Neurosci.*, 9, 7, 545–56, 2008 Jul.
64. Abbott, L.F., Farhi, E., Gutmann, S., The path integral for dendritic trees. *Biol. Cybern.*, 66, 1, 49–60, 1991 Nov.
65. Ascoli, G.A., Mobilizing the base of neuroscience data: the case of neuronal morphologies. *Nat. Rev. Neurosci.*, 7, 4, 318–24, 2006 Apr 1.
66. Migliore, M., Morse, T.M., Davison, A.P., Marenco, L., Shepherd, G.M., Hines, M.L., ModelDB: making models publicly accessible to support computational neuroscience. *Neuroinformatics*, 1, 135–9, 2003 Mar.
67. López-Muñoz, F., Baumeister, A.A., Hawkins, M.F., Álamo, C., The role of serendipity in the discovery of the clinical effects of psychotropic drugs: beyond of the myth. *Actas Esp. Psiquiatr.*, 40, 1, 34–42, 2012 Jan 1.
68. Talevi, A., Bellera, C.L., Di Ianni, M., Gantner, M., Bruno-Blanch, L.E., Castro, E.A., CNS drug development-lost in translation? *Mini-Rev. Med. Chem.*, 12, 10, 959–70, 2012 Sep 1.
69. Margineanu, D.G., Systems biology, complexity, and the impact on antiepileptic drug discovery. *Epilepsy Behav.*, 38, 131–42, 2014 Sep 1.
70. Margineanu, D.G., Systems (network) pharmacology for brain functionality restoration, in: *Homeostatic Control of Brain Function*, p. 231, 2015 Nov 3.
71. Geerts, H., Spiros, A., Roberts, P., Carr, R., Has the time come for predictive computer modeling in CNS drug discovery and development? *CPT: Pharmacometrics Syst. Pharmacol.*, 1, 11, 1–4, 2012 Nov.
72. Geerts, H., Spiros, A., Roberts, P., Carr, R., Quantitative systems pharmacology as an extension of PK/PD modeling in CNS research and development. *J. Pharmacokinet. Pharmacodyn.*, 40, 257–65, 2013 Jun.
73. Geerts, H., Roberts, P., Spiros, A., Potkin, S., Understanding responder neurobiology in schizophrenia using a quantitative systems pharmacology model: application to iloperidone. *J. Psychopharmacol.*, 29, 4, 372–82, 2015 Apr.
74. Emmett, M.R., Kroes, R.A., Moskal, J.R., Conrad, C.A., Priebe, W., Laezza, F., Meyer-Baese, A., Nilsson, C.L., Integrative biological analysis for neuropsychopharmacology. *Neuropsychopharmacology*, 39, 1, 5–23, 2014 Jan.
75. Ascoli, G.A., Donohue, D.E., Halavi, M., NeuroMorpho. Org: a central resource for neuronal morphologies. *J. Neurosci.*, 27, 35, 9247–51, 2007 Aug 29.
76. Teoh, H., Fowler, L.J., Bowery, N.G., Effect of lamotrigine on the electrically-evoked release of endogenous amino acids from slices of dorsal horn of the rat spinal cord. *Neuropharmacology*, 34, 10, 1273–8, 1995 Oct 1.

77. Brookmeyer, R., Gray, S., Kawas, C., Projections of Alzheimer's disease in the United States and the public health impact of delaying disease onset. *Am. J. Public Health*, 88, 9, 1337–42, 1998 Sep.
78. Rodda, J., Morgan, S., Walker, Z., Are cholinesterase inhibitors effective in the management of the behavioral and psychological symptoms of dementia in Alzheimer's disease? A systematic review of randomized, placebo-controlled trials of donepezil, rivastigmine and galantamine. *Int. Psychogeriatr.*, 21, 5, 813–24, 2009 Oct.
79. Pulikkal, B.P., Marunnan, S.M., Bandaru, S., Yadav, M., Nayarisseri, A., Sureshkumar, S., Common SAR Derived from Linear and Non-linear QSAR Studies on AChE Inhibitors used in the Treatment of Alzheimer's Disease. *Curr. Neuropharmacol.*, 15, 8, 1093–9, 2017 Nov 1.
80. Marunnan, S.M., Pulikkal, B.P., Jabamalairaj, A., Bandaru, S., Yadav, M., Nayarisseri, A., Doss, V.A., Development of MLR and SVM aided QSAR models to identify common SAR of GABA uptake herbal inhibitors used in the treatment of Schizophrenia. *Curr. Neuropharmacol.*, 15, 8, 1085–92, 2017 Nov 1.
81. Malhotra, S., Mathew, O.K., Sowdhamini, R., DOCKSCORE: a webserver for ranking protein-protein docked poses. *BMC Bioinf.*, 16, 1, 1–6, 2015 Dec.
82. Kaushik, S. and Sowdhamini, R., Structural analysis of prolyl oligopeptidases using molecular docking and dynamics: insights into conformational changes and ligand binding. *PLoS One*, 6, 11, e26251, 2011 Nov 23.
83. Kanagarajadurai, K., Malini, M., Bhattacharya, A., Panicker, M.M., Sowdhamini, R., Molecular modeling and docking studies of human 5-hydroxytryptamine 2A (5-HT2A) receptor for the identification of hotspots for ligand binding. *Mol. Biosyst.*, 5, 12, 1877–88, 2009.
84. Hodos, R.A., Kidd, B.A., Shameer, K., Readhead, B.P., Dudley, J.T., *In silico* methods for drug repurposing and pharmacology. *Wiley Interdiscip. Rev. Syst. Biol. Med.*, 8, 3, 186–210, 2016 May.
85. Shameer, K. and Readhead BDudley, J.T., Computational and experimental advances in drug repositioning for accelerated therapeutic stratification. *Curr. Top. Med. Chem.*, 15, 1, 5–20, 2015 Jan 1.
86. Shameer, K. and Sowdhamini, R., Functional repertoire, molecular pathways and diseases associated with 3D domain swapping in the human proteome. *J. Clin. Bioinf.*, 2, 1, 1–6, 2012 Dec.
87. Shameer, K., Shingate, P.N., Manjunath, S.C., Karthika, M., Pugalenthi, G., Sowdhamini, R., 3DSwap: curated knowledgebase of proteins involved in 3D domain swapping. *Database*, 2011, bar042, 2011 Jan 1.
88. Shameer, K., Pugalenthi, G., Kumar Kandaswamy, K., Sowdhamini, R., 3dswap-pred: prediction of 3D domain swapping from protein sequence using Random Forest approach. *Protein Pept. Lett.*, 18, 10, 1010–20, 2011 Oct 1.
89. Shameer, K., Pugalenthi, G., Kandaswamy, K.K., Suganthan, P.N., Archunan, G., Sowdhamini, R., Insights into protein sequence and structure-derived

features mediating 3D domain swapping mechanism using support vector machine based approach. *Bioinf. Biol. Insights*, 4, BBI–S4464, 2010 Jan.
90. Xu, K., Schadt, E.E., Pollard, K.S., Roussos, P., Dudley, J.T., Genomic and network patterns of schizophrenia genetic variation in human evolutionary accelerated regions. *Mol. Biol. Evol.*, 32, 5, 1148–60, 2015 May 1.
91. Ruderfer, D.M., Charney, A.W., Readhead, B., Kidd, B.A., Kähler, A.K., Kenny, P.J., Keiser, M.J., Moran, J.L., Hultman, C.M., Scott, S.A., Sullivan, P.F., Polygenic overlap between schizophrenia risk and antipsychotic response: a genomic medicine approach. *Lancet Psychiatry*, 3, 4, 350–7, 2016 Apr 1.
92. Aarthy, M., Panwar, U., Selvaraj, C., Singh, S.K., Advantages of structure-based drug design approaches in neurological disorders. *Curr. Neuropharmacol.*, 15, 8, 1136–55, 2017 Nov 1.
93. Griffith, K.S., Lewis, L.S., Mali, S., Parise, M.E., Treatment of malaria in the United States: a systematic review. *Jama*, 297, 20, 2264–77, 2007 May 23.
94. Pisarev, V.B., Butenko, A.M., Petrov, V.A., Grigor'eva, N.V., Larichev, V.F., Hepatotropic action of the virus of West Nile fever. *Vopr. Virusol.*, 50, 1, 37–8, 2005 Jan 1.
95. Pewter, S.M., Huw Williams, W., Haslam, C., Kay, J.M., Neuropsychological and psychiatric profiles in acute encephalitis in adults. *Neuropsychol. Rehabil.*, 17, 4–5, 478–505, 2007 Aug 1.
96. Manoharan, M., Ng Fuk Chong, M., Vaïtinadapoulé, A., Frumence, E., Sowdhamini, R., Offmann, B., Comparative genomics of odorant binding proteins in Anopheles gambiae, Aedes aegypti, and Culex quinquefasciatus. *Genome Biol. Evol.*, 5, 1, 163–80, 2013 Jan 1.
97. Sparks, J.T., Bohbot, J.D., Dickens, J.C., Olfactory disruption: toward controlling important insect vectors of disease. *Prog. Mol. Biol. Transl. Sci.*, 130, 81–108, 2015 Jan 1.
98. Huys, Q.J., Maia, T.V., Frank, M.J., Computational psychiatry as a bridge from neuroscience to clinical applications. *Nat. Neurosci.*, 19, 3, 404–13, 2016 Mar.
99. Mohammadi, M.R., Khaleghi, A., Nasrabadi, A.M., Rafieivand, S., Begol, M., Zarafshan, H., EEG classification of ADHD and normal children using non-linear features and neural network. *Biomed. Eng. Lett.*, 6, 66–73, 2016 May.
100. Khaleghi, A., Sheikhani, A., Mohammadi, M.R., Nasrabadi, A.M., Vand, S.R., Zarafshan, H., Moeini, M., EEG classification of adolescents with type I and type II of bipolar disorder. *Australas. Phys. Eng. Sci. Med.*, 38, 551–9, 2015 Dec.
101. Khaleghi, A., Mohammadi, M.R., Shahi, K., Nasrabadi, A.M., Computational neuroscience approach to psychiatry: a review on theory-driven approaches. *Clin. Psychopharmacol. Neurosci.*, 20, 1, 26, 2022 Feb 2.
102. Moeini, M., Khaleghi, A., Mohammadi, M.R., Characteristics of alpha band frequency in adolescents with bipolar II disorder: a resting-state QEEG study. *Iran. J. Psychiatry*, 10, 1, 8, 2015.

103. Khaleghi, A., Birgani, P.M., Fooladi, M.F., Mohammadi, M.R., Applicable features of electroencephalogram for ADHD diagnosis. *Res. Biomed. Eng.*, 36, 1–1, 2020 Mar.
104. Mohammadi, M.R. and Khaleghi, A., Transsexualism: A different viewpoint to brain changes. *Clin. Psychopharmacol. Neurosci.*, 16, 2, 136, 2018 May.
105. Wiecki, T.V., Antoniades, C.A., Stevenson, A., Kennard, C., Borowsky, B., Owen, G., Leavitt, B., Roos, R., Durr, A., Tabrizi, S.J., Frank, M.J., A computational cognitive biomarker for early-stage Huntington's disease. *PLoS One*, 11, 2, e0148409, 2016 Feb 12.
106. Brodersen, K.H., Deserno, L., Schlagenhauf, F., Lin, Z., Penny, W.D., Buhmann, J.M., Stephan, K.E., Dissecting psychiatric spectrum disorders by generative embedding. *Neuroimage Clin.*, 4, 98–111, 2014 Jan 1.

2

Neuropharmacology in the Molecular Epoch

Neelakanta Sarvashiva Kiran[1], Chandrashekar Yashaswini[1], and Bhupendra G. Prajapati[2,3*]

[1]*Department of Biotechnology, School of Applied Sciences, REVA University, Bengaluru, Karnataka, India*
[2]*Shree. S. K. Patel College of Pharmaceutical Education and Research, Ganpat University, Kherva, Gujarat, India*
[3]*Department of Industrial Pharmacy, Faculty of Pharmacy, Silpakorn University, Nakhon Pathom, Thailand*

Abstract
Through the development of medications for neurologic illnesses during the last 50 years, the field of pharmacology known as neuropharmacology has progressively developed. Neuropharmacology has gained a solid foundation because of discoveries in integral neurosciences, notably those involving receptors and neurotransmitters, and various novel medications have been developed as a result. The last decade of the 20th century marked the beginning of the molecular neuropharmacology period as molecular biology began to influence neuropharmacology. Approximately, 1% of most traditional drugs can cross the blood–brain barrier and reach the central nervous system, which prevents many diseases of the central nervous system, including stroke, neurodegenerative diseases, and brain tumors, from having successful therapies. The creation of innovative methods for delivering drugs is a subject of significant curiosity and ongoing study because the majority of therapies for illnesses of the central nervous system are unsuccessful. Future treatments may be more efficient and safer because of a number of emerging approaches that show potential for efficient central nervous system medication delivery. Improved medication delivery to the central nervous system is crucial

**Corresponding author:* bhupen27@gmail.com; ORCID: https://orcid.org/0000-0001-8242-4541
Neelakanta Sarvashiva Kiran: ORCID: https://orcid.org/0000-0002-2404-3254
Chandrashekar Yashaswini: ORCID: https://orcid.org/0000-0001-7415-748X

Bhupendra Prajapati, Alok Tripathi, Rishabha Malviya and Lucy Mohapatra (eds.) Computational Neuropharmacology: Fundamentals and Clinical Aspects, (31–66) © 2025 Scrivener Publishing LLC

because it would enable enhanced control of CNS disorders and result in better treatments for those suffering from CNS diseases. In this chapter, we provide an in-depth understanding of various drug delivery strategies, including evaluation of potential therapeutic uses, restrictions, its mechanism of action, and instances of ongoing clinical and preclinical research.

***Keywords*:** Neuropharmacology, molecular biology, blood–brain barrier, neurodegenerative, drug delivery

List of Abbreviations

3D	Three Dimensional
5-HT	Serotonin
AEDs	Anti-Epileptic Drugs
AI	Artificial Intelligence
AMPA	Alpha-Amino-3-Hydroxy-5-Methyl-4-Isoxazolepropionic Acid
cAMP	Cyclic Adenosine Monophosphate
CNS	Central Nervous System
CRISPR	Clustered Regularly Interspaced Short Palindromic Repeats
DAG	Diacylglycerol
DNA	Deoxyribonucleic Acid
GABA	Gamma Aminobutyric Acid
GPCRs	G Protein Coupled Receptors
HPLC	High Performance/Pressure Liquid Chromatography
IP3	Inositol Triphosphate
LGIC	Ligand Gated Ion Channels
mAChR	Muscarinic Acetylcholine Receptor
MRI	Magnetic Resonance Imaging
nAChR	Nicotinic Acetylcholine Receptor
NMDA	N-Methyl-d-Aspartate
NMR	Nuclear Magnetic Resonance
NSAIDs	Nonsteroidal Anti-Inflammatory Drugs
PET	Positron Emission Tomography
PKA	Protein Kinase A
RNA	Ribonucleic Acid
RNAi	Ribonucleic Acid Interference
SPECT	Single-Photon Emission Computed Tomography
SSRIs	Selective Serotonin Reuptake Inhibitors

2.1 Introduction

The intriguing and important field of pharmacology known as neuropharmacology explores the complex interactions between medications and the nervous system. At its foundation, it is the study of how different chemicals, or "drugs," interact with neurons to impact brain activity and behaviour. Information is sent through the nervous system, which is made up of the central and peripheral nervous systems, through intricate networks of specialized cells called neurons. Neuropharmacologists seek to comprehend how medications affect these neurons, control neurotransmission, and alter mental activity [1].

The idea of neurotransmission, or the process by which neurons interact with other cells and with one other through chemical messengers called neurotransmitters, is crucial to this area. These neurotransmitters are implicated in a wide variety of neurological and psychiatric illnesses and are crucial for many different brain processes and behaviours [1, 2]. Serotonin, dopamine, norepinephrine, acetylcholine, and γ-aminobutyric acid [GABA] are important neurotransmitters having distinct roles and effects on the brain. Drugs known as neuropharmacological agents work on the nervous system to alter certain neurotransmitter concentrations, receptor activities, or neuronal functions. These medications are essential in the treatment of a wide range of neurological and psychiatric disorders, including schizophrenia, epilepsy, depression, and Alzheimer's disease. Researchers can create pharmaceuticals that are more effective and have fewer adverse effects by better understanding how these compounds interact with neurons and their targets. Additionally, the study of receptors is crucial to neuropharmacology [3, 4]. Drugs can interact with the particular receptors that neurotransmitters attach to on the surface of neurons, increasing or suppressing their activity. Researchers can create novel therapeutic strategies by learning more about the processes behind medication action by examining these interactions [5].

Another topic covered by neuropharmacology is how repeated drug use can result in tolerance, dependence, and addiction. Understanding the neurochemical underpinnings of addiction aids in the development of methods to prevent drug usage and enhance the results of therapy for those suffering from substance dependency. In summary, neuropharmacology is a crucial area of research that unites pharmacology and neurology. Neuropharmacologists work towards the creation of new treatments for neurological and mental illnesses by examining how pharmaceuticals interact with the nervous system [2, 3, 6]. Their work advances our knowledge of how the brain works and creates new opportunities for enhancing the quality of life for people dealing with diverse neurological conditions.

2.2 History of Neuropharmacology

The field of neuropharmacology has a long, rich history that has seen incredible growth and development [7]. The field has roots in ancient cultures when opium and other herbal remedies were utilized to numb pain and bring about altered states of awareness. However, real scientific investigation into how medications affect the neurological system started in the 19th century as chemistry and physiology advanced. The basis for more organized studies into the nervous system's reaction to certain substances was established by the discovery of morphine in 1804 and the separation of additional alkaloids from plants [8].

Significant turning points in the development of contemporary neuropharmacology occurred throughout the 20th century. The evolution of the area was considerably aided by improvements in methodologies like electrophysiology and our understanding of neurotransmitters and their receptors. Acetylcholine's identification as a neurotransmitter by Otto Loewi in the 1920s proved the chemical transmission of nerve impulses and offered significant new information on neural communication. Future investigations into the role of neurotransmitters were enabled by this important discovery [9]. Chlorpromazine, the first antipsychotic medication, was introduced in the 1940s, ushering in a significant advancement in the treatment of mental illnesses and shedding light on the function of dopamine in schizophrenia. Serotonin was subsequently identified as a neurotransmitter in the 1950s, which opened up new perspectives for understanding and managing neurological illnesses such as mood disorders. Selective serotonin reuptake inhibitors (SSRIs), including fluoxetine (Prozac), were created in the 1960s and completely changed how depression was treated. These medications marked a move to focused therapies, concentrating in particular on neurotransmitter systems. The thalidomide disaster, however, stained this time period and demonstrated the need for strict safety standards and drug testing, particularly for medications that target the neurological system [9–12]. With the development of neuroimaging methods like MRI and PET scans, neuropharmacology made tremendous strides in the late 20th and early 21st centuries. With the use of these instruments, researchers were better able to examine medication effects and visualize brain activity. In the field of neuropharmacology, the

21st century also saw a trend towards precision medicine and tailored medicines. Drugs created by researchers to target certain molecules have improved the effectiveness and personalization of therapy for neurological illnesses [13].

The study of novel medication treatments, the creation of neuroprotective compounds, and the exploration of fresh methods to treat neurodegenerative illnesses and mental health issues are all ongoing areas of neuropharmacology research. As our knowledge of the nervous system increases and technology develops, neuropharmacology holds the potential of further advancements that will improve our capacity to treat and manage a variety of neurological diseases. The discipline has gone a long way from its early days [9, 10, 13].

2.3 Neurochemical Interactions

To fully appreciate the potential advancements that neuropharmacology may make in medicine, one must have a thorough understanding of neuronal communication and neurotransmitter/receptor interactions (Figure 2.1). In order to facilitate quick depolarization and the production of action potentials, neurons, which are excitable cells, have ion channels on their surface membrane [14]. These channels allow charged particles to move in and out. The axon terminal receives this electrical signal from the dendrites, and calcium influx causes the release of neurotransmitters from vesicles into the synapses. In order to send impulses between neurons, neurotransmitters are essential. They can undergo re-uptake or breakdown processes after being released into the synapse, bind to receptors on the post-synaptic neuron, or both. Neurological problems can result from any disturbance to the connections between neurotransmitters and receptors, which control the strength and length of the signal transmission. Post-synaptic receptors may be divided into two categories: G-protein-coupled receptors (GPCRs) and ligand-gated ion channels (LGICs). In order to quickly convert a chemical signal into an electrical one, LGIC receptors allow ions to enter the cell promptly after attaching to a neurotransmitter [15, 16]. On the other hand, upon neurotransmitter binding, GPCRs start slower intracellular biochemical activities, which result in a variety of cellular changes, including modifications to the biochemistry, physiology, and gene expression. The interactions between neurotransmitters and receptors are important therapeutic targets in the discipline of neuropharmacology. The binding mechanism is either enhanced or inhibited by medications, which modifies neural transmission [17]. Antidepressants,

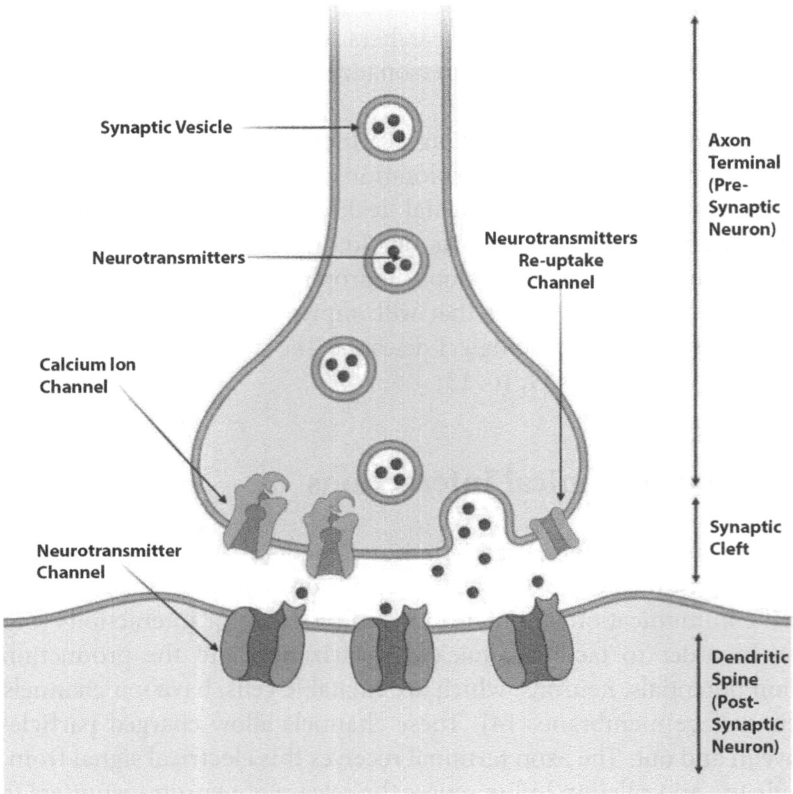

Figure 2.1 Representation of mechanism of neuronal communication (or) synapsis.

for instance, can lessen the symptoms of depression by increasing the availability of neurotransmitters in the synapse by preventing their re-uptake. Contrarily, antipsychotic medications may focus on certain receptors to lower dopamine activity in schizophrenia patients. Neuropharmacologists can create novel pharmaceuticals to more effectively treat a wide range of neurological and psychiatric illnesses by studying the mechanics of neurotransmission and how drugs might change these processes. The quality of life for those who suffer from illnesses including depression, anxiety, epilepsy, Alzheimer's disease, and more may be enhanced by developments in this subject. Furthermore, a greater comprehension of the molecular underpinnings of human behaviour and mind brings up interesting opportunities for the future creation of novel treatment modalities [14, 18].

2.4 Molecular Pharmacology of Neuronal Receptors

The molecular basis of neuropharmacology primarily depends on the various forms of receptors that play a vital role in the neural functioning, development and cognition. Table 2.1 provides an understanding on various neuronal receptors at molecular level and its role and functions in the field of neuropharmacology. The fundamentals of neuropharmacology research primarily focuses on six vital receptors: Acetylcholine, Serotonin, Dopamine, GABA, Norepinephrine and Glutamate receptors (Figure 2.2).

2.4.1 Acetylcholine Receptors

Acetylcholine is an essential neurotransmitter that plays a role in a number of physiological functions, including as muscular contraction, cognition, and control of the autonomic nervous system [31]. The postsynaptic membrane of cholinergic neurons and the neuromuscular junction both include the acetylcholine receptor, also referred to as the cholinergic receptor. Acetylcholine diffuses through the synaptic cleft and binds to the cholinergic receptor on the postsynaptic membrane when it is released from the presynaptic neuron into the synapses. The receptor's ion channel pore opens as a result of the binding event's conformational change. Based on how they react to various medicines and where in the nervous system they are found, acetylcholine receptors are divided into two basic types: nicotinic receptors and muscarinic receptors [32].

The neuromuscular junction, the autonomic ganglia, and the central nervous system all contain nicotinic receptors. They are referred to as "nicotinic" because nicotine, a substance present in tobacco, has the ability to activate them. By attaching to nicotinic receptors, nicotine imitates the effect of acetylcholine by causing the ion channel to open and enabling positively charged ions, such as sodium and calcium, to enter the postsynaptic cell. The cell membrane is depolarized by this inflow of ions, which causes an excitatory reaction. As a result, at the neuromuscular junction, muscles contract, and autonomic ganglia send signals to the effector organs. However, persistent nicotinic receptor activation can cause desensitization, which over time decreases the receptor's receptivity to nicotine and acetylcholine [33, 34]. The neuromuscular junction, the autonomic ganglia, and the central nervous system all contain nicotinic receptors. They are referred to as "nicotinic" because nicotine, a substance present in tobacco, has the ability to activate them. By attaching to nicotinic receptors, nicotine imitates the effect of acetylcholine by causing the

Table 2.1 Diverse neuronal receptors targeted in neuropharmacology.

Neurotransmitters	Receptor type	Importance	Effects on human body	Mechanism of action	Drugs mimicking them	Drugs blocking them	References
Acetylcholine	Nicotinic (nAChR)	Essential for muscle contraction	Muscle movement, learning, and memory	Ligand-gated ion channels	Nicotine	Curare, alpha-bungarotoxin	[19]
	Muscarinic (mAChR)	Regulates various autonomic functions	Heart rate, digestion, and glandular secretions	G-protein-coupled receptors	Muscarine, Pilocarpine	Atropine, Scopolamine	[20]
Serotonin	5-HT (various subtypes)	Mood regulation, sleep/wake cycle	Mood, sleep, appetite, and pain perception	G-protein-coupled receptors	SSRIs (e.g., Prozac, Zoloft)	Antipsychotics (e.g., Chlorpromazine)	[21]
Dopamine	D1-like (D1, D5)	Reward and pleasure pathways	Motor control, motivation, mood, and cognition	G-protein-coupled receptors	Dopamine agonists (e.g., Bromocriptine)	Typical antipsychotics (e.g., Haloperidol)	[22]
	D2-like (D2, D3, D4)	Inhibition of reward pathways	Emotional regulation and coordination	G-protein-coupled receptors	Dopamine antagonists (e.g., Haloperidol)	Dopamine agonists (e.g., Ropinirole)	[23]

(Continued)

Table 2.1 Diverse neuronal receptors targeted in neuropharmacology. (*Continued*)

Neurotransmitters	Receptor type	Importance	Effects on human body	Mechanism of action	Drugs mimicking them	Drugs blocking them	References
GABA	GABA-A	Major inhibitory neurotransmitter	Reduces neuronal excitability	Ligand-gated ion channels	Benzodiazepines (e.g., Valium)	Bicuculline, Flumazenil	[24]
	GABA-B	Modulates synaptic transmission	Reduces neurotransmitter release	G-protein-coupled receptors	Baclofen	CGP-52432, Saclofen	[25]
Norepinephrine	Alpha (α1, α2)	Regulates stress response	Constriction of blood vessels and alertness	G-protein-coupled receptors	Phenylephrine, Clonidine	Phentolamine, Yohimbine	[26]
	Beta (β1, β2, β3)	Modulates heart and lung functions	Increased heart rate and bronchodilation	G-protein-coupled receptors	Isoproterenol, Albuterol	Propranolol, Atenolol	[27]
Glutamate	AMPA	Mediates fast excitatory signals	Enhances synaptic transmission	Ligand-gated ion channels	AMPA receptor agonists	NBQX, GYKI-52466	[28]
	NMDA	Involved in learning and memory	Long-term potentiation and synaptic plasticity	Ligand-gated ion channels	NMDA receptor agonists	MK-801, Ketamine	[29]
	Metabotropic	Modulates synaptic activity	Regulates neuronal excitability	G-protein-coupled receptors	L-AP4, DCG-IV	L-CCG-I, MPEP	[30]

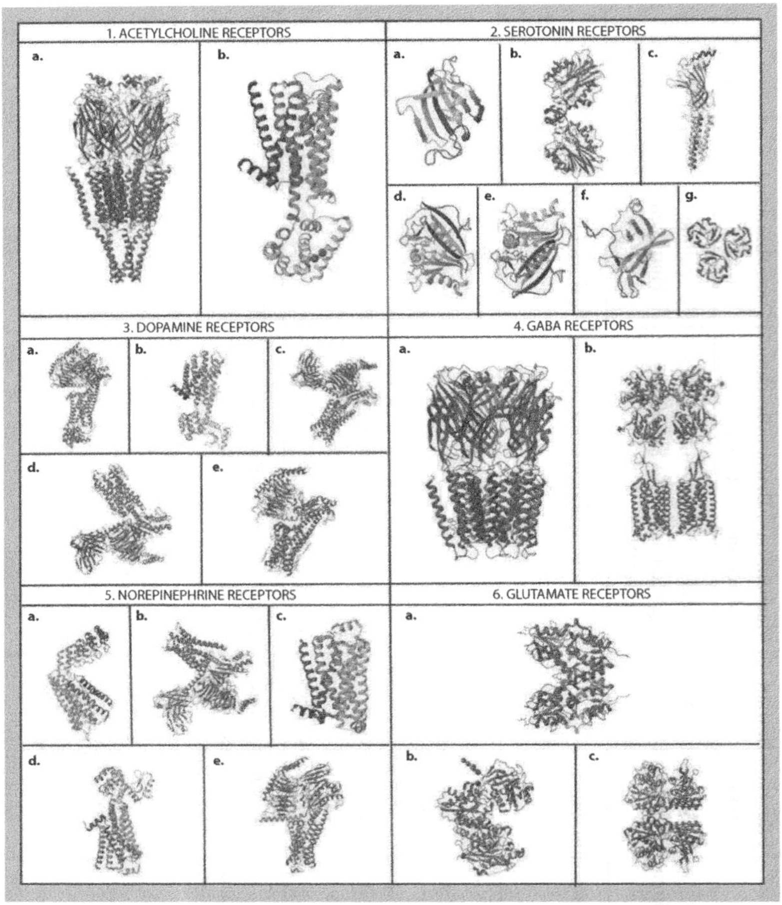

Figure 2.2 Protein structure of various neurotransmitter receptors (1a: Nicotinic Acetylcholine Receptor (https://doi.org/10.2210/pdb2BG9/pdb); 1b: Muscarinic Acetylcholine Receptor (https://doi.org/10.2210/pdb5ZHP/pdb); 2a: 5-HT1 (https://doi.org/10.2210/pdb5HT1/pdb); 2b: 5-HT2 (https://doi.org/10.2210/pdb5HT2/pdb); 2c: 5-HT3 (https://doi.org/10.2210/pdb6HIN/pdb); 2d: 5-HT5 (https://doi.org/10.2210/pdb5HT5/pdb); 2e: 5-HT4 (https://doi.org/10.2210/pdb5HT4/pdb); 2f: 5-HT6 (https://doi.org/10.2210/pdb5HT6/pdb); 2g: 5-HT7 (https://doi.org/10.2210/pdb5HT7/pdb); 3a: D1 (https://doi.org/10.2210/pdb7JOZ/pdb); 3b: D2 (https://doi.org/10.2210/pdb6CM4/pdb); 3c: D3 (https://doi.org/10.2210/pdb8IRT/pdb); 3d: D4 (https://doi.org/10.2210/pdb8IRU/pdb); 3e: D5 (https://doi.org/10.2210/pdb8IRV/pdb); 4a: GABA-A (https://doi.org/10.2210/pdb6CDU/pdb); 4b: GABA-B (https://doi.org/10.2210/pdb7C7S/pdb); 5a: α1 (https://doi.org/10.2210/pdb7B6W/pdb); 5b: α2 (https://doi.org/10.2210/pdb7EJ8/pdb); 5c: β1 (https://doi.org/10.2210/pdb2YCX/pdb); 5d: β2 (https://doi.org/10.2210/pdb6E67/pdb); 5e: β3 (https://doi.org/10.2210/pdb7DH5/pdb); 6a: AMPA (https://doi.org/10.2210/pdb6YK5/pdb); 6b: NMDA (https://doi.org/10.2210/pdb5EWL/pdb); 6c: Metabotropic Receptor (https://doi.org/10.2210/pdb1ISR/pdb)).

ion channel to open and enabling positively charged ions, such as sodium and calcium, to enter the postsynaptic cell [35]. The cell membrane is depolarized by this inflow of ions, which causes an excitatory reaction. As a result, at the neuromuscular junction, muscles contract, and autonomic ganglia send signals to the effector organs. However, persistent nicotinic receptor activation can cause desensitization, which over time decreases the receptor's receptivity to nicotine and acetylcholine [36]. The heart, smooth muscles, and glands are among the autonomic nervous system's many locations where muscarinic receptors are mostly present. Their name comes from the chemical muscarine, which has the ability to activate these receptors [37, 38]. Acetylcholine stimulates intracellular signalling pathways through G-proteins when it interacts to muscarinic receptors, which controls a range of biological reactions. Depending on the target tissue and the subtype of muscarinic receptor, these reactions can either be excitatory or inhibitory. Muscarinic receptor activation, for instance, lowers heart rate in the heart while increasing or decreasing constriction or relaxation in smooth muscles [39].

Clinical consequences of pharmacological drugs that target cholinergic receptors can be profound. For instance, medications that selectively block or modify nicotinic receptors can be used as muscle relaxants during surgery or to help people stop smoking. On the other hand, medications that specifically activate or inhibit muscarinic receptors are used to treat a variety of diseases, such as asthma, overactive bladder, and cardiac arrhythmias [40].

2.4.2 Serotonin Receptors

Serotonin, commonly referred to as 5-hydroxytryptamine (5-HT), is largely produced in the raphe nuclei of the brainstem and functions as a neurotransmitter [41]. By attaching to certain serotonin receptors found on the surface of target cells, such as neurons and other cells in the central and peripheral nervous systems, it has an impact. Serotonin receptors come in a variety of subtypes that are categorized into seven main classes (5-HT1 to 5-HT7), each of which has a distinct function and location throughout the body and brain. These receptors can be generally divided into two groups: ligand-gated ion channels and G-protein-coupled receptors (GPCRs). GPCRs make up the vast bulk of serotonin receptors [42, 43]. Serotonin activates a cascade of intracellular signalling processes through G-proteins when it binds to a GPCR. As molecular switches, these G-proteins convey the signal from the receptor to a number of effector proteins inside the cell. The signalling route can result in either excitatory or inhibitory responses, depending on the particular GPCR subtype and the cellular environment. One second messenger that controls several cellular functions is cyclic adenosine monophosphate (cAMP), which is produced when certain 5-HT receptors are activated. Serotonin receptors that are ligand-gated ion channels include the 5-HT3 receptor [44–46]. When serotonin activates these receptors, ion channels quickly open, enabling positively charged ions like sodium and calcium to enter the cell. The postsynaptic membrane becomes depolarized as a result, producing an excitatory reaction. Antiemetic medications specifically target the 5-HT3 receptor since it is well known for its role in mediating nausea and vomiting and is used to treat these symptoms [47].

Serotonin receptor-targeting medications can have a variety of physiological and cognitive impacts on the body and brain. For instance, selective serotonin reuptake inhibitors (SSRIs), which are frequently used to treat depression and anxiety disorders, increase serotonin availability in the synapses and enhance its effects on post-synaptic receptors by inhibiting serotonin reuptake. Serotonin transporters, which are in charge of reabsorbing serotonin into the presynaptic neuron, are the main targets of SSRIs [48, 49]. These medications have the potential to reduce depression symptoms over time by extending serotonin's availability in the synaptic cleft. To accomplish desired effects, other medications may target particular subtypes of serotonin receptors. For example, 5-HT3 receptor antagonists can be used to reduce chemotherapy-induced nausea, whereas medicines that selectively activate certain 5-HT receptor subtypes can be used to treat migraines [50].

2.4.3 Dopamine Receptors

Dopamine is a critical neurotransmitter involved in many different brain activities, including reward, motivation, movement, and emotional control, thus understanding how medications interact with dopamine receptors is important for understanding neuropharmacology [51]. Specifically in the substantia nigra and ventral tegmental area of the brain, dopamine is produced and functions as a signalling chemical by attaching to dopamine receptors on target cells. Dopamine receptors may be divided into two groups: D1-like receptors (D1 and D5), and D2-like receptors (D2, D3, and D4). These receptors belong to the G-protein-coupled receptor (GPCR) family and are connected to intracellular G-proteins that, upon receptor activation, mediate signalling pathways [52].

Cyclic adenosine monophosphate (cAMP) levels rise in cells when D1-like receptors are activated. As a result, protein kinase A (PKA) is activated, which sets off a series of intracellular processes that increase cellular excitability and have many additional consequences [53, 54]. D1-like receptors are broadly distributed in areas of the brain associated with motivation and cognition, and they are mostly excitatory. Inhibitory effects on intracellular signalling pathways are produced by D2-like receptors. Dopamine lowers cAMP levels and inhibits PKA activity when it interacts to D2-like receptors. Cellular excitability is reduced as a result, and neurotransmission is inhibited. In the parts of the brain responsible for motor control and emotional processing, D2-like receptors are widely distributed [55].

Drugs that target dopamine receptors can have a big impact on how the brain and behaviour work. Such chemicals as cocaine and amphetamines have the potential to become addicted because they raise dopamine levels in the brain, which can result in emotions of pleasure and reward [56, 57]. These medications work by preventing dopamine reuptake, preventing its breakdown, or enhancing its release from presynaptic neurons. On the other hand, a number of neurological and psychiatric conditions can be treated using medications that specifically target particular dopamine receptor subtypes. For instance, D2-like receptors are the main target of antipsychotic drugs, which function by preventing their activation. By doing this, these medications help lessen the brain's excessive dopamine activity, thereby treating the symptoms of schizophrenia and other psychotic diseases [58–60].

2.4.4 γ-Aminobutyric Acid Receptors

Gamma-aminobutyric acid (GABA) is the main inhibitory neurotransmitter in the central nervous system, making the GABA receptor an essential component of neuropharmacology. It is essential for controlling brain activity because it reduces neuronal excitability and avoids overstimulation [61]. The enzyme glutamate decarboxylase converts glutamate, another neurotransmitter, into GABA. In the synapse, where it interacts to certain receptors on the postsynaptic membrane, GABA is released once it has been synthesized. GABA receptors come in two primary categories: both GABA-A and GABA-B receptors [62].

As ligand-gated ion channels, GABA-A receptors permit the flow of ions only when GABA binds to their particular binding sites. The GABA-A receptor undergoes a conformational change upon GABA binding, which causes a chloride ion channel to open. The postsynaptic membrane becomes hyperpolarized by the inflow of chloride ions into the cell, which reduces its propensity to produce an action potential [63]. By lowering neuron excitability, the inhibitory impact of GABA-A receptor activation contributes to the stability and equilibrium of the nervous system. GABA-B receptors work through intracellular signalling pathways, in contrast to GABA-A receptors, which are metabotropic receptors. The activation of G-proteins by GABA upon binding to GABA-B receptors results in the regulation of intracellular enzymes and ion channels. When GABA-B receptors are activated, potassium channels can open, causing potassium ions to flow out and the cell membrane to become hyperpolarized, which inhibits neurons [64, 65].

GABA-A receptor-targeting medications frequently augment the inhibitory effects of GABA. For instance, a family of medications known as benzodiazepines bind to certain locations on the GABA-A receptor, enhancing that receptor's affinity for GABA. This leads to higher inhibitory effects and increased chloride ion channel opening, which have the relaxing, sedating, and antianxiety effects. Muscle spasms, sleep difficulties, and anxiety disorders are all often treated with benzodiazepines [66–68]. There are several uses for pharmacological drugs that target GABA-B receptors. For instance, the muscle relaxant baclofen, a GABA-B receptor agonist, is used to treat spasticity in disorders including multiple sclerosis and spinal cord injury [69, 70].

2.4.5 Norepinephrine Receptors

A crucial neurotransmitter and hormone, norepinephrine (sometimes referred to as noradrenaline), is essential for the sympathetic nervous system's fight-or-flight response. The enzyme dopamine beta-hydroxylase converts dopamine into norepinephrine, which is then released into the synapses where it interacts to certain receptors on target cells [71]. Adrenergic receptors- α and β-adrenergic receptors—are the two primary subtypes of norepinephrine receptors, which are also known as adrenergic receptors. Each subtype is further subdivided into many subgroups, referred to as α1, α2, and β1 and β2 receptors, among other designations, depending on their unique characteristics and signalling mechanisms [72].

Alpha-adrenergic receptors come in two primary subtypes: α1 and α2. Inositol trisphosphate (IP3) and diacylglycerol (DAG) are produced when phospholipase C enzymes are activated as a result of the activation of α1-adrenergic receptors. These second messengers cause a rise in intracellular calcium levels and the contraction of smooth muscle through a sequence of intracellular signalling processes. Vascular smooth muscles have a large number of α1 receptors, and their activation causes vasoconstriction, which raises blood pressure [73, 74]. Adenylyl cyclase is inhibited by the activation of α2-adrenergic receptors, which results in less cyclic adenosine monophosphate (cAMP) being produced. Negative feedback regulation occurs as a result of decreasing intracellular calcium levels and inhibition of norepinephrine release. The brain has α2 receptors that control the release of neurotransmitters, mood, and other mental processes. Subtypes β1 and β2 of beta-adrenergic receptors are further separated. The stimulation of adenylyl cyclase by the activation of β1 receptors results in an increase in cAMP generation. This causes the cardiac muscle cells to contract more readily and beat faster [75]. Drugs used to treat disorders like heart failure and arrhythmias specifically target β1 receptors, which are largely present in the heart. In addition to raising cAMP levels, β2 receptor activation causes smooth muscles in the blood arteries, the respiratory tract, and other organs to relax and enlarge. Albuterol is a popular example of a β2 adrenergic agonist that is used to treat asthma and other respiratory diseases [76–78].

The body and the sympathetic nervous system are both susceptible to a variety of side effects from pharmacological drugs that target norepinephrine receptors. Many illnesses, including hypertension, heart failure, and asthma, are treated with medications that precisely bind to and activate or inhibit certain subtypes of norepinephrine receptors [79]. These drugs can modify sympathetic nervous system responses and help control physiological processes related to the "fight-or-flight" response by altering the activation of norepinephrine receptors [80].

2.4.6 Glutamate Receptors

Since glutamate is the main excitatory neurotransmitter in the central nervous system, glutamate receptors are an important component of neuropharmacology [81]. Precursor molecules are converted into glutamate, which is then released into the synaptic cleft and binds to certain receptors on the postsynaptic membrane. Glutamate receptors come in a variety of forms, with ionotropic and metabotropic glutamate receptors being the two primary kinds [82].

Ionotropic glutamate receptors are ligand-gated ion channels, which means that when glutamate binds to their particular binding sites, ions can flow through. Alpha-amino-3-hydroxy-5-methyl-4-isoxazolepropionic acid (AMPA), *N*-methyl-D-aspartate (NMDA), and kainate receptors are the three primary subtypes of ionotropic glutamate receptors. When AMPA receptors are activated, sodium ions flood the cell, causing rapid excitatory synaptic transmission [83, 84]. These receptors are crucial for functions like learning and memory because they mediate fast communication between neurons. In order for calcium ions to enter the cell and trigger postsynaptic depolarization (the removal of a magnesium barrier), NMDA receptors must bind both glutamate and a co-agonist, such as glycine. NMDA receptors are essential for synaptic plasticity, learning, and memory due to this particular gating mechanism. Excessive NMDA receptor activation, on the other hand, can result in excitotoxicity, a condition in which the receptor is overstimulated and results in cellular harm and death. Although less well understood than AMPA and NMDA receptors, kainate receptors have a variety of roles in synaptic transmission and plasticity [85, 86]. Metabotropic glutamate receptors are G-protein-coupled receptors without ion channels that, when glutamate binds to them, modify intracellular signalling pathways. Metabotropic glutamate receptors come in a variety of subtypes, each with a special purpose. Numerous physiological processes, including as synaptic plasticity, the control of neurotransmitter release, and the excitability of neurons, are regulated by these receptors [87, 88].

Drugs that target glutamate receptors can have a big impact on how the brain and behaviour work. For instance, anaesthetics that block NMDA receptors, such as ketamine, have been investigated for their possible antidepressant benefits. On the other hand, medications like ampakins that promote the activity of certain glutamate receptors are being researched for their potential to increase cognition and memory [89, 90].

2.5 Neuropharmacological Drugs

In order to treat a variety of neurological and psychiatric problems, physicians use pharmaceuticals known as neuropharmacologicals, which target the nerve system. Based on their modes of action and therapeutic benefits, these medications may be roughly categorized into several kinds (Table 2.2).

Antidepressants are a significant group of neuropharmacological medications that are frequently used to treat mood disorders including depression and anxiety. Antidepressants with well-known names include fluoxetine (Prozac) and sertraline (Zoloft), which are selective serotonin reuptake inhibitors (SSRIs). Serotonin is a neurotransmitter that affects mood, emotion, and cognition. SSRIs function by preventing serotonin from being reabsorbed. These medications lessen depression symptoms by

Table 2.2 List of various approved neuropharmacological drugs.

Chemical name	Class of drug	Commercial name	Mechanism of action	Medicinal usage	Side effects	References
Aspirin	Nonsteroidal anti-inflammatory drugs (NSAIDs)	Bayer Aspirin	Inhibits COX enzymes, reducing prostaglandin production and inflammation	Analgesic (pain relief), anti-inflammatory, antipyretic (fever reduction)	Gastric irritation, bleeding tendencies	[91]
Acetaminophen	Analgesic and antipyretic	Tylenol	Works centrally to inhibit prostaglandin synthesis in the brain	Analgesic (pain relief), antipyretic (fever reduction)	Liver toxicity (with overdose)	[92]
Sertraline	Selective serotonin reuptake inhibitor (SSRI)	Zoloft	Blocks the reuptake of serotonin, increasing its availability	Antidepressant	Nausea, sexual dysfunction, insomnia	[93]

(Continued)

Table 2.2 List of various approved neuropharmacological drugs. *(Continued)*

Chemical name	Class of drug	Commercial name	Mechanism of action	Medicinal usage	Side effects	References
Alprazolam	Benzodiazepine	Xanax	Enhances the effects of GABA, reducing brain activity	Anxiolytic (anti-anxiety), sedative, anticonvulsant	Drowsiness, dependence, withdrawal symptoms	[94]
Risperidone	Atypical antipsychotic	Risperdal	Antagonizes dopamine and serotonin receptors	Antipsychotic, bipolar disorder, autism-related irritability	Weight gain, movement disorders	[95]
Levodopa	Dopamine precursor	Sinemet	Converted to dopamine in the brain, replenishing its levels	Parkinson's disease treatment, restores dopamine levels	Nausea, dyskinesia (abnormal movements)	[96]
Memantine	NMDA receptor antagonist	Namenda	Blocks NMDA receptors, reducing excitotoxicity in the brain	Alzheimer's disease treatment	Dizziness, confusion, headaches	[94]

reducing serotonin reuptake, which increases serotonin availability in the synapse and improves neurotransmission [97, 98]. Antipsychotics are an important group of neuropharmacological medications that are used to treat psychotic conditions including schizophrenia. In this group, atypical antipsychotics like risperidone (Risperdal) and aripiprazole (Abilify) are often recommended medications. Dopamine and serotonin receptors are the primary targets of atypical antipsychotics in the brain. They diminish psychotic symptoms such hallucinations and delusions by acting as partial agonists or antagonists, regulating dopamine and serotonin signalling. Antipsychotic medications assist in controlling the bizarre behaviours and thinking processes connected to psychotic diseases by modulating these neurotransmitter systems. Another crucial class of neuropharmacological medicines is anti-epileptic medications (AEDs). These medications are used to treat seizure disorders such as epilepsy. AEDs include, for instance, valproic acid (Depakote) and carbamazepine (Tegretol). Ion channels and receptors implicated in neuronal excitability are modulated by AEDs. They provide seizure control for people with epilepsy by reducing excessive neuronal firing and seizure activity [6, 99, 100]. AEDs assist prevent or reduce the incidence of seizures by controlling the electrical activity in the brain, enabling patients to live more stable and useful lives. Another class of neuropharmacological medications used to treat anxiety disorders and the symptoms of acute anxiety is called anxiolytics. Anxiolytics are often used benzodiazepines like diazepam (Valium) and alprazolam (Xanax). Gamma-aminobutyric acid (GABA) has inhibitory properties, and by binding to particular locations on GABA-A receptors, benzodiazepines increase these effects. This causes an increase in the inflow of chloride ions, which has sedative and anxiolytic effects. Anxiolytics assist people in managing symptoms of anxiety by encouraging relaxation and lowering excessive anxiety [101, 102].

These illustrations show the wide variety of neuropharmacological medications that are offered as well as their possible therapeutic effects on the nervous system. With continued research aimed at creating more specialized and efficient drugs for neurological and mental diseases, the subject of neuropharmacology is still developing [103, 104]. To guarantee their safe and proper usage, taking into account the person's unique condition and medical history, it is essential to use these medications sparingly and under medical supervision. Neuropharmacological treatments have the potential to considerably better the lives of people with a variety of neurological and mental diseases by reducing symptoms and boosting general wellbeing.

2.6 Impact of Biotechnology of Neuropharmacology

Neuropharmacology has been significantly impacted by biotechnology, which has transformed how scientists investigate, comprehend, and create remedies for neurological illnesses. The term "biotechnology" describes the utilization of biological systems, processes, or creatures to create new goods or technologies. Biotechnology has aided developments in a variety of fields, including drug discovery and personalized treatment, in the field of neuropharmacology [105]. The creation of new drugs is one of the major effects of biotechnology on neuropharmacology. Novel medications that target particular biochemical pathways implicated in neurological illnesses have been made possible through biotechnology. Researchers can more effectively find prospective drug candidates by using methods including high-throughput screening, combinatorial chemistry, and computer modelling. Genetically modified animal models and other biotechnological technologies have been crucial in evaluating the effectiveness and drug safety [5]. Table 2.3 provides an in-depth information about the various fields of biotechnology and is role in the development in the basis and foundation of neuropharmacology.

Additionally, the development of biologics for neuropharmacology has been aided by biotechnology. Drugs called biologics are created from cells or live creatures. The use of monoclonal antibodies, a kind of biologic, in the treatment of neurological diseases including multiple sclerosis and Alzheimer's disease has shown considerable promise. These antibodies can target certain proteins implicated in disease processes, making the treatment approach more focused and exact [130]. Neuropharmacology is another area where gene therapy has an influence. Gene therapy, which involves changing a patient's genes to treat or prevent disease, has become more feasible because to advances in biotechnology. Gene therapy has the possibility to fix or replace defective genes, offering long-term therapeutic advantages, in neurological illnesses brought on by genetic abnormalities. Early clinical studies have demonstrated promising results in illnesses including Parkinson's disease and other hereditary neuropathies [131]. In addition, neuropharmacology has benefited from the identification of biomarkers—molecular indications of illness existence, progression, or response to therapy. These biomarkers make it possible to identify diseases early, keep track of how well treatments are working, and forecast patient outcomes. Biomarkers aid in the creation of personalized medicine strategies, which allow for the customization of medicines to the requirements of specific patients, when combined with genetic data and other

Table 2.3 Various biotechnological fields and tools and their impact on neuropharmacology.

Biotechnological field	Tools and techniques	Impact on neuropharmacology	References
Genomics	DNA sequencing	Identifying genetic variations linked to neurological disorders	[106]
	Microarray analysis	Studying gene expression patterns in the brain and response to drugs	[107]
	CRISPR-Cas9	Creating animal models with specific genetic mutations for drug testing	[108]
Proteomics	Mass spectrometry	Identifying and characterizing proteins in the brain and their interactions	[99]
	Protein microarrays	Studying protein-protein interactions relevant to neuropharmacology	[109]
	Protein crystallography	Understanding protein structures, aiding in drug design and development	[110]

(*Continued*)

Table 2.3 Various biotechnological fields and tools and their impact on neuropharmacology. (*Continued*)

Biotechnological field	Tools and techniques	Impact on neuropharmacology	References
Transcriptomics	RNA sequencing	Analyzing gene expression and identifying non-coding RNAs relevant to drug targets	[111]
	Single-cell transcriptomics	Examining gene expression in individual neurons to understand cellular diversity	[112]
	RNA interference (RNAi)	Silencing specific genes to study their functions in the brain	[113]
Metabolomics	Mass spectrometry	Profiling small molecules and metabolites in the brain, offering insights into drug metabolism and pathways	[114]
	Nuclear magnetic resonance (NMR)	Studying metabolic changes in response to drug treatments	[115]
	High-performance liquid chromatography (HPLC)	Separating and quantifying neurotransmitters, helping to study their regulation and role in drug action	[116]

(*Continued*)

Table 2.3 Various biotechnological fields and tools and their impact on neuropharmacology. (*Continued*)

Biotechnological field	**Tools and techniques**	**Impact on neuropharmacology**	**References**
High-Throughput Screening	Automated screening platforms	Rapidly testing thousands of compounds for potential drug candidates	[117]
	Fluorescence-based assays	Assessing drug-receptor interactions and cellular responses in a high-throughput manner	[118]
	Ion channel screening platforms	Identifying compounds that modulate ion channels relevant to neurological diseases	[119]
Imaging Techniques	Positron emission tomography (PET)	Visualizing neurotransmitter activity and receptor occupancy in the brain	[120]
	Functional magnetic resonance imaging (fMRI)	Monitoring brain activity in response to pharmacological interventions	[121]
	Single-photon emission computed tomography (SPECT)	Assessing brain blood flow and neuroreceptor binding for drug development	[122]

(*Continued*)

Table 2.3 Various biotechnological fields and tools and their impact on neuropharmacology. (*Continued*)

Biotechnological field	Tools and techniques	Impact on neuropharmacology	References
Brain Organoids	3D cell cultures	Creating miniature brain-like structures to study drug responses in a more realistic environment	[123]
	Organ-on-a-chip technology	Simulating brain tissue in microfluidic devices to evaluate drug effects	[124]
Optogenetics	Light-activated proteins (e.g., channelrhodopsin)	Controlling specific neurons with light to understand brain circuits and drug effects	[125]
	Halorhodopsin	Inhibiting neuronal activity optically to study the effects of neuromodulators	[126]
Bioinformatics	Data analysis software	Analyzing large-scale genomic, proteomic, and transcriptomic data related to neuropharmacology	[127]
	Network biology tools	Understanding complex interactions between genes and proteins involved in neurological disorders	[128]
	Machine learning algorithms	Predicting drug-target interactions and identifying potential new drug candidates	[129]

patient-specific information. Furthermore, biotechnology has developed neuroimaging methods that shed light on the anatomy and function of the brain, such as functional magnetic resonance imaging (fMRI) and positron emission tomography (PET). These instruments have greatly improved our knowledge of brain illnesses and how medications affect certain parts of the brain, assisting in the creation of specialized treatments [132, 133].

Furthermore, medication delivery methods for neuropharmacology have used nanoscale ideas from biotechnology. Drugs can be delivered to certain brain areas more efficiently by using nanoparticles that have been designed to pass the blood-brain barrier. This method of focused medicine administration reduces negative effects and increases treatment effectiveness [134]. Overall, biotechnology has had a profound influence on neuropharmacology. The process of finding new drugs has been sped up, new therapy strategies have been made easier to design, and our understanding of neurological illnesses has increased. The prospect for even more novel and efficient therapies for neurological illnesses remains encouraging as biotechnological developments proceed, giving people living with these disorders hope for improved outcomes and a higher quality of life [121].

2.7 Future Research and Perspectives

The area of neuropharmacology has an exciting trajectory ahead of it since new developments in science and technology are expected to completely transform it. The future of neuropharmacology is anticipated to be shaped by a number of important areas, presenting fresh opportunities for improving treatment results for neurological and psychiatric illnesses as well as for studying the brain and designing tailored medicines [135]. The ongoing use of personalized medicine in neuropharmacology is one of the most promising future directions. Researchers are learning more about the genetic elements that affect a person's reaction to drugs because to advances in genomics and the availability of extensive patient data. Clinicians will be able to customize prescription regimens to each patient's unique genetic profile thanks to this personalized approach to therapy, maximizing therapeutic advantages while minimizing negative effects [136, 137].

The creation of inventive medication delivery methods is another field with a lot of potential. Researchers are investigating ways to deliver medications directly to certain brain cells or areas using nanotechnology. Bypassing the blood-brain barrier and delivering drugs more precisely, nanoparticles can improve therapeutic efficacy and lower the chance of systemic adverse effects [5, 105, 138]. This may completely alter how neurological disorders including brain tumors, neurodegenerative illnesses, and epilepsy are treated. The development of brain mapping and neuroimaging techniques will also continue to be crucial in the study of neuropharmacology. Innovative methods like functional connectomics and cutting-edge imaging modalities will provide previously unattainable insights into the complex neuronal networks of the brain and how they malfunction in a variety of illnesses. This more thorough comprehension of brain connections will make it easier to find new medication targets and create more precision [139]. Furthermore, the development of biologics and gene treatments is anticipated to accelerate in the field of neuropharmacology. Numerous neurological and psychiatric diseases have a high potential for improvement with the use of monoclonal antibodies and gene-editing techniques like CRISPR. While gene treatments can fix genetic abnormalities that are the cause of some neurological illnesses, biologics have the benefit of offering highly focused medicines that can modify particular pathways implicated in disease processes [103, 140].

Finally, developments in artificial intelligence (AI) and machine learning will influence neuropharmacology. These technologies have the speed and capacity to forecast medication interactions, analyze massive volumes of data, and find novel therapeutic ideas. AI-driven methodologies will speed up the development of new neuropharmacological therapies and reduce the time and money needed for their development [141].

2.8 Conclusion

In conclusion, neuropharmacology is crucial for understanding the intricate relationships between neurotransmitters and receptors in the nervous system that affect cognition and behaviour in people. The creation of several medications that specifically target particular neurotransmitter systems has been made possible by the historical development and milestones in this field, revolutionizing the management of neurological and psychiatric illnesses. Neuropharmacological medications have shown promise in treating a range of illnesses and enhancing patients' lives through modulating acetylcholine and serotonin receptors as well as dopamine and GABA receptors. The development of targeted medicines, gene editing methods, and personalized medical strategies have all been made possible by the integration of biotechnology, which has improved drug discovery even more. The discipline of neuropharmacology has a bright future ahead of it as research advances, thanks to cutting-edge technologies like AI-driven drug discovery, nanotechnology-based drug delivery systems, and precision medicine. Neuropharmacology is well positioned to continue playing a crucial role in advancing our understanding of the brain and creating novel treatments for neurological and psychiatric disorders, which will ultimately improve patient outcomes and pave the way for a more promising future for both neuroscience and medicine.

Acknowledgments

The authors greatly acknowledge and convey our sincere gratitude to REVA University and Ganpat University for providing an opportunity and platform for research.

References

1. Barnhill, J., *et al.*, Pharmacology, Psychopharmacology, and Adverse Drug Reactions, in: *Neurodevelopmental Pediatrics: Genetic and Environmental Influences*, pp. 713–729, Springer, Cham, 2023.
2. Mastinu, A., *et al.*, The bright side of psychedelics: Latest advances and challenges in neuropharmacology. *Int. J. Mol. Sci.*, 24, 2, 1329, 2023.
3. Ferraiolo, M. and Hermans, E., The complex molecular pharmacology of the dopamine D2 receptor: Implications for pramipexole, ropinirole, and rotigotine. *Pharmacol. Ther.*, 245, 108392, 2023.
4. Casadó, V. and Casadó-Anguera, V., *What are the current trends in G protein-coupled receptor targeted drug discovery?*, pp. 1–6, Taylor & Francis, 2023.
5. Premoli, M., *et al.*, Mouse and rat ultrasonic vocalizations in neuroscience and neuropharmacology: State of the art and future applications. *Eur. J. Neurosci.*, 57, 12, 2062–2096, 2023.
6. Salmanzadeh, H., *et al.*, Neuropharmacology of human TERA2. cl. SP12 stem cell-derived neurons in ultra-long-term culture for antiseizure drug discovery. *Front. Neurosci.*, 17, 1182720, 2023.
7. Lodge, D., The history of the pharmacology and cloning of ionotropic glutamate receptors and the development of idiosyncratic nomenclature. *Neuropharmacology*, 56, 1, 6–21, 2009.
8. Gould, R.W., *et al.*, Effects of nicotinic acetylcholine receptor agonists on cognition in rhesus monkeys with a chronic cocaine self-administration history. *Neuropharmacology*, 64, 479–488, 2013.
9. Katz, N.K. and Barohn, R.J., The history of acetylcholinesterase inhibitors in the treatment of myasthenia gravis. *Neuropharmacology*, 182, 108303, 2021.
10. Tricklebank, M. and Daly, E., *The serotonin system: history, neuropharmacology, and pathology*, Academic Press, Cambridge, 2019.

11. Richard Green, A., Neuropharmacology of 5-hydroxytryptamine. *Br. J. Pharmacol.*, 147, S1, S145–S152, 2006.
12. Bockaert, J., *et al.*, 5-HT4 receptors: history, molecular pharmacology and brain functions. *Neuropharmacology*, 55, 6, 922–931, 2008.
13. Hirota, K. and Lambert, D.G., Ketamine; history and role in anesthetic pharmacology. *Neuropharmacology*, 216, 109171, 2022.
14. Ziminski, J.J., *et al.*, Microstructural and neurochemical plasticity mechanisms interact to enhance human perceptual decision-making. *PLoS Biol.*, 21, 3, e3002029, 2023.
15. Smith, D.M., *et al.*, The neurochemical bases of verbal suggestion and hypnosis, *The Routledge International Book of Clinical Hypnosis*, Taylor & Francis Group, Oxfordshire, 2023.
16. Rodríguez-Nieto, G., *et al.*, Organization of neurochemical interactions in young and older brains as revealed with a network approach: Evidence from proton magnetic resonance spectroscopy (1H-MRS). *Neuroimage*, 266, 119830, 2023.
17. McNealy, K.R., Weyrich, L., Bevins, R.A., The Co-use of Nicotine and Prescription Psychostimulants: A Review of Their Behavioral and Neuropharmacological Interactions. *Drug Alcohol Depend.*, 248, 109906, 2023.
18. Duo, L., *et al.*, Sleep disorders in chronic pain and its neurochemical mechanisms: a narrative review. *Front. Psychiatry*, 14, 1157790, 2023.
19. Spinella, M., *The psychopharmacology of herbal medicine: plant drugs that alter mind, brain, and behavior*, Mit Press, Cambridge, 2001.
20. Abraham, G., The importance of muscarinic receptors in domestic animal diseases and therapy: Current and future perspectives. *Vet. J.*, 208, 13–21, 2016.
21. Pourhamzeh, M., *et al.*, The roles of serotonin in neuropsychiatric disorders. *Cell. Mol. Neurobiol.*, 42, 6, 1671–1692, 2022.
22. Kaur, S., *et al.*, Pharmacology of dopamine and its receptors, in: *Frontiers in Pharmacology of Neurotransmitters*, pp. 143–182, 2020.
23. Ayano, G., Dopamine: receptors, functions, synthesis, pathways, locations and mental disorders: review of literatures. *J. Ment. Disord. Treat.*, 2, 120, 2, 2016.
24. Chebib, M. and Johnston, G.A.R., GABA-activated ligand gated ion channels: medicinal chemistry and molecular biology. *J. Med. Chem.*, 43, 8, 1427–1447, 2000.
25. Xu, M.-Y. and Wong, A.H.C., GABAergic inhibitory neurons as therapeutic targets for cognitive impairment in schizophrenia. *Acta Pharmacol. Sin.*, 39, 5, 733–753, 2018.
26. Perez, D.M., α1-Adrenergic receptors in neurotransmission, synaptic plasticity, and cognition. *Front. Pharmacol.*, 11, 581098, 2020.
27. Rosenthal, J.S. and Yuan, Q., Constructing and tuning excitatory cholinergic synapses: the multifaceted functions of nicotinic acetylcholine receptors in

Drosophila neural development and physiology. *Front. Cell. Neurosci.*, 15, 720560, 2021.
28. Köles, L., Wirkner, K., Illes, P., Modulation of ionotropic glutamate receptor channels. *Neurochem. Res.*, 26, 925–932, 2001.
29. Szénási, G., *et al.*, 2, 3-benzodiazepine-type AMPA receptor antagonists and their neuroprotective effects. *Neurochem. Int.*, 52, 1-2, 166–183, 2008.
30. Vikelis, M. and Mitsikostas, D.D., The role of glutamate and its receptors in migraine. *CNS Neurol. Disord. Drug Targets (Formerly Curr. Drug Targets-CNS Neurol. Disord.)*, 6, 4, 251–257, 2007.
31. Xu, J., *et al.*, Structural and dynamic insights into supra-physiological activation and allosteric modulation of a muscarinic acetylcholine receptor. *Nat. Commun.*, 14, 1, 376, 2023.
32. Terry Jr., A.V., Jones, K., Bertrand, D., Nicotinic acetylcholine receptors in neurological and psychiatric diseases. *Pharmacol. Res.*, 191, 106764, 2023.
33. Richter, K. and Grau, V., Signaling of nicotinic acetylcholine receptors in mononuclear phagocytes. *Pharmacol. Res.*, 191, 106727, 2023.
34. Power, S.K., Venkatesan, S., Lambe, E.K., Xanomeline restores endogenous nicotinic acetylcholine receptor signaling in mouse prefrontal cortex. *Neuropsychopharmacology*, 48, 4, 671–682, 2023.
35. Papke, R.L., Quadri, M., Gulsevin, A., Silent agonists for α7 nicotinic acetylcholine receptors. *Pharmacol. Res.*, 190, 106736, 2023.
36. Hone, A.J. and McIntosh, J.M., Nicotinic acetylcholine receptors: Therapeutic targets for novel ligands to treat pain and inflammation. *Pharmacol. Res.*, 190, 106715, 2023.
37. Fontana, I.C., Kumar, A., Nordberg, A., The role of astrocytic α7 nicotinic acetylcholine receptors in Alzheimer disease. *Nat. Rev. Neurol.*, 19, 5, 278–288, 2023.
38. Elgoyhen, A.B., The α9α10 acetylcholine receptor: a non-neuronal nicotinic receptor. *Pharmacol. Res.*, 190, 106735, 2023.
39. Chrestia, J.F., *et al.*, Regulation of nicotinic acetylcholine receptors by post-translational modifications. *Pharmacol. Res.*, 190, 106712, 2023.
40. Allen, N.M., *et al.*, The emerging spectrum of fetal acetylcholine receptor antibody-related disorders (FARAD). *Brain*, 146, awad153, 2023.
41. Shahraki, A., Selent, J., Kolb, P., On the construction of LIECE models for the serotonin receptor 5-HT $$ _ {\text {2A}} $$ R. *J. Comput.-Aided Mol. Des.*, 37, 1–11, 2023.
42. Salvan, P., *et al.*, Serotonin regulation of behavior via large-scale neuromodulation of serotonin receptor networks. *Nat. Neurosci.*, 26, 1, 53–63, 2023.
43. Pizzi, S.D., *et al.*, Spatial Correspondence of LSD-Induced Variations on Brain Functioning at Rest With Serotonin Receptor Expression. *Biol. Psychiatry: Cognit. Neurosci. Neuroimaging*, 8, 7, 768–776, 2023.
44. Peroutka, S.J. and Sleight, A.J., Central serotonin receptors: Functional correlates and clinical relevance, in: *Role Of Serotonin In Psychiatric Disorders*, pp. 8–26, Routledge, Oxfordshire, 2023.

45. Pei, Y., *et al.*, Structural insight into the selective agonist ST1936 binding of serotonin receptor 5-HT6. *Biochem. Biophys. Res. Commun.*, 671, 327–334, 2023.
46. Moskaliuk, V.S., *et al.*, On Associations between Fear-Induced Aggression, Bdnf Transcripts, and Serotonin Receptors in the Brains of Norway Rats: An Influence of Antiaggressive Drug TC-2153. *Int. J. Mol. Sci.*, 24, 2, 983, 2023.
47. Mercurio, S., *et al.*, Serotonin Receptors and Their Involvement in Melanization of Sensory Cells in Ciona intestinalis. *Cells*, 12, 8, 1150, 2023.
48. Gumpper, R.H. and Roth, B.L., SnapShot: psychedelics and serotonin receptor signaling. *Cell*, 186, 1, 232–232, 2023.
49. Chaumont-Dubel, S., *et al.*, Impact of 5-HT6 Receptor Subcellular Localization on Its Signaling and Its Pathophysiological Roles. *Cells*, 12, 3, 426, 2023.
50. Calderón, J.C., *et al.*, General Metadynamics Protocol To Simulate Activation/Deactivation of Class A GPCRs: Proof of Principle for the Serotonin Receptor. *J. Chem. Inf. Model.*, 63, 10, 3105–3117, 2023.
51. Yin, N., *et al.*, Dopamine receptor-mediated roles on retinal ganglion cell hyperexcitability and injury in experimental glaucoma. *Cell. Signalling*, 109, 110781, 2023.
52. Xu, P., *et al.*, Structural genomics of the human dopamine receptor system. *Cell Res.*, 33, 1–13, 2023.
53. Neumann, J., *et al.*, Role of Dopamine in the Heart in Health and Disease. *Int. J. Mol. Sci.*, 24, 5, 5042, 2023.
54. Kawahata, I. and Fukunaga, K., Endocytosis of dopamine receptor: Signaling in brain. *Prog. Mol. Biol. Transl. Sci.*, 196, 99–111, 2023.
55. Ji, G.-J., *et al.*, Neuroplasticity-related genes and dopamine receptors associated with regional cortical thickness increase following electroconvulsive therapy for major depressive disorder. *Mol. Neurobiol.*, 60, 3, 1465–1475, 2023.
56. Hao, H., Song, L., Zhang, L., Wolfram syndrome 1 regulates sleep in dopamine receptor neurons by modulating calcium homeostasis. *PLos Genet.*, 19, 7, e1010827, 2023.
57. Gerfen, C.R., Segregation of D1 and D2 dopamine receptors in the striatal direct and indirect pathways: An historical perspective. *Front. Synaptic Neurosci.*, 14, 1002960, 2023.
58. Cookson, J., Pimm, J., Reynolds, G., Partial agonists of dopamine receptors: clinical effects and dopamine receptor interactions in combining aripiprazole with a full antagonist in treating psychosis. *BJPsych Adv.*, 29, 2, 158–165, 2023.
59. Chhabra, S., *et al.*, Striatal increase of dopamine receptor 2 density in idiopathic and syndromic mouse models of autism spectrum disorder. *Front. Psychiatry*, 14, 1110525, 2023.
60. Carney, A.E., Clarke, C., Pratt, W.E., Administration of Neuropeptide Y into the rat nucleus accumbens shell, but not core, attenuates the motivational

impairment from systemic dopamine receptor antagonism by α-flupenthixol. *Neurosci. Lett.*, 797, 137069, 2023.
61. Zhang, C., *et al.*, Structural Diversity Design, Synthesis, and Insecticidal Activity Analysis of Ester-Containing Isoxazoline Derivatives Acting on the GABA Receptor. *J. Agric. Food. Chem.*, 71, 7, 3184–3191, 2023.
62. Wagner, A., Yan, Z., Kulka, M., A Human Microglial Cell Line Expresses γ-Aminobutyric Acid (GABA) Receptors and Responds to GABA and Muscimol by Increasing Production of IL-8. *Neuroglia*, 4, 3, 172–187, 2023.
63. Tian, J. and Kaufman, D.L., The GABA and GABA-Receptor System in Inflammation, Anti-Tumor Immune Responses, and COVID-19. *Biomedicines*, 11, 2, 254, 2023.
64. Sharma, P., *et al.*, Endocytosis of GABA receptor: Signaling in nervous system. *Prog. Mol. Biol. Transl. Sci.*, 196, 125–139, 2023.
65. Park, C.W., *et al.*, Sleep-promoting activity of amylase-treated Ashwagandha (Withania somnifera L. Dunal) root extract via GABA receptors. *J. Food Drug Anal.*, 31, 2, 278, 2023.
66. Omoumi, S., *et al.*, Corticosterone injection into the infralimbic prefrontal cortex enhances fear memory extinction: Involvement of GABA receptors and the extracellular signal-regulated kinase. *Physiol. Behav.*, 265, 114156, 2023.
67. Lv, M., *et al.*, Serotonin/GABA receptors modulate odor input to olfactory receptor neuron in locusts. *Front. Cell. Neurosci.*, 17, 1156144, 2023.
68. Li, Y., *et al.*, Design, Synthesis, and Insecticidal Activity of Novel Isoxazoline Diacylhydrazine Compounds as GABA Receptor Inhibitors. *J. Agric. Food. Chem.*, 71, 17, 6561–6569, 2023.
69. Guo, L., *et al.*, Convergent resistance to GABA receptor neurotoxins through plant–insect coevolution. *Nat. Ecol. Evol.*, 7, 9, 1–13, 2023.
70. Bailey, A.M., *et al.*, Changes in social, sexual, and hedonic behaviors in rats in response to stress and restoration by a negative allosteric modulator of α5-subunit containing GABA receptor. *Behav. Brain Res.*, 452, 114554, 2023.
71. Zhang, S., *et al.*, Hyperglycemia Induces Tear Reduction and Dry Eye in Diabetic Mice through the Norepinephrine–α1 Adrenergic Receptor–Mitochondrial Impairment Axis of Lacrimal Gland. *Am. J. Pathol.*, 193, 7, 913–926, 2023.
72. Yan, H., *et al.*, Fast-specific fluorescent probes to visualize norepinephrine signaling pathways and its flux in the epileptic mice brain. *J. Am. Chem. Soc.*, 145, 5, 3229–3237, 2023.
73. Shibao, C.A., Norepinephrine transporter inhibitors in autonomic disorders, in: *Primer on the Autonomic Nervous System*, pp. 733–736, Elsevier, Cambridge, 2023.
74. Reitman, M.E., *et al.*, Norepinephrine links astrocytic activity to regulation of cortical state. *Nat. Neurosci.*, 26, 4, 579–593, 2023.

75. Knittel, J., *et al.*, Monoamine Oxidase A Contributes to Serotonin—But Not Norepinephrine-Dependent Damage of Rat Ventricular Myocytes. *Biomolecules*, 13, 6, 1013, 2023.
76. Jin, M., *et al.*, Norepinephrine/β2-Adrenergic Receptor Pathway Promotes the Cell Proliferation and Nerve Growth Factor Production in Triple-Negative Breast Cancer. *J. Breast Cancer*, 26, 3, 268, 2023.
77. Grosman-Rimon, L., *et al.*, Relationships among norepinephrine levels, exercise capacity, and chronotropic responses in heart failure patients. *Heart Fail. Rev.*, 28, 1, 35–45, 2023.
78. Fage, N., *et al.*, Norepinephrine and Vasopressin in Hemorrhagic Shock: A Focus on Renal Hemodynamics. *Int. J. Mol. Sci.*, 24, 4, 4103, 2023.
79. Dong, Q., *et al.*, Overactivation of Norepinephrine–β2-Adrenergic Receptor Axis Promotes Corneal Neovascularization. *Invest. Ophthalmol. Visual Sci.*, 64, 3, 20–20, 2023.
80. Chen, C.S., *et al.*, Dopamine and norepinephrine differentially mediate the exploration-exploitation tradeoff, bioRxiv, pp. 2023–01, 2023.
81. Simon, A.A., Navarro-Retamal, C., Feijó, J.A., Merging Signaling with Structure: Functions and Mechanisms of Plant Glutamate Receptor Ion Channels. *Annu. Rev. Plant Biol.*, 74, 415–452, 2023.
82. Sehatpour, P., *et al.*, Dose-dependent augmentation of neuroplasticity-based auditory learning in schizophrenia: A double-blind, placebo-controlled, randomized, target engagement clinical trial of the NMDA glutamate receptor agonist d-serine. *Biol. Psychiatry*, 94, 2, 164–173, 2023.
83. Schackert, F.K., *et al.*, Mechanism of calcium permeation in a glutamate receptor ion channel. *J. Chem. Inf. Model.*, 63, 4, 1293–1300, 2023.
84. Nicoletti, F., *et al.*, GPCR interactions involving metabotropic glutamate receptors and their relevance to the pathophysiology and treatment of CNS disorders. *Neuropharmacology*, 235, 109569, 2023.
85. Kwan, C., *et al.*, Metabotropic glutamate receptors in Parkinson's disease, in: *International Review of Neurobiology*, pp. 1–31, Elsevier, Cambridge, 2023.
86. Kukaj, T., *et al.*, Kinetic fingerprinting of metabotropic glutamate receptors. *Commun. Biol.*, 6, 1, 104, 2023.
87. Huh, E., *et al.*, Coevolutionary signals in metabotropic glutamate receptors capture residue contacts and long-range functional interactions. *J. Biol. Chem.*, 299, 4, 2023.
88. Grenzi, M., *et al.*, Long-distance turgor pressure changes induce local activation of plant glutamate receptor-like channels. *Curr. Biol.*, 33, 6, 1019–1035, 2023.
89. Clennell, B., *et al.*, Ultrasound modulates neuronal potassium currents via ionotropic glutamate receptors. *Brain Stimul.*, 16, 2, 540–552, 2023.
90. Chen, T.-S., *et al.*, The Role of Glutamate Receptors in Epilepsy. *Biomedicines*, 11, 3, 783, 2023.

91. Sepehrmanesh, Z., *et al.*, The effects of combined sertraline and aspirin therapy on depression severity among patients with major depressive disorder: a randomized clinical trial. *Electron. Physician*, 9, 11, 5770, 2017.
92. Hasanjani, H.R.A. and Sohrabi, M.R., Artificial neural networks (ANN) for the simultaneous spectrophotometric determination of fluoxetine and sertraline in pharmaceutical formulations and biological fluid. *Iran. J. Pharm. Res. IJPR*, 16, 2, 478, 2017.
93. Javed, S. and Abdi, S., Chronic and Interventional Pain Management in Patients With Cancer, in: *Perioperative Care of the Cancer Patient*, pp. 466–472, Elsevier, Amsterdam, 2023.
94. Papadopoulou, L., *et al.*, Neuropsychiatric symptoms in dementia. The added value of cannabinoids. Are they a safe and effective choice? Case series with cannabidiol 3%. *Ann. Case Rep.*, 7, 799, 2022.
95. Pirttilä, A., *et al.*, Hospitalization and the Risk of Initiation of Antipsychotics in Persons With Parkinson's Disease. *J. Am. Med. Dir. Assoc.*, 24, 9, 1290–1296, 2023.
96. Samanta, D., Management of alternating hemiplegia of childhood: a review. *Pediatr. Neurol.*, 103, 12–20, 2020.
97. Varshney, S., *et al.*, A Comprehensive Review on the Neuropharmacological Effects of Antidepressants in Various Prospects. *Current Pharm. Biotechnol.*, 25, 3, 247–267, 2023.
98. Vandeputte, M.M., *et al.*, Comparative neuropharmacology of structurally distinct non-fentanyl opioids that are appearing on recreational drug markets worldwide. *Drug Alcohol Depend.*, 249, 109939, 2023.
99. Rudra, S., *et al.*, Neuropharmacological and antiproliferative activity of Tetrastigma leucostaphyllum (Dennst.) Alston: Evidence from *in-vivo, in-vitro* and *in-silico* approaches. *Saudi Pharm. J.*, 31, 6, 929–941, 2023.
100. Kuropka, P., Zawadzki, M., Szpot, P., A narrative review of the neuropharmacology of synthetic cathinones—Popular alternatives to classical drugs of abuse. *Hum. Psychopharmacol. Clin. Exp.*, 38, 3, e2866, 2023.
101. Kaur, H., Neuropharmacology of neuro-degenerative,-logical,-psychiatric disorders. *Front. Pharmacol.*, 14, 1223200, 2023.
102. Gonzalez-Rivera, M.L., *et al.*, *In Vivo* Neuropharmacological Effects of Neophytadiene. *Molecules*, 28, 8, 3457, 2023.
103. Gao, Y., *et al.*, Neuropharmacological insight into preventive intervention in posttraumatic epilepsy based on regulating glutamate homeostasis. *CNS Neurosci. Ther.*, 29, 9, 2430–2444, 2023.
104. Brofiga, M., *et al.*, Modularity and neuronal heterogeneity: Two properties that influence *in vitro* neuropharmacological experiments. *Front. Cell. Neurosci.*, 17, 1147381, 2023.
105. Salehi, A., *et al.*, Neuropharmacological Potential of Diterpenoid Alkaloids. *Pharmaceuticals*, 16, 5, 747, 2023.

106. Zhao, T., *et al.*, ABCB1 Gene Polymorphisms Are Closely Associated With Drug-Resistant Epilepsy: Evidence Based on 377 Subjects in Chinese Pediatric Patients. *Clin. Neuropharmacol.*, 46, 4, 10–1097, 2023.
107. Anwar, F., *et al.*, Analysis of Interaction Between Odorant Receptors and Flexible Spike of SARS CoV-2-Key to Loss of Smell. *Curr. Neuropharmacol.*, 21, 1, 151–159, 2023.
108. Wen, Y.-X., *et al.*, Oxytocinergic neurons, but not oxytocin, are crucial for male penile erection. *Neuropharmacology*, 235, 109576, 2023.
109. Gao, F., *et al.*, Fear extinction induced by activation of PKA ameliorates anxiety-like behavior in PTSD mice. *Neuropharmacology*, 222, 109306, 2023.
110. Saha, S. and González-Maeso, J., The crosstalk between 5-HT2AR and mGluR2 in schizophrenia. *Neuropharmacology*, 230, 109489, 2023.
111. Serafini, R., *et al.*, Nucleus Accumbens Single Nucleus RNA Sequencing Identifies Cell Subtype-Specific Maladaptations after Prolonged Spared Nerve Injury, ASPET, 2023.
112. Li, Y., *et al.*, Protective Effects of Ozanimod in Cisplatin-induced Neuropathic Pain in Mice: Insights from Single-cell Transcriptomic Profiling in Spinal Cord Tissues, ASPET, 2023.
113. Ratna, D., Mondal, A.C., Mallick, B.N., Modulation of dopamine from ventral tegmental area neurons by the LC-REM-OFF and PPT-REM-ON neurons in REMS regulation in freely moving rats. *Neuropharmacology*, 237, 109621, 2023.
114. Zou, Y., *et al.*, Measuring Brain Temperature in Youth Bipolar Disorder Using a Novel Magnetic Resonance Imaging Approach: A Proof-of-concept Study. *Curr. Neuropharmacol.*, 21, 6, 1355–1366, 2023.
115. Papageorgiou, M.P., *et al.*, Deciphering the Metabolome under Stress: Insights from Rodent Models. *Curr. Neuropharmacol.*, 22, 5, 884–903, 2023.
116. de la Cabeza Fernández, M., *et al.*, Neuropharmacological Effects in Animal Models and HPLC-Phytochemical Profiling of Byrsonima crassifolia (L.) Kunth Bark Extracts. *Molecules*, 28, 2, 764, 2023.
117. Ernst, A., *et al.*, Identification of side effects of COVID-19 drug candidates on embryogenesis using an integrated zebrafish screening platform, bioRxiv, pp. 2023–06, 2023.
118. Ungvári, O. and Bakos, É., Fluorescence-based methods for studying activity and drug-drug interactions of hepatic solute carrier and ATP binding cassette proteins involved in ADME-Tox. *Biochem. Pharmacol.*, 209, 115448, 2023.
119. Jung, S.-C., Zhou, T., Ko, E.-A., Age-dependent expression of ion channel genes in rat. *Korean J. Physiol. Pharmacol. Off. J. Korean Physiol. Soc. Korean Soc. Pharmacol.*, 27, 1, 85–94, 2023.
120. Leroy, C., *et al.*, A pharmacological imaging challenge based on 11C-buprenorphine PET-MRI to explore the response to opioids in humans. *Eur. J. Nucl. Med. Mol. Imaging*, 50, 10, 1–2, 2023.

121. Battaglia, S., *et al.*, Neuropharmacological modulation of N-methyl-D-aspartate, noradrenaline and endocannabinoid receptors in fear extinction learning: Synaptic transmission and plasticity. *Int. J. Mol. Sci.*, 24, 6, 5926, 2023.
122. Pinna, A., Parekh, P., Morelli, M., Serotonin 5-HT1A receptors and their interactions with adenosine A2A receptors in Parkinson's disease and dyskinesia. *Neuropharmacology*, 226, 109411, 2023.
123. Polit, L.D., *et al.*, Recommendations, guidelines, and best practice for the use of human induced pluripotent stem cells for neuropharmacological studies of neuropsychiatric disorders. *Neurosci. Appl.*, 2, 101125, 2023.
124. Tamargo-Rubio, I., *et al.*, Human induced pluripotent stem cell–derived liver-on-a-chip for studying drug metabolism: the challenge of the cytochrome P450 family. *Front. Pharmacol.*, 14, 1223108, 2023.
125. Ko, T., *et al.*, A Photoactivated Protein Degrader for Optical Control of Synaptic Function, bioRxiv, pp. 2023–02, 2023.
126. Bansal, A., Shikha, S., Zhang, Y., Towards translational optogenetics. *Nat. Biomed. Eng.*, 7, 4, 349–369, 2023.
127. Biney, R.P., *et al.*, Neuropharmacological Assessment of the Methanolic Stem Back Extract of Anopyxis klaineana (Pierre) Engl. (Rhizophoraceae) in Mice. *Phytomed. Plus*, 3, 3, 100470, 2023.
128. Chang-Halabi, Y., *et al.*, Crosstalking interactions between P2X4 and 5-HT3A receptors. *Neuropharmacology*, 236, 109574, 2023.
129. Grigoriou, S., *et al.*, Comparison of dyskinesia profiles after L-DOPA dose challenges with or without dopamine agonist coadministration. *Neuropharmacology*, 237, 109630, 2023.
130. Noorani, B., *et al.*, Advanced Microfluidic Vascularized Tissues as Platform for the Study of Human Diseases and Drug Development. *Curr. Neuropharmacol.*, 21, 3, 599, 2023.
131. Gozzo, L., Spina, E., Drago, F., Innovative treatments for neuro-psychiatric diseases, Frontiers Media SA, p. 1247681, 2023.
132. Garcia-Durillo, M. and Frenguelli, B.G., Antagonism of P2X7 receptors enhances lorazepam action in delaying seizure onset in an *in vitro* model of status epilepticus. *Neuropharmacology*, 239, 109647, 2023.
133. Cardoner, N., *et al.*, Impact of Stress on Brain Morphology: Insights into Structural Biomarkers of Stress-related Disorders. *Curr. Neuropharmacol.*, 22, 5, 935–962, 2023.
134. Beesley, S. and Kumar, S.S., The tN-methyl-d-aspartate receptor: Making the case for d-Serine to be considered its inverse co-agonist. *Neuropharmacology*, 238, 109654, 2023.
135. Zia, F.Z., *et al.*, Are psychedelic medicines the reset for chronic pain? Preliminary findings and research needs. *Neuropharmacology*, 233, 109528, 2023.

136. Xi, D., *et al.*, National Institutes of health psilocybin research speaker series: State of the science, regulatory and policy landscape, research gaps, and opportunities. *Neuropharmacology*, 230, 109467, 2023.
137. Widyarti, S., *et al.*, Dysfunctional energy and future perspective of low dose H2O2 as protective agent in neurodegenerative disease. *Heliyon*, 9, 7, 2023.
138. Saxena, K., Chang, K., Sani, G., Pediatric Bipolar Disorder: Evolution in Clinical and Biological Markers and Future Perspectives. *Curr. Neuropharmacol.*, 21, 6, 1300–1301, 2023.
139. Kundu, S. and Singh, S., *What happens in TBI?* A *wide talk on animal models and future perspective. Curr. Neuropharmacol.*, 21, 5, 1139–1164, 2023.
140. Joksimović, S. and Joksimović, S., Advanced approaches for selective investigation of neuronal function and circuitry: the future of developing novel therapeutic strategies in neuropharmacology? *Arch. Pharm.*, 73, Notebook 2, 109–125, 2023.
141. Heal, D.J., *et al.*, Psychedelics: Threshold of a Therapeutic Revolution, *Neuropharmacology*, 236, 109610, Elsevier, 2023.

3

Basics of Theoretical Neuroscience

Anil P. Dewani[1]*, Deepak S. Mohale[1], Alok S. Tripathi[2] and Naheed Waseem A. Sheikh[3]

[1]Department of Pharmachemistry, P. Wadhwani College of Pharmacy, Yavatmal, Maharashtra, India
[2]Department of Pharmacology, ERA University Lucknow, Uttar Pradesh, India
[3]Department of Pharmacology, Hitech College of Pharmacy, Chandrapur, Maharashtra, India

Abstract
Theoretical neuroscience also called as computational neuroscience or understanding the information contained in neural signals is one goal of mathematical neuroscience. The study of signals is done via modeling the several structural scales of the nervous system, including the circuit, biophysical and system levels. Theoretical neuroscience describes as how the computational methods can be used more precisely in both theoretical and mathematical ways for understanding neural phenomena occurring at different levels of neuronal organization. This chapter discuses about neuronal structure its properties, neuronal networks. The chapter will also discuss about neuronal signaling, neuronal encoding, decoding in neuronal networks, various neuronal network models involved in progression of machine learning and learning and synaptic plasticity.

Keywords: Neuroscience, computational, brain, neural networks

List of Abbreviations

CNNs	Convolutional neural networks
EEG	Electroencephalography
ECoG	Electrocorticography
fMRI	Functional Magnetic Resonance Imaging

**Corresponding author*: anildewani007@gmail.com

Bhupendra Prajapati, Alok Tripathi, Rishabha Malviya and Lucy Mohapatra (eds.) Computational Neuropharmacology: Fundamentals and Clinical Aspects, (67–86) © 2025 Scrivener Publishing LLC

LTP Long-term Potentiation
RNNs Recurrent neural networks

3.1 Introduction

Theoretical neuroscience also known as computational neuroscience is a field that has seen a significant rise in previous three decades [1–3]. Theoretical neuroscience involves computer assisted theoretical and mathematical processes to interpret neural phenomenon occurring at various levels of neural organization. It also tries to understand how brain computes [4–6]. Neuroscience studies approaches involved at molecular and cellular levels to human psychology and psychophysics [7, 8]. Theoretical neuroscience promotes links between these levels by creating a solid representation of what has been learned. It also attempts to create links between various levels of description and looks for principles that connect these ideas [9, 10]. In this chapter, we present details about neuronal structure, basic properties of neurons and neuronal networks. The chapter also briefed about neuronal signaling, neuronal encoding, decoding in neuronal networks, various neuronal network models involved in progression of machine learning and learning and synaptic plasticity.

3.1.1 Neurons and Neuronal Network

3.1.1.1 Neurons

Among all the other cells in the body, neurons are special in that they have the capacity to transmit impulses quickly over considerable distances. The neurons accomplish this by producing distinctive electrical pulses known as action potentials. By spiking in varied temporal patterns, neurons represent and convey information. The measurement and characterization of action potential representations of sensory characteristics, such as light or sound intensity, or motor activities, such as the direction of an arm movement, is fundamental to the study of brain coding [11–13].

Studying the relationship between stimuli and response is possible from two opposing perspectives. The pathway from input to reaction is referred to as neural encoding. In order to build models that attempt to anticipate how neurons will react to different stimuli, for instance, we can catalogue how neurons react to a wide range of stimuli. The aim is to rebuild a stimulus, or certain elements of that stimulus, from the spike sequences it evokes, which is referred to as neural decoding [14–16].

3.1.1.2 Neural Network

A branch of machine learning known as a "neural network" takes its cues from the structure and operation of the human brain. It is a computer model made up of interconnected nodes that may either be thought of as artificial neurons or units that are arranged in layers. These layers are segregated into 3 types: input layer, hidden layers, and output layer [17].

In order for a neural network to learn, its weights and biases must be modified based on the input data and the desired results. Typically, backpropagation is used to do this, which uses an optimization algorithm to reduce the difference between the output as projected and the actual output. The learning process aims to find the optimal values for the network's parameters, allowing it to accurately predict based on fresh, unforeseen data [18].

Neural networks have shown remarkable performance in a variety of applications, such as speech and image recognition, computer vision, and natural language processing, game playing, recommendation systems, and much more. The advancements in neural network architectures and the availability of large datasets have played a crucial part in the resurgence of artificial intelligence and machine learning in recent years. Some popular Feedforward neural networks, convolutional neural networks (CNNs), recurrent neural networks (RNNs), and transformer networks are many forms of neural networks. Each type is tailored to handle specific types of data and tasks more effectively [19].

The neural network has various components, which are as follows [19]:

Input Layer: The neural network's initial layer is where the raw input data is fed. A feature of the input data is represented by each node in the input layer.
Hidden Layers: There may be one or more hidden layers between the input and output layers. Each hidden layer consists of multiple nodes, and these nodes are interconnected to the nodes in the adjacent layers.
Output Layer: the neural network's last layer, which generates the output predictions or decisions based on the processed information from the hidden layers.

3.2 Properties of Neurons and Neuronal Signaling

3.2.1 Properties of Neurons

Neurons are specialized cells that play an essential role in the nervous system, enabling the transmission of electrical signals and facilitating communication between different parts of the body [20–24]. They serve a number of purposes and are the fundamental components of the nervous system, including sensory perception, motor control, cognitive processes, and memory formation. Here are some key components and properties of neurons (Figure 3.1).

- **Excitability:** Neurons can receive and respond to stimuli from their environment or other neurons. When stimulated, they generate electrical impulses or action potentials.
- **Cell Body (Soma):** The nucleus and other organelles necessary for cell maintenance and function are found in the cell body, which is the major component of the neuron.
- **Dendrites:** Branching extensions known as dendrites extend from the cell body. They send impulses towards the cell body after receiving incoming signals from other neurons.
- **Axon:** The neuron's long, thin axon is the extension that sends electrical impulses from the cell body away. A myelin sheath that protects axons can hasten the conduction of electrical signals.
- **Synapses:** At specialized junctions known as synapses, neurons communicate with one another. Neurotransmitters are

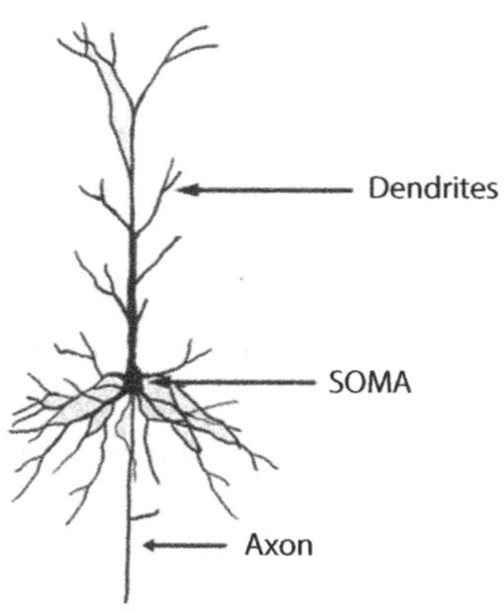

Figure 3.1 Key components of neurons.

released into the synaptic cleft to alert the following neuron (the postsynaptic terminal) when an action potential reaches the end of an axon (the presynaptic terminal).

- **Neurotransmitters:** These are chemical messengers that neurons release into the synapse in order to communicate with other neurons or target cells, such as muscles or glands.
- **Action Potential:** A neuron produces an electrical impulse known as an action potential when it is sufficiently stimulated. This electrical impulse moves through the axon and enables long-distance communication between neurons.
- **All-or-None Response:** The reaction of the action potential is an all-or-none one, meaning that it either happens completely or not at all. Once a certain threshold of stimulation is reached, the neuron fires an action potential with a constant magnitude.
- **Integration:** Neurons can receive input from multiple sources through their dendrites, and these inputs are integrated at the cell body. An action potential is produced if the total input surpasses the threshold.
- **Plasticity:** Neurons can undergo changes in their structure and function in response to experiences and learning. This property is known as neuroplasticity and is fundamental to memory formation and adaptation.
- **Different Types:** There are various types of neurons in the nervous system, including sensory neurons, motor neurons, and interneurons, each with specific roles in transmitting and processing information.

3.2.2 Neuronal Signaling

Neuronal signaling, also known as neural signaling or neuronal communication, refers to the process by which neurons (nerve cells) in the nervous system communicate with each other [25–29]. This communication is essential for the functioning of the brain and the transmission of information throughout the nervous system. Neuronal signaling occurs through both electrical and chemical processes and involves the transmission of signals from one neuron to another, as well as the transmission of signals from neurons to other types of cells, such as muscle cells or gland cells.

The key elements of neuronal signaling include:

- **Electrical signaling:** Neurons are electrically excitable cells. They can generate and transmit electrical signals known as action potentials or spikes. When a neuron receives sufficient excitatory input, its cell membrane depolarizes, causing a rapid change in electrical potential across the membrane. If this depolarization hits a specific threshold, an action potential triggers. The action potential then propagates along the neuron's axon, allowing it to transmit signals over long distances.

- **Synaptic transmission (chemical signaling):** Most communication between neurons occurs at specialized junctions called synapses. Neurotransmitters are released into the synaptic cleft, a tiny space between the presynaptic and postsynaptic neurons, when an action potential contacts a neuron's presynaptic terminal (also known as the axon terminal). Neurotransmitters are chemical messengers that carry the signal across the synapse. They bind to certain postsynaptic neuron receptors, resulting in changes in its electrical potential, either depolarizing it (excitatory) or hyperpolarizing it (inhibitory).
- **Neuromodulation:** In addition to classic neurotransmitters, neurons can release neuromodulators, which can modify the strength of synaptic connections or the excitability of neurons. Neuromodulators play a significant role in regulating neural circuit dynamics and are involved in various brain functions, including learning, memory, and emotional processing.
- **Feedback Loops:** Neuronal signaling often involves feedback loops, where the output from one set of neurons acts as input to another set of neurons, leading to recurrent or ongoing neural activity. These feedback loops are essential for complex information processing and control in the brain.
- **Signal Integration:** Neurons receive input from multiple sources, and they integrate these inputs to determine whether to generate an action potential. This process of signal integration allows neurons to perform complex computations and respond to changing environmental conditions.

3.3 Recording Neuronal Responses

Recording neuronal responses is a crucial technique in neuroscience that allows researchers to study the activity of particular neurons or groups of neurons [30–47]. These recordings provide valuable insights into the functioning of the brain and how it processes information. There are various methods used to record neuronal responses, depending on the experimental goals and the scale of recording required. Some common techniques include:

- **Extracellular Single-Unit Recording:** This method involves inserting a fine electrode (usually made of glass or metal) into the brain tissue to detect the electrical activity of individual neurons. The electrode can be positioned near the desired neuronal population, and the electrical signals generated by the firing of these neurons are amplified, filtered, and recorded.
- **Electroencephalography:** EEG records the electrical activity of the brain by putting number of electrodes to the scalp a number of electrodes to the scalp. It measures the summed activity of large groups of neurons and is useful for studying overall brain states and rhythms.
- **Electrocorticography:** ECoG is similar to EEG but involves placing electrodes directly on the surface of the brain (cortical surface). It provides a higher spatial resolution than EEG and is used in certain clinical and research settings.
- **Intracellular Recording:** In this method, an electrode is inserted into a single neuron to measure its membrane potential changes more directly. Intracellular recordings offer high precision but are more technically challenging and limited to specific experimental setups.
- **Calcium Imaging:** This technique uses genetically encoded calcium indicators to detect changes in intracellular calcium levels, which are associated with neuronal activity. Calcium imaging can record the activity of multiple neurons simultaneously and is often used in studies involving intact brain tissue or animal models.
- **Voltage-Sensitive Dye Imaging: As l**ike calcium imaging, voltage-sensitive dyes are used to monitor changes in the membrane potential of neurons. This technique provides high spatial resolution but is generally limited to *in vitro* preparations.
- **Functional Magnetic Resonance Imaging (fMRI):** fMRI measures changes in blood oxygen levels in different brain regions, which indirectly reflects neural activity. It offers good spatial resolution and is often used in human studies.

3.4 Neural Encoding and Neuronal Decoding

3.4.1 Neuronal Encoding

Neuronal encoding is the process by which sensory information from the external world is transformed into neural signals within the nervous system [48–56]. It entails comprehending how the activity of specific neurons or populations of neurons encodes and represents stimuli. Two important techniques used in studying neural encoding, particularly in the context of visual information, are reverse correlation and visual receptive fields.

Reverse Correlation: Reverse correlation, also known as reverse correlation analysis or spike-triggered averaging, is a mathematical technique used to characterize the receptive fields of neurons. Receptive fields are specific regions in the sensory space that influence the activity of a neuron when stimulated. In visual systems, a receptive field refers to the region of the visual field that manipulates the firing rate of a particular neuron when presented with visual stimuli.

The reverse correlation technique involves presenting a series of random stimuli (e.g., visual patterns) to a neuron while recording its responses. By analyzing the correlation between the presented stimuli and the neuron's responses, researchers can infer the neuron's receptive field properties. Specifically, they identify the features of the stimuli that trigger the neuron to fire more or less frequently.

- **Visual Receptive Fields:** Visual receptive fields are critical in understanding how neurons process visual information. In the visual system, each neuron, particularly in the early stages of visual processing (e.g., in the retina or visual cortex), responds to specific visual features within its receptive field.

For example, in the visual cortex, some neurons may have receptive fields that respond best to lines oriented in a specific direction, while others may be more sensitive to particular colors or spatial frequencies. By studying the receptive fields of neurons in the visual system, researchers can gain insights into how visual information is encoded and represented in the brain.

Visual receptive fields can have different shapes and sizes depending on the neuron's location in the visual pathway and the specific visual feature to which it is sensitive. By mapping these receptive fields, researchers can create receptive field profiles for individual neurons or neural populations, helping them better understand how the brain processes visual stimuli.

In the study of neuroscience, receptive field mapping and reverse correlation are both crucial technologies that have greatly advanced our understanding of how neurons encode sensory information, particularly in the visual system. These techniques have broader applications beyond vision and are used to study encoding mechanisms in various sensory modalities.

3.4.2 Neural Decoding

Neural decoding, also known as brain decoding or brain-machine decoding, is a process in neuroscience and neuroengineering that aims to interpret the activity patterns of neurons or brain regions to extract meaningful information or make predictions about cognitive states, intentions, or sensory perceptions [57–61].

The basic idea behind neural decoding is to analyze the neural activity recorded from the brain and decode or translate it into meaningful information that can be used to understand brain function, enable communication with external devices, or restore lost sensory or motor functions. This is typically done using advanced statistical and machine learning techniques.

The process of neural decoding (Figure 3.2) can be summarized as follows:

- **Neural Data Acquisition:** Neural decoding requires recording neural activity from the brain. This can be done using various techniques, such as electroencephalography (EEG),

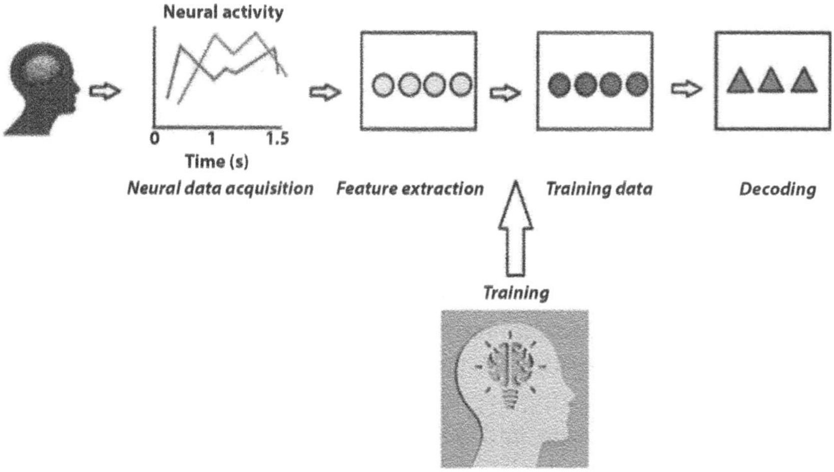

Figure 3.2 Schematic process of neuronal decoding.

intracranial electrocorticography (ECoG), functional magnetic resonance imaging (fMRI), or single-neuron recordings using electrodes.

- **Feature Extraction:** The recorded neural data needs to be preprocessed to extract relevant features that carry information. These features could be the firing rates of individual neurons, frequency bands in EEG signals, or patterns of brain activity in fMRI data.
- **Training Data Collection:** To build a decoding model, it is necessary to collect data while the subject performs specific tasks or experiences certain sensory stimuli. For example, if the goal is to decode visual information, the subject may be presented with different visual stimuli while their brain activity is recorded.
- **Decoding Model Training:** Support vector machines and neural networks are examples of machine learning algorithms, or Bayesian classifiers, are used to build a decoding model. The model is trained using the neural data collected during the training tasks/stimuli, along with the corresponding behavioral or sensory information. Once the decoding model is trained, it can be used to predict or decode brain activity in real-time. New neural data collected during different tasks or stimuli can be fed into the model, which then predicts the corresponding cognitive state, intention, or sensory perception.

3.5 Neuronal Network Models

Neural circuitry is characterized by high levels of synaptic connection. A normal mammalian neuron, for instance, gets hundreds of synaptic inputs. Utilizing simulations and analyses, network models enable us to investigate the computational potential of such connectedness.

Decoding in neuronal networks is a very powerful tool, also it is helpful in studying model neural networks as it gives us a brief information about activity of population. Computational models in theoretical neuroscience are essential for generating testable predictions, understanding neural mechanisms, and exploring hypotheses about brain function. They often require parameter tuning and validation against experimental data to ensure their accuracy and relevance.

Computational models for theoretical neuroscience

- **Spiking neural network models [62–66]:** These models simulate individual neurons as spiking units that generate action potentials (spikes) in response to inputs (Figure 3.3). The timing and pattern of spikes encode information and communication between neurons. Examples of spiking neural network models include the Hodgkin-Huxley model and the integrate-and-fire model.
- **Firing rate models [67–70]:** In contrast to spiking models, rate-based models consider the average firing rate of neurons rather than individual spikes. These models are computationally less demanding and are often used to study large-scale neural networks and their emergent properties.
- **Recurrent Models [71–73]:** Recurrent models (Figure 3.4) are a kind of artificial neural network with recurrent connections that allow information to persist over time. They can capture temporal dynamics and are often used to model working memory, decision-making, and sequence processing in the brain.

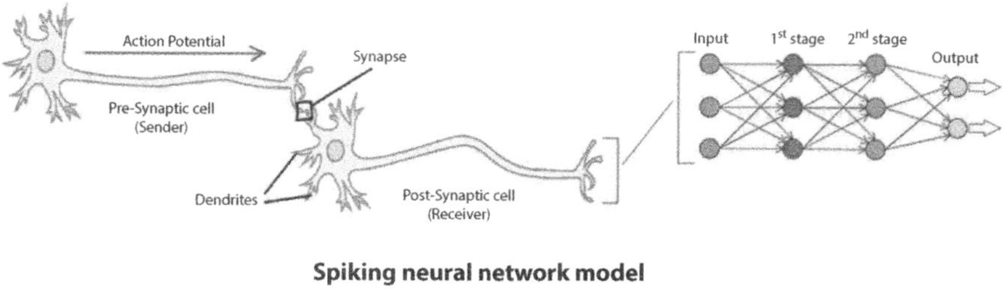

Figure 3.3 Schematic diagram showing spiking neural network model.

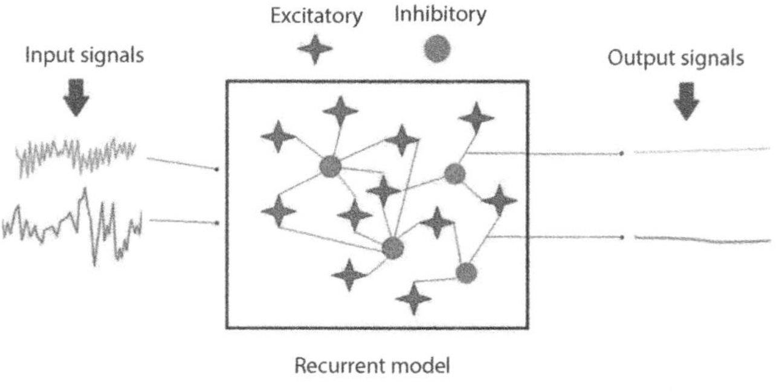

Figure 3.4 Schematic diagram showing recurrent neuronal model.

- **Feed forward model [74–78]:** Feed forward model is an artificial neural network model which features linear networks and do not involve formation of cycle. It is opposite to recurrent model which features cycled pathways. It is one of the simplest forms of network model featuring processing of information in a linear one way.

3.6 Learning and Synaptic Plasticity

Learning and synaptic plasticity are essential concepts in neuroscience that help us understand how the brain adapts and changes in response to experiences and environmental stimuli. The ability of synapses, which are the connections between neurons, to change their strength and efficacy over time is referred to as synaptic plasticity. This phenomenon is crucial for learning and memory processes in the brain [79–83].

There are two main types of synaptic plasticity:

- **Long-term Potentiation [84–88]**: long term potentiation is a process by which synaptic connections are strengthened, leading to increased efficiency in transmitting signals between neurons. It occurs when there is repeated and sustained stimulation of a synapse, and it is often considered a cellular basis for learning and memory. Synapses undergo modifications during LTP, including an increase in the number of neurotransmitter receptors and the development of new synaptic connections.
- **Long-term Depression:** The opposite of long-term potentiation is long-term depression. It causes synaptic connections to deteriorate due to prolonged low-frequency stimulation or lack of stimulation. Long-term Depression plays a crucial role in neural network refinement and eliminating unnecessary connections, which is important for information processing and memory consolidation [89–91].

Learning involves the acquisition and storage of new information or skills. It frequently involves the creation of fresh synaptic connections or the alteration of already existing ones. Specific brain pathways are active when we learn something new, and synaptic plasticity mechanisms come into play to strengthen the relevant connections, allowing us to retain that information for future use [79–83].

Different factors can influence synaptic plasticity and learning:

- **Neurotransmitters [92–94]:** Chemicals released by neurons, such as glutamate, play an important role in long term potentiation and long-term depression by facilitating or inhibiting synaptic changes.
- **Receptor activation [95–97]:** Neurotransmitter receptors on the postsynaptic membrane are crucial for transmitting signals during long term potentiation and long-term depression.
- **Neural activity [98–100]:** Repeated or intense neural activity strengthens synapses through long term potentiation, while infrequent or low-intensity activity can weaken synapses through long term depression.
- **Calcium influx [101–105]:** Calcium ions play a significant role in the signaling pathways that mediate synaptic plasticity.
- **Neurotrophic factors [106–109]:** These molecules support the growth, survival, and maintenance of neurons, influencing synaptic plasticity.
- **Brain-derived neurotrophic factor [106–109]:** A protein that promotes neuronal survival and supports the growth and differentiation of new synapses.

The interplay between long term potentiation and long-term depression is critical for maintaining a balance in neural circuits, preventing excessive excitation or inhibition, and ensuring efficient information processing and memory storage.

3.7 Conclusion

Theoretical neuroscience provides a framework for understanding the complex workings of the brain. By combining experimental data, mathematical modeling, and computational simulations, theoretical neuroscientists aim to uncover the fundamental principles of brain function. The concepts discussed in this chapter lay the foundation for further exploration into the fascinating field of theoretical neuroscience.

References

1. Izhikevich, E.M., *Dynamical Systems in neuroscience: The geometry of excitability and bursting*, MIT Press, London, 2010.
2. Anderson, B., *Computational neuroscience and cognitive modelling. A student's introduction to methods and procedures*, Sage Publ. Ltd, Beverley Hills, 2014.
3. Arbib, M., *The handbook of brain theory and neural networks*, MIT Press, Cambridge, 2002.
4. Bechtel, W., *Philosophy of science: An overview for cognitive science*, Psychology Press, New York, 2013.
5. Green, S., Şerban, M., Scholl, R., Jones, N., Brigandt, I., Bechtel, W., Network analyses in systems biology: new strategies for dealing with biological complexity. *Synthese*, 195, 4, 1751–77, 2018.
6. Levenstein, D., Alvarez, V.A., Amarasingham, A., Azab, H., Gerkin, R.C., Hasenstaub, A., On the role of theory and modeling in neuroscience, arXiv preprint arXiv:200313825, 2020.
7. Montague, P.R., Dolan, R.J., Friston, K.J., Dayan, P., Computational psychiatry. *Trends Cognit. Sci.*, 16, 1, 72–80, 2012.
8. Rieke, F., Warland, D., Van Steveninck, R.D.R., Bialek, W.S., *Spikes: exploring the neural code*, MIT press, Cambridge, 1999.
9. Insel, T.R., Landis, S.C., Collins, F.S., The NIH brain initiative. *Science*, 340, 6133, 687–8, 2013.
10. Kass, R.E., Amari, S.-I., Arai, K., Brown, E.N., Diekman, C.O., Diesmann, M., Computational neuroscience: Mathematical and statistical perspectives. *Annu. Rev. Stat. Appl.*, 5, 183–214, 2018.
11. Churchland, P.S. and Sejnowski, T.J., *The Computational Brain*, MIT Press, London, 1992.
12. Harris-Warrick, R.M., Marder, E., Selverston, A.I., Moulins, M., *Dynamic Biological Networks: The Stomatogastric Nervous System*, MIT Press, London, 1992.
13. Gardner, D., *The Neurobiology of Neural Networks*, MIT Press, London, 2003.
14. Koch, C. and Davis, J.L., *Large-Scale Neuronal Theories of the Brain*, MIT Press, London, 2003.
15. Idan, S., John, R., Gordon, M.S., *The Theoretical Foundations of Dendritic Function: Selected Papers of Wilfrid Rall with Commentaries*, The MIT Press, London, England, 1995.
16. James, C.H., Joel, L.D., David, G.B., *Models of Information Processing in the Basal Ganglia*, MIT press, USA, 1995.
17. Hopfield, J.J., Neural networks and physical systems with emergent collective computational properties. *Proc. Nat. Acad. Sci. (USA)*, 79, 2554–2558, 1982.
18. Abdi, H., Valentin, D., Edelman, B., *Neural Networks*, SAGE Publication Inc., Thousand Oaks, CA, 1999.

19. Haykin, S., *Neural Networks: A Comprehensive Foundation*, Macmillan Publishing, New York, 1994.
20. Friesen, W.O. and Jonathon, A.F., Properties of Neurons', in: *NeuroDynamix II: Concepts of Neurophysiology Illustrated by Computer Simulations*, New York, 2009; online edn, Oxford Academic, 2010.
21. Arshavskii, Y.I., Berkinblit, M.B., Kovalev, S.A., Smolyaninov, V.V., Chailakhyan, L.M., The role of dendrites in the functioning of nerve cells. *Dokl. Akad. Nauk SSSR*, 163, 994–997, 1965.
22. Barnwell, G.M. and Cerimele, B.J., A mathematical model of the effects of spatio-temporal patterns of dendritic input potentials on neuronal somatic potentials. *Kybernetik*, 10, 144–155, 1972.
23. Binstock, L. and Goldman, L., Current and voltage clamped studies on Myxicola giant axons. *J. Gen. Physiol.*, 54, 730–740, 1969.
24. Hagen, E., Magnusson, S.H., Ness, T.V., Halnes, G., Babu, P.N., Linssen, C., Morrison, A., Einevoll, G.T., Brain signal predictions from multi-scale networks using a linearized framework. *PLoS Comput. Biol.*, 18, 8, e1010353, 2022.
25. Sullivan, A.M., Neuronal Signaling: an introduction. *Neuronal Signal.*, 1, 1, NS20160025, 2016.
26. Russo, S.J., Dietz, D.M., Dumitriu, D., Morrison, J.H., Malenka, R.C., Nestler, E.J., The addicted synapse: mechanisms of synaptic and structural plasticity in nucleus accumbens. *Trends Neurosci.*, 33, 6, 267–76, 2010.
27. Ebert, D.H. and Greenberg, M.E., Activity-dependent neuronal signalling and autism spectrum disorder. *Nature*, 493, 7432, 327–37, 2013.
28. Suomi, F. and McWilliams, T.G., Autophagy in the mammalian nervous system: a primer for neuroscientists. *Neuronal Signal.*, 3, 3, NS20180134, 2019.
29. Paolini, A.G., Morgan, S.J., Kim, J.H., Auditory fear conditioning alters neural gain in the cochlear nucleus: a wireless neural recording study in freely behaving rats. *Neuronal Signal.*, 4, 4, NS20200009, 2020.
30. Herrmann, C.S. and Debener, S., Simultaneous recording of EEG and BOLD responses: a historical perspective. *Int. J. Psychophysiol.*, 67, 3, 161–8, 2008.
31. Duru, A.D., Erdogan, S.B., Kasikci, I., Bayram, A., Ademoglu, A., Demiralp, T., Investigaton of the neuronal efficacy and EEG source power under steady-state visual stimulation. *Annu. Int. Conf. IEEE Eng. Med. Biol. Soc.*, 2011, 6576–9, 2011.
32. Toda, H., Suzuki, T., Sawahata, H., Majima, K., Kamitani, Y., Hasegawa, I., Simultaneous recording of ECoG and intracortical neuronal activity using a flexible multichannel electrode-mesh in visual cortex. *Neuroimage*, 54, 1, 203–12, 2011.
33. Piantoni, G., Hermes, D., Ramsey, N., Petridou, N., Size of the spatial correlation between ECoG and fMRI activity. *Neuroimage*, 242, 118459, 2021.
34. Bugeon, S., Duffield, J., Dipoppa, M., Ritoux, A., Prankerd, I., Nicoloutsopoulos, D., Orme, D., Shinn, M., Peng, H., Forrest, H., Viduolyte, A., Reddy, C.B.,

Isogai, Y., Carandini, M., Harris, K.D., A transcriptomic axis predicts state modulation of cortical interneurons. *Nature*, 607, 7918, 330–338, 2022.
35. Xu, S., Yang, H., Menon, V., Lemire, A.L., Wang, L., Henry, F.E., Turaga, S.C., Sternson, S.M., Behavioral state coding by molecularly defined paraventricular hypothalamic cell type ensembles. *Science*, 370, 6514, eabb2494, 2020.
36. Aramuni, G. and Griesbeck, O., Chronic calcium imaging in neuronal development and disease. *Exp. Neurol.*, 242, 50–6, 2013.
37. Pachitariu, M., Stringer, C., Harris, K.D., Robustness of Spike Deconvolution for Neuronal Calcium Imaging. *J. Neurosci.*, 38, 37, 7976–7985, 2018.
38. Li, L., Durand-de Cuttoli, R., Aubry, A.V., Burnett, C.J., Cathomas, F., Parise, L.F., Chan, K.L., Morel, C., Yuan, C., Shimo, Y., Lin, H.Y., Wang, J., Russo, S.J., Social trauma engages lateral septum circuitry to occlude social reward. *Nature*, 613, 7945, 696–703, 2023.
39. Momose-Sato, Y. and Sato, K., Voltage-sensitive dye imaging during functional development of the embryonic nervous system: a brief review with special thanks to Professor Larry Cohen. *Neurophotonics*, 2, 2, 021009, 2015.
40. Zhang, H.K., Yan, P., Kang, J., Abou, D.S., Le, H.N., Jha, A.K., Thorek, D.L., Kang, J.U., Rahmim, A., Wong, D.F., Boctor, E.M., Loew, L.M., Listening to membrane potential: photoacoustic voltage-sensitive dye recording. *J. Biomed. Opt.*, 22, 4, 45006, 2017.
41. Kozyrev, V., Eysel, U.T., Jancke, D., Voltage-sensitive dye imaging of transcranial magnetic stimulation-induced intracortical dynamics. *Proc. Natl. Acad. Sci. U.S.A.*, 111, 37, 13553–8, 2014.
42. Inaji, M., Sato, K., Momose-Sato, Y., Ohno, K., Voltage-sensitive dye imaging analysis of functional development of the neonatal rat corticostriatal projection. *Neuroimage*, 54, 3, 1831–9, 2011.
43. Hordacre, B., Moezzi, B., Ridding, M.C., Neuroplasticity and network connectivity of the motor cortex following stroke: A transcranial direct current stimulation study. *Hum. Brain Mapp.*, 39, 8, 3326–3339, 2018.
44. Johnson, L.C., Richards, T.L., Archbold, K.H., Landis, C.A., Functional magnetic resonance imaging in nursing research. *Biol. Res. Nurs.*, 8, 1, 43–54, 2006.
45. Schomer, D.L., Bonmassar, G., Lazeyras, F., Seeck, M., Blum, A., Anami, K., Schwartz, D., Belliveau, J.W., Ives, J., EEG-Linked functional magnetic resonance imaging in epilepsy and cognitive neurophysiology. *J. Clin. Neurophysiol.*, 17, 1, 43–58, 2000.
46. Bell, A.H., Malecek, N.J., Morin, E.L., Hadj-Bouziane, F., Tootell, R.B., Ungerleider, L.G., Relationship between functional magnetic resonance imaging-identified regions and neuronal category selectivity. *J. Neurosci.*, 31, 34, 12229–40, 2011.
47. Ma, Z., Cao, P., Sun, P., Lu, Z., Li, L., Chen, Y., Chai, X., Negative hemodynamic response without neuronal inhibition investigated by combining optical imaging and electrophysiological recording. *Neurosci. Lett.*, 637, 161–167, 2017.

48. Williams, A.M., Angeloni, C.F., Geffen, M.N., Sound Improves Neuronal Encoding of Visual Stimuli in Mouse Primary Visual Cortex. *J. Neurosci.*, 43, 16, 2885–2906, 2023.
49. Backen, T., Treue, S., Martinez-Trujillo, J.C., Encoding of Spatial Attention by Primate Prefrontal Cortex Neuronal Ensembles. *eNeuro*, 5, 1, ENEURO.0372-16.2017, 2018.
50. Averbeck, B.B., Latham, P.E., Pouget, A., Neural correlations, population coding and computation. *Nat. Rev. Neurosci.*, 7, 5, 358–66, 2006.
51. Kirsch, J.A., Vlachos, I., Hausmann, M., Rose, J., Yim, M.Y., Aertsen, A., Güntürkün, O., Neuronal encoding of meaning: establishing category-selective response patterns in the avian 'prefrontal cortex'. *Behav. Brain Res.*, 198, 1, 214–23, 2009.
52. Cubero, R.J., Marsili, M., Roudi, Y., Multiscale relevance and informative encoding in neuronal spike trains. *J. Comput. Neurosci.*, 48, 1, 85–102, 2020.
53. Yang, Y. and Mailman, R.B., Strategic neuronal encoding in medial prefrontal cortex of spatial working memory in the T-maze. *Behav. Brain Res.*, 343, 50–60, 2018.
54. Xia, J., Marks, T.D., Goard, M.J., Wessel, R., Diverse coactive neurons encode stimulus-driven and stimulus-independent variables. *J. Neurophysiol.*, 124, 5, 1505–1517, 2020.
55. Huang, C., Resnik, A., Celikel, T., Englitz, B., Adaptive Spike Threshold Enables Robust and Temporally Precise Neuronal Encoding. *PLoS Comput. Biol.*, 12, 6, e1004984, 2016.
56. Rieke, F., Warland, D., van Steveninck, R.R., Bialek, W., *Spikes: Exploring the Neural Code*, MIT press, USA, 1997.
57. Mancinelli, S. and Lodato, S., Decoding neuronal diversity in the developing cerebral cortex: from single cells to functional networks. *Curr. Opin. Neurobiol.*, 53, 146–155, 2018.
58. Tabata, H. and Nagata, K., Decoding the molecular mechanisms of neuronal migration using in utero electroporation. *Med. Mol. Morphol.*, 49, 2, 63–75, 2016.
59. Abbott, L.F., Decoding neuronal firing and modelling neural networks. *Q. Rev. Biophys.*, 27, 3, 291–331, 1994.
60. Zhang, Y., Aghajan, Z.M., Ison, M., Lu, Q., Tang, H., Kalender, G., Monsoor, T., Zheng, J., Kreiman, G., Roychowdhury, V., Fried, I., Decoding of human identity by computer vision and neuronal vision. *Sci. Rep.*, 13, 1, 651, 2023.
61. Quian Quiroga, R. and Panzeri, S., Extracting information from neuronal populations: information theory and decoding approaches. *Nat. Rev. Neurosci.*, 10, 3, 173–85, 2009.
62. Chen, L. and Campbell, S.A., Exact mean-field models for spiking neural networks with adaptation. *J. Comput. Neurosci.*, 50, 4, 445–469, 2022.
63. Pietrzak, P., Szczęsny, S., Huderek, D., Przyborowski, Ł., Overview of Spiking Neural Network Learning Approaches and Their Computational Complexities. *Sensors (Basel)*, 23, 6, 3037, 2023.

64. Hu, L., Liu, Y., Qiu, W., A Deep Spiking Neural Network Anomaly Detection Method. *Comput. Intell. Neurosci.*, 6391750, 2022.
65. Ghosh-Dastidar, S. and Adeli, H., Spiking neural networks. *Int. J. Neural Syst.*, 19, 4, 295–308, 2009.
66. Zhao, J., Yang, J., Wang, J., Wu, W., Spiking Neural Network Regularization With Fixed and Adaptive Drop-Keep Probabilities. *IEEE Trans. Neural Netw. Learn. Syst.*, 33, 8, 4096–4109, 2022.
67. Schaffer, E.S., Ostojic, S., Abbott, L.F., A complex-valued firing-rate model that approximates the dynamics of spiking networks. *PLoS Comput. Biol.*, 9, 10, e1003301, 2013.
68. Keeley, S., Byrne, Á., Fenton, A., Rinzel, J., Firing rate models for gamma oscillations. *J. Neurophysiol.*, 121, 6, 2181–2190, 2019.
69. Fasoli, D., Faugeras, O., Panzeri, S., A formalism for evaluating analytically the cross-correlation structure of a firing-rate network model. *J. Math. Neurosci.*, 5, 6, 2015.
70. Apfaltrer, F., Ly, C., Tranchina, D., Population density methods for stochastic neurons with realistic synaptic kinetics: firing rate dynamics and fast computational methods. *Network*, 17, 4, 373–418, 2006.
71. Barak, O., Recurrent neural networks as versatile tools of neuroscience research. *Curr. Opin. Neurobiol.*, 46, 1–6, 2017.
72. Tang, M., Salvatori, T., Millidge, B., Song, Y., Lukasiewicz, T., Bogacz, R., Recurrent predictive coding models for associative memory employing covariance learning. *PLoS Comput. Biol.*, 19, 4, e1010719, 2023.
73. Harkin, E.F., Shen, P.R., Goel, A., Richards, B.A., Naud, R., Parallel and Recurrent Cascade Models as a Unifying Force for Understanding Subcellular Computation. *Neuroscience*, 489, 200–215, 2022.
74. Luz, Y. and Shamir, M., Oscillations via Spike-Timing Dependent Plasticity in a Feed-Forward Model. *PLoS Comput. Biol.*, 12, 4, e1004878, 2016.
75. Sinha, D.B., Ledbetter, N.M., Barbour, D.L., Spike-timing computation properties of a feed-forward neural network model. *Front. Comput. Neurosci.*, 8, 5, 2014.
76. Karimi-Rouzbahani, H., Bagheri, N., Ebrahimpour, R., Hard-wired feedforward visual mechanisms of the brain compensate for affine variations in object recognition. *Neuroscience*, 349, 48–63, 2017.
77. Frolov, N., Maksimenko, V., Lüttjohann, A., Koronovskii, A., Hramov, A., Feed-forward artificial neural network provides data-driven inference of functional connectivity. *Chaos*, 29, 9, 091101, 2019.
78. Burns, T.F., Classic Hebbian learning endows feed-forward networks with sufficient adaptability in challenging reinforcement learning tasks. *J. Neurophysiol.*, 125, 6, 2034–2037, 2021.
79. Magee, J.C. and Grienberger, C., Synaptic Plasticity Forms and Functions. *Annu. Rev. Neurosci.*, 43, 95–117, 2020.

80. Neves, G., Cooke, S.F., Bliss, T.V., Synaptic plasticity, memory and the hippocampus: a neural network approach to causality. *Nat. Rev. Neurosci.*, 9, 1, 65–75, 2008.
81. Martin, S.J., Grimwood, P.D., Morris, R.G., Synaptic plasticity and memory: an evaluation of the hypothesis. *Annu. Rev. Neurosci.*, 23, 649–711, 2000.
82. Porter, J.T. and Sepulveda-Orengo, M.T., Learning-induced intrinsic and synaptic plasticity in the rodent medial prefrontal cortex. *Neurobiol. Learn. Mem.*, 169, 107117, 2020.
83. Niu, W., Duan, Y., Kang, Y., Cao, X., Xue, Q., Propofol improves learning and memory in post-traumatic stress disorder (PTSD) mice via recovering hippocampus synaptic plasticity. *Life Sci.*, 293, 120349, 2022.
84. Lüscher, C. and Malenka, R.C., NMDA receptor-dependent long-term potentiation and long-term depression (LTP/LTD). *Cold Spring Harb. Perspect. Biol.*, 4, 6, a005710, 2012.
85. Bin Ibrahim, M.Z., Benoy, A., Sajikumar, S., Long-term plasticity in the hippocampus: maintaining within and 'tagging' between synapses. *FEBS J.*, 289, 8, 2176–2201, 2022.
86. Dringenberg, H.C., The history of long-term potentiation as a memory mechanism: Controversies, confirmation, and some lessons to remember. *Hippocampus*, 30, 9, 987–1012, 2020.
87. Hernandez, A., Tan, C., Mettlach, G., Pozo, K., Plattner, F., Bibb, J.A., Cdk5 Modulates Long-Term Synaptic Plasticity and Motor Learning in Dorsolateral Striatum. *Sci. Rep.*, 6, 29812, 2016.
88. Cobar, L.F., Yuan, L., Tashiro, A., Place cells and long-term potentiation in the hippocampus. *Neurobiol. Learn. Mem.*, 138, 206–214, 2017.
89. Hirano, T., Long-term depression and other synaptic plasticity in the cerebellum. *Proc. Jpn. Acad. Ser. B Phys. Biol. Sci.*, 89, 5, 183–95, 2013.
90. Zhuo, M. and Hawkins, R.D., Long-term depression: a learning-related type of synaptic plasticity in the mammalian central nervous system. *Rev. Neurosci.*, 6, 3, 259–77, 1995.
91. Shim, H.G., Jang, D.C., Lee, J., Chung, G., Lee, S., Kim, Y.G., Jeon, D.E., Kim, S.J., Long-Term Depression of Intrinsic Excitability Accompanied by Synaptic Depression in Cerebellar Purkinje Cells. *J. Neurosci.*, 37, 23, 5659–5669, 2017.
92. Speranza, L., di Porzio, U., Viggiano, D., de Donato, A., Volpicelli, F., Dopamine: The Neuromodulator of Long-Term Synaptic Plasticity, Reward and Movement Control. *Cells*, 10, 4, 735, 2021.
93. McCarberg, B. and Peppin, J., Pain Pathways and Nervous System Plasticity: Learning and Memory in Pain. *Pain Med.*, 20, 12, 2421–2437, 2019.
94. Puścian, A., Winiarski, M., Łęski, S., Charzewski, Ł., Nikolaev, T., Borowska, J., Dzik, J.M., Bijata, M., Lipp, H.P., Dziembowska, M., Knapska, E., Chronic fluoxetine treatment impairs motivation and reward learning by affecting neuronal plasticity in the central amygdala. *Br. J. Pharmacol.*, 178, 3, 672–688, 2021.

95. Baez, M.V., Cercato, M.C., Jerusalinsky, D.A., NMDA Receptor Subunits Change after Synaptic Plasticity Induction and Learning and Memory Acquisition. *Neural Plast.*, 2018, 5093048, 2018.
96. Handler, A., Graham, T.G.W., Cohn, R., Morantte, I., Siliciano, A.F., Zeng, J., Li, Y., Ruta, V., Distinct Dopamine Receptor Pathways Underlie the Temporal Sensitivity of Associative Learning. *Cell*, 178, 1, 60–75.e19, 2019.
97. Senkov, O., Andjus, P., Radenovic, L., Soriano, E., Dityatev, A., Neural ECM molecules in synaptic plasticity, learning, and memory. *Prog. Brain Res.*, 214, 53–80, 2014.
98. Tsilibary, E., Tzinia, A., Radenovic, L., Stamenkovic, V., Lebitko, T., Mucha, M., Pawlak, R., Frischknecht, R., Kaczmarek, L., Neural ECM proteases in learning and synaptic plasticity. *Prog. Brain Res.*, 214, 135–57, 2014.
99. Chen, C. and Tonegawa, S., Molecular genetic analysis of synaptic plasticity, activity-dependent neural development, learning, and memory in the mammalian brain. *Annu. Rev. Neurosci.*, 20, 157–84, 1997.
100. Xie, Y.F., Jackson, M.F., Macdonald, J.F., Optogenetics and synaptic plasticity. *Acta Pharmacol. Sin.*, 34, 11, 1381–5, 2013.
101. Inglebert, Y. and Debanne, D., Calcium and Spike Timing-Dependent Plasticity. *Front. Cell. Neurosci.*, 15, 727336, 2021.
102. Cascella, R. and Cecchi, C., Calcium Dyshomeostasis in Alzheimer's Disease Pathogenesis. *Int. J. Mol. Sci.*, 22, 9, 4914, 2021.
103. Honnuraiah, S. and Narayanan, R., A calcium-dependent plasticity rule for HCN channels maintains activity homeostasis and stable synaptic learning. *PLoS One*, 8, 2, e55590, 2013.
104. Zhai, J., Navakkode, S., Yeow, S.Q.Z., Krishna-K, K., Liang, M.C., Koh, J.H., Wong, R.X., Yu, W.P., Sajikumar, S., Huang, H., Soong, T.W., Loss of Ca_V1.3 RNA editing enhances mouse hippocampal plasticity, learning, and memory. *Proc. Natl. Acad. Sci. U.S.A.*, 119, 32, e2203883119, 2022.
105. Wang, H. and Zhang, M., The role of Ca^{2+}-stimulated adenylyl cyclases in bidirectional synaptic plasticity and brain function. *Rev. Neurosci.*, 23, 1, 67–78, 2012.
106. Leal, G., Comprido, D., Duarte, C.B., BDNF-induced local protein synthesis and synaptic plasticity. *Neuropharmacology*, 76 Pt C, 639–56, 2014.
107. Colucci-D'Amato, L., Speranza, L., Volpicelli, F., Neurotrophic Factor BDNF, Physiological Functions and Therapeutic Potential in Depression, Neurodegeneration and Brain Cancer. *Int. J. Mol. Sci.*, 21, 20, 7777, 2020.
108. Lu, B., Nagappan, G., Lu, Y., BDNF and synaptic plasticity, cognitive function, and dysfunction. *Handb. Exp. Pharmacol.*, 220, 223–50, 2014.
109. Leal, G., Bramham, C.R., Duarte, C.B., BDNF and Hippocampal Synaptic Plasticity. *Vitam. Horm.*, 104, 153–195, 2017.

4

In Silico Modeling of Drug–Receptor Interactions for Rational Drug Design in Neuropharmacology

Princy Shrivastav[1], Bhupendra Prajapati[2,3], Chandni Chandarana[4] and Parixit Prajapati[1]*

[1]Department of Pharmaceutical Chemistry, SSR College of Pharmacy, Sayli-Silvassa Road, Sayli, Silvassa, Union Territory of Dadra and Nagar Haveli & Daman Diu, India
[2]Department of Pharmaceutics and Pharmaceutical Technology, Shree S.K. Patel College of Pharmaceutical Education & Research, Faculty of Pharmacy, Ganpat University, Mehsana, Gujarat, India
[3]Department of Industrial Pharmacy, Faculty of Pharmacy, Silpakorn University, Nakhon Pathom, Thailand
[4]Department of Pharmaceutical Quality Assurance, SSR College of Pharmacy, Sayli-Silvassa Road, Sayli, Silvassa, Union Territory of Dadra and Nagar Haveli & Daman Diu, India

Abstract

Since conventional therapies for neurological disorders are ineffective, developing potential medicines is a difficult task. The goals of current research trends are to find high-quality leads and to reduce the expenditures involved. Consequently, there is a lot of interest in optimizing *in silico* screening, as well as high-throughput screening. These methods support the rational creation of powerful treatments that have multi-targeting effects, greater efficacies, and fewer adverse effects. Furthermore, a broad range of molecular and structural properties of chemical compounds have been used to predict their physicochemical properties and to evaluate their binding affinities to the target. The chapter covers a wide range of subjects, including target identification, toxicity prediction, target-based chemical space categorization, target-based lead design, and virtual screening. We end by discussing the techniques' limitations and their wide potential for future applications.

**Corresponding author*: parixitprajapati@gmail.com; ORCID: 0000-0001-9323-0179

Bhupendra Prajapati, Alok Tripathi, Rishabha Malviya and Lucy Mohapatra (eds.) Computational Neuropharmacology: Fundamentals and Clinical Aspects, (87–126) © 2025 Scrivener Publishing LLC

Keywords: Neuropharmacology, computer-aided drug design, drug–receptor interaction, virtual screening

Abbreviations

ADMET	Absorption, distribution, metabolism, excretion, and toxicity
BBB	Blood-brain barrier
CADD	Computer-aided drug design
GPCR	G protein-coupled receptor
HTS	High-throughput screening
IFP	Interaction fingerprint
LBDD	Ligand-based drug design
MaSIF	Molecular surface interaction fingerprinting
NAM	Negative allosteric modulator
NMR	Nuclear magnetic resonance
PAM	Positive allosteric modulator
QSAR	Quantitative structure activity relationship
RTK	Receptor tyrosine kinase
SAR	Structure-activity relationship
SBDD	Structure-based drug design
SERT	Serotonin transporter
SIFt	Structural Interaction Fingerprint
SSRI	Selective serotonin reuptake inhibitor

4.1 Introduction

The process of modern drug development is rather well defined, interdisciplinary, expensive, and time consuming. Target identification and validation, lead discovery (finding relevant chemicals that modify the biological function of the receptor), lead optimization, and clinical trials can all be used to generally describe the process of discovery of drug.

Since many years, pharmaceutical companies have used a chance approach to assess the therapeutic activity of both nature-originated and manufactured drugs. The industrialization of the process of discovery of drug, best exemplified by the emergence of high-throughput screening (HTS) in the late 1990s, was fueled by the failure to identify target-specific leads, a dearth of target-focused biological activity data, and rapid advancements in combinatorial and parallel chemistry.

The development of HTS from its infancy in the late 1980s to today's standards has been a process. Using robots, automated liquid handling equipment, hardware control software, robust data processing, and a variety of sensitive detectors, it may be broadly defined as the process of quickly and consistently testing huge collections of substances. Both the techniques employed for HTS and the amount of data gathered from a screen have developed over time. In the beginning, high-throughput screening involved analyzing the effects of a single compound acting at a single target at a single concentration. This was done by using radioligand binding to see if the compound bound to the target or kinetic or end-point assays to see if it might inhibit a specific enzymatic reaction [1].

Numerous novel biological targets are being created through developments in the areas of cell biology, molecular biology, and biochemistry, which are made possible by advancements in genomics and proteomics. Structural biology and rational drug design techniques have considerable promise to speed up the development of new medicines. The process of discovering novel drugs and developing them is a lengthy, expensive, dangerous process with few commercial counterparts. For this reason, procedures for computer-aided drug design (CADD) are frequently applied in pharmacy sector to quicken the procedure. Using computational methods during the lead optimization stage of drug development provides a significant financial advantage. On average, a considerable percentage of the time and money needed to bring a drug in market goes towards the production as well as testing of lead analogs. To reduce the number of compounds that need to be produced and tested *in vitro* while covering a larger chemical space, hit-to-lead optimization benefits from the application of computational approaches.

Computational approaches to drug design are based on the idea that biological effects are mediated by interactions between pharmacologically active medicines and their macromolecular targets, primarily nucleic acids or proteins. These interactions are largely influenced by hydrophobic contact, electrostatic force, the surfaces of molecules, and hydrogen bond formation. When evaluating and predicting the interaction of two molecules, several factors are mostly taken into consideration. Figure 4.1 depicts the basic flow of *in silico* drug design process.

CADD, which also documents the design process, is a method of creating products. By providing in-depth drawings of a product's components, operations, dimensions, and tolerances along with the suitable conventions for the product in question, it can hasten the production process. It can be used to create two- or three-dimensional diagrams, and any point of view, including from the inside looking out, can be achieved by rotating the diagram. Chemical

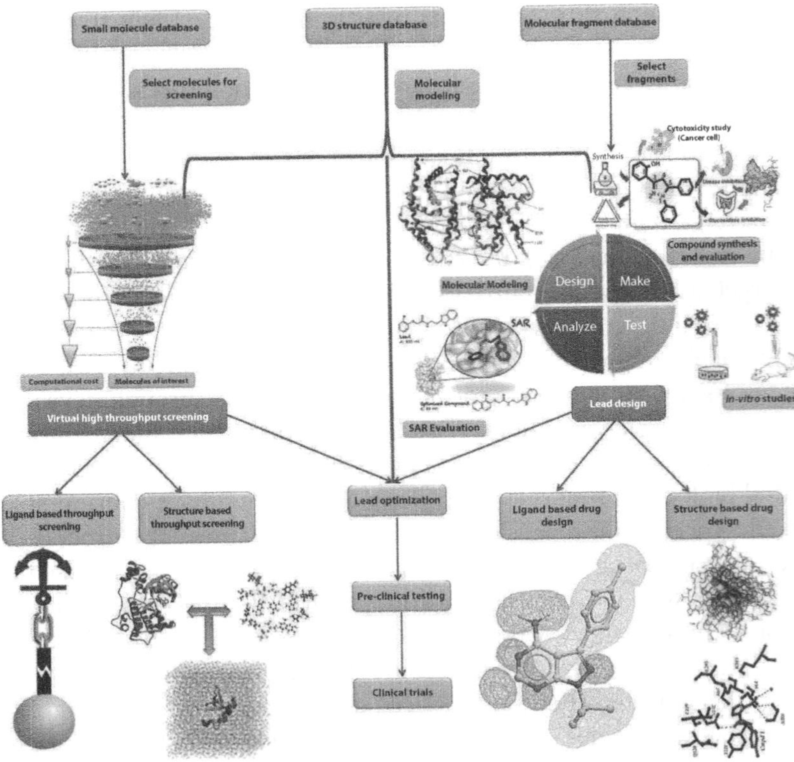

Figure 4.1 Processes involved in *in silico* drug design.

synthesis or natural resources can be used to create test substances. Due to lack of efficacy, poor activity, toxicity, carcinogenicity, complexity of synthesis, insufficient efficiency, etc., these compounds may be disregarded as potentials. Only one of a million researched compounds may be released onto the market as a result, and the cost of developing a new medication has increased to several million dollars. Due to the high governmental standards for its execution, it is doubtful that the final steps of drug testing will take less time or cost less money. Therefore, the stages of ligand discovery and optimization are the focus of major efforts to improve the effectiveness of drug development.

Through the use of bioinformatics tools, *in silico* methods can assist in discovering pharmacological targets. They can also be used to develop and optimize candidate molecules. They can also be used to analyze the target structure for potential binding sites. The foundation of both structure-based drug design (SBDD) as well as ligand-based drug design (LBDD) today is the use of computers and computational approaches, which permeate all phases of discovery of drug [2].

Understanding the 3D structure of the biomolecular target is essential to the SBDD method for discovering novel medicines. The chemical structure of a medicine must be created and improved in order to correspond to the biological target. The SBDD process typically involves the following steps: sequencing of the human genome, target protein extraction and purification, structure determination of therapeutically significant proteins using approaches like homology modeling, nuclear magnetic resonance (NMR), and X-ray crystallography, database preparation of the active compounds, identification of the druggable target polypeptide and its binding site, design and optimization of the chemical substances [3].

The atomic resolution structures of the target proteins have historically benefited SBDD, a popular approach in rational drug development. When a target's experimental structure is unknown, it could be possible to predict the target's homology by examining the experimental structure of a related protein. Biopharmaceuticals, such as peptides and therapeutic antibodies, are

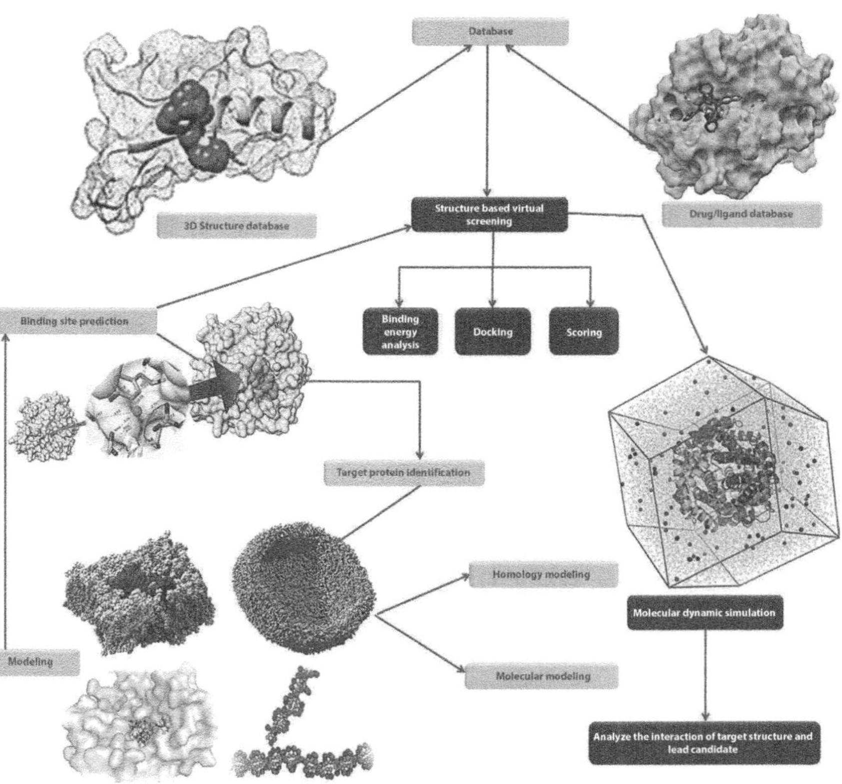

Figure 4.2 Structure-based drug design workflow for identifying potential drug candidates.

another important type of pharmaceuticals in addition to small molecules. Additionally, computational methods have been created to improve the affinities, selectivities, and stabilities of these protein-based therapies [4]. Refer to Figure 4.2 for SBDD workflow for identifying potential drug candidates.

When the target structure is unavailable but a small number of known small molecules can influence the target, the LBDD technique is utilized in drug discovery. Starting with either a sole compound or a collection of compounds that are well-known to be effective against a target, LBDD then designs suitable analogues to enhance potency and other crucial features based on the understanding of structure-activity relationships (SARs). The medicinal chemistry team and the computational chemistry team work closely together to manage the projects needing LBDD activities [5].

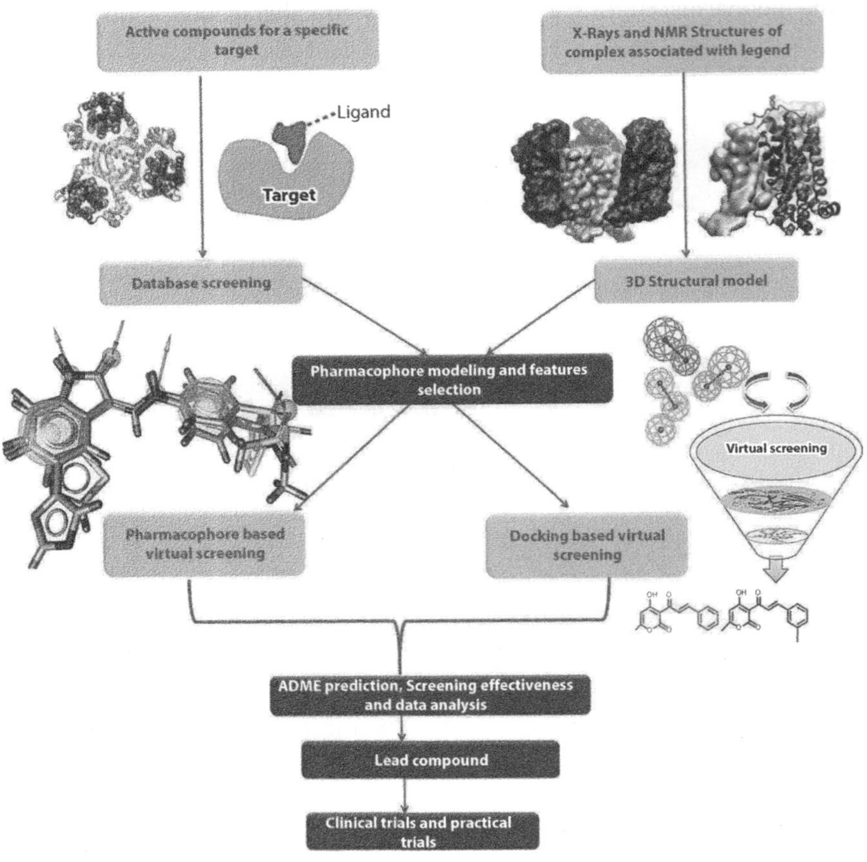

Figure 4.3 Ligand-based drug design workflow for identifying potential drug candidates.

Site point connection approaches, pharmacophore mapping, molecular shape analysis, molecular interaction fields, molecular dynamics, and machine learning are six key classifications that can be used to categorize LBDD [5]. The pharmacophore model developed by LBDD depends on the knowledge of existing compounds that bind to the protein target of interest and outlines the basic structural requirements of a molecule needed to bind to the target [5, 6].

Several pharmacophoric methods, alignment strategies based on ligand shape and electrostatic similarity, scaffold hopping and fragment replacement tools, matched molecular pair analysis, Free-Wilson models, local QSAR models, and machine learning with project data are among the features of LBDD. These instruments can be employed to forecast potency versus the desired target and/or ADMET.

By examining molecules that interact with the target, LBDD is an indirect method for facilitating the development of pharmacologically active drugs. Utilizing ligands is a fairly effective method of creating novel chemical substances [6, 7]. Figure 4.3 depicts LBDD workflow for identifying potential drug candidates.

4.2 Drug–Receptor Interactions

Due to electrostatic forces including dispersion forces, hydrogen bonds, and Coulombic attractions, molecules have a natural affinity towards one another. Biological processes, such as the activation of cellular activity, biocatalysis of chemical reactions, and neutralization of foreign toxins, all depend on the noncovalent interactions that arise from this affinity. Ligands and receptors trade interactions with solute and solvent molecules for interacting with one another to start these processes.

All biological receptor–natural ligand interactions have a number of characteristics, including selectivity, affinity, saturation, and physiological response. Specificity is the property of biological receptors that strongly bind to a single natural ligand. With a competitive-binding test, ligand specificity can be easily ascertained. The quantity of ligand that has been bound to a receptor is determined when there are more potential ligands present. If the receptor is indeed selective for the first ligand, the presence of the other ligands has no impact on the amount of ligand bound.

The second characteristic, affinity, refers to non-covalent interactions between molecules. For example, polypeptides adhere to glass a lot, which is partially due to polar surfaces interacting. The surface of a cell is also quite polar, which is mostly caused by the large amount of carbohydrates that extend from membrane proteins and membrane lipids. Because of this, every protein has a specific degree of affinity for cell surfaces. Receptor–ligand interactions differ from other noncovalent interactions between molecules due to their high affinity.

Saturation also has a significant impact. A receptor becomes saturated with a lot of ligands since it only has a limited number of binding sites. There is no longer a rise in binding with increasing ligand concentration when all of the binding sites are occupied, and instead, a curvilinear relationship between the concentration of bound ligand and that of total ligand is shown. Last but not least, a significant receptor–ligand interaction results in a physiological reaction. Furthermore, binding and physiological reactions ought to follow one another in terms of correlation and parallelism [8].

4.2.1 Overview of Receptors and Their Role in Drug Action

Large protein molecules called receptors are found in cell membrane or wall and receive chemical information from other molecules like medicines, hormones, or neurotransmitters that are found outside the cell. These outside substances attach to cell receptors and activate them, causing the cell to release an internal chemical or electrical signal. In response to this signal, the cell does certain actions, such as causing us to experience pain [9].

An agonist is a substance that attaches to a particular receptor and increases the activity of a cell process. A cell's physiological response to a particular stimulus is called an agonist. They could exist naturally or artificially. Endorphins are one such naturally occurring agonist of opioid receptors. Apart from that, heroin and morphine are synthetic opioid agonists [10].

Receptors dictate how specifically a drug's activity is regulated. The form, size, and electrical charge of a drug's molecules all affect how well it can bind to a receptor. Because of the thorough isolation and characterization of several drug receptors, a precise understanding of the molecular basis of pharmacological action is now attainable.

Receptor activation controls cellular biochemical processes such as enzyme activity, DNA transcription, protein phosphorylation, and ion conductance. Neurotransmitters, hormones, and medicines are examples of molecules that bind to receptors and are referred to as ligands. The binding could be specific and reversible. A ligand can either activate or inactivate a receptor, and activation can alter how well a cell performs. Each ligand may interact with various receptor subtypes [9, 10].

Drugs rarely show 100% specificity for a single receptor or subtype, but the majority do. The regulation of physiological processes like contraction and secretion typically involves numerous receptor-mediated mechanisms and a number of phases.

In conclusion, by accepting chemical information from other molecules and producing a biochemical or electric signal inside the cell, receptors play a critical part in how drugs operate. A drug's capacity to attach to a receptor depends on size, shape, and electrical charge of its molecules. Each ligand may interact with different subtypes of receptors, and the binding can be both selective and reversible. Few medicines show 100% selectivity for a single receptor or subtype, but the majority do [9].

4.2.2 Types of Receptor and Ligand Interactions

One of the important principles of how living systems work is highly specific molecular recognition, which is involved in all of the most important biological processes, including transmembrane transport of various substances, enzymatic reactions, antigen-antibody interactions, and hormone-receptor interactions, among others. The concept of molecular recognition is crucial in supramolecular chemistry. It can be characterized as a process of selective ligand binding by a receptor and high affinity interaction. The ligands are often quite straightforward organic or synthesized compounds. Receptors, on the other hand, are complex, three-dimensional, water-soluble macromolecules that are membrane-anchored or membrane-embedded. The biological activity of many biomolecules and the effects of therapeutic drugs are governed by the physiological reactions that are induced by ligand-stimulated changes in cell receptor state. By creating relatively strong non-covalent bonds by ionic, hydrophobic, electrostatic, or Van der Waals forces, the ligand and receptor come together to form a complementary pair. The nature of this interaction is determined by the biopartners' structural and energetic compatibility. The development of receptor–ligand couples is a difficult process that may go through numerous stages. Sometimes the ligands have several recognition sites, which makes it possible for them to attach to a group of receptors. The affinity between the receptor and the ligand can be significantly increased by this multivalent binding. A series of chemical processes start when a ligand binds to its complementary receptor and produces a signal. In the cell, the high affinity receptor–ligand combinations create a highly complex network of interconnected molecules. While distinct receptors can activate diverse downstream signaling pathways through a single signaling molecule, a single receptor may convey a variety of signals depending on its ligand partners. In order for a cell to communicate with its neighbors and the rest of the body, ligand-receptor interactions are essential. They initiate dynamic processes like death, cell proliferation, and migration, but they also keep cells in homeostasis and ensure that all cell systems are functioning properly [11].

Receptor–ligand interactions play a crucial role in various biological processes. Some common types of receptor–ligand interactions have been discussed in the subsequent sections of the chapter.

4.2.2.1 Enzyme–Substrate Interaction

Enzymes are proteins with specific roles that speed up chemical processes in the body. An enzyme-substrate complex is formed when the active site of enzyme and a substrate molecule work together. This complex subsequently catalyzes a chemical reaction to produce a product. This contact is often short and reversible. Substances bind to enzymes similarly to how ligands interact to proteins. The substrates, however, undergo a chemical shift when they bind to an enzyme, which turns them from substrates into products. The substrate interacts with the amino acids of the enzyme to establish a bond that enables the enzyme to function at the binding site, sometimes referred to as the active site since it has amino acids that both bind the substrate and aid in its conversion to product. In the enzyme active site, amino acids serve two distinct functions, which occasionally cross over. Some of the amino acids in the active site are responsible for tying up the substrate, while others are in charge of promoting the chemical process [12, 13].

4.2.2.2 Ligand-Gated Ion Channels

Integral membrane proteins called ligand-gated ion channels have a hole that allows specific ions to penetrate the plasma membrane under control. They participate in basic brain functions like learning, memory, and attention. These channels regulate the binding of neurotransmitters and the opening of channels in response to the binding of ligands, which regulates synaptic transmission between a neuron and a muscle or between two neurons. The majority of ligand-gated ion channels are pentameric, with each subunit consisting of four transmembrane helices that make up the transmembrane domain and an extracellular, beta sheet sandwich-type ligand binding region. They are also referred to as ionotropic receptors, and they control ion channel function as well as the channel's location in the membrane and all over the cell.

In certain subcellular localizations, a large number of proteins that bind to the different subunits of ligand-gated ion channel receptors interact with ligand-gated ion channels. These interfering proteins can change how an ion channel functions and where it is located in the membrane and within the cell [14, 15].

4.2.2.3 *Hormone–Receptor Interaction*

Endocrine glands are the organs responsible for producing hormones in the body. Hormones often bind to particular receptor proteins on or inside of cells. A signaling cascade that results from this binding trigger specific cellular processes. The extremely selective binding of hormone to its receptor is produced by the accumulation of weak interactions between hydrophilic, hydrophobic residues, neutral, charged, polar, or non-polar. A signaling pathway is activated when a ligand attaches to the receptor, which regularly changes shape and may bind more ligands. Even very low hormone concentrations can cause a significant physiological response due to the highly specialized and wealthy interactions between hormones and their receptors. Cells in various body tissues may contain the same type of receptor, which could result in slightly different reactions. As a result, both the hormone and the target cell are necessary for a response to be triggered by a hormone. When the amount of a hormone is consistently decreased, target cells upregulate to make more receptors. Because of this mechanism, cells can now react to the current hormone more sensitively. Cells can also modify the receptors' sensitivity to particular hormones. There are numerous ways in which two or more hormones might interact to change how cells react [16, 17].

4.2.2.4 *Ligand-Activated Transcription Factors*

DNA-binding transcription factors that are ligand-activated govern the transcription of particular gene sets. They do this by binding to ligands. They belong to the nuclear receptor class, which is a superfamily of ligand-inducible transcription factors made up of several domains. Lipid-soluble compounds like retinoic acid and steroid hormones are examples of the ligands that activate these transcription factors.

Recent research has focused on the structural features of nuclear receptors' ligand-dependent transcription activation. They undergo conformational changes after binding to ligands, and these changes in shape cause them to attract cofactor complexes to improve target gene transcription. Nuclear receptors' ligand-dependent increase of transcription is mediated by cofactor complexes, one of which is the DRIP complex.

The signal-sensing domain of transcription factors can be triggered in a number of ways, including by ligand interaction. The majority of transcription factors interact with one another; through dimerization, many big transcription factor families create intricate homotypic or heterotypic interactions. A number of transcription factors must bind to DNA regulatory regions for gene transcription to take place. These intermediary proteins, such as cofactors, then effectively recruit the preinitiation complex and RNA polymerase. The type II nuclear receptors are examples of ligand-activated transcription factors that regulate a variety of cellular processes, such as inflammation, metabolism, and development [18–20].

4.2.2.5 Receptor Tyrosine Kinase Interaction

A subclass of tyrosine kinases known as receptor tyrosine kinases (RTKs) is essential for controlling a wide range of intricate biological processes, such as metabolism, differentiation, motility, and cell growth [21]. RTKs are triggered by ligand-induced oligomerization, which commonly dimerizes and juxtaposes the cytoplasmic tyrosine kinase domains [22]. When RTKs are active, several intracellular proteins involved in signal transduction can bind on their cytoplasmic tails. of the cytoplasmic tails of RTK receptors, phosphorylated tyrosines are bound by a specific region of these proteins known as SH2. RTKs influence changes through interacting with other proteins and phosphorylating the proteins that are their substrates [23]. The connections can be stable, as is the case when the TKD's active site is in contact with the C-terminal tail of the RON (MST-1R), MET, and TEK receptors. This prevents substrate access and stabilizes an inactive conformation that strongly inhibits kinase activity [21]. Signal transmission is corrupted and a variety of pathological effects result from aberrations in regulation of the composition of RTK-mediated lipid micro-domains, which is controlled by adaptor polypeptide interaction. RTK mutations cause the activation of a variety of signaling cascades that have multiple impacts on polypeptide expression and are crucial in the initiation and development of many types of cancer [24].

4.2.2.6 G Protein-Coupled Receptor Interaction

In eukaryotes, G protein-coupled receptors (GPCRs) are a broad and diversified class of membrane receptors that control a wide range of physiological and cellular processes by reacting to a variety of extracellular stimuli, such as light, neurotransmitters, hormones, and odors [25]. When a GPCR attaches to an external signaling molecule, the GPCR undergoes a conformational shift that sets off an interaction between the GPCR and a neighboring G protein. Alpha, beta, and gamma are the three subunits that make up the G protein. The alpha subunit is attached to GDP when the G protein is inactive, but when the GPCR activates it, it releases its bond with GDP and forms a new one with GTP, which allows the α subunit to separate from the β and γ subunits. A physiological response can then result from the alpha subunit's interaction with and control of a variety of downstream effectors, including enzymes and ion channels [26]. There is evidence to suggest that GPCRs can interact with one another to create dimers and higher-order oligomers in addition to binding G proteins and arrestins. Single-molecule microscopy, computational modeling, and microplate readers have all been used to investigate the intricate interactions between GPCRs and G-proteins. In order to create new medications that can control GPCR function and cure prevalent diseases like diabetes or heart failure, it is crucial to understand these interactions [25, 27].

4.2.3 Factors Affecting Drug–Receptor Interactions

Several factors can influence drug receptor interactions. Some of the key factors are discussed below [28]:

a. Receptor Conformation: Different conformational states of receptors are possible, and drug interaction can cause the receptor to change its conformation. The strength and length of the receptor–ligand interaction can be influenced by the capacity of drug to bind to and stabilize a particular receptor conformation.
b. Drug's Chemical Structure: The capacity of a drug to interact with particular receptors depends on its chemical composition. Size, shape, charge, and functional groups, among other things, can affect a drug's affinity and selectivity for a certain receptor.
c. Concentration of Drug and Receptor: The likelihood and magnitude of their interaction can be affected by the drug's and the receptor's concentrations. The likelihood of binding and the total reaction are both enhanced by higher doses of either the medication or the receptor.
d. Competitive and Non-competitive Interactions: Drugs may fight each other or endogenous ligands for a receptor's attention. When a drug and an endogenous ligand fight for the

same binding site on the receptor, a competitive interaction is created. On the other hand, non-competitive interactions entail binding to many places on the receptor, indirectly changing how it functions.

e. Pharmacokinetics: Drug receptor interactions can be impacted by drug pharmacokinetics, which includes drug absorption, distribution, metabolism, and excretion. The concentration of the medication at the receptor site and, subsequently, its pharmacological effects, can be influenced by variables such as bioavailability, plasma protein binding, metabolism, and elimination rate.
f. Tolerance and Sensitization: Long-term use of some medications can cause tolerance, in which the receptor gradually loses responsiveness to the drug. On the other side, sensitization happens when the receptor gets more receptive to the medication. The efficiency and response of the medicine may be affected by these adaptive alterations in receptor function.
g. Genetic Variations: Drug receptor interactions may be impacted by genetic differences between people. Drug binding and response may be impacted by polymorphisms in drug receptor genes that change the shape or expression of the receptor.
h. Affinity and Binding Strength: The degree of the binding relationship is referred to as a drug's affinity for its receptor. Because they attach to receptors with high affinity, medicines have longer-lasting effects. The affinity and binding strength are influenced by variables like the quantity and quality of the links formed between the drug and receptor.
i. Baseline receptor density: The number of receptors present can be increased or decreased by other drugs, supplements, foods, aging, genetic mutations, and disorders.
j. Steric features of a drug: To trigger a response at a specific receptor, the medication must have stereoselective properties.
k. Enantiomers, asymmetry, and chirality: These ideas are crucial for improving the way that medications interact with receptors. While each enantiomer has some physical and molecular characteristics, their pharmacological effects can vary depending on how and where they act.
l. Isosterism: Isosteres are collections of atoms with properties that are similar to those of a molecule in terms of size,

electronegativity, or stereochemistry. Isomerism is the scientific word for the presence of such groups in molecules.

m. Physiological functions: Multiple receptor-mediated pathways are typically involved in controlling physiological activities, and there may be several steps between the initial molecular interaction between drug and receptor and the final response.

Every medication has a preferred receptor, but it may also bind to additional receptors with equal or lower affinity than its chosen receptor. Because a single medicine can bind to several receptors and trigger undesirable biological reactions, side effects are a direct outcome of inadequate selectivity. Additionally, when a drug binds to a receptor, it can cause the receptor to become active to produce the desired biological effect (agonist) or it can merely occupy the receptor to prevent an agonist from accessing its binding site on the receptor, obliterating the agonist response (antagonist).

4.3 *In Silico* Methods for Modeling Drug–Receptor Interactions

In silico modeling of drug–receptor communications involves using computational methods to simulate and predict how drugs interact with receptors in the body. This strategy can assist researchers in making choices and simulating nearly all aspects of drug discovery and development [29]. There are several methods for *in silico* modeling of drug–receptor interactions, including molecular docking, molecular dynamic simulations, QSAR modeling, and pharmacophore modeling [29–32].

For *in silico* modeling of drug–receptor interactions, molecular docking is a widely utilized technique. It entails simulating the interaction of a ligand with a receptor in three dimensions in order to determine the binding affinity of a ligand to a receptor. Molecular docking, for instance, was used in a study to forecast how naturally occurring ligands will attach to the oestrogen receptor alpha, a target for the therapy of breast cancer. According to the study, a number of natural substances had a high affinity for the receptor and may one day be used as anti-breast cancer medications [30].

On the basis of drug chemical structure and protein sequences, machine learning-based algorithms have also been suggested for predicting drug–target interaction networks. These approaches involve building a classification model that can predict whether a given drug will interact with a given target protein based on their chemical and structural properties [31].

Before conducting time-consuming and costly *in vitro* and *in vivo* tests, researchers can uncover new chemical entities and optimize their properties with the aid of *in silico* drug–receptor interaction modeling, which can be a powerful tool for drug discovery and development [32, 33]. Table 4.1 gives a brief summary of various *in silico* modeling techniques used.

Table 4.1 A brief summary of *in silico* techniques used for modeling drug–receptor interactions.

Sr. no.	*In silico* modeling technique	Description	References
1.	Molecular Docking	Predicts the ligand's affinities and binding orientations to a receptor's active site.	[2]
2.	Molecular Dynamics Simulations	Studies the dynamics and stability of drug–receptor complexes by simulating how atoms behave and move over time.	[34]
3.	QSAR (Quantitative Structure-Activity Relationship)	Predicting the potency of novel compounds by relating a ligand's chemical structure to its biological activity.	[2]
4.	Pharmacophore Modeling	Determines which ligands have the required chemical characteristics to bind to the target receptor.	[2]
5.	Interaction Fingerprints	The 3D character of interactions in molecular complexes, generally generated between target and ligand, is encapsulated in interaction fingerprints, which are vector representations.	[37]
6.	Receptor-guided Alignment	Receptor-guided alignment is a method used in 3D-QSAR studies to align molecules based on their binding to a receptor.	[43]

4.3.1 Molecular Docking

The computational modeling of the structure of complexes made up of two or more interacting molecules is known as molecular docking. The goal of molecular docking is to predict three-dimensional structure. Docking is essential to the drugs' logical design. In order to lower the free energy of the overall system, molecular docking aims to optimize the shape of the polypeptide and its ligand as well as their relative orientation. The activation of fundamental biomolecular events, such as interactions between medications and proteins, nucleic acids, and enzymes, is greatly aided by molecular recognition.

There are many subjects related to docking theory. The first is called CDOCKER, and it uses full CHARMm forcefield-based docking along with preliminary ligand placement that is chosen at random. The next is LibDock, a quick docking method based on hotspots or properties of binding sites. Another notion, known as LigandFit, is based on a preliminary shape match to the binding site. Finally, MCSS employs CHARMm to dock fragments using a special technique known as Multiple Copy Simultaneous Search (MCSS), which is computationally efficient.

It is possible for medicines and receptors to interact at the atomic level. To fully understand how and why pharmacological substances attach to protein targets, we need to understand the atomic-level biophysical and biochemical properties of the drug and its target. One excellent source for this is the Swiss PDB (protein data bank). It has the ability to predict critical physico-chemical properties like polarity and hydrophobicity, which have a big impact on how drugs bind to proteins [2].

4.3.2 Molecular Dynamics Simulations

Computer simulation techniques called "molecular dynamics simulations" are used to examine the actual motions of atoms and molecules. The simulations must be chosen carefully to ensure that the calculation can be completed in an acceptable amount of time. These considerations include simulation size, timestep, and overall time duration. The CPU-intensive step in a conventional MD simulation is the evaluation of the potential as a function of the internal coordinates of the particles. The non-bonded or non-covalent portion of this energy evaluation is the most expensive.

A computer model of the system must first be built using information from NMR, crystallography, or homology modeling in order to represent a molecular system in molecular dynamics simulations. The locations of these atoms are changed in line with Newton's equations of motion after the forces exerted on each of the system's atoms have been determined. The procedure is then carried out again after moving forward in the simulation time. On computer clusters or supercomputers, molecular dynamics simulations are generally run in parallel with dozens or even hundreds of processors.

Macromolecular systems can now be understood using a mature technique called molecular dynamics simulations. From simulating a small number of atoms, they have advanced to simulating biologically significant systems, such as huge macromolecular complexes like entire proteins, proteins embedded in membranes, or nucleosomes or ribosomes in solution with accurate solvent representations [34].

New frameworks for molecular simulations have been created as a result of recent developments in deep learning, such as TorchMD, which uses empirical potentials to provide a mechanistic description of molecules [35]. A potent simulation technique to examine the dynamics of the conformational space where both the ligand and the receptor are present is molecular dynamics simulation [36].

4.3.3 Quantitative Structure–Activity Relationship Modeling

Quantitative structure–activity relationship (QSAR) modeling is the process of creating prediction models of biological activities in connection to structural and molecular data from a chemical library. The term quantitative structure-property relationship (QSPR) is used to describe how the idea of QSAR, which is typically used for drug discovery and development, has gained widespread applicability for correlating molecular information with both biological activities and other physicochemical properties. Typical molecular features that are used to account for topology, steric effects, hydrophobicity, and electrical properties can be established either theoretically through computational chemistry or physically through testing. Then, a specific set of data sets is subjected to statistical or machine learning approaches for data pre-processing and data modeling.

In QSAR, equations that connect distinct sets of data that have been obtained via experimental research or generated through theoretical study are developed using a collection of mathematical techniques known as regression analysis. The data are fed into the proper computer programme, which when executed outputs an equation that reflects the line that best fits the data. The line of best fit to the data in a regression analysis would be used to calculate the values of m and c [2].

4.3.4 Pharmacophore Modeling

An abstract description of the molecular properties required for molecular identification and for a biological macromolecule to recognize a ligand is called a pharmacophore. Typical pharmacophoric properties include cationic charge, anionic charge, hydrogen bond donors and acceptors, aromatic rings, and hydrophobic centroids. Pharmacophore techniques have evolved over the past century to become one of the most important tools in the drug discovery process. The development of numerous ligand- and structure-based approaches has improved pharmacophore modeling. To create pharmacophore models, one can employ both the ligand-based and structure-based techniques. Using the ligand-based technique, a set of physiologically active compounds are superimposed, and the chemical properties that are essential to their bioactivity are identified. In the structure-based approach, potential locations of interaction between the ligands and macromolecular targets are looked for. In virtual screening, pharmacophore techniques have been widely used [2]. Table 4.2 lists out various software used in *in silico* methods for modeling drug–receptor interactions.

Table 4.2 List of various softwares used in *in silico* methods for modeling drug–receptor interactions.

Sr. no.	Software name	Description	Website
Databases			
1.	Zinc15 Database	A freely accessible database of compounds available commercially for virtual screening.	https://zinc15.docking.org/
2.	ChEMBL	Database of bioactive substances with drug-like properties.	https://www.ebi.ac.uk/chembl/

(*Continued*)

Table 4.2 List of various softwares used in *in silico* methods for modeling drug–receptor interactions. (*Continued*)

Sr. no.	Software name	Description	Website
3.	Bingo	A data cartridge for storage and searching solution for chemical information.	https://lifescience.opensource.epam.com/bingo/index.html
4.	JChem for Office	Integrates handling of chemical structure, analysing data, visualizing and reporting of data within Microsoft Office.	https://chemaxon.com/jchem-for-office
5.	ChemDiff	An Indigo-based tool to compare two files with numerous structures visually and identify duplications.	https://lifescience.opensource.epam.com/indigo/chemdiff.html
6.	Protein Data Bank (PDB)	A database containing major biological molecules' 3D structural information, like nucleic acids and proteins.	https://www.rcsb.org/search
7.	Binding MOAD (Mother of All Database)	A part of PDB that includes each and every exceptional ligand-protein binding example.	http://www.bindingmoad.org/

(*Continued*)

Table 4.2 List of various softwares used in *in silico* methods for modeling drug–receptor interactions. (*Continued*)

Sr. no.	Software name	Description	Website
8.	PDBbind	Database for experimentally measured binding affinity data for all biomolecular complexes deposited in PDB.	http://www.pdbbind.org.cn/
9.	TTD	A database that offers details on the targeted diseases, pathways, known and investigated therapeutic nucleic acid and polypeptide targets, and the medications that are aimed towards each of these targets.	https://db.idrblab.net/ttd/
Molecular Dynamics Simulation Software Tools			
10.	CHARMM	Molecular simulation package.	https://www.charmm.org/
11.	GROMACS	Molecular dynamics simulation package.	http://www.gromacs.org/
12.	Amber	Assisted Model Building with Energy Refinement.	https://ambermd.org/

(*Continued*)

Table 4.2 List of various softwares used in *in silico* methods for modeling drug–receptor interactions. (*Continued*)

Sr. no.	Software name	Description	Website
13.	Desmond	Molecular dynamics software from D.E. Shaw Research.	https://www.deshawresearch.com/resources_desmond.html
14.	VMD	Visualization and analysis of molecular dynamics.	http://www.ks.uiuc.edu/Research/vmd/
15.	NAMD	Scalable Molecular Dynamics.	https://www.ks.uiuc.edu/Research/namd/
Protein Structure Prediction Software Tools			
16.	Modeler	Protein structure homology modeling tool.	https://salilab.org/modeller/
17.	I-TASSER (Iterative Threading ASSEmbly Refinement)	Hierarchical approach to prediction of polypeptide structure and structure-based function annotation.	https://zhanggroup.org/I-TASSER/
18.	LOMETS	Prediction of polypeptide structure and structure-based function annotation using a meta-server approach.	https://zhanggroup.org/LOMETS/

(*Continued*)

Table 4.2 List of various softwares used in *in silico* methods for modeling drug–receptor interactions. (*Continued*)

Sr. no.	Software name	Description	Website
19.	SWISS-MODEL	A fully automated protein structure homology-modeling server.	https://swissmodel.expasy.org/
20.	Rosetta	Protein structure prediction and design.	https://www.rosettacommons.org/
Protein Structure Analysis Software Tools			
21.	MED-SuMo	Software to compare and superpose any 3D interaction molecular surface across the PDB.	http://www.medit.fr/
22.	TRAPP	TRAnsient Pockets in Proteins with Druggability Calculation.	https://trapp.h-its.org/
23.	CAVER	a program for examining and visualizing the channels and tunnels seen in protein structures.	https://caver.cz/
24.	CASTp	Computed Atlas of Surface Topography of proteins.	http://sts.bioe.uic.edu/castp/index.html?2was

(*Continued*)

Table 4.2 List of various softwares used in *in silico* methods for modeling drug–receptor interactions. (*Continued*)

Sr. no.	Software name	Description	Website
25.	Pocketome	A comprehensive catalog of druggable binding sites' conformational ensembles that can be determined by experimentation based on co-crystal structures found in the PDB.	http://www.pocketome.org/
26.	3DLigandSite	An automated process for prediction of ligand binding site.	http://www.sbg.bio.ic.ac.uk/~3dligandsite/index.cgi
Molecular Docking Software Tools			
27.	Autodock Vina	A popular molecular docking software.	http://vina.scripps.edu/
28.	DOCK	A molecular docking program.	http://dock.compbio.ucsf.edu/
29.	GOLD	Genetic Optimization for Ligand Docking.	https://www.ccdc.cam.ac.uk/solutions/csd-discovery/components/gold/
30.	SwissDock	Protein-ligand docking via web services.	http://www.swissdock.ch/

(*Continued*)

Table 4.2 List of various softwares used in *in silico* methods for modeling drug–receptor interactions. (*Continued*)

Sr. no.	Software name	Description	Website
31.	DockingServer	A user-friendly, web-based interface that manages every step of molecular docking, starting with protein and ligand setup.	https://www.dockingserver.com/web
32.	1-Click Docking	Free web-based tool for molecular docking.	https://mcule.com/apps/1-click-docking/
Ligand-Based Virtual Screening Software Tools			
33.	PharmaGist	A web server that can identify pharmacophores, or the layout of characteristics that permits a chemical to bind with a certain target receptor.	http://bioinfo3d.cs.tau.ac.il/PharmaGist/
34.	SwissSimilarity	Interface to perform ligand-based virtual screening of chemical libraries.	http://www.swisssimilarity.ch/
Structure-Based Virtual Screening Software Tools			
35.	PatchSearch	Allows local nonsequential searching for similar regions, called patches, on protein surfaces.	https://bioserv.rpbs.univ-paris-diderot.fr/services/PatchSearch/

(*Continued*)

Table 4.2 List of various softwares used in *in silico* methods for modeling drug–receptor interactions. (*Continued*)

Sr. no.	Software name	Description	Website
36.	Swiss TargetPrediction	Webtool to estimate the most probable macromolecular targets of a small molecule.	http://www.swisstargetprediction.ch/
37.	CSNAP	A network-based approach for automated compound target identification.	https://services.mbi.ucla.edu/CSNAP/html/search9.php
Bioisosteric Replacement and Scaffold Hopping			
38.	SwissBioisostere	Database with details on molecular substitutes and how well they work in biochemical tests.	http://www.swissbioisostere.ch/

4.3.5 Interaction Fingerprints

The 3D character of interactions in molecular complexes, generally generated between a target and a ligand, is encapsulated in interaction fingerprints, which are vector representations. They have numerous uses in drug development efforts, including structure-based virtual screening (SBVS) and machine learning [37]. When creating an interaction fingerprint, two groups of atoms are frequently considered to be involved in an interaction if they follow certain geometrical constraints based on angles and distances. Drug discovery and design can benefit greatly from the analysis and prediction of protein-ligand binding interactions provided by interaction fingerprints [38]. There are several types of interaction fingerprints, including structural interaction fingerprints, ProLIF, geometric deep learning, and machine learning.

4.3.5.1 *Structural Interaction Fingerprint (SIFt)*

For simulating and studying 3D polypeptide-ligand binding interactions, it provides a quick and precise method. It produces an interaction fingerprint by reducing the three-dimensional structural binding information from a polypeptide-ligand complex to a one-dimensional binary string. The "structural interaction profile" of the complex, which may be used to organize, assess, and visualize the vast quantity of information contained in the complex, can be considered to represent each fingerprint [39].

SIFt has various uses in drug development initiatives, including machine learning and SBVS. For instance, De Graaf *et al.* rescored the outcomes of virtual screening on a GPCR using the Tanimoto similarity between the interaction fingerprint (IFP) of a crystallographic reference and that of docking poses [37].

4.3.5.2 *ProLIF*

For molecular complexes comprised of ligands, proteins, and other compounds, ProLIF is a Python package that creates interaction fingerprints. It is a tool that can be used to fingerprint the molecular interactions between molecules and is compatible with many different molecular structure files, including docking and experimental structures. Support for molecular dynamics trajectories is one of the key ways that ProLIF differs from other freely accessible interaction fingerprint programmes. A GitHub repository hosts ProLIF's installation files. There are several tutorials in the library that address typical use-case scenarios. Code snippets are also provided as support for the documentation to show how the library can be integrated with other data-analysis tools for a more user-friendly experience. Researchers working in the domains of molecular dynamics, virtual screening, docking, and structural biology can benefit from using ProLIF [37].

4.3.5.3 *Geometric Deep Learning*

Protein molecular surfaces can be utilized to learn interaction signatures using geometric deep learning. Molecular surface interaction fingerprinting (MaSIF), a theoretical framework built on geometric deep learning, collects fingerprints crucial for particular biomolecular interactions. A convolutional neural network (CNN) that learns the characteristics of the patches is used to overlap radial patches in order to characterize the molecular surface data. Visual analysis may find it challenging to understand these signatures, but extensive datasets can teach us about them.

The patterns of chemical and geometrical properties that make up a polypeptide's fingerprint its forms of interactions with other biomolecules.

Based simply on structure, MaSIF has been used to forecast interactions between proteins and other macromolecules. Additionally, it has been utilized to document crucial features of novel protein interactions and molecular recognition. The paradigm not only provides a method for analyzing interaction networks in biological processes at the systems level, but it may also open up prospects in important biotechnological fields like biomaterials, biosensing, or drug design.

In conclusion, geometric deep learning provides a method for discovering interaction signatures on protein molecular surfaces. The MaSIF conceptual framework employs geometric deep learning to identify fingerprints required for specific biomolecular interactions. The modalities of interactions between a protein and other biomolecules are characterized by patterns of chemical and geometric properties known as protein fingerprints. In addition to giving a way to examine interaction networks in biological processes at the systems level, MaSIF has been used to predict interactions between polypeptides and other macromolecules based solely on structure [40].

4.3.5.4 Machine Learning

A method that uses structural interaction fingerprints for prediction and explanation of binding of ligands to RNA [41]. In one study, researchers combined machine learning and a pharmacophore-based interaction fingerprint (Pharm-IF) to examine its application in *in silico* screening [42].

Overall, machine learning in interaction fingerprints is a promising area of research that has shown success in prediction and explanation of binding of ligands to RNA. It has the potential to improve virtual screening and drug discovery.

4.3.6 Receptor-Guided Alignment

Receptor-guided alignment is a method used in 3D-QSAR studies to align molecules based on their binding to a receptor [43]. To create a prognostic model for the activity of the molecules, it is necessary to determine its optimal alignment with the receptor. Most of the time, the alignment is accomplished by flexible docking of the molecules into the binding area of the receptor [43].

The inhibitory models of HER-2 kinase and EGFR inhibitors can be studied using receptor-guided alignment. In these investigations, composite alignment techniques were created using a variety of binding modes discovered by docking. When compared to models from alignments III as well as IV, those from alignments I and II had a superior capacity for prediction. Alignments I and II yielded equal 3D-QSAR models for the EGFR kinase results, with Alignment II marginally outperforming I, but Alignments I and II created superior predictive models for the HER-2 data [43–45].

Receptor-guided alignment has also been used in the design of more potent insulin-like growth factor-1 receptor (IGF-1R) inhibitors. The method has been shown to yield better 3D-QSAR models than other alignment schemes [46].

Overall, receptor-guided alignment is a powerful tool for developing predictive models of molecular activity based on their binding to a receptor. It is commonly used in 3D-QSAR studies to align molecules based on their binding to a receptor and to develop predictive models for their activity.

4.4 Applications of *In Silico* Modeling in Neuropharmacology

A computational technique called "*in silico* modeling" has been created and is frequently used to develop and test pharmacological hypotheses [47]. In neuropharmacology, *in silico* modeling can be used for prediction of the efficacy of drugs on the nervous system, identify potential drug targets, and optimize drug design [48]. Here are some specific applications of *in silico* modeling in neuropharmacology:

Virtual ligand screening and profiling: large databases of chemicals can be screened using *in silico* approaches to find potential drug candidates for a given target. In order to find the molecules that are most likely to attach to a particular target, it entails scanning vast databases of chemicals using computer algorithms. This method is regarded as a valuable target-based virtual since it fosters a favorable approximation of the predicted conformers.

Another *in silico* technique used to forecast a compound's pharmacological characteristics is virtual ligand profiling. It entails the use of computer algorithms to forecast a compound's physicochemical characterization, absorption, distribution, metabolism, excretion, and hazardous characteristics [29]. The time and expense involved with conventional drug discovery techniques may be decreased as a result.

Prediction of drug efficacy and toxicity: Due to the enormous libraries of compounds that high-throughput screening and combinatorial chemistry are now able to manage, identification of probable toxicity is advised even before synthesis. As a result, predictive toxicology is required. There are several *in silico* methods for predicting toxicity that are described. For a variety of toxicity outcomes, QSARs, usually linked to certain chemical classes, have long been employed for this purpose. However, QSARs are also used to predict the toxicity of very different libraries. However, these QSARs are frequently of the classification type, which means that they just predict whether a molecule is dangerous or not and do not indicate the level of toxicity [48, 49]. Early on in the process of developing a medicine, this can assist in identifying potential safety concerns.

Identification of drug targets: A considerable bottleneck is created at the foundational stage of drug development, which is the finding and validation of pharmacological targets of interest for further downstream procedures. This is especially true for polygenic illnesses. Consequently, a variety of computer techniques have been created to support experimental drug development procedures [50]. It is possible to identify prospective drug targets in the nervous system using *in silico* approaches [47]. This can help to guide drug discovery efforts and identify new therapeutic targets.

Optimization of drug design: *In silico* modeling can be used to optimize the design of drugs to improve their efficacy and reduce their toxicity. Given the rapid development of *in silico* techniques, it is reasonable to assume that biomedical research in virtual reality will ultimately lead to significant changes in the pharmaceutical research landscape by streamlining the road to personalized medicine, streamlining the drug development process, and lowering the number of animal experiments. As a result, there may be a decrease in the likelihood of negative effects and an increase in the success rate of medication development [47, 51].

Overall, by cutting the time and expense involved with conventional methods, increasing how often drugs are developed successfully, and lowering risk of negative effects, *in silico* modeling has the probability to revolutionize drug development in neuropharmacology.

4.5 Case Studies

In silico approaches have been used in neuropharmacology to identify novel drug compounds and accelerate the drug discovery process. They were used to identify a compound that could target a specific receptor involved in the regulation of mood [52], a specific ion channel involved in the regulation of neuronal excitability [47], and a specific enzyme involved in the metabolism of dopamine [47].

Databases, homology models, QSARs, pharmacophores, and other molecular modeling approaches are examples of *in silico* techniques utilized in neuropharmacology [29, 47]. These methods have been successful in identifying new drug compounds and accelerating the drug discovery process in neuropharmacology, illustrating how computational techniques contribute to the discovery and design of probable drug candidates in the field.

Despite several successes, *in silico* drug discovery in neuropharmacology also has several challenges that need to be addressed. The behavior of a drug in a biological system may not always be precisely predicted by *in silico* methods, which rely on computational models and simulations. Therefore, it is essential to validate the results obtained from *in silico* methods experimentally [47].

The nervous system is highly complex, and the interactions between different components are not fully understood. Therefore, it is challenging to model the interactions between drugs and the nervous system accurately [53]. Also, *in silico* methods require huge amounts of data to build accurate models. However, the availability of data on neuropharmacology is limited, which can hinder the development of accurate models [54]. Additionally, *in silico* methods require significant computational power and resources, which can be expensive and time-consuming to obtain [52].

To overcome these challenges, researchers are developing new computational tools and algorithms that can improve the efficiency and accuracy of CADD in neuropharmacology [53]. Additionally, collaborations between computational scientists and experimentalists can help bridge the gap between *in silico* predictions and experimental validation [47].

4.5.1 Virtual Screening for Anti-Alzheimer's Drugs

Bellver *et al.* used SBVS method to find putative inhibitors of methyltransferase G9a. Following *in vitro* and *in vivo* testing on the top candidates, one (F) was discovered to be a promising G9a inhibitor with a fresh and powerful structure. Additionally, it was discovered that the candidate decreased H3 K9me2 in a Caenorhabditis elegans strain, which reduced the Alzheimer's disease (AD) pathology [55].

Lakra *et al.* looked for new prospective medication candidates as anti-Alzheimer medicines using pharmacophore-based virtual screening. Their goal was to find compounds that could prevent the enzyme cholinesterase from breaking down acetylcholine, a neurotransmitter crucial for memory and learning. The research discovered a number of possible drug candidates that may be enhanced for the eventual discovery of brand-new anti-Alzheimer's medications [56].

Additionally, virtual screening has been performed to look for fresh, effective anti-Alzheimer medications. The study done by Basile emphasized the benefits of virtual screening, which makes it possible to choose substances with the best physicochemical, pharmacokinetic, and pharmacodynamic features at a very cheap cost [57].

Another study by Khalid *et al.* employed virtual screening to find drugs with neuroprotective and anti-aggregatory properties in AD that are biaryl scaffold focused. The research discovered a number of substances that shown promising outcomes both *in vivo* and *in vitro* [58].

Virtual screening has also been used to search for pyridazine derivatives for AD treatment. The study identified several compounds that showed good SAR and drug-likeness [59].

Then a study was conducted to screen a sizable chemical database for substances that might block the activity of beta-secretase (BACE1), an enzyme involved in the formation of amyloid β plaques, using molecular docking and virtual screening techniques. Several new BACE1 inhibitors with potential therapeutic efficacy were found as a result of this [60].

Last but not least, Zhang *et al.* used virtual screening to isolate chemical components from traditional Chinese medicine formulations that were multi-target anti-AD. A number of chemical components with three or more possible AD-related targets were discovered by the investigation, and they could be further researched as prospective medication candidates [61].

4.5.2 Designing Antagonists of Dopamine D3 Receptor for Addiction Treatment

The dopamine D3 receptor is known to be impacted by behaviors connected to addiction. The designing of antagonists of selective D3 receptor has benefited from *in silico* techniques like pharmacophore modeling and molecular docking. By predicting the binding affinities and interactions of drug candidates with receptors, these computational techniques enabled researchers to optimize the chemical structures of drug candidates, which resulted in the creation of viable drug addiction treatments. Studies on *in silico* drug repositioning have been done to find dopamine modulators that may be useful in treating schizophrenia. Utilizing the crystal structure of the D3 receptor, an investigation of fragment-based drug design has been conducted [62–64].

4.5.3 Structure-Based Design of Selective Serotonin Reuptake Inhibitors

Drugs in this class are used to treat depression and anxiety. A protein-based pharmacophore was created in one investigation to find new structure-based design of selective serotonin reuptake inhibitors (SSRIs). The serotonin transporter (SERT), a target for SSRIs, served as the basis for the pharmacophore development [65].

Another study created a small chemical library by modifying the well-known SSRI paroxetine using a structure-based drug discovery methodology. Three of these inhibitors were crystallized in association with GPCR kinase 2 (GRK2) in order to gain knowledge of the structural factors that determine the inhibitors' efficacy and selectivity [66].

Based on the architecture of existing SSRIs and a pharmacophore model of the human SERT receptor, candidates for new SSRIs were developed in a computational modeling study [68]. Using LBDD and SBDD, a paroxetine analogue was created, synthesized, and tested for its biological activity in a different investigation [67].

Singh *et al.* discovered conformationally selective inhibitors of the SERT using a structure-based approach [68]. A different study used a structure-based pharmacophore model (SBPM) to find new SSRIs by docking conformations of known SSRIs [69].

4.5.4 Prediction of Blood–Brain Barrier Permeability

Blood–brain barrier (BBB) prevents many medications from entering the brain. The BBB permeability of potential therapeutic candidates may now be predicted *in silico*, allowing researchers to give priority to substances that have a better probability of crossing the BBB. As the blood-brain barrier can prevent many potentially helpful pharmaceuticals from reaching the brain, this is a crucial stage in the development of new medications. These hypotheses encourage the development of drugs that target the central nervous system and increase the possibility that effective novel neuropharmacological treatments will be found [70, 71].

The *in-silico* BBB permeability prediction has been subject of numerous investigations. Some of these researches have employed resampling and machine learning techniques to forecast the chemical permeability. In other investigations, models for forecasting BBB permeability were created using artificial neural networks [70].

One review article summarized recent advancements in *in silico* models of BBB permeation and reviewed the datasets currently available for *in silico* modeling [71]. Another review article discussed the need for an *in-silico* BBB permeability prediction model dedicated to CNS positron emission tomography tracer development [72].

4.5.5 Design of Allosteric Modulators for G Protein-Coupled Receptors

Drugs known as allosteric modulators bind to allosteric pockets rather than the orthosteric site where endogenous ligands bind, which is a distinct location. By doing so, they might indirectly modify receptor activation [73]. Positive allosteric modulators (PAMs) as well as negative allosteric modulators (NAMs) are the two categories into which allosteric modulators can be separated. While NAMs function as an allosteric antagonist, PAMs increase agonist affinity [74, 75].

In silico methods can be used to design allosteric modulators for GPCRs [73, 76, 77]. For instance, a computational fragment-based approach was used to identify allosteric modulators for the metabotropic glutamate receptor 5 (mGlu5). Persechino *et al.* thoroughly examined the GPCR structures that were already known in complex with small-molecule allosteric ligands to get structural knowledge for *in silico* drug discovery [73].

4.6 Conclusion

A complicated system is computationally simulated in *in silico* modeling, which takes the form of equations or rules. The capabilities to qualitatively and quantitatively assess various treatments on certain diseases and test a bigger set of distinct situations are provided by *in silico* models. Building *in silico* models is made simpler by quickly evolving experimental and analytical methods that deliver information-rich, high-throughput data. *In silico* modeling gives a framework for the description and comprehension of biological systems by formulating equations that express biological knowledge. *In silico* experiment findings serve as the basis for future investigation. *In silico* modeling has a number of benefits, such as the ability to predict phenomena using a hypothesis generator, the ability to assess unknown chemicals by *in silico* predictions or Adverse Outcome Pathway (AOP)-based *in silico* models based on known chemicals and structural similarity characteristics, and the ability to explore the chemical space to predict key events, key event relationships, metabolites, events at the molecular level, and/or adverse outcomes, and bi Additionally, *in silico* modeling can aid in the creation of more effective therapeutics for infectious diseases as well as in the study of the dynamics of infectious diseases.

Researchers can improve their understanding of biological processes, fasten the discovery of novel drugs, and make more informed judgments on the creation of novel therapies by combining the advantages of *in silico* approaches and experimental studies. The use of virtual populations via *in silico* models is still relatively new and is still in the early stages of clinical research, which presents one of the major problems of incorporating them into drug development. Furthermore, *in vitro/vivo* procedures frequently place restrictions on the idea of simulating human disorders.

In conclusion, the discipline of *in silico* modeling of drug–receptor interactions is expanding quickly and has the possibility to revolutionize the process of drug development. The number of *in vitro* tests needed can be greatly decreased by using *in silico* approaches to design medications that are more efficient and have fewer adverse effects.

References

1. Phatak, S.S., Stephan, C.C., Cavasotto, C.N., High-throughput and *in silico* screenings in drug discovery. *Expert Opin. Drug Discov.*, 4, 9, 947–959, 2009.
2. Hoque, I., Chatterjee, A., Bhattacharya, S., Biswas, R., An Approach of Computer-Aided Drug Design (CADD) Tools for *In Silico* Pharmaceutical Drug Design and Development. *Int. J. Adv. Res. Biol. Sci.*, 4, 2, 60–71, 2017.
3. Anderson, A.C., Review The Process of Structure-Based Drug Design. *Chem. Biol.*, 10, 787–797, 2003.
4. Aplin, C., Milano, S.K., Zielinski, K.A., Pollack, L., Cerione, R.A., Evolving Experimental Techniques for Structure-Based Drug Design. *J. Phys. Chem. B*, 126, 35, 6599–6607, 2022.
5. Aparoy, P., Reddy, K.K., Reddanna, P., Structure and Ligand Based Drug Design Strategies in the Development of Novel 5-LOX Inhibitors. *Curr. Med. Chem.*, 19, 3763–3778, 2012.
6. Acharya, C., Coop, A., Polli, J.E., Mackerell, A.D., Recent Advances in Ligand-Based Drug Design. Relevance and Utility of the Conformationally Sampled Pharmacophore Approach. *Curr. Comput.-Aided Drug Des.*, 7, 10–22, 2011.
7. March-Vila, E., Pinzi, L., Sturm, N., Tinivella, A., Engkvist, O., Chen, H., Rastelli, G., On the integration of *in silico* drug design methods for drug repurposing. *Front. Pharmacol.*, 8, May, 1–7, 2017.
8. Attie, A.D. and Raines, R.T., Analysis of Receptor-Ligand Interactions. *J. Chem. Educ.*, 72, 119–124, 1995.
9. Holloway, F.A. and Peirce, J.M., Fundamental Psychopharmacology, in: *Comprehensive Clinical Psychology*, pp. 175–205, 1998.

10. Marc, J., Pharmacogenetics of drug receptors. *Electron. J. Int. Fed. Clin. Chem. Lab. Med.*, 19, 48–53, 2008.
11. Guryanov, I., Fiorucci, S., Tennikova, T., Receptor-ligand interactions: Advanced biomedical applications. *Mater. Sci. Eng., C*, 68, 890–903, 2016.
12. Haburcak, R., Shi, J., Du, X., Yuan, D., Xu, B., Ligand-Receptor Interaction Modulates the Energy Landscape of Enzyme-Instructed Self-Assembly of Small Molecules. *J. Am. Chem. Soc.*, 138, 47, 15397–15404, 2016.
13. Wolfenden, R., Conformational Aspects of Inhibitor Design. Enzyme± Substrate Interactions in the Transition State. *Bioorg. Med. Chem.*, 7, 647–652, 1999.
14. Pacheco, M.A., Receptor Types and Subtypes, in: *The Comprehensive Pharmacology Reference*, pp. 1–5, 2011.
15. Li, S., Wong, A.H.C., Liu, F., Ligand-gated ion channel interacting proteins and their role in neuroprotection. *Front. Cell. Neurosci.*, 125, 8, 1–5, 2014.
16. Rodbard, D., Mathematics of Hormone-Receptor Interaction, in: *Receptors for reproductive hormones*, pp. 289–326, 1973.
17. Wells, J.A., Cunningham, B.C., Fuh, G., Lowman, H.B., Bass, S.H., Mulkerrin, M.G., Ultsch, M., Devos, A.M., The molecular basis for growth hormone-receptor interactions. *Recent Prog. Hormone Res.*, 48, 1, 253–275, 1993.
18. Dos Santos, C.E.M., Miranda, R.G., de Oliveira, D.P., Dorta, D.J., Challenges and opportunities for integrating *in silico* models and adverse outcomes pathways to set and relate new biomarkers. *Water (Basel)*, 12, 2–10, 2020.
19. Rachez, C., Lemon, B.D., Suldan, Z., Bromleigh, V., Gamble, M., Na, A.M., Erdjument-Bromage, H., Tempst, P., Freedman, L.P., Ligand-dependent transcription activation by nuclear receptors requires the DRIP complex. *Nature*, 398, 824–828, 1999.
20. Penvose, A., Keenan, J.L., Bray, D., Ramlall, V., Siggers, T., Comprehensive study of nuclear receptor DNA binding provides a revised framework for understanding receptor specificity. *Nat. Commun.*, 10, 1, 1–15, 2019.
21. Du, Z. and Lovly, C.M., Mechanisms of receptor tyrosine kinase activation in cancer. *Mol. Cancer*, 17, 1, 1–3, 2018.
22. Hubbard, S.R. and Miller, W.T., Receptor tyrosine kinases. mechanisms of activation and signaling. *Curr. Opin. Cell Biol.*, 19, 2, 117–123, 2007.
23. Salokas, K., Liu, X., Öhman, T., Chowdhury, I., Gawriyski, L., Keskitalo, S., Varjosalo, M., Physical and functional interactome atlas of human receptor tyrosine kinases. *EMBO Rep.*, 23, 6, 1–28, 2022.
24. Rohwedder, A., Knipp, S., Roberts, L.D., Ladbury, J.E., Composition of receptor tyrosine kinase-mediated lipid micro-domains controlled by adaptor protein interaction. *Sci. Rep.*, 11, 1, 1–10, 2021.
25. Calebiro, D., Koszegi, Z., Lanoiselée, Y., Miljus, T., O'brien, S., G Protein-Coupled Receptor-G Protein Interactions: A Single-Molecule Perspective. *Physiol. Rev.*, 101, 3, 857–906, 2021.
26. Mahoney, J.P. and Sunahara, R.K., Mechanistic insights into GPCR–G protein interactions. *Curr. Opin. Struct. Biol.*, 41, 247–254, 2016.

27. Heifetz, A., Morao, I., Babu, M.M., James, T., Southey, M.W.Y., Fedorov, D.G., Aldeghi, M., Bodkin, M.J., Townsend-Nicholson, A., Characterizing Interhelical Interactions of G-Protein Coupled Receptors with the Fragment Molecular Orbital Method. *J. Chem. Theory Comput.*, 16, 4, 2814–2824, 2020.
28. Gomez-Jeria, J.S., Modeling the Drug-Receptor Interaction in Quantum Pharmacology, in: *Molecules in Physics, Chemistry, and Biology*, vol. 4, pp. 215–231, 1989.
29. Ekins, S., Mestres, J., Testa, B., *In silico* pharmacology for drug discovery. Methods for virtual ligand screening and profiling. *Br. J. Pharmacol.*, 152, 1, 9–20, 2007.
30. Maruthanila, V.L., Elancheran, R., Roy, N.K., Bhattacharya, A., Kunnumakkara, A.B., Kabilan, S., Kotoky, J., *In silico* Molecular Modelling of Selected Natural Ligands and their Binding Features with Estrogen Receptor Alpha. *Curr. Comput.-Aided Drug Des.*, 15, 1, 89–96, 2018.
31. Li, Z., Han, P., You, Z.H., Li, X., Zhang, Y., Yu, H., Nie, R., Chen, X., *In silico* prediction of drug-target interaction networks based on drug chemical structure and protein sequences. *Sci. Rep.*, 7, 1, 1–31, 2017.
32. Brogi, S., Ramalho, T.C., Kuca, K., Medina-Franco, J.L., Valko, M., Editorial. *In silico* Methods for Drug Design and Discovery. *Front. Chem.*, 8, 1–5, 2020.
33. Peletier, L.A. and Gabrielsson, J., New Equilibrium Models of Drug-Receptor Interactions Derived from Target-Mediated Drug Disposition. *AAPS J.*, 20, 4, 1–15, 2018.
34. Hospital, A., Goñi, J.R., Orozco, M., Gelpí, J.L., Molecular dynamics simulations: Advances and applications. *Adv. Appl. Bioinform. Chem.*, 8, 37–47, 2015.
35. Doerr, S., Majewski, M., Pérez, A., Krämer, A., Clementi, C., Noe, F., Giorgino, T., De Fabritiis, G., TorchMD: A Deep Learning Framework for Molecular Simulations. *J. Chem. Theory Comput.*, 17, 4, 2355–2363, 2021.
36. Polanski, J., Chemoinformatics, in: *Comprehensive Chemometrics: Chemical and Biochemical Data Analysis*, pp. 459–506, 2009.
37. Bouysset, C. and Fiorucci, S., ProLIF. a library to encode molecular interactions as fingerprints. *J. Cheminf.*, 13, 1, 1–9, 2021.
38. Da, C. and Kireev, D., Structural protein-ligand interaction fingerprints (SPLIF) for structure-based virtual screening. Method and benchmark study. *J. Chem. Inf. Model.*, 54, 9, 2555–2561, 2014.
39. Deng, Z., Chuaqui, C., Singh, J., Structural Interaction Fingerprint (SIFt). A Novel Method for Analyzing Three-Dimensional Protein-Ligand Binding Interactions. *J. Med. Chem.*, 47, 2, 337–344, 2004.
40. Gainza, P., Sverrisson, F., Monti, F., Rodolà, E., Boscaini, D., Bronstein, M.M., Correia, B.E., Deciphering interaction fingerprints from protein molecular surfaces using geometric deep learning. *Nat. Methods*, 17, 2, 184–192, 2020.
41. Szulc, N.A., Mackiewicz, Z., Bujnicki, J.M., Stefaniak, F., Structural interaction fingerprints and machine learning for predicting and explaining binding of small molecule ligands to RNA. *Brief Bioinform.*, 24, 4, 1–30, 2023.

42. Sato, T., Honma, T., Yokoyama, S., Combining machine learning and pharmacophore-based interaction fingerprint for *in silico* screening. *J. Chem. Inf. Model.*, 50, 1, 170–185, 2010.
43. Kamath, S. and Buolamwini, J.K., Receptor-Guided Alignment-Based Comparative 3D-QSAR Studies of Benzylidene Malonitrile Tyrphostins as EGFR and HER-2 Kinase Inhibitors. *J. Med. Chem.*, 46, 22, 4657–4668, 2003.
44. Zong, F., Long, C., Hu, W., Chen, S., Dai, W., Xiao, Z.-X., Cao, Y., Abalign. a comprehensive multiple sequence alignment platform for B-cell receptor immune repertoires. *Nucleic Acids Res.*, 51, W1, 17–24, 2023.
45. Radaev, S., Zou, Z., Huang, T., Lafer, E.M., Hinck, A.P., Sun, P.D., Ternary complex of transforming growth factor-β1 reveals isoform-specific ligand recognition and receptor recruitment in the superfamily. *J. Biol. Chem.*, 285, 19, 14806–14814, 2010.
46. Muddassar, M., Pasha, F.A., Hwan, W.C., Kyung, H.Y., Chang, H.O., Seung, J.C., The receptor guided 3D-QSAR method is a powerful tool to design more potent IGF-1R inhibitors, in: *Proceedings of the Frontiers in the Convergence of Bioscience and Information Technologies*, pp. 48–52, 2007.
47. Ekins, S., Mestres, J., Testa, B., *In silico* pharmacology for drug discovery: Applications to targets and beyond. *Br. J. Pharmacol.*, 152, 1, 21–37, 2007.
48. Sacan, A., Ekins, S., Kortagere, S., Applications and limitations of *in silico* models in drug discovery. *Methods Mol. Biol.*, 910, 87–124, 2012.
49. Dearden, J.C., *In silico* prediction of drug toxicity. *J. Comput.-Aided Mol. Des.*, 17, 119–127, 2003.
50. Agamah, F.E., Mazandu, G.K., Hassan, R., Bope, C.D., Thomford, N.E., Ghansah, A., Chimusa, E.R., Computational/*in silico* methods in drug target and lead prediction. *Brief Bioinform.*, 21, 5, 1663–1675, 2020.
51. Noori, H.R. and Spanagel, R., *In silico* pharmacology. drug design and discovery's gate to the future. *In Silico Pharmacol.*, 1, 1, 1–2, 2013.
52. Shaker, B., Ahmad, S., Lee, J., Jung, C., Na, D., *In silico* methods and tools for drug discovery. *Comput. Biol. Med.*, 137, 1–15, 2021.
53. Meqbil, Y.J. and van Rijn, R.M., Opportunities and Challenges for *In Silico* Drug Discovery at Delta Opioid Receptors. *Pharmaceuticals*, 15, 7, 1–15, 2022.
54. Medina-Franco, J.L., Grand Challenges of Computer-Aided Drug Design: The Road Ahead. *Front. Drug Discov.*, 1, 1–4, 2021.
55. Bellver-Sanchis, A., Singh Choudhary, B., Companys-Alemany, J., Ávila-López, P.A., Martínez Rodríguez, A.L., Brea Floriani, J.M., Malik, R., Pallàs, M., Pérez, B., Griñán-Ferré, C., Structure-Based Virtual Screening and *in vitro* and *in vivo* Analyses Revealed Potent Methyltransferase G9a Inhibitors as Prospective Anti-Alzheimer's Agents. *ChemMedChem*, 17, 13, 1–10, 2022.
56. Lakra, N., Matore, B.W., Banjare, P., Singh, R., Singh, J., Roy, P.P., Pharmacophore based virtual screening of cholinesterase inhibitors. search of new potential drug candidates as antialzheimer agents. *In Silico Pharmacol.*, 10, 1, 1–14, 2022.

57. Basile, L., Virtual screening in the search of new and potent anti-alzheimer agents, in: *Neuromethods*, vol. 132, pp. 107–37, Humana Press Inc., New York, U.S.A., 2018.
58. Khalid, S., Zahid, M.A., Ali, H., Kim, Y.S., Khan, S., Biaryl scaffold-focused virtual screening for anti-aggregatory and neuroprotective effects in Alzheimer's disease. *BMC Neurosci.*, 19, 1, 1–11, 2018.
59. Zerroug, A., Belaidi, S., BenBrahim, I., Sinha, L., Chtita, S., Virtual screening in drug-likeness and structure/activity relationship of pyridazine derivatives as Anti-Alzheimer drugs. *J. King Saud Univ. Sci.*, 31, 4, 595–601, 2019.
60. Chidambaram, K., Identification of BACE-1 Inhibitors against Alzheimer's Disease through E-Pharmacophore-Based Virtual Screening and Molecular Dynamics Simulation Studies: An Insilco Approach. *Life*, 13, 4, 1–13, 2023.
61. Zhang, B., Zhao, J., Wang, Z., Guo, P., Liu, A., Du, G., Identification of Multi-Target Anti-AD Chemical Constituents From Traditional Chinese Medicine Formulae by Integrating Virtual Screening and *In Vitro* Validation. *Front. Pharmacol.*, 12, 1–19, 2021.
62. Mejia-Gutierrez, M., Vásquez-Paz, B.D., Fierro, L., Maza, J.R., *In Silico* Repositioning of Dopamine Modulators with Possible Application to Schizophrenia: Pharmacophore Mapping, Molecular Docking and Molecular Dynamics Analysis. *ACS Omega*, 6, 23, 14748–14764, 2021.
63. Keck, T.M., Burzynski, C., Shi, L., Newman, A.H., Beyond small-molecule SAR: Using the dopamine d3 receptor crystal structure to guide drug design. *Adv. Pharmacol.*, 69, 267–300, 2014.
64. Maramai, S., Gemma, S., Brogi, S., Campiani, G., Butini, S., Stark, H., Brindisi, M., Dopamine D3 receptor antagonists as potential therapeutics for the treatment of neurological diseases. *Front. Neurosci.*, 10, 1–16, 2016.
65. Manepalli, S., Geffert, L.M., Surratt, C.K., Madura, J.D., Discovery of novel selective serotonin reuptake inhibitors through development of a protein-based pharmacophore. *J. Chem. Inf. Model.*, 51, 9, 2417–2426, 2011.
66. Waldschmidt, H.V., Homan, K.T., Cato, M.C., Cruz-Rodríguez, O., Cannavo, A., Wilson, M.W., Song, J., Cheung, J.Y., Koch, W.J., Tesmer, J.J.G., Larsen, S.D., Structure-Based Design of Highly Selective and Potent G Protein-Coupled Receptor Kinase 2 Inhibitors Based on Paroxetine. *J. Med. Chem.*, 60, 7, 3052–3069, 2017.
67. Jaramillo, D.N., Millán, D., Guevara-Pulido, J., Design, synthesis and cytotoxic evaluation of a selective serotonin reuptake inhibitor (SSRI) by virtual screening. *Eur. J. Pharm. Sci.*, 183, 1–9, 2023.
68. Singh, I., Seth, A., Billesbølle, C.B., Braz, J., Rodriguiz, R.M., Roy, K., Bekele, B., Craik, V., Huang, X.-P., Boytsov, D., Pogorelov, V.M., Lak, P., O'Donnell, H., Sandtner, W., Irwin, J.J., Roth, B.L., Basbaum, A.I., Wetsel, W.C., Manglik, A., Shoichet, B.K., Rudnick, G., Structure-based discovery of conformationally selective inhibitors of the serotonin transporter. *Cell*, 186, 10, 2160–2175, 2023.

69. Zhou, Z.L., Liu, H.L., Wu, J.W., Tsao, C.W., Chen, W.H., Liu, K.T., Ho, Y., Combining structure-based pharmacophore and *in silico* approaches to discover novel selective serotonin reuptake inhibitors. *Chem. Biol. Drug Des.*, 82, 6, 705–717, 2013.
70. Wang, Z., Yang, H., Wu, Z., Wang, T., Li, W., Tang, Y., Liu, G., *In Silico* Prediction of Blood–Brain Barrier Permeability of Compounds by Machine Learning and Resampling Methods. *ChemMedChem*, 13, 20, 2189–2201, 2018.
71. Jain, T. and Kumar, P., *In Silico* Prediction of Blood-Brain Barrier Permeability of Chemical Compounds through Molecular Feature Modeling, ArXiv, 1–17, 2022.
72. Clark, D.E., *In silico* prediction of blood–brain barrier permeation. *Drug Discov. Today*, 8, 927–933, 2003.
73. Persechino, M., Hedderich, J.B., Kolb, P., Hilger, D., Allosteric modulation of GPCRs: From structural insights to *in silico* drug discovery. *Pharmacol. Ther.*, 237, 1–5, 2022.
74. Basith, S., Cui, M., Macalino, S.J.Y., Park, J., Clavio, N.A.B., Kang, S., Choi., S., Exploring G protein-coupled receptors (GPCRs) ligand space via chemin-formatics approaches: Impact on rational drug design. *Front. Pharmacol.*, 9, 1–26, 2018.
75. Mohanty, S.K., Mittal, A., Gaur, A., Solanki, S., Kumar, S., Duari, S., Arora, S., Gautam, V., Dixit, N., Ghosh, T., Sengupta, D., Artificial Intelligence uncovers Evolutionarily Conserved Intracellular Allosteric Modulators of GPCR-Gα interface, BioRxiv, 1–53, 2023.
76. Bian, Y., Feng, Z., Yang, P., Xie, X.Q., Integrated *in Silico* Fragment-Based Drug Design: Case Study with Allosteric Modulators on Metabotropic Glutamate Receptor 5. *AAPS J.*, 19, 4, 1235–1248, 2017.
77. Lu, S. and Zhang, J., Small Molecule Allosteric Modulators of G-Protein-Coupled Receptors: Drug-Target Interactions. *J. Med. Chem.*, 62, 1, 24–45, 2019.

5

Computational Attitudes in Counselling Psychology

Bharat Mishra[1]*, Farha Deeba Khan[2], Archita Tiwari[3] and Anitta Joseph[4]

[1]Institute of Pharmacy, Dr. Shakuntala Misra National Rehabilitation University, Lucknow, Uttar Pradesh, India
[2]RGS College of Pharmacy, Lucknow, Uttar Pradesh, India
[3]Khwaja Moinuddin Chishti Language University, Lucknow, Uttar Pradesh, India
[4]RSQ JRec, Manuscript Services, Springer Nature, Pune, Maharashtra, India

Abstract

Computational neuroscience is a branch of neuroscience which employs mathematical models, computer simulations, theoretical analysis, and abstractions of the brain to understand the principles that govern the development, structure, physiology and cognitive abilities of the nervous system. The main goal of computational neuroscience is to build biologically realistic mathematical models and to test how they match and predict experimental data by using computer simulations. A wide range of computational models in fact have been developed to explore the alterations of the mechanisms involved in neurotransmission following the emergence of neurological pathologies such as rank order, clockwork orange, visual attention, model of area 3b, decision making model, clock driven vs event driven stimulus, NEURON, NEST stimulator, Brian stimulator, Reproducible science using Python software etc. A large no of experimental methods has been tuned to dissect the mechanisms taking part in the neurotransmission process. The identification and characterization of potential pharmacological targets in neurology and psychiatry is a fundamental problem at the intersection between medicinal chemistry and the neurosciences. The ultimate goal of computational neuroscience is to explain how electrical and chemical signals are used in the brain to represent and process information. It explains the biophysical mechanisms of computation in neurons, computer simulations of neural circuits, and models of

**Corresponding author*: bmishra_pharmacy@dsmnru.ac.in; bharatekansh@gmail.com; ORCID: 0000-0001-8761-3499
Archita Tiwari: ORCID: 0009-0001-7067-0304

Bhupendra Prajapati, Alok Tripathi, Rishabha Malviya and Lucy Mohapatra (eds.) Computational Neuropharmacology: Fundamentals and Clinical Aspects, (127–158) © 2025 Scrivener Publishing LLC

learning. Computational neuroscience and bioinformatics can play a major role in this functional connection. Robust quantitative models exist describing all major active membrane properties under endogenous and exogenous chemical control. These include voltage-dependent ionic channels (sodium, potassium, calcium, etc.), synaptic receptor channels (e.g. glutamatergic, GABAergic, cholinergic), and G protein–coupled signaling pathways (protein kinases, phosphatases, and other enzymatic cascades) Computational methods such as Hodgkin - Huxley formalism ADME/T evaluation, molecular modeling, logical modeling, network modeling, pathway analysis, multi-scale systems pharmacology platforms and virtual patient for QSP have been developed. The availability of electroencephalography (EEG) and magnetoencephalography (MEG) data that reflect spatio-temporal neural activity, together with hemodynamic imaging such as positron emission tomography (PET) and functional magnetic resonance imaging (fMRI) obtained from human subjects, fostered the emergence of new modeling techniques that integrate different phenomena occurring at different spatiotemporal scales. The parallel progresses in the experimental and ncomputational neuroscience, together with the improvement of high-performing computing prelude to an exponential increase in the development of computational models aiming to investigate the mechanisms underlying neurological diseases. Computational perspective provides compelling examples of how simulations can elucidate, explain, and predict the effect of chemical agonists, antagonists, and modulators in the nervous system. The main objective of this chapter is to evaluate electrical patterns in neurohumoral transmission and signal transduction using the above-mentioned computational models and methods. Effects of various drugs and neuromodulators on synaptic transmission and activity patterns of neurons influencing behavior consciousness and mood. This study could provide new strategies for screening new drugs and help search for novel target specific drugs in different neurological illness. Computational approach is cost effective compared with traditional methods and it can be applied to several neurological and psychiatric diseases.

Keywords: Computational neuropharmacology, neurohumoral transmission, neurological disorders

List of Abbreviations

AI	Artificial Intelligence
AR	Augmented Reality
AU	Action unit
AUC	Area under the Curve
CBT	Cognitive Behavioral Therapy
CCBT	Computerized Cognitive Behavioral Therapy
CCTM	Classical Computational Theory of Mind

CERT	Computer Expression Recognition Toolbox
CNN	Convolutional neural network
COVID-19	Coronavirus disease 2019
CPT	Cognitive Processing Therapy
CPT	Cognitive Processing Therapy for PTSD
EEG	Electroencephalography
fMRI	Functional Magnetic Resonance Imaging
GAD	Generalized anxiety disorder
GOMS	Goals, Operators, Methods, and Selection
IRT	Item-response theory
MEG	Magnetoencephalography
NEST	Neural Simulation Tool
NLP	Natural Language Processing
PET	Positron Emission Tomography
PTSD	Post-Traumatic Stress Disorder
PTSD	Post-Traumatic Stress Disorder
QSP	Quantitative System Pharmacology
REBT	Rational Emotive Behavior Therapy
ROC	Receiver Operating Characteristic
SVM	Support vector machine
VR	Virtual Reality
VRET	Virtual Reality Exposure Therapy
WCST	Wisconsin Card Sorting Test

5.1 Introduction

Not long ago, a journalist quoted an interesting fact about computers and the human mind. When the computer was invented, they described it as the 'electronic brain' or 'thinking machine'. As computers evolved to be a part and parcel of our lives, the metaphor has turned upside down. Conversely, the brain is now described as a computer; an information-processing machine made of a billion neurons in place of transistors [1].

Scientists have dedicated decades of research to understand the complexity of the human mind. In 1943, Warren McCulloch and Walter Pitts put forward the thought that the human mind is a computational system. This notion is known as the Classical Computational Theory of Mind (CCTM). It says that the human mind is a computing system where the neurons in the brain act as the processing units. They proposed that both cognition and consciousness are a form of computation [2].

In the world of computer science, a definition of computer is anything which can receive input and produce output. Alan Turing proposed a computing system called a Turing machine. It is a theoretical model which does any computation by taking all the input and producing the output according to our instructions. This machine is used to describe what a computer basically is, in the world of computer science [2]. The Classical Computational Theory of Mind states that the human mind is like the Turing machine. The mental processes like reasoning, problem-solving, decision-making, cognition, consciousness etc. are similar to the computations executed by a computer [2].

5.1.1 Counselling Psychology

As per the American Psychological Association, Counseling Psychology is a specialty in professional psychology which uses a wide range of techniques to enable people to improve their quality of life, prevent mental distress, help cope with traumatic events and thus improve their overall life [7]. Counselling psychologists are professionals who use various psychological therapies to prevent or decrease problems related to emotional, physical, health, social or identity aspects.

5.1.2 Indications for Psychological Therapy

Psychological therapies are treatments which help people experiencing a wide range of mental challenges. Such therapies help one to identify the root cause of the psychological distress, overcome it and help to lead a happy, healthier life. People trying to cope with stressful events, traumatic experiences, suffering from medical illness, loss of loved ones or dealing with mental health conditions could benefit from psychological therapies [8, 9].

5.1.2.1 Types of Psychological Therapies

Several types of therapies exist, and the choice is made depending on the type of illness the patient experiences [8, 9].

1. Cognitive behavioral therapy or CBT: This therapy makes people aware of their detrimental patterns of thinking and behaviors and helps them to substitute them with healthy thinking patterns. It includes practicing newly acquired skills in a real-world scenario. CBT is beneficial in treating

depression and anxiety disorders, eating disorders and traumatic events.

2. Interpersonal therapy or IBT: This form of therapy helps patients to explore the underlying causes of their troubles like unresolved grief, sudden changes in social or work roles etc. They learn how to communicate and express their emotions in a healthy way. This therapy is commonly used to treat depression.
3. Dialectical behaviour therapy: It is a type of CBT often used to treat people having chronic suicidal thought patterns and borderline personality disorders, eating disorders as well as post-traumatic stress disorders. People are taught new skills which help them to take personal responsibility for changing their behaviour patterns.
4. Psychodynamic therapy: This approach is based on the idea that our present behaviours are rooted in past experiences, especially those of childhood years. It helps to recognize those experiences and bring them into one's awareness. By being aware of them, a person can begin to work on the deep-seated patterns and improve their quality of life.
5. Psychoanalysis: It is an intensive form of psychodynamic therapy. This type of therapy requires a close partnership between the patient and the therapist. By interacting with the therapist, the patients learn more about themselves and help them to identify and change their behaviours.
6. Supportive therapy: The patient is guided and encouraged to develop their own resources. The patients build better self-esteem, reduce anxiety, help to cope more effectively with their surroundings and improve their social functioning.

Psychology is the scientific study of the human mind and behaviour. There are various psychological therapies to treat the disorders of the human mind. But they have the following drawbacks [11]:

1. Psychotherapy fails because both patient and the therapist are unwilling to put in adequate effort. Patients might disagree with the therapist's diagnosis of the problem altogether. They show reluctance to continue with the subsequent appointments for therapy. They also fail to do the homework suggested by the therapists.

2. The patient might hold maladaptive beliefs about the therapist and vice versa. Such negatively biased beliefs may hinder a healthy relationship between the patient and therapist. In therapy, such relationships are necessary for the patient to have trust in his therapist and thus make the therapy an overall success.
3. Psychological therapies can have negative or side effects as much as positive effects. Patients might not have a full idea of how the therapy works and might hold unrealistic expectations about the results.
4. Psychological therapies fail because it assumes neurocognitive resources which the patients might not have.
5. Therapists may have the wrong model of illness.
6. Psychological therapies fail with some patients even though they have acquired new knowledge from the therapy.

The emerging field of computational psychology investigates the computations done by the human brain to satisfy human needs in a state of both health as well as disorder. Computational psychology helps researchers to understand the human mind, its working mechanism and thus find answers which are of clinical importance.

5.1.3 Why Do We Need Computer Models in the Field of Psychology?

There are four areas where computational psychology can help in psychological therapies [10]:

1. Better understanding of the mechanism of mental illness: Mental illness is a result of underlying neural mechanisms and to treat them, it is important to have the knowledge of this biological mechanism. Understanding the biological aspects of mental illness can help in better chances for prevention and therapeutic interventions.
2. Subtyping of disease: The ignorance of the underlying biological mechanisms of mental illness has hindered the proper diagnosis and treatment of such disorders. Subtyping involves identifying similar people with similar disease conditions. This practice is essential to develop effective clinical protocols to accurately diagnose and treat the patient. In

cases like mental disorders, there is a great heterogeneity of the diseases and most of the time, the underlying biology is unknown. This has led to difficulty in subtyping. The computational models will be able to solve this heterogeneity and provide a better understanding of the disease states.

3. Prediction of mental health status: Computational models will be beneficial in accurately predicting the mental health status before the development of the disease or during the disease process. Such predictions might allow early prevention of the disease, especially in the early psychosis field where the identification of high-risk states provides valuable insights for prevention.
4. Treatment stratification: The prediction of which patient will benefit from what kind of treatment is traditionally done by a slow trial and error process. Computational models can be a valuable tool for selecting the specific treatment for mental disease or predicting the treatment outcomes.

Psychologists conduct experiments and formulate various theories to explain the human mind. Theories are hypotheses backed by evidence. Most of the time, these theories fail to reproduce the results in psychology. In spite of these, numerous theories come and go without being testified. This has led to a theory crisis in psychology. One of the major reasons why building a theory in psychology is difficult is because of the subject matter.

Together with that, there are three reasons why developing a theory in psychology has become hard [3]:

1. The lack of strong pieces of evidence or phenomena leading to an unlimited number of theories.
   In psychology, we need theories to explain the evidence or phenomena. Sometimes, formulating a theory which explains the entire pieces of phenomena becomes difficult. For example, in biology, we have Darwin's theory of evolution. It fits well with the phenomena we observe in the world, that it has literally become impossible to come up with a new theory of evolution. This is an example of a robust phenomenon. The problem in psychology is that in most cases, there is no robust phenomenon. Due to this lack of robust phenomenon, there is no restriction in coming up with numerous theories to explain a phenomenon. Due to insufficient evidence, these theories cannot be tested for their truthfulness. The end result is a pile

of theories which remain untested, thus weakening the base of theoretical psychology [3].

2. The lack of validity testing of psychological constructs
Psychological constructs are variables which are used to describe elements of human experience. They cannot be directly measured. For example, self-esteem. Construct validity means how well something measures what it was designed to evaluate. How consistently it measures what it was designed to measure, is called reliability. But most scientists do reliability tests than validity tests for constructs because it is relatively easy and has well-described parameters. Due to this, there are many unvalidated constructs in psychology now [3].
Usually in science or physics, once a concept is defined (e.g: electron) it is constantly challenged and changed until it reaches a near-perfect stage. This is called epistemic iteration. But in psychology, this rarely happens because psychological concepts like depression are deeply entrenched in society that it has almost become impossible to challenge those ideas and make improvements [3].
3. Inability to form a connection between psychological variables
In biology, the DNA theory explains the mechanism of inheritance and Darwin's theory of evolution explains the mechanism of evolution through natural selection. The problem with psychological theories is that they cannot describe the causes accurately. In an experiment, changing a psychological variable often requires a change in a non-psychological aspect. For example, we cannot directly change the thoughts of a person without changing an external factor. And when the external factor is changed, multiple psychological aspects are altered. So, it becomes difficult to find the exact mechanism of action of a psychological phenomenon [3].

Psychology's focus on finding explanations for the causes of behaviours has led to the formation of numerous research theories and methodologies. As enlisted above, these theories lack fail to explain the evidence; lack construct validity and are unable to draw a connection between cause and effect. In the next part, we will see the limitations of methodologies in psychological therapy.

5.1.4 Limitations of Methodologies in Psychological Therapy [4]

In psychological therapy, it is important to establish a relationship between cause and effect. In words, identifying the mechanism of action is crucial in psychological therapy. The most reliable experimental method until now is the randomized controlled trials. It is considered the 'gold standard' to elucidate the mechanism of action. Unlike clinical trials or biology experiments, the issue in psychological experiments is that it is difficult to change one variable without affecting the other. Psychological interventions are complex, and it is difficult to find the exact relationship between cause and effect in such psychological experiments.

1. Current research methodologies used in psychological therapy do not consider the feedback relationship between behavior and sensory input of an individual.
2. The current research methods lack theories to explain the dynamic and continuous variation in the behaviour of an individual. Instead, they explain the average behaviour or try to explain the behaviour in 'parts'.
3. The research methods use group statistics to understand what is happening on an individual level. However, it can lead to incorrect conclusions when it comes to using group data to explain individual behaviour.
4. The mechanism of psychological therapy is difficult to be presented as a diagram or as a working model. Most of the mechanisms are described only verbally. This makes it difficult to test against real-world data. The progress of physics and engineering disciplines is mostly because they could produce functional models which could be tested and verified against real-world data. A similar approach could be used in psychological therapy to build functional models.

It is evident from the above cases that, the theory-building in psychology is always prone to errors since it is done by humans [2]. This is where computational modelling comes into the picture. As we saw earlier, the human mind is a computation system and can be simulated using computational models. This can lead to more reliable and error-proof theories in psychology.

Computational modelling involves the use of computers to simulate complex systems like the human mind, using mathematics, physics and computer science. To generate a computational model, it is inevitable to convert the psychological attribute into mathematical terms [5]. Computational modelling in psychology means describing a psychological theory in mathematical terms. Various computational models such as the generative model, Bayesian inference, predictive investigative coding, hierarchical model, cortical architecture, Wisconsin Card Sorting Test, the Attentional-Updating Model Cognitive Reinforcement-Learning Model, and Parallel Reinforcement-Learning Model, etc. are used in psychology.

The following are the advantages of computational modelling in psychology

1. It helps in constructing robust and valid theory building [3].
2. It increases the reproducibility of research in psychology [3].
3. It helps to focus on prediction rather than explanation, which promotes a greater appreciation of human behaviour [6].
4. It enables an understanding of the mechanism behind psychological behaviours [6].
5. The robust theory-building process finally leads to a strong foundation for theoretical psychology, helping the field to make further progress [3].

Computational models use mathematical equations to structure the information and they form associations between the input and output variables [10]. Input variables include neural activity, self-reported outcomes etc. and output variables include behaviour, psychiatric diagnosis and treatment response. Such models can even quantify the output variables and enables simulation to better understand how they work [11].

5.1.5 Evolution of Technology in Counselling Psychology

The evolution of technology in counselling psychology has been a transformative journey that has significantly impacted how mental health services are delivered. Over the years, technological advancements have introduced innovative tools and platforms that complement traditional therapeutic approaches, enhancing accessibility, effectiveness, and reach. The evolution can be divided into several key stages:

In the 1960s and 1970s, early computer technologies were introduced into counselling psychology. Computer programs were developed to aid therapists in data collection, assessment, and treatment planning [12]. These computer-assisted tools were a precursor to more sophisticated applications that emerged later. With the widespread adoption of the internet in the 1990s, opportunities for remote communication and information exchange expanded. Counsellors began exploring email counselling and chat-based therapy, enabling patients to receive support from the comfort of their homes. This marked the beginnings of telehealth in counselling psychology. In the early 2000s, web-based platforms and mobile applications started gaining traction in the mental health field [13]. These platforms offered self-help resources, psychoeducation, mood tracking, and interactive exercises, empowering individuals to engage in self-guided interventions and self-monitoring. Advancements in virtual reality (VR) and augmented reality (AR) technologies brought new possibilities for exposure therapy and immersive therapeutic experiences. VR-based environments allowed patients to confront and manage anxiety-provoking situations in a controlled setting, fostering therapeutic progress. The integration of artificial intelligence (AI) and natural language processing (NLP) in counselling psychology introduced chatbots and virtual agents capable of engaging in human-like conversations. These AI-driven tools could deliver personalized support, conduct assessments, and provide immediate feedback to users. The increasing availability of data and the development of data analytics techniques have enabled researchers and practitioners to gain insights into large datasets related to mental health [14]. This has led to a better understanding of various psychological conditions and the optimization of interventions. Over time, dedicated online therapy platforms and digital mental health services have emerged. These platforms connect patients with licensed therapists through secure video conferencing and messaging, expanding the reach of mental health services beyond geographical limitations.

Modern counselling psychology now emphasizes the integration of technology with traditional therapeutic modalities. Therapists are encouraged to embrace technological tools that can complement their practice, promoting more effective interventions and better client outcomes. Throughout this evolution, technology in counselling psychology has encountered challenges related to ethics, data privacy, therapist training, and therapeutic efficacy. However, ongoing research, professional guidelines, and the development of best practices have helped to address these concerns and ensure responsible and ethical use of technology in therapy [14]. As the field continues to evolve, it is essential for practitioners to stay informed about new technological advancements and embrace innovative approaches that can enhance their therapeutic practice.

5.1.6 Emergence of Computational Attitudes in Counselling Psychology

The emergence of computational attitudes in counselling psychology refers to the increasing integration of technology, particularly computational tools and methodologies, into therapeutic practice. It represents a shift in the field towards embracing technological advancements to enhance the delivery of mental health services and improve client outcomes [15]. The term "computational attitudes" reflects the mindset of therapists and researchers who actively seek to incorporate technology into their work, leveraging its potential to complement and enhance traditional therapeutic approaches [16]. The following factors have contributed to the rise of computational attitudes in counselling psychology:

Rapid advancements in technology, such as artificial intelligence, natural language processing, virtual reality, and data analytics, have provided new opportunities for the mental health field. These technologies offer innovative ways to deliver interventions, conduct assessments, and enhance therapeutic processes. The widespread ownership of digital devices, such as smartphones and computers, has made it easier for individuals to engage with mental health resources and interventions through digital platforms [17]. This increased accessibility has opened up new avenues for reaching a broader population. The demand for convenient and remote mental health services has grown in recent years. Computational attitudes enable therapists to offer online therapy, telehealth sessions, and self-help resources, catering to patients' preferences and busy lifestyles. Empirical evidence demonstrating the efficacy of technology-assisted interventions has contributed to the adoption of computational attitudes. Research studies have shown positive outcomes for various digital mental health interventions, reinforcing the value of integrating technology in therapy. Computational tools enable personalized interventions, tailoring treatment plans to individual needs. They analyze large mental health datasets, providing valuable insights and supporting evidence-based practice [18]. Computational attitudes bridge the treatment gap, increasing access to mental health services for underserved areas. Cultural acceptance of technology in mental health care is growing, with therapists and patients embracing digital tools. Training programs are incorporating computational attitudes, equipping future therapists with technology skills. Global events, like the COVID-19 pandemic, accelerated technology adoption for remote care. Computational attitudes hold promise in transforming mental health services but require ethical considerations, data privacy, and ongoing research for responsible integration.

5.1.7 Current Trends and Challenges

In the field of counseling psychology, there are several prevailing trends and challenges. As of that time, the integration of technology is a prominent trend, with telehealth, online therapy platforms, mobile mental health applications, virtual reality interventions, and AI-driven chatbots being widely utilized for support and guidance [17, 19]. Additionally, there is a growing emphasis on cultural competence, diversity, and inclusion. Therapists are encouraged to be more aware of their patients' diverse backgrounds and experiences, providing culturally sensitive and responsive care. Another significant trend is the shift towards preventive and early intervention approaches. Efforts are being made to identify and address mental health concerns at their early stages to prevent them from escalating. Positive psychology has also been embraced in the field, aiming to promote well-being, strengths, and positive emotions alongside addressing psychological distress. Therapists are incorporating positive psychology techniques to enhance resilience and life satisfaction. Moreover, mindfulness-based interventions and self-compassion practices have gained popularity [20]. These approaches help patients cultivate present-moment awareness and self-kindness, leading to reduced stress and improved overall mental well-being. Lastly, the rise of social media and online platforms has given rise to various online support communities and mental health advocacy groups. These communities offer peer support and valuable resources to individuals facing mental health challenges. It is important to note that newer developments might have occurred since the time of these observations.

5.2 Theoretical Foundations of Computational Attitude

5.2.1 Cognitive and Behavioral Theories in Computational Counselling

In computational counselling, cognitive and behavioral theories play a significant role in shaping the development and application of technology-driven interventions. These theories form the foundation for understanding human behavior, cognition, and emotional processes, and they are essential in guiding the design and implementation of computational tools aimed at promoting psychological well-being. Here's how cognitive and behavioral theories are integrated into computational counselling:

5.2.1.1 *Cognitive Theories*

Cognitive behavioural therapy (CBT) is one of the most influential cognitive theories in counselling. It emphasizes the role of thoughts, beliefs, and cognitive processes in shaping emotions and behaviors. In computational counselling, CBT principles are integrated into chatbots and virtual agents that engage in cognitive restructuring, challenging negative thoughts, and offering coping strategies [21].

Cognitive Processing Therapy (CPT) is often used to treat post-traumatic stress disorder (PTSD) and involves addressing cognitive distortions related to traumatic events. In computational counselling, CPT principles can be applied through online platforms that help individuals process and challenge traumatic memories [22].

Rational emotive behavior therapy (REBT) focuses on identifying and challenging irrational beliefs that contribute to emotional distress. In computational counselling, REBT principles can be embedded in chatbots or mobile apps that guide patients through disputing irrational thoughts and adopting more rational beliefs [23].

5.2.1.2 *Behavioral Theories*

Behavioral activation is used to treat depression and involves engaging in rewarding and meaningful activities to improve mood and motivation. In computational counselling, mobile apps can be designed to track and encourage patients to participate in positive activities. Exposure therapy is an effective treatment for anxiety disorders, involving gradual exposure to feared stimuli to reduce anxiety responses. Virtual reality applications are employed in computational counselling to create virtual environments for exposure therapy in a safe and controlled manner.

Operant conditioning principles, such as reinforcement and punishment, can be integrated into behavior change apps that provide rewards or incentives to encourage desired behaviors and discourage unwanted ones [24].

5.2.1.3 *Integrative Approaches*

In computational counselling, many interventions are designed based on integrative approaches that combine elements from various cognitive and behavioral theories. For example, an AI-driven mental health platform might utilize CBT strategies for anxiety, behavioral activation for depression, and mindfulness techniques for stress management.

Overall, cognitive and behavioral theories serve as the theoretical framework for developing technology-driven interventions in counselling. These theories guide the selection of therapeutic strategies, content delivery, and the incorporation of personalized elements in computational tools. By integrating cognitive and behavioral principles, computational counselling aims to enhance therapeutic outcomes and provide accessible and effective mental health support. However, it's essential to ensure that ethical considerations, privacy protection, and the quality of therapeutic relationships are maintained when employing computational approaches in counselling psychology [25].

5.2.2 Key Technologies in Computational Counselling

Artificial Intelligence (AI) and Natural Language Processing (NLP) are two interrelated technologies that have significantly influenced the field of counselling psychology. These technologies have been applied to create innovative interventions, improve communication between therapists and patients, and enhance the overall therapeutic experience [26]. Here's an overview of AI and NLP in counselling psychology:

5.2.2.1 Artificial Intelligence and Natural Language Processing

5.2.2.1.1 Artificial Intelligence (AI)

AI refers to the simulation of human intelligence in machines that can perform tasks typically requiring human intelligence, such as reasoning, learning, problem-solving, and decision-making. In counselling psychology, AI is used to create virtual agents, chatbots, and other intelligent systems that can interact with patients and deliver therapeutic support. Some key applications of AI in counselling psychology include: AI-powered virtual therapists are virtual agents that use natural language to engage in conversation with patients. They can provide psychoeducation, assessments, interventions, and emotional support. AI analyzes client data and interactions to personalize interventions, increasing their relevance and effectiveness. It also analyzes the emotional tone in patients' language to identify signs of distress, depression, or anxiety, aiding therapists in understanding emotional states and tracking changes. AI-driven predictive analytics identify patterns in client data and predict potential mental health issues, enabling timely intervention and preventive strategies [26].

5.2.2.1.2 Natural Language Processing (NLP)

NLP is a subfield of AI that focuses on enabling computers to understand, interpret, and generate human language. In counselling psychology, NLP facilitates communication between patients and AI-driven systems, enabling more natural and empathetic interactions. Some applications of NLP in counselling psychology include:

NLP allows AI-driven systems to engage in human-like conversations with patients, offering a more personalized and supportive therapeutic experience [22]. NLP helps virtual agents interpret and comprehend patients' language, allowing them to respond appropriately to their needs and concerns. NLP can be used to analyze text-based data, such as chat transcripts or journal entries, to extract valuable insights and track therapeutic progress. NLP techniques can determine the sentiment and emotional tone of patients' language, helping therapists assess emotional states and tailor interventions accordingly. NLP can be utilized to facilitate communication with patients who speak different languages, breaking down language barriers in therapy.

The integration of AI and NLP in counselling psychology holds great promise for enhancing the accessibility, effectiveness, and efficiency of mental health services. However, ethical considerations are crucial when using these technologies to ensure client privacy, data security, and responsible use of AI-driven interventions. Therapists must carefully balance the benefits of technology with the maintenance of the therapeutic relationship and the humanistic elements of counselling practice [27, 28].

5.2.2.1.3 Virtual Reality and Augmented Reality Applications

Virtual Reality (VR) and Augmented Reality (AR) are immersive technologies that have found applications in the field of counselling psychology. These technologies create simulated environments or augment real-world settings to offer unique and effective therapeutic experiences. Here's an overview of VR and AR applications in counselling psychology:

The use of VR and AR applications in counselling psychology has shown promising results in improving treatment outcomes, enhancing client engagement, and providing novel therapeutic experiences. However, therapists must ensure that patients are comfortable and informed about the use of these technologies, and ethical considerations related to privacy, informed consent, and data security should be carefully addressed. Additionally, therapists should be adequately trained to utilize VR and AR technologies effectively and ethically in their practice [29].

5.2.2.1.4 Mobile Applications and Digital Mental Health Platforms

Mobile applications and digital mental health platforms have revolutionized the accessibility and delivery of mental health services. These technologies provide individuals with convenient and immediate access to mental health resources, interventions, and support. Mobile applications and digital mental health platforms have made mental health support more accessible, flexible, and cost-effective. However, ethical considerations related to data privacy, informed consent, and therapeutic boundaries must be carefully addressed. Additionally, it's essential to ensure that these technologies complement rather than replace traditional face-to-face therapy when needed, and that patients have access to qualified and licensed mental health professionals [30].

5.2.2.1.5 Computational Models in Counselling Psychology

In counseling psychology, computational models play a crucial role in the application of computational attitudes. These models utilize mathematical

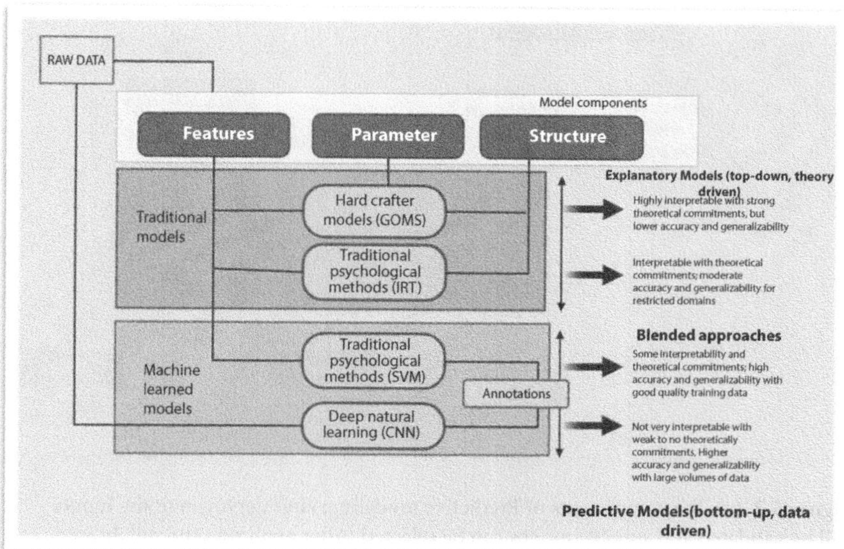

Figure 5.1 The top four computer modelling strategies. The methods can be mixed without conflict. Annotations only apply to model training. GOMS: Goals, Operators, Methods, and Selection; IRT: ltem-response theory; SVM: Support vector machine; CNN: Convolutional neural network. In handcrafted models the parameters are prespecified, and for both types of traditional model the structure is prespecified. Annotations provided by outside to guide the machine-learning process by feeding a supervisory signal. Model-training phase is needed for supervising the machine learning [50].

and computational techniques to simulate and understand complex psychological processes, inform therapeutic decisions, and improve treatment outcomes. Here are some ways in which computational models are used in counseling psychology and the main approach of computational methods depicts in Figure 5.1.

5.2.2.1.5.1 PREDICTIVE MODELS

Computational models can be used to predict client responses to various therapeutic interventions. By analyzing historical data from previous patients with similar characteristics and treatment histories, therapists can develop predictive models that estimate the likelihood of success for different therapeutic approaches [31]. This can help therapists tailor treatment plans to individual patients' needs and improve the chances of positive outcomes Figure 5.2.

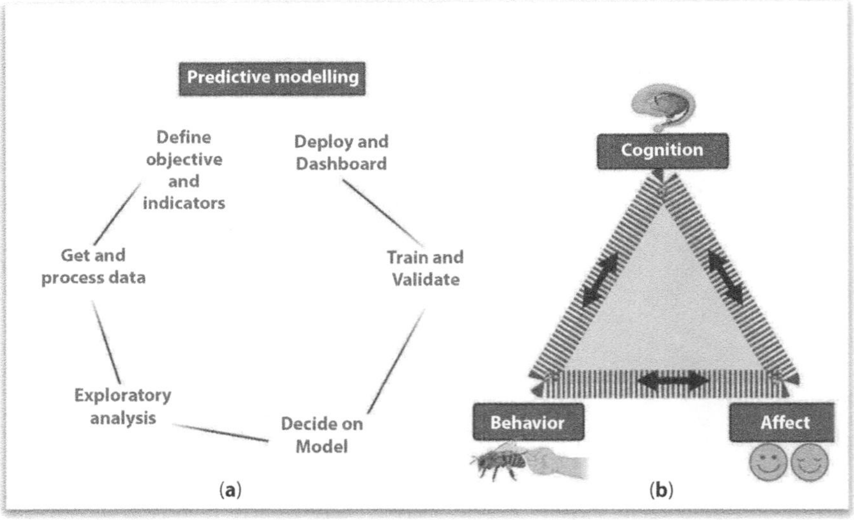

Figure 5.2 (a) The components of Predictive modeling: after deployment the inputs will be validated and suitable modes can be selected. After analyzing the suitability, the data will be processed. After the completion of the process the final objectives and indicators will be defined. (b) Cognition-affect-behavior triangle, as used in CBT. Each part has related links to the other two components. For example, a thought can promote similar emotions and behaviors, and an emotion can also promote similar thoughts and behaviors [51].

5.2.2.1.5.2 Virtual Reality Simulations

Virtual reality (VR) simulations often rely on computational models to create realistic and dynamic environments for exposure therapy and social skills training. These models mimic real-life scenarios and interactions, allowing patients to practice coping strategies in a controlled and safe environment. The development of VR simulations models involves a combination of 3D modeling, computer graphics, physics simulations, and human-computer interaction. Advancements in VR technology, such as motion tracking and haptic feedback, have significantly improved the realism and interactivity of these simulations [29].

5.2.2.1.5.3 Cognitive Models

Computational cognitive models are used to represent and understand cognitive processes, such as memory, decision-making, and problem-solving. These models can help therapists identify cognitive biases or dysfunctional thought patterns in patients and develop targeted interventions to address them. Cognitive models help in understanding the cognitive deficits resulting from brain injuries or neurological disorders. By comparing observed cognitive impairments with predictions from models, researchers can pinpoint areas of cognitive processing affected by the brain damage. Figure 5.3 is showing the cognitive models for mental addition fluency [32].

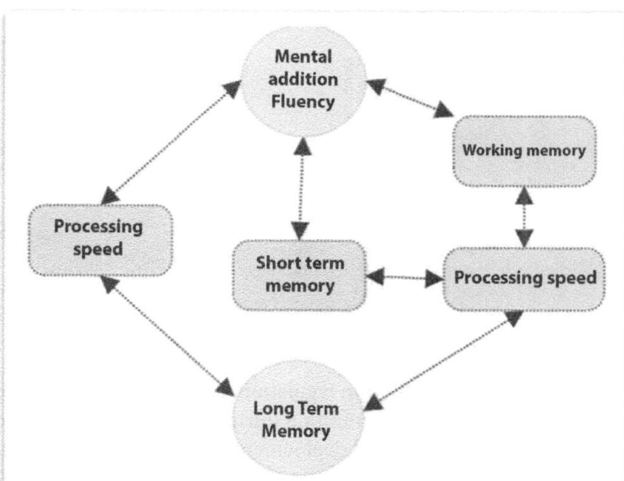

Figure 5.3 Cognitive models for mental addition fluency [52]. Processing speeds of models can define the mental addition fluency. It improves and detects the working memory, short term memory and long-term memory.

5.2.2.1.5.4 Reinforcement Learning

Computational reinforcement learning models can be applied to behavioral interventions. By providing positive reinforcement for desired behaviors and ignoring or discouraging unwanted behaviors, therapists can help patients learn new adaptive behaviors and decrease maladaptive ones [33].

5.2.2.1.5.5 Dynamic Systems Models

Dynamic systems models help therapists understand how psychological processes change over time and in response to different influences. These models can assist in predicting the course of therapy and identifying critical points where interventions may be most effective [34].

5.2.2.1.5.6 Emotion Regulation Models

Computational models of emotion regulation can help therapists understand how patients regulate their emotions and develop interventions to

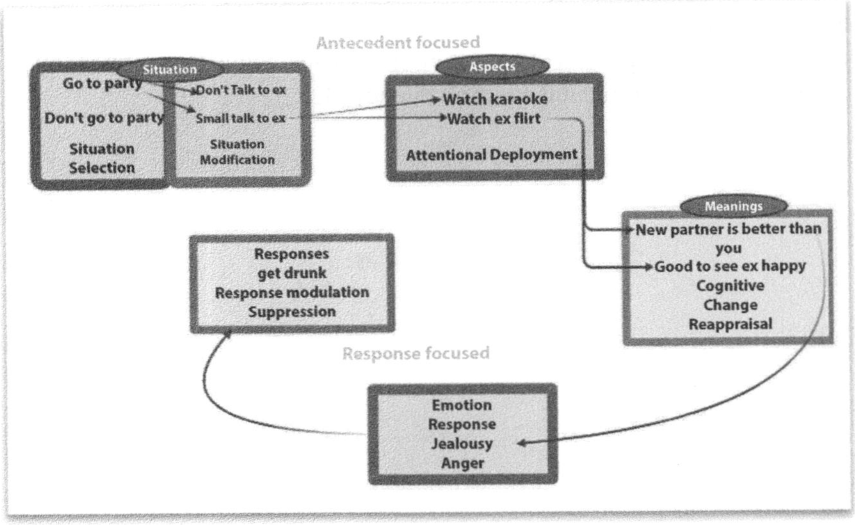

Figure 5.4 The process model of emotion regulation proposed by gross (1998) [53]. Based on many primary points of concentration during emotion regulation: context selection, situation alteration, attentional deployment, cognitive change, and reaction modification. Emotion regulation is the process of modulating emotion in order to change both what and how emotions are experienced [53]. Emotion regulation is concerned with the modulation of emotion in order to change what and how emotions are experienced. The major points of focus during emotion regulation are situation selection, situation modification, attentional deployment, cognitive change, and response modification.

improve emotional regulation skills. These models may also assist in predicting emotional responses in specific situations [35]. They are relevant in various fields, including psychology, psychiatry, counseling, and neuroscience, and they provide valuable insights into emotion-related disorders and interventions aimed at improving emotion regulation skills Figure 5.4.

5.2.2.1.5.7 Network Models

Network models in counseling psychology analyze the relationships between different psychological constructs and symptoms. These models can help therapists identify central symptoms or issues and design interventions to target them effectively. Network models provide powerful tools for analyzing complex systems and understanding their structure and behavior. They allow researchers to uncover patterns, identify central elements or hubs, measure network robustness, and predict system dynamics under different conditions.

5.2.2.1.5.8 Data-Driven Treatment Decisions

Computational models can process large datasets and integrate various sources of information to inform treatment decisions. By analyzing data from assessments, physiological measures, and client self-reports, therapists can make data-driven decisions to guide therapy.

5.2.2.1.5.9 Chat Bots and Virtual Assistants

Computational models underpin the functioning of chatbots and virtual assistants that offer support and psychoeducation to patients outside of therapy sessions. These models allow virtual assistants to understand client input and provide appropriate responses. The choice of the model depends on the specific use case, available data, computational resources, and desired level of sophistication in the chatbot's responses. As AI and natural language processing continue to advance, chatbots and virtual assistants are becoming increasingly capable of engaging in human-like conversations and providing valuable support across various domains [36].

5.2.2.1.5.10 Generative Model

A generative model is a type of statistical model that is used to simulate or generate new data that is similar to the training data it was trained on. The goal of a generative model is to capture the underlying distribution of the training data, allowing it to generate new samples that exhibit similar patterns and characteristics as the original data. Generative models have numerous applications, including image and audio synthesis, data augmentation for machine learning tasks, anomaly detection, and data imputation in missing data scenarios. They are also used in creative applications, such as generating art and music [37].

5.2.2.1.5.11 Bayesian Inference

Bayesian inference is a method of statistical inference that uses Bayes' theorem to update the probability of a hypothesis or model based on new evidence or data. It is a fundamental concept in Bayesian statistics and is widely used in various fields, including machine learning, data analysis, and scientific research. Bayesian inference has several advantages over classical (frequentist) statistics. It provides a coherent framework for incorporating prior knowledge into the analysis, which can be particularly useful when dealing with small sample sizes. Bayesian methods also allow for the quantification of uncertainty in parameter estimates, making them more interpretable [38].

5.2.2.1.5.12 The Wisconsin Card Sorting Test (WCST)

WCST is a widely used neuropsychological assessment tool that is designed to assess executive function and cognitive flexibility. It is used to evaluate frontal lobe dysfunction and is particularly useful in identifying cognitive impairments related to conditions like traumatic brain injury, dementia, and various neuropsychiatric disorders. The Wisconsin Card Sorting Test provides valuable information about a person's executive functioning and cognitive adaptability [39]. Impairments in these areas can be indicative of certain neurological or psychiatric conditions, making the test an essential tool in clinical assessments as shown in Figure 5.5.

Overall, computational models in counseling psychology facilitate evidence-based and personalized treatment approaches, enhance therapist decision-making, and contribute to a deeper understanding of psychological processes. Integrating computational attitudes and models into counseling practice can lead to more effective and efficient mental health care. However, it is crucial to use these tools responsibly, ethically, and in conjunction with traditional therapeutic approaches to ensure the best possible outcomes for patients.

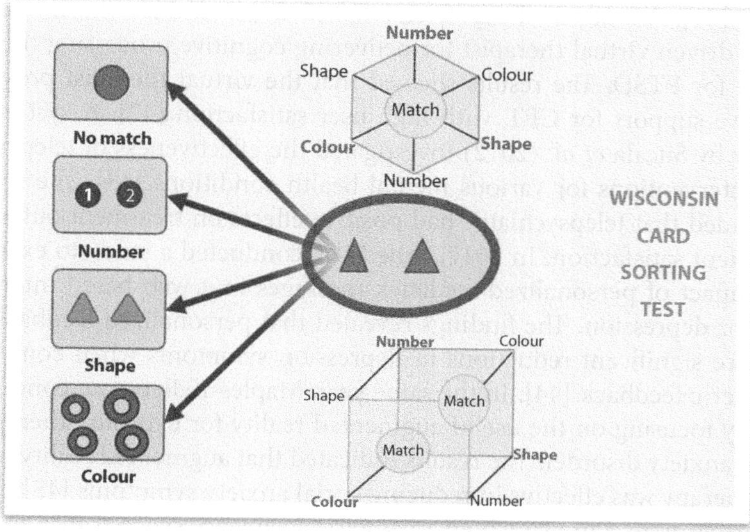

Figure 5.5 Wisconsin Card sorting test, used to detect cognitive deficits, appears to fit the structure of the new mathematics [54]. For the task the participant has a target card that defines the rule they are looking for. Below that card, there are three more cards with variations in the criteria (i.e. representing different possibilities in shape, number and color). The participant has to guess what the target criteria is in order to get correct feedback response.

5.3 Empirical Evidence and Efficacy of Computational Counselling

In 2019, Andersson and colleagues conducted a meta-analysis investigating the effectiveness of Computerized Cognitive Behavioral Therapy (CCBT) for depression and anxiety disorders. Their findings revealed that CCBT interventions were successful in reducing symptoms of depression and anxiety, demonstrating effect sizes comparable to face-to-face CBT. Rothbaum *et al.* (2014) carried out a randomized controlled trial to assess the efficacy of Virtual Reality Exposure Therapy (VRET) for veterans with Post-Traumatic Stress Disorder (PTSD) [40]. The study demonstrated that VRET significantly decreased PTSD symptoms, and these improvements were sustained during follow-up assessments. In 2017, Firth *et al.* conducted a systematic review examining the impact of mobile apps on mental health outcomes. The review showed that mobile apps targeting depression, anxiety, and stress exhibited moderate to large effect sizes in reducing symptoms, suggesting their potential as a valuable tool in improving mental health [41]. A study by Fitzpatrick *et al.* (2023) examined the efficacy of an AI-driven virtual therapist for delivering cognitive processing therapy (CPT) for PTSD. The results showed that the virtual therapist provided effective support for CPT, with high user satisfaction [42]. A systematic review by Sucala *et al.* (2012) investigated the effectiveness of telepsychiatry interventions for various mental health conditions [43]. The review concluded that telepsychiatry had positive effects on treatment outcomes and client satisfaction. In 2017, Erbe *et al.* conducted a study to examine the impact of personalized feedback messages in a web-based intervention for depression. The findings revealed that personalized feedback led to more significant reductions in depression symptoms when compared to generic feedback [44]. In the same year, Maples-Keller *et al.* conducted a study focusing on the use of augmented reality for exposure therapy in social anxiety disorder. The results indicated that augmented reality exposure therapy was effective in reducing social anxiety symptoms [45].

Furthermore, a growing body of empirical evidence suggests that psychological therapy induces neural reorganization. Longitudinal functional MRI studies investigating Cognitive Behavioral Therapy (CBT) have identified strengthened connections between the prefrontal cortex and limbic regions, which are consistent with contextualizing and regulating emotions and potential threats [46, 47] Additionally, the extent of these CBT-induced changes may predict the degree of remission experienced several years after therapy [48]. Several MRI studies have also reported correlations between the prediction error signal and activity in the ventral striatum and frontal cortices during instrumental learning [49]. These findings contribute to our understanding of the neural mechanisms underlying the effects of therapeutic interventions. The field is continuously evolving, and new studies are being published regularly. It's important to note that while many studies show promising results, more research is needed to establish the long-term efficacy and potential limitations of these interventions. Additionally, factors such as user engagement, adherence, and the role of therapeutic alliance in technology-assisted interventions are areas of ongoing investigation in the research community.

5.3.1 Case Studies: Exemplifying Computational Attitudes in Practice

Computational attitudes in counselling psychology were gaining traction, and case studies exemplifying the integration of technology in practice were starting to emerge. Here are a couple of case studies that illustrate how computational attitudes can be applied in real-world therapeutic settings:

5.3.2 Case Study: AI-Driven Virtual Therapist for Anxiety Management

A 32-year-old professional experiencing generalized anxiety disorder (GAD). John has been struggling with excessive worry, restlessness, and physical symptoms of anxiety. He has difficulty managing his thoughts and emotions, impacting his work and personal life. John's therapist introduces an AI-driven virtual therapist as a supplementary intervention to his traditional face-to-face therapy. The virtual therapist is a chatbot equipped with natural language processing and AI capabilities. Patient interacts with the virtual therapist through a mobile app. The virtual therapist engages in conversation with John, asking about his thoughts and feelings and offering evidence-based CBT techniques. The chatbot provides psychoeducation on anxiety, relaxation exercises, and thought challenging exercises to help patient reframe negative thoughts. The virtual therapist tracks patient's mood and symptom changes over time and provides personalized feedback based on his interactions. It also reminds patient to practice coping strategies between therapy sessions. Patient finds the virtual therapist helpful in managing his anxiety between face-to-face therapy sessions. He reports feeling more empowered in challenging his negative thoughts and coping with stressful situations. The combination of face-to-face therapy and the AI-driven virtual therapist leads to significant improvements in patient 's anxiety symptoms and overall well-being [50].

5.3.3 Case Study: Virtual Reality Exposure Therapy (VRET) for Specific Phobia

A 26-year-old woman struggling with a severe fear of heights (acrophobia). Patient experiences intense anxiety and panic attacks when exposed to heights. Her fear of heights is limiting her ability to enjoy certain activities and is causing distress in her daily life. Patient's therapist incorporates Virtual Reality Exposure Therapy (VRET) as part of her treatment plan. Using a virtual reality headset, Patient is exposed to virtual environments that simulate various heights, such as standing on a tall building rooftop or a bridge. The therapist guides patient through these virtual scenarios, gradually increasing the intensity of exposure as she becomes more comfortable. The therapist monitors patient's anxiety levels during VRET sessions and tracks her progress over time. Patient also uses a mood tracking app to record her anxiety levels outside of therapy. Through VRET, patient learns to confront her fear of heights in a safe and controlled environment. Over the course of several sessions, her anxiety decreases significantly, and she gains confidence in her ability to cope with height-related situations. Patient starts engaging in activities that were previously avoided due to her acrophobia. These case studies demonstrate how computational attitudes can be integrated into counselling psychology to enhance therapeutic outcomes and improve patients' overall well-being. By leveraging technology, therapists can provide personalized interventions, extend treatment beyond session hours, and address specific client needs more effectively [51].

5.3.4 Case Study: Discriminating Between Genuine and Faked Expressions of Pain in the Lab

In 2014, Bartlett *et al.* conducted a case study to compare the ability of humans and a machine learning model to differentiate between genuine and faked expressions of pain [52]. For their analysis, the researchers employed the Computer Expression Recognition Toolbox (CERT) to examine facial expressions of 25 stimulus subjects during a cold-pressor pain induction task. During the task, genuine pain was induced by submerging the subjects' arms in ice water, while faked pain expressions were simulated as they submerged their arms in warm water. CERT was used to extract time series of facial action unit (AU) activations from the videos, representing various facial muscle movements associated with pain expressions. The team then computed descriptors to model the temporal dynamics of each AU throughout the 1-minute video sequence, capturing how facial muscle movements changed over time during genuine and faked pain expressions. To distinguish between genuine and faked pain expressions, the researchers employed a support vector machine (SVM), a supervised machine learning classifier. The SVM was trained on the temporal descriptors to learn patterns that could differentiate between the two types of expressions. The model's performance was evaluated using a leave-one-subject-out cross-validation approach, wherein the model was trained on data from all subjects except one and then tested on the held-out subject. This process was repeated for each subject to ensure a rigorous evaluation. The accuracy of the machine learning model was measured using the area under the ROC curve (AUC). The ROC curve plots true-positive rates against false-positive rates at various decision thresholds. An AUC of 1.0 indicates perfect discrimination, while an AUC of 0.5 represents a chance model. The results showed that the machine learning model achieved an AUC of 0.91, which was both statistically and substantially higher than chance. This suggests that the model effectively distinguished between genuine and faked expressions of pain with a high level of accuracy. The study's findings imply that machines, in this case, the SVM model using facial expression analysis, outperformed humans in discriminating between real and faked pain expressions. This has implications for applications in pain assessment, deception detection, and emotion recognition, where machine learning models may offer valuable insights and potentially surpass human capabilities. However, it is essential to note that the study focused specifically on pain expressions and may not necessarily generalize to other emotional expressions or contexts [52].

5.4 Ethical and Legal Considerations

When using computational attitudes in counseling psychology, it's crucial to address ethical and legal considerations. Patients must be fully informed about the technology's purpose, benefits, and risks, and obtaining informed consent is essential. Protecting sensitive client data through robust security measures and clarifying data ownership are important. Regularly assessing and addressing biases in computational models ensures fair treatment for all patients. Therapists should make the processes transparent and ensure patients understand the role of technology in therapy. Adequate training and competence in using computational approaches are necessary. Technology should enhance therapy, not replace human interaction. Evaluating evidence and complying with relevant laws are vital. By adhering to these considerations, therapists can responsibly integrate computational attitudes in counseling psychology, providing effective and ethical care for their clients [53].

5.5 Future Directions and Possibilities

The integration of computational attitudes in counseling psychology holds great potential to transform mental health care. Exciting future directions include personalized treatment plans based on individual data, advancements in Virtual Reality for immersive therapy, AI-driven assessments for deeper insights, and early detection of mental health issues through data analysis. AI-powered virtual assistants can offer 24/7 support, while data-driven decisions improve treatment strategies. Biometric data from wearables can provide objective insights, and telehealth services can extend care to remote areas. Ethical AI frameworks are essential to ensure fairness and transparency. Computational attitudes can also enhance interdisciplinary collaboration and therapeutic games. Responsible implementation, prioritizing ethics, privacy, and human connection, is crucial for realizing the full potential of computational attitudes in counseling psychology [27, 54]. Collaboration between mental health professionals, technologists, and researchers will be key in shaping this promising future.

5.6 Conclusion

The integration of computational attitudes into counseling psychology represents a groundbreaking development with immense promise for transforming mental health care. This book chapter has explored various aspects of this evolution, including the theoretical foundations, key technologies, computational models, empirical evidence, and case studies exemplifying computational attitudes in practice. Ethical and legal considerations have been highlighted to ensure responsible use of technology, and future directions and possibilities have been discussed. As computational attitudes continue to evolve, they offer personalized interventions, data-driven insights, and increased accessibility to mental health services. The integration of technology in counseling psychology has gained cultural acceptance, and training programs are equipping future therapists with essential computational skills. The COVID-19 pandemic has further underscored the value of technology in providing remote care and continuous support during emergencies. As we embrace these advancements, it is essential to address ethical concerns, data privacy, and continue research to ensure the responsible and effective integration of computational attitudes in therapeutic practice. In conclusion, the emergence of computational attitudes in counseling psychology holds great promise in enhancing therapeutic outcomes, improving accessibility, and ultimately advancing the field of mental health care. As we move forward, the responsible use of technology, combined with the expertise and empathy of therapists, will create a future where mental health care is not only more effective but also more accessible and tailored to the unique needs of each individual.

References

1. Simon, 2018, https://simons.berkeley.edu/news/mind-brain-computer.
2. Plato, 2015, https://plato.stanford.edu/entries/computational-mind/#Bib.
3. Eronen, M.I. and Bringmann, L.F., The theory crisis in psychology: How to move forward. *Perspect. Psychol. Sci.*, 16, 4, 779–88, 2021.

4. Mansell, W. and Huddy, V., Why do we need computational models of psychological change and recovery, and how should they be designed and tested? *Front. Psychiatry*, 11, 624, 2020.
5. Nibib,2020,https://www.nibib.nih.gov/science-education/science-topics/computational-modeling.
6. Yarkoni, T. and Westfall, J., Choosing prediction over explanation in psychology: Lessons from machine learning. *Perspect. Psychol. Sci.*, 12, 6, 1100–22, 2017.
7. Hays, D.G., Introduction to counseling outcome research and evaluation. *Couns. Outcome Res. Eval.*, Taylor & Francis, 1, 1–7, 2010.
8. Nesbitt, T.S., The evolution of telehealth: where have we been and where are we going, in: *The Role of Telehealth in an Evolving Health Care Environment: Workshop Summary*, Institute of Medicine, Washington, DC, 2012.
9. Kamran Ul haq, A., Khattak, A., Jamil, N., Naeem, M.A., Mirza, F., Data analytics in mental healthcare. *Sci. Program.*, 2020, 1–9, 2020.
10. Groves, P., Kayyali, B., Knott, D., Van Kuiken, S., The 'big data' revolution in healthcare: Accelerating value and innovation, 2016.
11. Miner, A.S., Fleming, S.L., Haque, A., Fries, J.A., Althoff, T., Wilfley, D.E., *et al.*, A computational approach to measure the linguistic characteristics of psychotherapy timing, responsiveness, and consistency. *npj Mental Health Res.*, 1, 1, 19, 2022.
12. Srivastava, K., Chaudhury, S., Dhamija, S., Prakash, J., Chatterjee, K., Digital technological interventions in mental health care. *Ind. Psychiatry J.*, 29, 2, 181, 2020.
13. Bell, I.H., Nicholas, J., Alvarez-Jimenez, M., Thompson, A., Valmaggia, L., Virtual reality as a clinical tool in mental health research and practice. *Dialogues Clin. Neurosci.*, 22, 169–177, 2022.
14. Philippe, T.J., Sikder, N., Jackson, A., Koblanski, M.E., Liow, E., Pilarinos, A., *et al.*, Digital health interventions for delivery of mental health care: systematic and comprehensive meta-review. *JMIR Ment. Health*, 9, 5, e35159, 2022.
15. Ryff, C.D., Positive psychology: Looking back and looking forward. *Front. Psychol.*, 13, 840062, 2022.
16. Resick, P.A., Galovski, T.E., Uhlmansiek, M.O., Scher, C.D., Clum, G.A., Young-Xu, Y., A randomized clinical trial to dismantle components of cognitive processing therapy for posttraumatic stress disorder in female victims of interpersonal violence. *J. Consult. Clin. Psychol.*, 76, 2, 243, 2008.
17. Morland, L.A., Mackintosh, M.A., Greene, C.J., Rosen, C.S., Chard, K.M., Resick, P., *et al.*, Cognitive processing therapy for posttraumatic stress disorder delivered to rural veterans via telemental health: A randomized noninferiority clinical trial. *J. Clin. Psychiatry*, 75, 5, 3193, 2014.
18. Dryden, W., Rational emotive behavior therapy, in: *Encyclopedia of cognitive behavior therapy*, pp. 321–4, 2005.
19. Karimpour-Vazifehkhorani, A., Rudsari, A.B., Rezvanizadeh, A., Kehtary-Harzang, L., Hasanzadeh, K., Behavioral activation therapy on reward

seeking behaviors in depressed people: An experimental study. *J. Caring Sci.*, 9, 4, 195, 2020.
20. Harris, K.M., An integrative approach to health. *Demography*, 47, 1–22, 2010.
21. Kellogg, K.C. and Sadeh-Sharvit, S., Pragmatic AI-augmentation in mental healthcare: Key technologies, potential benefits, and real-world challenges and solutions for frontline clinicians. *Front. Psychiatry*, 13, 2057, 2022.
22. Bohr, A. and Memarzadeh, K., The rise of artificial intelligence in healthcare applications, in: *Artificial Intelligence in healthcare*, pp. 25–60, Elsevier, Denmark, 2020.
23. Duvvuri, V., Guan, Q., Daddala, S., Harris, M., Kaushik, S., Predicting Depression Symptoms from Discord Chat Messaging Using AI Medical Chatbots, in: *Proceedings of the 2022 6th International Conference on Machine Learning and Soft Computing*, pp. 111–9, 2022.
24. Dascal, J., Reid, M., IsHak, W.W., Spiegel, B., Recacho, J., Rosen, B., *et al.*, Virtual reality and medical inpatients: a systematic review of randomized, controlled trials. *Innov. Clin. Neurosci.*, 14, 1–2, 14, 2017.
25. Chandrashekar, P., Do mental health mobile apps work: evidence and recommendations for designing high-efficacy mental health mobile apps. *Mhealth*, 4, 1–4, 2018.
26. Dolce, P., Marocco, D., Maldonato, M.N., Sperandeo, R., Toward a machine learning predictive-oriented approach to complement explanatory modeling. an application for evaluating psychopathological traits based on affective neurosciences and phenomenology. *Front. Psychol.*, 11, 446, 2020.
27. Barsalou, L.W., *Cognitive psychology: An overview for cognitive scientists*, Psychology Press, New York, 2014.
28. Eckstein, M.K., Wilbrecht, L., Collins, A.G.E., What do reinforcement learning models measure? Interpreting model parameters in cognition and neuroscience. *Curr. Opin. Behav. Sci.*, 41, 128–37, 2021.
29. Homer, J.B. and Hirsch, G.B., System dynamics modeling for public health: background and opportunities. *Am. J. Public Health*, 96, 3, 452–8, 2006.
30. Grecucci, A., Messina, I., Amodeo, L., Lapomarda, G., Crescentini, C., Dadomo, H., *et al.*, A dual route model for regulating emotions: Comparing models, techniques and biological mechanisms. *Front. Psychol.*, 11, 930, 2020.
31. Ahmed, A., Hassan, A., Aziz, S., Abd-Alrazaq, A.A., Ali, N., Alzubaidi, M., *et al.*, Chatbot features for anxiety and depression: A scoping review. *Health Inf. J.*, 29, 1, 14604582221146720, 2023.
32. Haines, N., Kvam, P.D., Irving, L.H., Smith, C., Beauchaine, T.P., Pitt, M.A., *et al.*, Theoretically informed generative models can advance the psychological and brain sciences: Lessons from the reliability paradox, 2020.
33. Wagenmakers, E.J., Marsman, M., Jamil, T., Ly, A., Verhagen, J., Love, J., *et al.*, Bayesian inference for psychology. Part I: Theoretical advantages and practical ramifications. *Psychon. Bull. Rev.*, 25, 35–57, 2018.

34. Kopp, B., Lange, F., Steinke, A., The reliability of the Wisconsin card sorting test in clinical practice. *Assessment*, 28, 1, 248–63, 2021.
35. Rothbaum, B.O., Price, M., Jovanovic, T., Norrholm, S.D., Gerardi, M., Dunlop, B., *et al.*, A randomized, double-blind evaluation of D-cycloserine or alprazolam combined with virtual reality exposure therapy for posttraumatic stress disorder in Iraq and Afghanistan War veterans. *Am. J. Psychiatry*, 171, 6, 640–8, 2014.
36. Firth, J., Torous, J., Nicholas, J., Carney, R., Rosenbaum, S., Sarris, J., Can smartphone mental health interventions reduce symptoms of anxiety? A meta-analysis of randomized controlled trials. *J. Affect. Disord.*, 218, 15–22, 2017.
37. Fitzpatrick, S.S., Liebman, R.E., Monson, C.M., Resick, P.A., Latent emotion profiles of PTSD and specific emotions predicting differential therapy outcomes in a dismantling study of cognitive processing therapy. *J. Anxiety Disord.*, 95, 102681, 2023.
38. Sucala, M., Schnur, J.B., Constantino, M.J., Miller, S.J., Brackman, E.H., Montgomery, G.H., The therapeutic relationship in e-therapy for mental health: a systematic review. *J. Med. Internet Res.*, 14, 4, e110, 2012.
39. Erbe, D., Eichert, H.C., Riper, H., Ebert, D.D., Blending face-to-face and internet-based interventions for the treatment of mental disorders in adults: systematic review. *J. Med. Internet Res.*, 19, 9, e306, 2017.
40. Maples-Keller, J.L., Bunnell, B.E., Kim, S.J., Rothbaum, B.O., The use of virtual reality technology in the treatment of anxiety and other psychiatric disorders. *Harv. Rev. Psychiatry*, 25, 3, 103, 2017.
41. Mason, L., Peters, E., Kumari, V., Functional connectivity predictors and mechanisms of cognitive behavioural therapies: A systematic review with recommendations. *Aust. New Z. J. Psychiatry*, 50, 4, 311–21, 2016.
42. Mason, L., Peters, E.R., Dima, D., Williams, S.C., Kumari, V., Cognitive behavioral therapy normalizes functional connectivity for social threat in psychosis. *Schizophr. Bull.*, 42, 3, 684–92, 2016.
43. Mason, L., Peters, E., Williams, S.C., Kumari, V., Brain connectivity changes occurring following cognitive behavioural therapy for psychosis predict long-term recovery. *Transl. Psychiatry*, 7, 1, e1001–e1001, 2017.
44. Garrison, J., Erdeniz, B., Done, J., Prediction error in reinforcement learning: a meta-analysis of neuroimaging studies. *Neurosci. Biobehav. Rev.*, 37, 7, 1297–310, 2013.
45. Zhang, W., Paudel, D., Shi, R., Liang, J., Liu, J., Zeng, X., *et al.*, Virtual reality exposure therapy (VRET) for anxiety due to fear of COVID-19 infection: A case series. *Neuropsychiatr. Dis. Treat.*, 16, 2669–75, 2020.
46. Albakri, G., Bouaziz, R., Alharthi, W., Kammoun, S., Al-Sarem, M., Saeed, F., *et al.*, Phobia exposure therapy using virtual and augmented reality: a systematic review. *Appl. Sci.*, 12, 3, 1672, 2022.

47. Bartlett, M.S., Littlewort, G.C., Frank, M.G., Lee, K., Automatic decoding of facial movements reveals deceptive pain expressions. *Curr. Biol.*, 24, 7, 738–43, 2014.
48. Lustgarten, S.D., Garrison, Y.L., Sinnard, M.T., Flynn, A.W.P., Digital privacy in mental healthcare: current issues and recommendations for technology use. *Curr. Opin. Psychol.*, 36, 25–31, 2020.
49. Memarian, B. and Doleck, T., Fairness, Accountability, Transparency, and Ethics (FATE) in Artificial Intelligence (AI), and higher education: A systematic review. *Comput. Educ.: Artif. Intell.*, 5, 100152, 2023.
50. D'Mello, S.K., Tay, L., Southwell, R., Psychological measurement in the information age: machine-learned computational models. *Curr. Dir. Psychol. Sci.*, 31, 1, 76–87, 2022.
51. Smith, R., Moutoussis, M., Bilek, E., Simulating the computational mechanisms of cognitive and behavioral psychotherapeutic interventions: insights from active inference. *Sci. Rep.*, 11, 1, 10128, 2021.
52. Kornev, A.N., Rakhlin, N., Grigorenko, E.L., Dyslexia from a cross-linguistic and cross-cultural perspective: The case of Russian and Russia. *Learn. Disabil. Contemp. J.*, 8, 1, 41–69, 2010.
53. Gross, J.J., The extended process model of emotion regulation: Elaborations, applications, and future directions. *Psychol. Inq.*, 26, 1, 130–7, 2015.
54. Grant, D.A. and Berg, E.A., Wisconsin card sorting test. *J. Exp. Psychol.*, 38, 404, 1993.

6

Computational Psychiatry: Addressing the Gap Between Pathophysiology and Psychopathology

Jignasha Derasari Pandya[1]* and Bhupendra Prajapati[2,3]

[1]Arihant School of Pharmacy & Bio-Research Institute, Gujarat Technological University, Ahmedabad, Gujarat, India
[2]Faculty of Pharmacy, Ganpat University, Gujarat, India
[3]Department of Industrial Pharmacy, Faculty of Pharmacy, Silpakorn University, Nakhon Pathom, Thailand

Abstract
Advancement in decoding of neuroscience keen on assistance for patients with psychological disorder offers huge challenges as it encompasses interaction between the brain, a multidimensional body part of human physique and likewise convoluted environs of numerous variables full of societal, folk, and realistic factors. This makes them intricately multifarious phenomena that hang on and interrelate with a big number of variables. The last few years have perceived a rebellion in the arena of medical psychology and psychiatry. Nowadays this has enhanced the understanding of the communal and dissimilar pathophysiological mechanisms of psychological illness to provide more effective, person-tailored cures. Yet some challenges like genotype of psychopathology, characterization of molecular pathology of each disorder, developing new theories, heterogeneity of psychiatric phenotype etc. allied with mental health issues remaining unanswered. Moreover, psychopathology is unable to assimilate ecology, information processing, subjective experience, therapeutic interventions, and relative understanding to certain clinical phenomena. Dealing with such complexities stresses more prominent systems to be developed with the help of machine learning and consequently computational psychiatry was born. It combines several levels and sorts of calculation with numerous forms of data to improve thoughtfulness, extrapolation and dealing

**Corresponding author*: jignashaderasari@gmail.com; ORCID: 0000-0001-7250-3748
Bhupendra Prajapati: ORCID: 0000-0001-8242-4541

Bhupendra Prajapati, Alok Tripathi, Rishabha Malviya and Lucy Mohapatra (eds.) Computational Neuropharmacology: Fundamentals and Clinical Aspects, (159–206) © 2025 Scrivener Publishing LLC

of cerebral disease. Largely it comprehends two corresponding approaches: data driven and theory driven.

In conclusion, it helps psychiatry accomplish exactitude and reproducibility, which can benefit the mental health field attain noteworthy progress. Hence, present review is the attempt to portray computational psychiatry as a precise, efficient and advanced statistical tool to address the gap between pathophysiology and psychopathology.

***Keywords*:** Computational psychiatry, mental illness, pathophysiology, psychopathology, machine learning

List of Abbreviations

DSM	(Diagnostic and Statistical Manual of Mental Disorders)
ICD	(International Classification of Diseases)
DALY	(Disability-Adjusted Life-Year)
PID	(Person-centred Integrative Diagnosis)
HiTOp	(Hierarchical Taxonomy of psychopathology)
RDoC	(Research Domain Criteria)
CP	(Computational Psychiatry)

6.1 Introduction

Human brain is the super computer ever and incredible invention created by lord. No one can imagine and predict how a brain works in various stimuli present in the outer and inner world due to its amazing functional ability and complexity both. The concept of mental illness and the advent of psychiatry are the offshoot of individual's personality which is the reflection of his or her mind. Fundamentally, there are three components which constitutes mind; *i.e.* id, ego and super ego. Though each part of the mind encompasses exclusive landscapes, they interrelate to form a whole, and each part makes a relative input to a person's deeds or it is called personality of human being. When a person fails to maintain balance among these, create mental issues or behavioral abnormality that finally leads to mental illness or psychological disorders, proposed by Sigmund Freud [1]. Here starts the role of an individual as a priest, sociologists, and philosophers in ancient times and as a clinicians, doctors, psychiatrist, psychologists, psychotherapist, medical officer *etc.* in modern times which necessitated the birth of intriguing field of "Psychiatry."

6.1.1 Psychiatry

The seeds of psychiatry were as old as the ancient Greek philosophy reported by the philosophers and physicians. A mental illness is associated with brain functioning and how it occurs is unknown till the discovery of pathophysiology. Thousands years ago, where medical science was not popular in the society, most of the clinicians, as well as laypersons, firmly believed that the main psychological disorders were triggered by demonic "control." As time passed, investigation on mental illness causes led to the development of other relevant therapies and revealed more erudite and recent system for the remedy of psychological disorders which is known as psychiatry [1].

6.1.1.1 Clinical Challenges

Although advancement made in present treatment over the last 100 years, thought and for mental disorders has continued slightly immobile [2–4]. Practical approach has been adopted for various other fields of medicines to overcome or prevent the severity of diseases, while the progress in psychiatric field remained at the therapeutic level. Even though progression of genetic and neurology branches speeded the growth of mental illness, the vibrant features of mental disorders has made this evolution hard and clinically challenging [5]. Some of these clinical challenges required to work up on for the smooth functioning and better outcomes of treatments. Few of them are discussed as below:

(a) **Nature of mental illness:** Psychological complaint is considered by not only environmental science but also the impact of various conservational variables which can make them erratic. As it is the study of all about brain which is very complex organ of human body.
(b) **Cost:** Treatment of mental illness is costlier than any other disease that is confirmed from the statistics of it, as it includes; medical treatment, food and lodging in asylum, some special kind of training, instruments/equipment's and therapy.
(c) **Dishonor:** Dishonoring acts a noteworthy starring role in the matters dogging psychological disease in health related professions. The identified persons with psychological diseases have to deal with both the illness itself and the undesirable implication raised by society on them. Compassion, distress & segregation, and despotism are three divergent

qualities that persons detected with a psychological illness frequently direct due to this stigma [5]. In so far, the most operative technique in contending this problem has been, and will remain to be, education and awareness.

(d) **Social acceptance, job retention, and treatment:** Still people in society cannot consider mental illness is like other medical illness and so these disgraceful attitudes create a major community problems. Also an intellect insufficiencies are naturally more of a matter with severe disorders like schizophrenia [6]. These problems pose a noteworthy impact on attention, working pattern, its recalling and processing efficiency, official functioning ability. Consequently, a leading exertion of diseased people group to protect their employment, to supervise as well as to renew their knowledge for enduring their routine existence being a human [7].

(e) **Drug addiction and substance abuse:** Moreover, comorbid addiction is conjoint in severe mental disorders; as study reports showing more than half of schizophrenic patients either addicted with drugs or misuse of them [8]. Although this leads to increase number of deaths associated with opioids overdosing and suicide. All these issues required medical treatment, generally antipsychotic drugs are popular in market but adverse drug reaction profile are so high for these drugs which limit the use of it, as well as initiates novel research and funding for discovery and development of new drugs and promising therapies. To combat this, precision medicine can be practiced with the help of unique genotyping and neuroimaging techniques for psychiatric care and diagnosis.

(f) **Nonconsistency, legitimacy, and clinical usefulness:** Since long psychological disorders have been treated with diagnostic tools like numerical and analytical manuals for mental illness and Global categorization of these disease which are commonly known as DSM and ICD guidelines. They are symptoms based tool which always in searching of certain causes of mental illness. Yet available evidences of these systems fails to segregate healthy and mentally ill people and so pathology of psychological diseases is baseless. Diagnostic treatment or approaches towards psychological disorders are unsuccessful to recognize definite causes for a particular category of mental diseases [9].

DSM guidelines gradually modified for regular update and its third version published in 1980, was the clear evidence of it. A more realistic methodology which operates present scientific and medical skill to outline mental illness could be assimilated into the existing model. This manual is just an initial guide but the study of causative factors of mental issues and their analysis could be more trust worthy and authorized [10].

6.1.1.2 Clinical Load and Diagnostic Complication

Cerebral, neural, and material use illnesses pay to a major share of ill health load. Well-being organisms can reply by applying well recognized, economically reasonable involvements, sound research programs to produce the best prevention and management alternatives [11]. Occurrence cost, mortality rate, infirmities, and deficiencies allied to the psychological disorders and other relevant costs are the major factors increasing clinical load and complexity of disease analysis. Mental illnesses are progressively accepted as prominent roots of disease load. One directive group known as Lancet stressed out the fact that psychological well-being is vital element required to spread worldwide as its human birth right.

This group also undertaken the responsibility of investigating the facilities available and their effective utilization to the affected population and to manage urgencies [12]. Each country has to transform their priorities related to health concerns to handle the mental fitness complications. Further, it is very crucial to deal with other consequences arose all through the administration of psychological diseases [13].

The comprehensive drain of sicknesses, damages, and threat factors Revision 2019 is a wide-ranging worldwide exertion that comprises the burden quantification of mental disorders. This revision cast off the disability-adjusted life-year (DALY) metric that deals the lacking between the existing fitness of the population and a normative usual life expectancy spent in throughout healthy life. It fabrication based on integration of fresh records and working advances. This learning makes easy and methodological assessment of the occurrence and load due to 369 diseases and injuries. During the period of 1990 to 2019, DALY metrics was indicated complete change in disease load [14–16].

Diagnostic Complications

Psychiatry has endeavored to device methodologies from precision medicine by looking for biomarkers to apprise mental health management for the betterment of diagnosis. This urges the requirement of precise approach in psychiatry field as existing treatments enabled preliminary success rates for patients of already defined analytical classes [17]. However, it is reverse to other branches of medicines, people with mental illness does not match apparently with any symptoms based category of diagnosis due to its manifold diagnoses (co-morbidity) [17]. Symptoms based nosology slows down the process of biomarker searching which are the diagnostic group and are likely to be unrelated in terms of aetiology and pathogenesis.

Diagnosis of mental illness is much more complicated as many neurobiological processes works simultaneously which may give rise to same type of symptoms or sometimes single procedure gives diverse signs reliant on their dealings with other bio-psychosocial factors [18].

Moreover, experimental complication is developing as a piercing display of the intangible and observed productivity of mental science. These complication requirements to be evaluated, implicit and expressed in responsiveness to its various facets and levels in order to inform adequately developed crucial clinical tools such as an effective diagnostic model.

A broad model relevant to this concern is being outlined under the term of person-centred integrative diagnosis (PID) [19, 20]. It is a variable term covering numerous levels and areas. Comorbidity is the strategic formula of medical intricacy that requires a health consideration. Feinstein, who coined comorbidity word first time. This term is extended into certain physiological conditions like pregnancy, trauma and child abuse. Comorbidity is the most common issue which is more frequent as age is growing of human being. Moreover, it has other associated effects of experimental maintenance that has lot influence on both analysis and wellbeing management.

6.1.1.3 *Treatment Development*

However, the research progress of mental illness is devastating, (~26% lifespan frequency global), there is a huge breach between the requirement for illness management and its establishment over the globe. Mental illness is quite different than any other forms of physical illness [21]. If these illness are cured, the achievement rates of intrusions are unsure, in grown-ups and in youths [22, 23]. Very few people who suffered from mental illness hardly get improvement from therapy. Also, the cure of narrow window outcomes does not assure enduring results. The uncertain conclusions of treatment proposes that yet the preservation work pattern of cerebral complaints and healing [24, 25] do not fully understood. Investigation to improve psychosomatic treatments is claimed to be "outrageously under-supported" [26]. Mental illness is much more prevalent in developed nations due to various reasons; among them one very common reason is that the proportion of funding and related investments spared for research is quite very low as compared to investment spent on scaling-up financial investments is high [25].

Scientific evolution of the study of mental illness starts from ancient science to psychiatry via pathophysiology and psychopathology with the amalgamation of neurobiology, network diagnosis system and finally reach to machine based psychiatry named computational psychiatry. In techno-advanced era, it is the utmost need of promising and outcomes oriented therapy with the involvement of techno-devices, modified, trans-diagnostic, vigorous classifications [23].

In a nutshell, journey and growth of mental illness is like zig-zag path of Zurasic Park adventure rides.

6.2 Roadmap of Conventional to Modern Evolution Towards Mental (Psychological) Illness

Since ancient times, the disorders of brain are prevalent but time to time they are known, understood and characterized by various different and evolving theories [27–29]. Here in this section, the evolution towards mental disorders is described starts from ancient conventional theories to modern advancement with medical treatment of them as depicted in Figure 6.1.

In Ancient Era (2100 BC) mental disorders word is not prevalent but something unusual societal activity observed in human being considered as mental symptoms and linked to either demonology or religious origins. Such things were believed to have need of cures like prayers, tiresome of certain pious stuffs, exorcisms and other sacred performs.

500–300 BC (Greek period) Hippocrates first time proposed that mental illness raised due to some bodily abnormalities which were the expressions of imbalance in body fluids like blood, bile pigments, phlegm and these were the basis of individual's psychopathology.

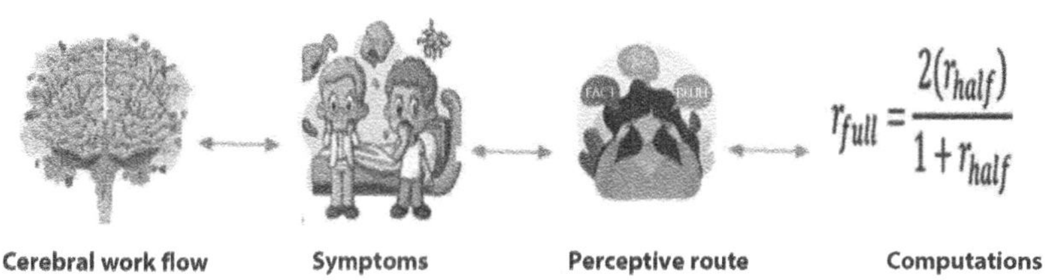

Figure 6.1 Evolution of mental illness from neuroscience, pathophysiology, psychopathology to computational psychiatry.

200–101 BC: This era was considered as an ancient china era. According to this mental issues arouse due to an inequality of the "yin" and "yang" forces which signified moral and wicked characteristics in an individual. The ancient Chinese can also be ascribed in evolving mental trials for assessments and bifurcated a groups of people for employment and categorize public servants.

Reawakening Era

The reawakening period can also be referred to as the period of more understanding and clarity about bodily changes happening in individuals. People were more concentrated about their ideology for mental illness and realises that all such mental ailments based on scientific ground rather than any evil spirit.

1500s: Paracelsus who was a physician and was introduced the word-homeopathy. He stated that all mental issues or any abnormal behavior had a biological base and tried to explain dysfunctional conduct as a biological process.

In the same era, Johann Weyer also disagreed with the ancient concept of spirituality as the only reasons of psychopathological activities. He constituted a refined and comprehensive arrangement scheme for numerous disorders like hysteria, delusions, paranoia, depression, and epilepsy.

In the 19th and 20th centuries, American psychologist **Lightner Witmer** was presented clinical psychology word first time in 1907. This can be explained as the scientific period which announced clinical psychology as a distinct division of psychology.

Literature witnessed about availability of first ever clinic in 1896 by Witmer for serving children with learning disabilities and by 1914, many more such types of clinics accessible for clinical psychology practice. During this period, this concept of psychology was well accepted and was practiced with more understanding and clarity of fundamentals like understanding of mental disorders, measurement of individual differences, emergence of scientific psychiatry, and psychological determinism.

Establishment of the field and modern era

Gradually developments of psychiatry led to existence of scientific psychiatry, treatment setups in terms of clinics, diagnostics methods and tools for analysis of individual's nature and capabilities and their interrelation with mental illness. Study of medical psychology was extended to pathophysiology of mental illness, psychopathology, and nowadays use of computational models also available in growing and latest development of psychiatry, i.e., Computational Psychiatry.

Conclusion

By the 1980s, clinical psychologists appreciated by increasing respect from the medical establishment as they grown hospital admitting civil liberties. As time passes from 1990 to 2000, clinical psychology field grown faster, large number of graduates and training centres established, even private practice was also started. At the same time, clinical psychiatry model fails to answer the reproducibility of the results which necessitates the new invention with the help of machine integrated approach in redefining psychiatry. In the late 20th century, novel concept of Computational Psychiatry was emerged with the hope of getting precise results for mental illness.

6.3 Pathophysiology of Mental Illness

Healthy lifestyle and maintenance of healthy mental status are the key elements for the balanced life which reflects into ones sound health and productivity in all means. In any case, if these balance will disturbed create numerous psychological fitness issues like depression, anxiety, eating disorders, personality disorders and many more mental illnesses. Various core elements as shown in Figure 6.2 are the major contributing factors for the regulation of mental health and its associated diseased conditions.

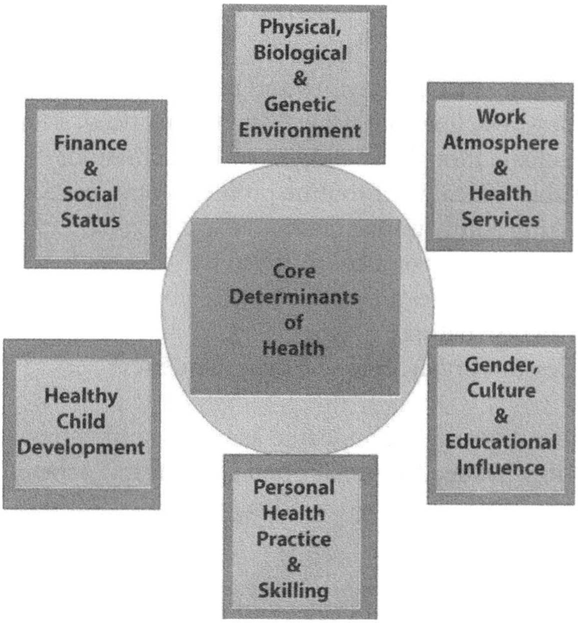

Figure 6.2 Effect of fundamental elements on mental health.

6.3.1 Overview of Mental Illness and Pathophysiology

Mental illness termed as disorders of mental health which is also known as psychological disorders that mainly affect the mood of persons and their abilities to think and causes behavioral changes and so they are known as mood disorders and cognitive disorders too [30]. This includes schizophrenia, depression, anxiety diseases, addictive nature and eating disorders. This health worries people have time to time with different intensities depending on the surrounding circumstances, but it will be problematic mental illness when the signs and symptoms of mental health leads to create stress and to disturb the thinking ability of them to execute even routine functions. Generally, these symptoms can be managed with medical treatment and counselling that is known as psychotherapy.

6.3.1.1 Signs and Symptoms

Signs and symptoms of psychological sickness can diverge, based on the type of mental sickness, surrounding situations and other aspects. These mainly include:

- Gloomy or unhappy feeling
- Chaotic thinking or concentration problem
- Unnecessary worries, or extreme guilty feelings
- High and low tides of mood
- Detachment from activities and people
- Restlessness, low energy level and feeling tired, sleeplessness
- Unrealistic thoughts-delusions or hallucinations
- Not able to deal with routine problems or stress
- Thinking ability impairment
- Drug addictions or alcohol abuse
- Frequent changes in eating habits, libido desires and self-destruction nature
- Aggression or fierceness, suicidal tendency

Occasionally these will be appeared as bodily difficulties, such as stomach ache, back-ache, headaches, or other inexplicable aches and pains.

6.3.1.2 Pathophysiology

Pathophysiology is originally Greek word combination of three words, i.e., "pathos" meaning distress, "physis" meaning nature & "logos" related to learning" that collectively states to the study of atypical variations in body tasks which initiates the mental health issues and stress. This exhaustive studies have identified major causes or origins responsible for the mental illness which are discussed as below.

6.3.1.2.1 Origins or Causes of Mental Illness

However, definite causes for mental illness is unknown that many literature has proved already. Such situations raised by combination of various living, emotional, and ecological elements. Few situations where brain circuits injury is also responsible for it. Here, some of them are summarized below.

6.3.1.2.1.1 Biological or Living or Genetic Elements

Few of the psychological problem due to unusual operation of nerve cell circuits or paths in brain, damage of these or brain injury leads to some of the cerebral issues.

Genetic Factors

Mostly such kind of cerebral illness having a propensity to spread or to pass in families, and the danger of emerging an illness is connected with biological relatedness of infected person [31, 32]. This outline of transmission sturdily advocates genetic interconnection. Twin studies constantly display that monozygotic twins are more likely to be concordant on each illness than dizygotic twins [33]. This alteration recommends that the connection of psychological illness with genetic factors is to a greater extent.

Mix genetic factors are the severe causative elements for quite less communal illnesses like autism, schizophrenia, bipolar disorder whereas depression and anxiety like few inherited disorders are of little illness and of more communal [34]. Learning of gene technology at molecular level practises as the gears to learn all about prevalence of psychological sickness. Research done since last couple of decades has brought amazing discoveries. Table 6.1 presents inherited variations among different psychosomatic disease (Table 6.1). All variants have slight effect & gradually increase effect with big sample sizes [35–37].

Polygenic hazard marks envisage psychological illness where genetic variants improve the disease conditions. This proposes many alternatives answerable for mental health progress related issues [38]. One more reliable discovery is that most communal and unusual genetic variants are non-specifically allied with a couple of mental disorders [39]. As a whole, more than 67% of genetic connections are there for mental diseases like schizophrenia, bipolar disorder and other depressive illness [40].

Table 6.1 Inherited deviations allied with psychological disease [43].

Particulars	Autism	Schizophrenia	Bipolar disorder	Depression	Ref [43]
Number of persons in largest genomic sample to date	13,088 samples and 16,664 controls	36,989 samples and 113,075 controls	7,481 samples and 9,250 controls	121,380 samples and 338,101 controls	
Number of inherited deviationsallied at genome-broad spectrum of statistical implication	5	126	17	17	
Odds ratio of the most strongly allied inherited variation	1.17	1.21	1.15	1.05	
Percentage variance described by common inherited variations across the genome	14%	23%	25%	5%	

6.3.1.2.1.2 Psychological Factors

Psychological factors associated with traumatic situations like sexual, emotional, and physical abuse of child; loss or separation of loved ones; negligence; poor ability to relate to others are also main causes of mental illness.

6.3.1.2.1.3 Environmental Factors

Surrounding is the biggest influencer on human psyche as human growth is sustained in it where lots of environmental factors contribute significantly for mental illness. These covers death or divorce, unorganized family life, moods of insufficiency, low self-assurance, nervousness, resentment, or solitude, frequently changes in job or school, sociocultural expectations, materialistic abuse by human beings.

It is impossible to segregate the effect of both heritable and ecological causes for the cause of emotional ailment reported by various literature since 1960s. In Table 6.2, various conservational aspects associated with different psychological diseases are listed.

R, rare proofs; M, modest proofs; V, various proofs; NA, not applicable.

Various common principles underlying which show the causes of mental illness are as follows:

1. The similar type of ecological contact increases the danger of different psychological illnesses. For example, city atmosphere recognized as a hazardous cause of schizophrenia, but a methodical scrutiny presented it as a peril of all forms of emotional conditions [41].

Table 6.2 Environmental factors allied with psychological diseases.

Risk factors	**Autism**	**Schizophrenia**	**Bipolar disorder**	**Depression**	**Ref [43]**
Gestational threat factors					
Infections	R	S	M	R	
Malnutrition		S	M	M	
Heavy metals	S	M			
Perinatal risk factors					
Preterm birth	M	M	M	M	
Season of birth	M	M	M	R	
Birth complications	S	S	0		
Childhood environment					
Urbanicity	V	V	R	R	
Poverty	M	V	R	V	
Maltreatment	NA	M	M	V	
Bullying	NA	M	R	V	
Drug use in adolescence					
Cannabis	NA	V	M	R	
Stimulants	NA	V	M	0	

2. Even frequent contact to different kinds of conservational resources causes same type of psychological complaint. Generally schizophrenia is more common in this type that grows due to prominent reasons like nutritional, vitamin and minerals deficiency into babies from mother and pathological toxicities, low socio-economic and minority prestige, metropolitan nurture and infant harm, as well as contact to stimulating substance, cannabis and tobacco [42].
3. Many people are resistant to any psychological symptoms, although they constantly under exposure of same kind of environmental factors.

Socioeconomic Factors like joblessness, small salary, scarcity, dues, poor or unbalanced existing conditions, educations are often linked to ecological factors accountable for psychosomatic diseases.

6.3.1.2.1.4 Gene–Environment Causation of Mental Illness
Some of the psychological sickness is associated by the combined effects of different causative parameters; especially interaction of genetic and ecological that is termed as gene-environment interaction (GxE). A global part of GxE with reference to psychological sickness is recommended due to inconsistency between the obtained outcomes of epidemiological readings and identical studies that is known as environment paradox [43, 44].

6.3.2 Evolving Background and Scientific Progress

Two main findings in psychiatry persist underestimated and are not replicated in most etiological exploration. The first one is the heterotypic endurance which show steadiness of pathology over the lifetime course. Although it has very clear about the mental illness existence from either in childhood or adulthood but its psychopathology characteristically fluctuates throughout life cycle of illness. For example, worry in infancy is a strong interpreter of psychological disorders in later life. However, the severity of illness and a methodical approach to the childhood symptoms and adult disorders [45]. During mental illness, most individuals pass through many investigative groupings enhances to the difficulties allied with research [46]. It necessitates the extensive study of emergence of various psychological disorders which is referred as psychopathology [46].

The second finding is that the psychological diseases life span differs among patients or individuals. Some of them are temporary whereas some are permanent. Mood disorders have been arranged as occasional and personality disorders have been hypothesized as persistent [47].

6.3.3 Limitation of Current Approaches

Based on available findings on psychiatry research, four major factors which are limiting the further growth of psychiatry research which are discussed as below:

1. Presumed facts.

Since the time of research, generally studies built up done on certain presumed facts like causation of mental illness and their influences although most of the notable discoveries were established before these presumed facts came in to the picture.

For example, lithium efficacy is responsible for bipolar disorder is discovered very early without any assumed information [48].

2. Data omissions.

The identification focused procedures and the consequent classes- vs.-dimensions causes the disregard of the growing framework and the chronological measurement of psychopathology.

3. Statistical power and quality of measurement.

Etiological study mainly requires huge archetypal trials with exact dimensions of hereditary dissimilarity, ecological contacts and psychopathology throughout the life cycle as countless inherited and ecological elements subsidize to utmost cases of psychological sickness. A single study statistic of environmental or high standard of inherited measurements or data of longitudinal study of high quality measurement of many individuals does not produce any accurate results instead of this, it is either dragged organized from various inconsistently evaluated samples or they have huge failure level and fewer correct extent. It is difficult to find out the exact causing factors responsible for psychological illness unless the experts of various fields like psychopathology, environment and genetics. The collaborative work could be helpful in this.

6.3.4 Basis for Discovery of Coherent Psychopathology

Development of psychological illness necessitates the in-depth study of psychopathology by cuddling the complications of many inherited, environmental and other related factors which causes mental sickness [49]. Advanced technology and practical understanding make this study quite accessible for further growth in treatment outcomes. It is crucial that an investigation work is intended in a way that make the most of the significant finding by evading the downsides of expectations, errors, insufficient quantity and statistical control.

To get better psychopathological outcomes for the treatments and to avoid major complications of mental illness, these kind of investigation under psychopathology should start in early phases of pregnancy, infantile or in puberty. Frequent evaluation of all these causative factors for a life span helps to find correct age at beginning, trail the progression and record chronological comorbidity to grow scientifically in psychopathology research.

6.4 Psychopathology

Psychopathology springs from two Greek words "psyche" sensing "soul," and "pathos" denotes "suffering." Presently, 'psychopathology' is assumed to mean the basis of psychological illnesses, how they progress and their indications [55].

It is the technical investigation of atypical psychological situations since century, has provided a contour for psychological complaints and directed experimental & systematic improvement in current investigations practiced for psychosomatic diseases. Influence of technical and scientific advancement sideline psychopathology today. This enduring attrition of psychiatric sensation is additionally raised by medical informality as well as overstretched well-being care and exploration. Accurate and vigilant skilling to evaluate psychopathology in a trained custom used to be a fundamental element of psychological well-being experts [50].

6.4.1 Advent of Various Approaches to Psychopathology

An evolution of the classification of psychological disease was started in the 19th century with the prehistoric and augmented advancing with the European nosologists. Earlier, this classification was based on two different systems and both were competing systems of classification. First system was based on symptoms which was termed as "symptomatological," and the other is based on the causes which is known as "causation-based or somatoetiological." As the time pass, later on causation based system was practiced in psychiatry field for the classification of mental illness. Although symptoms based system was not so promising to classify a specific mental illness due to symptoms similarity in different kind of diseases, causation based system of classification had also set a poor track record in this field. This often happens due to its failure to state any definite cause behind particular mental illness. That is why they both are seized from psychiatry and disappear into the diagnosis granary of other medicinal spheres [51].

6.4.2 Diagnostic Models of Psychopathology

However, the field of psychiatry has provided a variety of diverse prototypes of and tactics to emotional complaint, which have sometimes conveyed development in medical training, this have regularly also been complemented by disapproval from within and deprived of the field.

Continuous investigation on psychopathology gave birth of different types of diagnostic approaches which were bifurcated as A) Traditional diagnostic approach or Categorical diagnostic approach and B) Trans Diagnostic approach or Dimensional model of psychopathology.

6.4.2.1 Traditional Diagnostic Approaches

These approaches also termed as categorical diagnostic approach and they include two different systems to classify cerebral infection. One is Diagnostic and Statistical Manual of Mental Disorders (DSM) and the other one is International Classification of Diseases (ICD).

These suggests customary principles for the taxonomy of psychological sicknesses. These standards (that is, symptoms) are supposed to be expressions of a core article for diseases. This viewpoint is entitled the "communal cause approach" and has conquered psychological fitness medical exercise and investigation [52].

DSM is the system for the management of psychological disease was emerged by the effort of German psychologist; Emil Kraepelin as a diagnostic approach. Although system was emerged from the European country but European psychopathological knowledge futile to put it as a diagnostic system to the public domain. Then American Psychiatric Association has taken up this DSM series, as homespun American creation, start with Medical 203 in 1945, then steered by psychoanalytic visions over and done with various modified versions of DSM from I to III signified an enormous "revolving of the side" in nosology which finally exit in psychiatry. When this system reached to its modified version of DSM, i.e., DSM-V, still many questions existed about this scheme whether it is on right path or not.

Simultaneously, ICD was also popular in that era for mental illness classification. One lot of public adhere to this classification which was produced by WHO and known as traditional diagnostic model. Gradually it passed through various transitions and currently ICD-11 approach is working as a classification of mental illness [53, 54].

Abandon Old-style Investigative Taxonomy and Embrace a Wholly Dimensional Opportunity: Why?

Psychosomatic investigation is still stalwartly inclined by its key cataloguing arrangements, the DSM-5 and ICD-11 [55]. As per these two investigative cataloguing structures, psychosomatic nosology is understood built on multinomial taxonomic divisions. However, they are correspondingly confronted, and, sustained by progresses in device education, a tailored symptom-based tactic to accurate psychiatry is gradually more encouraged. Existing literature is the witness for the usage of these incongruous analyses as the core component of analysis. Although cataloguing system is available, its key restrictions significantly obscure the progress [56–58].

6.4.2.1.1 Condemnations of the Investigative (Diagnostic) Approach

Although Diagnostic models are widespread, suffer from certain limitations and criticisms; few of them are mentioned below [59–64].

1. There have been numerous reproaches of this tactic, counting the uncertain legitimacy, consistency, and usefulness of psychiatric identifies [65].
2. This is specious from the prominent heterogeneity in symptom outlines and progression paths in the interior the investigative groupings [61, 62].
3. Additionally, science of neurons, hereditary readings do not confirm biotic roots strongly [63].
4. One more imperative criticism is that the distinctive features of separate symptoms and their latent pivotal relationships are ignored [64].
5. Lastly, medical symptoms investigation for the categorization of such kind of sickness lead to daily complications.

In current times, classification of psychological illness has followed two main systems, i.e., DSM and ICD. Although successive versions of the DSM and ICD effectively practiced system among affected population, they have also heading that psychiatry arena is in calamity, and to support for exclusively novel models for analysis and assessment. In spite of the advancement in evidence-based pharmacotherapies and psychotherapies, there is huge breach exists in both research-practice and in health care [64].

6.4.2.2 *Dimensional Models of Psychopathology*

Associated challenges and limitations of the existing categorical or traditional psychiatric models urge the birth of dimensional model of psychopathology. Basically these models are categorized into three approaches; 1. Hierarchical Taxonomy of psychopathology (HiTOp) and 2. Research Domain Criteria (RDoC) program 3. Network approach.

6.4.2.2.1 HiTOp

An autonomous association of psychopathology investigators have developed this HiTOP program [66]. This program was a outcome of an enduring displeasure through the existing cataloguing schemes of mental illness and their incapability to effectively archetypal the organization of psychopathology. It is based on the main objective to deliver an empirical base, subjective approach of recent practicing diagnosis with complete dimensional association of psychopathology, disease patterns and indications to multivariate factor-analytic procedures. In recent times, many authors have meaningfully progressed visions in the quantifiable ranked construction of psychopathology [67–69].

HiTOP theorises this field in rank based organization that covers in various levels; diseases indications (level 1), their interior conditions/behaviors (level 2) features (level 3), and extensive ranges (level 4). Due to this, superordinate dominions may responsible for the interdependence of low-ranked constructions like interrelated character behaviors and psychopathological patterns. Wide range of it termed as broad spectra which is at level 4 incorporate adopting pathology, expressing pathology, thought syndrome (i.e., neurosis field illnesses), and pathological inwardness. Substitute features of the adopting range at level 3 comprise erotic difficulties, eating pathology, distress, and agony, while substitute features of the expressing range contain substance exploitation and rebellious conduct [66]. A ground-breaking revolution of the HiTOP is because of its combination of character qualities into the construction of psychopathology.

This is the fascinating figure of mark has established that character noticeably influences on the existence, reproduce, and tenacity of psychopathology [70], comorbidity among psychological sicknesses [71, 72]. HiTOP dimensions provide a clear and influential link for hereditary indicators, genetic susceptibility, neurobiological markers to give the disease sequence associated to psychopathology, and that to envisage effectiveness and generalizability of health managements [73–76].

6.4.2.2.2 RDoC

National Institute of Mental Health (NIMH) had initiated RDoC program as the existing psychiatric cataloguing scheme was not satisfactory to classify mental illness [77]. Due to lack of representing effective illness objects with the current system (DSM), the NIMH has supported with its own resources to genetic, neuroimaging, and cognitive sciences to propose an analytical pattern built on neurobiological and communicative phenotypes. Regardless of considerable hard work since the last 200 years, hereditary investigation and the neurosciences have principally unsuccessful to notice any consistent indicator that could be functional for scientific tests to support analysis and diagnosis [78]. This catastrophe is worthless to consider it as a future inside the medical science of neurons [79], and growth lacking as a part of the limitations to this category of system.

In dissimilarity to the typical DSM/ICD top-down methodology, that presents psychological illness oriented from marks and indications and then pursues to discover the etiopathology of concerned mental illness whereas RDoC inspires a bottom-up approach. As per this, scientists firstly describe the usual distribution of a certain attribute, that they inspect the mind system that trappings this purpose, and then the particular parameter which is responsible for dysfunction that leads to psychopathological indicators and conditions [78]. The project guides investigation in the direction of five neurobiological domains and these are allocated into numerous units related to the subject. Discrete pathophysiological appliances included under the same uncertain analytical DSM/ICD-based group are a main reason to less responsive to psychiatric medications [80]. Hence, RDoC establishes an encouraging biotic method that might assist to evade the difficulties allied with diagnostic heterogeneity. RDoC hence reasonably title that this neurobiological cataloguing system may significantly progress etiological study, encourage the expansion of new treatments, and effortlessness to detect genetic and neurobiological indicators for application in problem-solving scientific trials.

6.4.2.2.3 Network Approach

Network analysis is a fresh and different theoretic line of attack in the study of psychopathology and personality disorders [81–83]. This theoretic framework as an alternative to examine statistics, propose innovative customs of demonstrating and mimicking psychopathological practises, as well as initiating new skylines that enables to understand the cerebral complaints and their characterization in a diverse way [81–83] may assist to alter how psychopathology is implicit [82, 83]. This offers new responses to difficulties in psychopathology field which are already masterpieces. This network model of psychopathology hunts for analysing dynamic connections, occasionally of a fundamental type, between indications and signs. In this approach causal hidden variables would not be the reason of symptom co-variation, and nor would symptoms be compatible indicators of an original complaint. Symptoms do not therefore replicate primary mental illnesses.

Collection of symptoms cluster mutually influence each other and they have an autonomous causal influence and are not mere inactive significances of a public latent cause. That is why psychopathological disorders could be seen as a multifarious dynamic scheme of symptoms and signs and so network tactic postulates psychopathology as a vibrant scheme of causation based indications; it is not the psychological illness that aggravates the signs, but signs that aggravate or strengthen other signs.

Basic concepts [84, 85]: Networks comprise of bumps and boundaries. The bumps signify the substances or variables to be deliberated. The boundaries characterize the networks between bumps. In network analysis in psychopathology the bumps typically epitomize symptoms while the boundaries exemplify the links between them. Graphically, bumps are presented as "circles" and the boundaries are the "lines" joining them. Bumps may be variables apart from psychological indications, like shocking involvements, material misuse or everyday strain, to indicate just a few. They may also be additional kind of variable from dissimilar investigative stages for, e.g., hereditary or psychosomatically.

6.4.3 Comparative Outlook

[86–88] HiTOP, RDoC, and network approach, newly established, inventive and dimensional frame-works are finest comprehended as harmonizing systems, pointed next to substituting the scientifically blemished grouping structures towards investigation commitments. Both methodologies implement a ranked based completely dimensional arrangement where HiTOP emphases absolutely on psycho-pathological signs, patterns, and wide-ranging logical sphere, RDoC is principally intended by the side of exploratory inherited and neuron biology study. Hence, RDoC clasp-specific aptitude for proceeding neurobiological study, while HiTOP is primarily alarmed through reorganization experiment based psychiatric methods. While in DSM/ICD systems of diagnosis typically depends on discrete neurobiological etiopathologies and symptom summaries are also chief contributors. Psychiatric examination must, thus, intemperance old-style DSM/ICD nosology and accept effective and evidence-based dimensional systems if it needs to develop perceptions into the etiology of psychological illnesses.

Additionally, opponents of dimensional approaches classically strain that clinical exercise entails twofold diagnoses. Though, at this phase, HiTOP, network analysis and RDoC are just exploration agendas, so investigative objects are not essential. Once dimensional ideas are prepared for application in scientific exercise, this makes conversion easy from dimensional scores into categorical peculiarities which shows dimensional nature of psychopathology.

6.4.4 Major Psychological Disorders

Psychopathology majorly focused on following psychological disorders which required to treat with some novel, precise and personalized approaches for promising and better outcomes [89–91].

6.4.4.1 Schizophrenia

It is the utmost thoughtful and terrifying of altogether psychological diseases [89]. Raised grievance aggravates nervousness in each sectors where human being belong to. Active managements are available for health cure, yet patients and their relatives repeatedly catch it tough to contact noble care. The name schizophrenia originates from the initial statement that the sickness is characterized by "the cutting off or piercing of the mental tasks." This has run to the misconstruction that the sickness is considered by a "split character," People with schizophrenia stereotypically hear voices (acoustic hallucinations), which frequently criticise or misuse them.

As per global survey, it accounts for 1.2% of the total infirmity familiar life years universal and 2.7% of the years lived with incapacity globally. It characteristically grants in initial middle age or late teenage years. Men are more susceptible to experience a more thoughtful form of mental illness-negative symptoms than women and very rare chance of recovery which leads to poor results. Even, vulnerability to mental sickness of men is more frequent than women and is the most common in big cities & more prevalent among migrants. Environmental and social factors contribute significantly for this.

6.4.4.2 *Mood Disorders*

Mood disorders are communal, multifaceted, and one of the core grounds of illness globally [91]. There has been a growing acknowledgment that mental dysfunction is a dominant feature of most mood disorders and closely linked to the functional injury as a major cause of them. Hence, suitable evaluation methodology and handling of cognitive injury in mood diseases are the most important part to deal with them and to produce effective outcomes of treatment. To broaden the clarity of this and improve the treatment effectiveness, neurobiological and neuropsychological concepts of brain functions and underlying mechanism must be understood for particular mental illness. Moreover, tools must be developed to assess the cognitive insufficiencies. In addition, developing tools to measure cognitive deficits more accurately to advance the cataloguing of psychosomatic indications and progressions in psychiatry.

6.4.4.3 *Anxiety Disorders*

Nervousness complaints are the most widespread cerebral health situations which are just as disabling conditions [96]. However, these are less evident than any other psychological disorders like schizophrenia, depression, and bipolar disorder. The diagnosis system for such diseases are being constantly reviewed in terms of both dimensional and structural for clinical treatment and research to suggest new and effective categorization. Anxiety disorders set up the most prevailing subcategory of cerebral sicknesses. Regardless of their extensive occurrence, these illnesses have not acknowledged the same acknowledgement as other main diseases such as mood and psychotic illnesses; In such a case, primary health care plays an important role for the treatment but due to lack of proper dealing with them lead to poor efficiency, augmented morbidity, and mortality proportions.

6.4.4.4 *Eating Disorders*

Eating disorders (EDs) are of growing community well-being distress nowadays [92]. Around 1.69% of world's total population suffering from ED's by reason of body image fears, atypical eating, and weight-control behaviors. These stance one of the maximum mortality threats among psychological illnesses. The diversity in ED etiology along with social and individual identity characteristics plays a critical role in the alteration of demographics of the ED population which create the difference in perception of modifications among people. Multiplicity subjects such as sex, age, and origin are the main parameters used in research in recent times.

6.4.5 Concluding Observations of Psychological Disorders

Certain exceptional observations that have conquered precise recognition in modern eras and may meaningfully effect psychiatric workout and examination [93]. Various cerebral fitness methodology which projected changeovers from existing practices to innovative performs that redirect excitement or expectation. At the same time, each of this approach conveys only a limited view, and that any potential of a model variation for the field is not well stranded. Although considerable development done in psychiatric analysis and conduct, there is sizeable requirement to advance an assessment and interpolation.

6.5 Computational Psychiatry (CP)

6.5.1 Advent of Computational Psychiatry: Reshaping the Existing Nosology of Mental Illness

Although numerous efforts have been taken as a progressive steps in the journey of psychiatry field development for better clarity about what is mental illness and various causes responsible for it, the progress is stagnant [94]. As all those purposes have created diverse descriptions with several theoretic simulations, which are often not linked and deficient of multidimensional clarity of psychological health, emotional diseases, and psychopathology. These assimilating various planes and arrangements [95].

Moreover, the huge intricacy of the human intellect and the glitches in diagnosing and elucidation of the living, circumstantial and ecological reinforcements of psychological tasks. This makes an illustrative breach amid science of neuron and cerebral biology practice next to midway stages of depiction that essentially fix philosophies or/and theories connect clinical and molecular altitudes of psychological conditions. It is very apparent that many biological dysfunction may produce similar types of behavioral changes whereas sometimes one biological dysfunctioning may saw different pathological symptoms [96, 97]. Mostly, inadequate understanding of humanoid perceptive phenotypes has directed to the nonexistence of an actual connection between the phenomenological and the molecular events.

Due to this, repeated worries and queries about the sorting of psychiatric illnesses urges revolution in psychiatry field with the use of machines [98–100]. As psychiatry has always came across several debates, created manifold inner and outer catastrophes which interrogated the aforementioned legitimacy to claim it a skill for handling mental fitness [101–103]. These condemnations have concentrated largely on the rationality of their ideas and concepts [104], their analytical aptitude [105], the viewer's consistency, and the biomarkers non-availability for regulation of identifications, dealings and diagnosis [106, 107]. Furthermore, the obvious focus towards the growth of unpredictability of diverse illness [108]. It is very hard to manage such challenges without theoretic prototypes, intangible outlines and influential computational gears.

Although various practical presentations generate important and beneficial data to comprehend the rational developments and biological turbulences in mental illness [109–113], which are generally incomplete by various factors like applied, financial or moral factors. Hence, an archetype move is desired for such a dimensional and trans-diagnostic levels of understanding for mental illness. In a search of unique theory as promising tool to analyse the complex levels of mental illness, mathematical principles and machine based hypothesis are needed to be incorporated. The ultimate goal of this is to create more accurate, precise and reproducible model to answer multi-level analysis and to be validated thereby reshaping the nosology of mental illness is obvious [114].

In this framework, practicing computational psychiatry (CP) is the most appropriate alternative which cater all the answers to questions and controversies raised during the history of psychiatry and reshaping the nosology of psychiatry [115]. Currently accessible literature on computational psychiatry research is the witness of promising and sound scientific developments in psychiatry and allied branches of it.

6.5.2 Computational Psychiatry: Important Levels and Incorporated Models in Different Approaches

As CP is very challenging for research, special attention has been given. In this section, an overview of four levels and other incorporated models in different approaches are discussed in detail with cognitive function and dysfunction by computational analysis.

6.5.2.1 Important Computational Stages

These comprises of following levels [116]:

1. Stage 1, planned recognition or categorization of mental tasks which are required to be employed for gathering of performance statistics
2. Stage 2, the apt of computing prototypes to the recital statistics
3. Stage 3, factor assessment
4. Stage 4, recognition of groups and relationships to medical indication severity.

6.5.2.2 Computational Approaches

Intensification of computing and documenting records usage in almost science streams from the time of initial 21st century period (up to 2010), CP has established the latent to combat the lack of or challenges associated with the traditional procedures for the treatment of mental illness [117]. Various contrasts are there in CP prototypes (e.g., predictive vs descriptive, confirmatory vs. exploratory prototypes, generative vs discriminative) and each of them having specific characteristic to describe particular class of that models. Here, two distinct types of technical study approaches with their various purposes are summarized, i.e., data-driven vs. theory-driven approaches. The main goal of these approaches to explicate outlines in original data or facts "about" the unwell mind, then to form, authenticate or fabricate theories to provide details about mind.

6.5.2.2.1 Data-Driven Computational Approaches

Important goal mouth of this line of attack is to "code" investigational data available from numerical symmetries available from appliance knowledge. These approaches require very few former assumptions about the data is a major benefit of it. Yet, this makes data driven interpretation more challenging [118].

Machine learning, one of the common methods used in data driven approach which is utilized for biomarkers prediction. Amalgamation of machine learning with neuron imaging and medical data to reply queries for prognosis and treatment consequences. With this approach one study of selective serotonin re-uptake inhibitors treatment for depression was conducted by taking available scientific data and clinical variables to estimate the rate of lowering symptoms in depression [119]. This approach drastically reduced the number of clinical variables further to get selective variables for prediction. These were then enter into a machine learning system to estimate the reduction rate of depression symptoms in mentally ill patients [120]. Eventually, machine learning procedures could offer the root for the construction of gears to back scientific conclusion formation. Various examples available to support the facts about machine learning methodology in data-driven approach.

One more study is reported for dementia research. This mainly concentrate on illness recognition and estimation by using various other methods of this approach like recurrent neural networks based on neuroimaging, medical, and biotic data, backing vector machineries, verdict trees, Naïve Bayes classification system, bunching, or additional means [120–122]. Furthermore, literature of dementia reported the expansion of research in this approach by stating that the onset of dementia could be predicted with the development of related risk models years before the development of symptoms in patients with collective study of demographic and clinical antiquity data [123]. Data-driven presentations were also utilized to differentiate altered categories of various diseases, or to lower the numbers of trials required for identification.

Yet, translation of these efforts into medical exercise is quiet tricky and require more eco-friendly real-time methodology. Though this approach is able to convey disease mechanism, pathology or any subtypes of diseases [124], many of typical data-driven tactics do not clearly anticipate to give biological and pathological mechanism related to the diseased conditions. If these approaches for detection and prognosis we use not sufficient to offer unique treatment options and to apprehend the original intricacies of active nature of particular disease.

6.5.2.2.2 Theory-Based Computing Tactics

These tactics predominantly uses to explain the way by which any mental illness or neurobiological conditions will arise this is commonly known as a changed pathology or facts dealing out linked to the "origin" of psychiatric or neural situations. These approaches could be used in different types of models to explain different types of work phenomena of brain. These mainly include what central nervous systems sort out (i.e., descriptive models), functional definitions, work system could be operated on numerous stages of any idea pertaining to mental illness [125]. Hence, the purpose of theory-driven approaches is completely different as they stir the mind to search out possible reasons for the existence of any diseased conditions and how it works and grows.

All above replicas can be regarded as "simulated" patients which takes the intellectual function's defects on processor replications. Testing of these models along with other relevant data obtained from real patients help in the development of robotic understanding and form novel drug managements in-silico without experiments on animals and humans. Definitely, this could accelerate the translation effects and drug growth where unmet need satisfies.

6.5.2.2.3 Collective Approaches

The main goal of combined computational approaches in a single model (data-driven and theory-driven) has the great potential to explain about the various complicated forms of psychological conditions which could not be described with single approach.

This is principally essential for:

1. Delivering the brief about core mechanism and responsible cause connections acquired by theoretic models.
2. Lowering the chances of the existing scalability restrictions of theoretic models [126].

Hence, combined approaches increase translational advantages in cellular processes.

6.5.3 Obstacles and Resolutions to Implementation of Computational Psychiatry in Real World

Obstacles: Whenever a new terminology come on board, normally real-world worries, mostly seem to turn as obstacles in their execution in society. Such thing is very obvious in application of computational methods in psychiatry field. Few of them are very significant and described as below, i.e., time, ecological rationality, and practical implementation concerns.

Time: Computational modelling in psychiatry is machine based program which is quite unlike to clinical psychiatry where clinical symptoms assess the psychological behavior. Generally, assessment is based on set of various neuropsychological or psychophysical tests which helps to formulate reliable base for treatment of illness. Similarly, computational psychiatry uses computational modelling, mathematical equations, statistical data, software based program as a part of treatment for any specific mental illness.

However, accurate and precise results obtained but its validity necessitates to perform many trials which are time oriented and all these procedures usually consume time. In many cases, optimization of certain parameters is required for better treatment outcomes which also utilize time. In some of the emergency cases, decision is required to take very quickly. As a whole, timing factor is very crucial but in many cases, it is a big obstacle too for implementing computational psychiatry [127–131].

Ecological Validity: Ecological validity is now a days a big restriction as an experiences in computing practice for cerebral testing and behavior mostly do not replicate clinically related comfortable areas, backgrounds, and/or indications as they are practiced in the physical world [132]. For example, choices marked to get small amount of financial return or reducing minor shudders in the lab could reasonably utilize vary prospects than choices in real-world situations to evade dreaded circumstances or we can say to improve complete life gratification. Even lab performances also usually considered to segregate definite activities or cerebral expertise for effective measurements and their interpretation. Unfortunately, in real-world functioning, situations completely different where sometimes participant might utilize many abilities to solve a problem or multiple problems solution in just one set of ability [133–135]. However some neuropsychosomatic and computing chores, their factors connected straight to people skills and consequences, such connotations have been restricted to date [136–139].

Implementation Concerns: Along with above two, one of the big practical barrier is implementation of computational psychiatry in practice. Yet such system does not exist, which can easily assist the adoption of various trials needed to smooth functioning of computational psychiatry. CP is still struggling with sufficient computational framework to produce desired result for mental illness. Many times, investigators are fully equipped for refining, implementing and interpreting relevant models but computational framework is limited. Even in many cases, research questions are there but the standard comprehensible varieties of chores or replicas frequently do not exist. Furthermore, some investigators may not have expertise for such work also can perform task on computational models, but in that case development and validation process can be restricted [128].

Test–retest reliability: Generally, computational typical factors are required to be determined repeatedly to assess its reliability for experimental calculation. However, as many of these encompass different types of tactics which can often change. This can raise the point of test–retest reliability, and this cannot be the same at each level of assessment.

Resolutions: As CP is very vast field and very complex-in depth in understanding and implementing for a daily use. It's obvious that some concerns always there to be resolve. As said, when there is a problem, there is a solution. Here some of the practical solutions proposed for above obstacles are explained below.

Time: Timing issue can be resolved by redesigning of structure in which various estimation parameters and test size must be framed in such a way where overall task completes in minimum amount of time, reduced the numbers of trials required to take, battery designing like a game sequence to increase the tolerability of tasks together [140, 141].

Ecological validity: The execution of any idea or tasks related to psychological diseases in the form of game as a model of treatment to either the overall public or definite clusters of participants may have advantageous for common public with respect to better outcomes of therapy responses and improved patients compliances as it utilize minimum trials, multiple skills involvement, their utilization and influence for the treatment outcomes, and quite interesting method to diagnose the psychological problems. This strengthens the ecological validity of designed model by improving the simplicity of outcomes from these programmed tasks to medical results of interest. Through this ecologically valid form of tasks-game world patients or participants would be clever enough to make selections, to develop various tactics to solve a single problem. This provides a base for demonstrating of original factors correlated to people selection and actions which may replicate dormant conditions more appropriate for the creation of and preservation of various signs and diseases [135], but which have not been formerly restrained in economically effective means.

Implementation concerns: Computational tasks with installed cohesive battery is one of the best solution to sort out the concern of implementation as it combines multiple task in the form of game would offer a consistent setting, and could be planned to produce behavioral information of participant created on computational simulations automated into the software. These assimilated tasks in modular and modifiable form supports investigators to explore various levels of computations in their methodology. Digital tools are designed in such a way where they reduce time span of process, easy to handle multiple tasks, easy reach to participants by covering large population, reduced cost for validation and development of computational tasks and models and patient & researcher friendly.

Test–retest reliability: Tasks required to be assessed must be scrutinized for reliability before its addition in battery [139, 142]. One more probable solution is to have contributors' whole task several times before preparatory use of them for assessment increase reliability and improve clinical relevance. Another promising tactics is to use vary task content for assessment judgement structure will remain same so that each tasks though similar in result but each time new to participants.

6.5.4 Applications Area Where Computational Psychiatry Might Be Useful

Implementation of computing prototypes to analyse psychosomatic illness holds a great target to succeed following purposes that are habitually practiced in this branch [17, 143–146]. Few of them are summarized below.

Mechanism
Clinical psychiatry and available literature is still not assured about the cause of mental illness. Through computational frame work, attempt has been made to find out the root-cause responsible for mental illness and to detect the fault in brain working phenomena which further facilitates to develop promising biomarkers for identification, prevention, and therapeutic interference.

Subtyping
With the use of computational modelling, it would be easy to answer the heterogeneity of mental diseases as it mostly relies on unproven models like clustering, which aims to determine significant outlines in the data which can be assessed with external trials, like treatment consequences.

Status prediction
If any early symptoms of mental illness could be identified before its development in to diseased state, then mental illness can be prevented. Computation tools involvement in psychiatry able to predict health status of mentally ill people by identifying early stage of psychosis which might be helpful to stop its progression and its adverse effects and so cure must be assured [147, 148].

Treatment stratification although sets of pharmacological and non-pharmacological treatment options are available in psychiatry field, but significant size of population will not be getting treatment benefit. To find out which particular sets of treatment best suitable and assure the cure to particular patient is a very tedious and trial & error based procedure. Here the hope is CP models can advance the treatment options and so cure of mental illness by involving various therapeutic strategies.

6.5.5 Challenges Cum Limitations of Computational Psychiatry

Although psychiatry field is over flooded with computational tools which are all having similar in structural framework faces the same type of challenges to be resolve for smooth functioning of CP in assessment of mental illness. Few of them are discussed below.

Noisy data causes the impact on computational model's quality and reliability of data obtained which can create bias in treatment outcomes. This generally arises from uncontrolled or inconsistent data. This is the big challenge to resolve in CP to get quality and reliable data.
Missing and sparse data whenever computational data is in use, there is a possibility of missing data and or sparse data that is the major worry for archetypal construction. This necessitates supplementary processing for data rectifications in different types of studies in statistics. Generally these type of incomplete data leads to false or biased results which is again a big challenge in psychiatry [149].
Validation CP utilises computational models to assess the results for psychological disorders. When machine based system is in use, data authentication must be required for accurate results. That is why validation of data and system both require to improve system efficacy. This methodology is used to avoid exaggerated precision and so to get susceptible outcomes.
Small sample size and reproducibility Lab scale analysis generally requires small sample size for evaluation which tend to create biased and non-reproducible data with little arithmetic power [150, 151]. It is a challenging task to generate a device were data analysis is feasible for even a small sample size and that is possible only when it can be done online or smart data base collection with the use of digital device where tasks are installed in game type form [152].
Temporal dynamics Various data sources for psychological diseases symptoms, cerebral functions, intellectual variables, social media display vacillations and fluctuations on unlike time rulers. Time-based changing aspects might be illness related or fully self-determining. Because these changing aspects may not be distinguished while undergoing specific evaluation, to evaluate changing aspects to portray psychological well-being [153].

Generalizability: Generalizability is critical for the medical achievement of demonstrating exertions and also a fundamental trial particularly for the statistics vary widely from health realism. It is the skill to practice models out of facts utilized to progress the unique prototypical.

Implementation of transdiagnostic tools: However, practicing of transdiagnostic tools in clinical field have a great potential to current cosmological approaches are challenging. The complex nature of various mental illness and their behaviors do not assure about particular new computational batteries that provide holistic treatment via clinical assessment. Transdiagnostic tools must be reachable and easy to work in a low-resource situation in case of increasing their clinical potential. This would exclude the necessity of timewasting mental tasks and overpriced imaging that makes precision psychiatry more reachable.

In a nut shell, CP utilizing transdiagnostic computational phenotyping of psychological disease bridges the gap between pathophysiology and psychopathology which delivers exploratory and extrapolative tools in relieving specific psychological health indications.

6.6 Computational Psychiatry: An Advanced Version Links Pathology and Psychopathology

The existed history of psychiatric disorders is the witness of its foundation which is the combination of numerous interrelating stages, from inheritances to cells, intellect, conduct, the nearby atmosphere and neural routes, Moreover, huge amount of biotic and psychosomatic proof like the ENIGMA [154] and COCORO [155] ventures indicates latest progresses in knowledge gain to improve psychological disorders [156, 157]. Incorporation of mathematical procedures and hypothetical outlines to expand the safe curability scope of mental fitness, their routine usage towards investigation termed as branch of machinery based computing human wellbeing (CP), which is an imperative branch of psychiatry. The progress of psychiatry field starting from pathophysiology of mental illness, psychopathology to computational psychiatry is well narrated in various kinds of literature related to psychiatry [17, 116, 158–160].

To construct CP as a promising field in psychiatry research, many of researchers from various backgrounds like neuroscience, psychiatry, psychopathology and mathematical modelling are working together. Hence, to fast-track CP research, it is essential to link pathophysiology-neuroscience, psychopathology (traditional psychiatry) and computation.

6.6.1 Link to Neuroscience/Pathophysiology

Conventionally, patient's self-submitted subjective understandings and interactive opinions by clinicians have been utilized to gather observations for conceptualization and analysis of mental illness. But the current diagnosis system of pathophysiology faces numerous trials in terms of living acceptability [40, 161]. Besides this, psychologically ill patient's showing varied outcomes for the treatment options, variation in responses and the diseases procedures which are not consistent [162]. Research Domain Criteria (RDoC) structure has been established to schematize psychosomatic research depends on behavioral neuroscience that is free from any type of traditional disease type restriction [5, 163]. Even NIMH has given the RDoC matrix, which assimilates several points of evidence.

Psychiatric illnesses are measured on a scale from usual to unusual conditions of hypotheses and associated theories are ordered into a sphere. Moreover, they have listed units of many variables used in analysis at different levels. Linking of pathophysiology by fusion of RDoC along with computational approaches which are the part of CP will overcome the associated challenges mentioned above.

6.6.2 Link to Traditional Psychiatry/Psychopathology

At certain level, pathophysiology in terms of RDoC along with CP is not sufficient to shelter the old-style taxonomy and symptomatology of psychological complaints. As cited, psychopathology contributes a lot by incorporating symptomology to the study and investigative cataloguing measures due to its inter-reliability [40]. Thus, an incorporation of psychological warning sign and facts related neurons is wanted. Nevertheless, it is problematic for investigators to identify the symptoms which are important to study outside the domain of psychiatry.

6.6.3 Link to Computation

In this type of investigation generally commanding computational gears and theoretical bases are utilized to examine and operate high-dimensional, multimodal statistics groups which comprises medical, hereditary, interactive, neuroimaging, and other statistics forms gained by animal representations, humanoid tests, scientific information, and recreations. Even, this integrates dissimilar approaches based on theory and available data framework [40]. Moreover, several kinds of prototypes claim needed. Hence, integration of all these data, methodologies, theories and models is necessary to augment psychiatric research.

6.6.4 CP Addresses the Gap Between Pathophysiology and Psychopathology

CP shows a dynamic part in the advancement of pathology, psychopathology, management options and curative & defensive policies, as combined hard work in the disciplines of neuroscience, psychiatry, and mathematical modelling. Though, inadequate links between the perspectives of pathophysiology and psychopathology may hamper the psychological research expansion. Furthermore, CP addresses the gap between these two streams can be used to discover computational framework to a specific injury of mental functions. Additionally, CP is enabled to overcome the challenges associated separately with pathophysiology and psychopathology fields.

6.7 Conclusion

As time grows, it is very hard to assimilate the knowledge from diagonal readings deprived of a public-agreed calibration. For example, factor analysis methods to originate transdiagnostic indicators which are reliant on the self-report surveys [5]. Additionally, computational procedures also need substantial skill, and non-experts may discover the gears tough to steer. Although, diagnosis of mental illness based on behavior that can be categorized by computational factors with precise neurobiological links [164, 165], patients with parallel computational factors may not all react to a specific medicinal agent. This cluster of persons may also be varied and big investigational medication educations will be obligatory to create the diagramming among the fields of pathophysiology, psychopathology and computational psychiatry to advance person-specific managements. Hence, the study of computational psychiatry is the discovering unusual time scale in the brain through machines.

References

1. Neoadjuvant chemotherapy for locally advanced cervical cancer: a systematic review and meta-analysis of individual patient data from 21 randomised trials. *Eur. J. Cancer (Oxford, England: 1990)*, 39, 17, 2470–86, 2003.

2. Kazdin, A.E. and Blase, S.L., Rebooting psychotherapy research and practice to reduce the burden of mental illness. *Perspect. Psychol. Sci.*, 6, 1, 21–37, 2011.
3. Insel, T.R., Assessing the economic costs of serious mental illness. *Am. Psychiatric Assoc.*, 165, 6, 663–5, 2008, Jun.
4. Kong, C., Dunn, M., Parker, M., Psychiatric genomics and mental health treatment: setting the ethical agenda. *Am. J. Bioethics*, 17, 4, 3–12, 2017.
5. Cuthbert, B.N. and Insel, T.R., Toward the future of psychiatric diagnosis: the seven pillars of RDoC. *BMC Med.*, 11, 1, 1–8, 2013.
6. Trivedi, J.K., Cognitive deficits in psychiatric disorders: Current status. *Indian J. Psychiatry*, 48, 1, 10, 2006.
7. Gold, J.M., Goldberg, R.W., McNary, S.W., Dixon, L.B., Lehman, A.F., Cognitive correlates of job tenure among patients with severe mental illness. *Am. J. Psychiatry*, 159, 8, 1395–402, 2002.
8. Dixon, L.B., Holoshitz, Y., Nossel, I., Treatment engagement of individuals experiencing mental illness: review and update. *World Psychiatry*, 15, 1, 13–20, 2016.
9. Hyman, S.E., Psychiatric disorders: Grounded in human biology but not natural kinds. *Perspect. Biol. Med.*, 64, 1, 6–28, 2021.
10. Morris, S.E. and Cuthbert, B.N., Research Domain Criteria: cognitive systems, neural circuits, and dimensions of behavior. *Dialogues Clin. Neurosci.*, 14, 1, 29–37, 2022, Mar.
11. Whiteford, H.A., Ferrari, A.J., Degenhardt, L., Feigin, V., Vos, T., The global burden of mental, neurological and substance use disorders: an analysis from the Global Burden of Disease Study 2010. *PLoS One*, 10, 2, e0116820, 2015.
12. Eaton, W.W., Martins, S.S., Nestadt, G., Bienvenu, O.J., Clarke, D., Alexandre, P., The burden of mental disorders. *Epidemiol. Rev.*, 30, 1, 1–14, 2008.
13. Thornicroft, G. and Semrau, M., Health system strengthening for mental health in low-and middle-income countries: introduction to the Emerald programme. *BJPsych Open*, 5, 5, e66, 2019.
14. Collaborators, G.B.D. and Ärnlöv, J., Global age-sex-specific fertility, mortality, healthy life expectancy (HALE), and population estimates in 204 countries and territories, 1950–2019: a comprehensive demographic analysis for the Global Burden of Disease Study 2019. *Lancet*, 396, 10258, 1160–203, 2020.
15. Vos, T., Lim, S.S., Abbafati, C., Abbas, K.M., Abbasi, M., Abbasifard, M., *et al.*, Global burden of 369 diseases and injuries in 204 countries and territories, 1990–2019: a systematic analysis for the Global Burden of Disease Study 2019. *Lancet*, 396, 10258, 1204–22, 2020.
16. Collaborators GBDMD, Global, regional, and national burden of 12 mental disorders in 204 countries and territories, 1990–2019: a systematic analysis for the Global Burden of Disease Study 2019. *Lancet Psychiatry*, 9, 2, 137–50, 2022.

17. Huys, Q.J.M., Maia, T.V., Frank, M.J., Computational psychiatry as a bridge from neuroscience to clinical applications. *Nat. Neurosci.*, 19, 3, 404–13, 2016.
18. Stephan, K.E., Binder, E.B., Breakspear, M., Dayan, P., Johnstone, E.C., Meyer-Lindenberg, A., *et al.*, Charting the landscape of priority problems in psychiatry, part 2: pathogenesis and aetiology. *Lancet Psychiatry*, 3, 1, 84–90, 2016.
19. Mezzich, J.E. and Salloum, I.M., Towards innovative international classification and diagnostic systems: ICD-11 and person-centered integrative diagnosis, Wiley Online Library; *Acta. Psychiatr. Scand.*,116, 1, 1–5, Jul 2007.
20. Mezzich, J.E., Salloum, I.M., Cloninger, C.R., Salvador-Carulla, L., Kirmayer, L.J., Banzato, C.E.M., *et al.*, Person-centred integrative diagnosis: conceptual bases and structural model. *Can. J. Psychiatry*, 55, 11, 701–8, 2010.
21. Clark, D.M., Realizing the mass public benefit of evidence-based psychological therapies: the IAPT program. *Annu. Rev. Clin. Psychol.*, 14, 159–83, 2018.
22. Reynolds, S., Wilson, C., Austin, J., Hooper, L., Effects of psychotherapy for anxiety in children and adolescents: A meta-analytic review. *Clin. Psychol. Rev.*, 32, 4, 251–62, 2012.
23. Roefs, A., Fried, E.I., Kindt, M., Martijn, C., Elzinga, B., Evers, A.W.M., *et al.*, A new science of mental disorders: Using personalised, transdiagnostic, dynamical systems to understand, model, diagnose and treat psychopathology. *Behav. Res. Ther.*, 153, 104096, 2022.
24. Holmes, E.A., Craske, M.G., Graybiel, A.M., Psychological treatments: A call for mental-health science. *Nature*, 511, 7509, 287–9, 2014.
25. Riglin, L., Collishaw, S., Richards, A., Thapar, A.K., Maughan, B., O'Donovan, M.C., *et al.*, Schizophrenia polygenic risk score and psychotic risk detection–Authors' reply. *Lancet Psychiatry*, 4, 3, 188–9, 2017.
26. Ono, M., Bolland, S., Tempst, P., Ravetch, J.V., Role of the inositol phosphatase SHIP in negative regulation of the immune system by the receptor FeγRIIB. *Nature*, 383, 6597, 263–6, 1996.
27. Stoll, J., Müller, J.A., Trachsel, M., Ethical issues in online psychotherapy: A narrative review. *Front. Psychiatry*, 10, 993, 2020.
28. Hunsley, J. and Lee, C.M., *Introduction to clinical psychology*, John Wiley & Sons, NJ, USA, 2017.
29. Pomerantz, A.M., *Clinical psychology: Science, practice, and culture*, Sage Publications, Thousand Oaks, California, 2016.
30. Sharma, B., Neurochemical Signaling in Depression: Impact of Target-Based Pharmaceuticals. *Curr. Drug Ther.*, 18, 1, 2–14, 2023.
31. Gottesman, I.I., Laursen, T.M., Bertelsen, A., Mortensen, P.B., Severe mental disorders in offspring with 2 psychiatrically ill parents. *Arch. Gen. Psychiatry*, 67, 3, 252–7, 2010.
32. Rasic, D., Hajek, T., Alda, M., Uher, R., Risk of mental illness in offspring of parents with schizophrenia, bipolar disorder, and major depressive disorder:

a meta-analysis of family high-risk studies. *Schizophr. Bull.*, 40, 1, 28–38, 2014.
33. Polderman, T.J.C., Benyamin, B., De Leeuw, C.A., Sullivan, P.F., Van Bochoven, A., Visscher, P.M., *et al.*, Meta-analysis of the heritability of human traits based on fifty years of twin studies. *Nat. Genet.*, 47, 7, 702–9, 2015.
34. Gershon, E.S., Bipolar illness and schizophrenia as oligogenic diseases: implications for the future. *Biol. Psychiatry*, 47, 3, 240–4, 2000.
35. Rammos, A., Gonzalez, L.A.N., Schizophrenia Working Group of the Psychiatric Genomics C, Weinberger, D.R., Mitchell, K.J., Nicodemus, K.K., The role of polygenic risk score gene-set analysis in the context of the omnigenic model of schizophrenia. *Neuropsychopharmacology*, 44, 9, 1562–9, 2019.
36. Hyde, C.L., Nagle, M.W., Tian, C., Chen, X., Paciga, S.A., Wendland, J.R., *et al.*, Identification of 15 genetic loci associated with risk of major depression in individuals of European descent. *Nat. Genet.*, 48, 9, 1031–6, 2016.
37. Robinson, E.B., St Pourcain, B., Anttila, V., Kosmicki, J.A., Bulik-Sullivan, B., Grove, J., *et al.*, Genetic risk for autism spectrum disorders and neuropsychiatric variation in the general population. *Nat. Genet.*, 48, 5, 552–5, 2016.
38. Wray, N.R., Lee, S.H., Mehta, D., Vinkhuyzen, A.A.E., Dudbridge, F., Middeldorp, C.M., Research review: polygenic methods and their application to psychiatric traits. *J. Child Psychol. Psychiatry*, 55, 10, 1068–87, 2014.
39. Devlin, B., Kelsoe, J.R., Sklar, P., Daly, M.J., O'Donovan, M.C., Craddock, N., *et al.*, Genetic relationship between five psychiatric disorders estimated from genome-wide SNPs. *Nat. Genet.*, 45, 9, 984–94, 2013.
40. Cross-Disorder Group of the Psychiatric Genomics C, Identification of risk loci with shared effects on five major psychiatric disorders: a genome-wide analysis. *Lancet*, 381, 9875, 1371–9, 2013.
41. Vassos, E., Agerbo, E., Mors, O., Pedersen, C.B., Urban–rural differences in incidence rates of psychiatric disorders in Denmark. *Br. J. Psychiatry*, 208, 5, 435–40, 2016.
42. Davis, J., Eyre, H., Jacka, F.N., Dodd, S., Dean, O., McEwen, S., *et al.*, A review of vulnerability and risks for schizophrenia: Beyond the two hit hypothesis. *Neurosci. Biobehav. Rev.*, 65, 185–94, 2016.
43. Uher, R. and Zwicker, A., Etiology in psychiatry: embracing the reality of poly-gene-environmental causation of mental illness. *World Psychiatry*, 16, 2, 121–9, 2017.
44. Johnson, S.B., Riis, J.L., Noble, K.G., State of the art review: poverty and the developing brain. *Pediatrics*, 137, 4, 1–16, e20153075, 2016.
45. Stringaris, A. and Goodman, R., Longitudinal outcome of youth oppositionality: irritable, headstrong, and hurtful behaviors have distinctive predictions. *J. Am. Acad. Child Adolesc. Psychiatry*, 48, 4, 404–12, 2009.
46. Uher, R. and Rutter, M., Basing psychiatric classification on scientific foundation: problems and prospects. *Int. Rev. Psychiatry*, 24, 6, 591–605, 2012.

47. Zanarini, M.C., Frankenburg, F.R., Reich, D.B., Fitzmaurice, G., Attainment and stability of sustained symptomatic remission and recovery among patients with borderline personality disorder and axis II comparison subjects: a 16-year prospective follow-up study. *Am. J. Psychiatry*, 169, 5, 476–83, 2012.
48. Iniesta, R., Stahl, D., McGuffin, P., Machine learning, statistical learning and the future of biological research in psychiatry. *Psychol. Med.*, 46, 12, 2455–65, 2016.
49. Han, B., Pouget, J.G., Slowikowski, K., Stahl, E., Lee, C.H., Diogo, D., *et al.*, A method to decipher pleiotropy by detecting underlying heterogeneity driven by hidden subgroups applied to autoimmune and neuropsychiatric diseases. *Nat. Genet.*, 48, 7, 803–10, 2016.
50. Schultze-Lutter, F., Schmidt, S.J., Theodoridou, A., Psychopathology—a precision tool in need of re-sharpening. *Front. Psychiatry*, 9, 446, 2018.
51. Shorter, E., The history of nosology and the rise of the Diagnostic and Statistical Manual of Mental Disorders. *Dialogues Clin. Neurosci.*, 17, 1, 59–67, 2015.
52. Surís, A., Holliday, R., North, C.S., The Evolution of the Classification of Psychiatric Disorders. *Behav. Sci. (Basel, Switzerland)*, 6, 1, 1–10, 2016.
53. Bekhuis, E., Olde Hartman, T.C., Boschloo, L., Lucassen, P.L., A new approach to psychopathology: the example of depression. *Br. J. Gen. Pract. : J. R. Coll. Gen. Pract.*, 69, 680, 146–7, 2019.
54. Kawa, S. and Giordano, J., A brief historicity of the Diagnostic and Statistical Manual of Mental Disorders: issues and implications for the future of psychiatric canon and practice. *Philos. Ethics Humanit. Med. : PEHM*, 7, 2, 2012.
55. Hengartner, M.P. and Lehmann, S.N., Why Psychiatric Research Must Abandon Traditional Diagnostic Classification and Adopt a Fully Dimensional Scope: Two Solutions to a Persistent Problem. *Front. Psychiatry*, 8, 101, 2017.
56. Hyman, S.E., The diagnosis of mental disorders: the problem of reification. *Annu. Rev. Clin. Psychol.*, 6, 155–79, 2010.
57. Krueger, R.F. and Bezdjian, S., Enhancing research and treatment of mental disorders with dimensional concepts: toward DSM-V and ICD-11. *World Psychiatry*, 8, 1, 3–6, 2009.
58. Lilienfeld, S.O., DSM-5: Centripetal scientific and centrifugal antiscientific forces. *Clin. Psychol. Sci. Pract.*, 21, 3, 269, 2014.
59. Parker, G., Beyond major depression. *Psychol. Med.*, 35, 4, 467–74, 2005.
60. Regier, D.A., Narrow, W.E., Clarke, D.E., Kraemer, H.C., Kuramoto, S.J., Kuhl, E.A., *et al.*, DSM-5 field trials in the United States and Canada, Part II: test-retest reliability of selected categorical diagnoses. *Am. J. Psychiatry*, 170, 1, 59–70, 2013.
61. Penninx, B.W., Nolen, W.A., Lamers, F., Zitman, F.G., Smit, J.H., Spinhoven, P., *et al.*, Two-year course of depressive and anxiety disorders: results from

the Netherlands Study of Depression and Anxiety (NESDA). *J. Affective Disord.*, 133, 1-2, 76–85, 2011.
62. Fried, E.I. and Nesse, R.M., Depression is not a consistent syndrome: An investigation of unique symptom patterns in the STAR*D study. *J. Affective Disord.*, 172, 96–102, 2015.
63. Dowrick, C., *Beyond depression: a new approach to understanding and management*, Oxford University Press, USA, 2009.
64. Fried, E.I. and Nesse, R.M., Depression sum-scores don't add up: why analyzing specific depression symptoms is essential. *BMC Med.*, 13, 72, 2015.
65. Stein, D.J., Shoptaw, S.J., Vigo, D.V., Lund, C., Cuijpers, P., Bantjes, J., *et al.*, Psychiatric diagnosis and treatment in the 21st century: paradigm shifts versus incremental integration. *World Psychiatry*, 21, 3, 393–414, 2022.
66. Kotov, R., Krueger, R.F., Watson, D., Achenbach, T.M., Althoff, R.R., Bagby, R.M., *et al.*, The Hierarchical Taxonomy of Psychopathology (HiTOP): A dimensional alternative to traditional nosologies. *J. Abnorm. Psychol.*, 126, 4, 454–77, 2017.
67. Kotov, R., Ruggero, C.J., Krueger, R.F., Watson, D., Yuan, Q., Zimmerman, M., New dimensions in the quantitative classification of mental illness. *Arch. Gen. Psychiatry*, 68, 10, 1003–11, 2011.
68. Wright, A.G.C. and Simms, L.J., A metastructural model of mental disorders and pathological personality traits. *Psychol. Med.*, 45, 11, 2309–19, 2015.
69. Caspi, A., Houts, R.M., Belsky, D.W., Goldman-Mellor, S.J., Harrington, H., Israel, S., *et al.*, The p factor: one general psychopathology factor in the structure of psychiatric disorders? *Clin. Psychol. Sci.*, 2, 2, 119–37, 2014.
70. Hengartner, M.P., Ajdacic-Gross, V., Wyss, C., Angst, J., Rössler, W., Relationship between personality and psychopathology in a longitudinal community study: a test of the predisposition model. *Psychol. Med.*, 46, 8, 1693–705, 2016.
71. Hengartner, M.P., Kawohl, W., Haker, H., Rössler, W., Ajdacic-Gross, V., Big Five personality traits may inform public health policy and preventive medicine: Evidence from a cross-sectional and a prospective longitudinal epidemiologic study in a Swiss community. *J. Psychosom. Res.*, 84, 44–51, 2016.
72. Hengartner, M.P., Tyrer, P., Ajdacic-Gross, V., Angst, J., Rössler, W., Articulation and testing of a personality-centred model of psychopathology: evidence from a longitudinal community study over 30 years. *Eur. Arch. Psychiatry Clin. Neurosci.*, 268, 443–54, 2018.
73. Conway, C.C., Craske, M.G., Zinbarg, R.E., Mineka, S., Pathological personality traits and the naturalistic course of internalizing disorders among high-risk young adults. *Depression Anxiety*, 33, 1, 84–93, 2016.
74. Kendler, K.S., Aggen, S.H., Knudsen, G.P., Røysamb, E., Neale, M.C., Reichborn-Kjennerud, T., The structure of genetic and environmental risk factors for syndromal and subsyndromal common DSM-IV axis I and all axis II disorders. *Am. J. Psychiatry*, 168, 1, 29–39, 2011.

75. Hankin, B.L., Davis, E.P., Snyder, H., Young, J.F., Glynn, L.M., Sandman, C.A., Temperament factors and dimensional, latent bifactor models of child psychopathology: Transdiagnostic and specific associations in two youth samples. *Psychiatry Res.*, 252, 139–46, 2017.
76. Kendler, K.S., Prescott, C.A., Myers, J., Neale, M.C., The structure of genetic and environmental risk factors for common psychiatric and substance use disorders in men and women. *Arch. Gen. Psychiatry*, 60, 9, 929–37, 2003.
77. Lahey, B.B., Van Hulle, C.A., Singh, A.L., Waldman, I.D., Rathouz, P.J., Higher-order genetic and environmental structure of prevalent forms of child and adolescent psychopathology. *Arch. Gen. Psychiatry*, 68, 2, 181–9, 2011.
78. Insel, T., Cuthbert, B., Garvey, M., Heinssen, R., Pine, D.S., Quinn, K., *et al.*, Research domain criteria (RDoC): toward a new classification framework for research on mental disorders. *Am. Psychiatr. Assoc.*, 167, 7, 748–51, 2010.
79. Cuthbert, B.N., The RDoC framework: facilitating transition from ICD/DSM to dimensional approaches that integrate neuroscience and psychopathology. *World Psychiatry*, 13, 1, 28–35, 2014.
80. Reynolds Iii, C.F., Lewis, D.A., Detre, T., Schatzberg, A.F., Kupfer, D.J., The future of psychiatry as clinical neuroscience. *Acad. Med.: J. Assoc. Am. Med. Coll.*, 84, 4, 446, 2009.
81. Wong, E.H.F., Yocca, F., Smith, M.A., Lee, C.-M., Challenges and opportunities for drug discovery in psychiatric disorders: the drug hunters' perspective. *Int. J. Neuropsychopharmacol.*, 13, 9, 1269–84, 2010.
82. Borsboom, D. and Cramer, A.O.J., Network analysis: an integrative approach to the structure of psychopathology. *Annu. Rev. Clin. Psychol.*, 9, 91–121, 2013.
83. McNally, R.J., Can network analysis transform psychopathology? *Behav. Res. Ther.*, 86, 95–104, 2016.
84. Jones, P.J., Heeren, A., McNally, R.J., Commentary: A network theory of mental disorders. *Front. Psychol.*, 8, 1305, 2017.
85. Isvoranu, A.-M., van Borkulo, C.D., Boyette, L.-L., Wigman, J.T.W., Vinkers, C.H., Borsboom, D., *et al.*, A network approach to psychosis: pathways between childhood trauma and psychotic symptoms. *Schizophr. Bull.*, 43, 1, 187–96, 2017.
86. Klippel, A., Viechtbauer, W., Reininghaus, U., Wigman, J., van Borkulo, C., Merge, *et al.*, The cascade of stress: a network approach to explore differential dynamics in populations varying in risk for psychosis. *Schizophr. Bull.*, 44, 2, 328–37, 2018.
87. Müller, V.I., Cieslik, E.C., Serbanescu, I., Laird, A.R., Fox, P.T., Eickhoff, S.B., Altered brain activity in unipolar depression revisited: meta-analyses of neuroimaging studies. *JAMA Psychiatry*, 74, 1, 47–55, 2017.
88. Button, K.S., Ioannidis, J.P.A., Mokrysz, C., Nosek, B.A., Flint, J., Robinson, E.S.J., *et al.*, Power failure: why small sample size undermines the reliability of neuroscience. *Nat. Rev. Neurosci.*, 14, 5, 365–76, 2013.

89. Eklund, A., Nichols, T.E., Knutsson, H., Cluster failure: Why fMRI inferences for spatial extent have inflated false-positive rates. *Proc. Natl. Acad. Sci.*, 113, 28, 7900–5, 2016.
90. Geetha, P., Raksha, R.V., Ishak, J.M., Shanmugasundaram P. A Review on Generalized Anxiety Disorder. *Res. J. Pharm. Technol.*, 10, 11, 4043–6, 2017.
91. Picchioni, M.M. and Murray, R.M., Schizophrenia. *BMJ* (Clinical research ed), 335, 7610, 91–5, 2007.
92. Bystritsky, A., Khalsa, S.S., Cameron, M.E., Schiffman, J., Current diagnosis and treatment of anxiety disorders. *P T : Peer-Reviewed J. Formulary Manage.*, 38, 1, 30–57, 2013.
93. Halbeisen, G., Brandt, G., Paslakis, G., A Plea for Diversity in Eating Disorders Research. *Front. Psychiatry*, 13, 820043, 2022.
94. Khaleghi, A., Mohammadi, M.R., Shahi, K., Nasrabadi, A.M., Computational Neuroscience Approach to Psychiatry: A Review on Theory-driven Approaches. *Clin. Psychopharmacol. Neurosci.: Off. Sci. J. Korean Coll. Neuropsychopharmacol.*, 20, 1, 26–36, 2022.
95. Doi, K., Taniguchi, A., Nakai, Y., Kawamura, H., Higaki, Y., Yokoi, H., *et al.*, Decreased glucose effectiveness but not insulin resistance in glucose-tolerant offspring of Japanese non-insulin-dependent diabetic patients: a minimal-model analysis. *Metab. Clin. Exp.*, 46, 8, 880–3, 1997.
96. Stephan, K.E., Bach, D.R., Fletcher, P.C., Flint, J., Frank, M.J., Friston, K.J., *et al.*, Charting the landscape of priority problems in psychiatry, part 1: classification and diagnosis. *Lancet Psychiatry*, 3, 1, 77–83, 2016.
97. Talepasand, S., Mohammadi, M.R., Alavi, S.S., Khaleghi, A., Sajedi, Z., Akbari, P., *et al.*, Psychiatric disorders in children and adolescents: prevalence and sociodemographic correlates in Semnan Province in Iran. *Asian J. Psychiatry*, 40, 9–14, 2019.
98. Khaleghi, A., Mohammadi, M.R., Zandifar, A., Ahmadi, N., Alavi, S.S., Ahmadi, A., *et al.*, Epidemiology of psychiatric disorders in children and adolescents; in Tehran, 2017. *Asian J. Psychiatry*, 37, 146–53, 2018.
99. Mohammadi, M.R., Ahmadi, N., Khaleghi, A., Mostafavi, S.A., Kamali, K., Rahgozar, M., *et al.*, Prevalence and correlates of psychiatric disorders in a national survey of Iranian children and adolescents. *Iran. J. Psychiatry*, 14, 1, 1, 2019.
100. Mohammadi, M.R., Ahmadi, N., Khaleghi, A., Zarafshan, H., Mostafavi, S.-A., Kamali, K., *et al.*, Prevalence of autism and its comorbidities and the relationship with maternal psychopathology: a national population-based study. *Arch. Iran. Med.*, 22, 10, 546–53, 2019.
101. Phillips, J., Frances, A., Cerullo, M.A., Chardavoyne, J., Decker, H.S., First, M.B., *et al.*, The six most essential questions in psychiatric diagnosis: a pluralogue part 1: conceptual and definitional issues in psychiatric diagnosis. *Philos. Ethics Humanit. Med.*, 7, 1–29, 2012.
102. Phillips, J., Frances, A., Cerullo, M.A., Chardavoyne, J., Decker, H.S., First, M.B., *et al.*, The six most essential questions in psychiatric diagnosis: a

pluralogue part 3: issues of utility and alternative approaches in psychiatric diagnosis. *Philos. Ethics Humanit. Med.*, 7, 1–15, 2012.
103. Phillips, J., Frances, A., Cerullo, M.A., Chardavoyne, J., Decker, H.S., First, M.B., *et al.*, The six most essential questions in psychiatric diagnosis: a pluralogue. Part 4: general conclusion. *Philos. Ethics Humanit. Med. : PEHM*, 7, 14, 2012.
104. Salessi, S., Aporia of power: On the crises, science, and internal dynamics of the mental health field. *Eur. J. Philos. Sci.*, 7, 2, 175–200, 2017.
105. Aboraya, A., France, C., Young, J., Curci, K., LePage, J., The validity of psychiatric diagnosis revisited: the clinician's guide to improve the validity of psychiatric diagnosis. *Psychiatry (Edgmont)*, 2, 9, 48, 2005.
106. Venkatasubramanian, G. and Keshavan, M.S., Biomarkers in psychiatry-a critique. *Ann. Neurosci.*, 23, 1, 3–5, 2016.
107. McGorry, P., Keshavan, M., Goldstone, S., Amminger, P., Allott, K., Berk, M., *et al.*, Biomarkers and clinical staging in psychiatry. *World Psychiatry*, 13, 3, 211–23, 2014.
108 Kirmayer, L.J. and Crafa, D., What kind of science for psychiatry? *Front. Hum. Neurosci.*, 8, 435, 2014.
109. Zarafshan, H., Khaleghi, A., Mohammadi, M.R., Moeini, M., Malmir, N., Electroencephalogram complexity analysis in children with attention-deficit/hyperactivity disorder during a visual cognitive task. *J. Clin. Exp. Neuropsychol.*, 38, 3, 361–9, 2016.
110. Khaleghi, A., Zarafshan, H., Mohammadi, M.R., Visual and auditory steady-state responses in attention-deficit/hyperactivity disorder. *Eur. Arch. Psychiatry Clin. Neurosci.*, 269, 645–55, 2019.
111. Khaleghi, A., Pirzad Jahromi, G., Zarafshan, H., Mostafavi, S.A., Mohammadi, M.R., Effects of transcranial direct current stimulation of prefrontal cortex on risk-taking behavior. *Psychiatry Clin. Neurosci.*, 74, 9, 455–65, 2020.
112. Khaleghi, A., Zarafshan, H., Vand, S.R., Mohammadi, M.R., Effects of Non-invasive Neurostimulation on Autism Spectrum Disorder: A Systematic Review. *Clin. Psychopharmacol. Neurosci.: Off. Sci. J. Korean Coll. Neuropsychopharmacol.*, 18, 4, 527–52, 2020.
113. Mostafavi, S.A., Khaleghi, A., Mohammadi, M.R., Noninvasive brain stimulation in alcohol craving: A systematic review and meta-analysis. *Prog. Neuropsychopharmacol. Biol. Psychiatry*, 101, 109938, 2020.
114. Castro Martínez, J.C. and Santamaría-García, H., Understanding mental health through computers: An introduction to computational psychiatry. *Front. Psychiatry*, 14, 1092471, 2023.
115. Douven, I. and Schupbach, J.N., Probabilistic alternatives to Bayesianism: the case of explanationism. *Front. Psychol.*, 6, 459, 2015.
116. Wiecki, T.V., Poland, J., Frank, M.J., Model-based cognitive neuroscience approaches to computational psychiatry: clustering and classification. *Clin. Psychol. Sci.*, 3, 3, 378–99, 2015.

117. Kucikova, L., Danso, S., Jia, L., Su, L., Computational Psychiatry and Computational Neurology: Seeking for Mechanistic Modeling in Cognitive Impairment and Dementia. *Front. Comput. Neurosci.*, 16, 865805, 2022.
118. Goecks, J., Jalili, V., Heiser, L.M., Gray, J.W., How Machine Learning Will Transform Biomedicine. *Cell*, 181, 1, 92–101, 2020.
119. Chekroud, A.M., Zotti, R.J., Shehzad, Z., Gueorguieva, R., Johnson, M.K., Trivedi, M.H., *et al.*, Cross-trial prediction of treatment outcome in depression: a machine learning approach. *Lancet Psychiatry*, 3, 3, 243–50, 2016.
120. Cui, R. and Liu, M., RNN-based longitudinal analysis for diagnosis of Alzheimer's disease. *Comput. Med. Imaging Graph.: Off. J. Comput. Med. Imaging Soc.*, 73, 1–10, 2019.
121. Kuan, V., Fraser, H.C., Hingorani, M., Denaxas, S., Gonzalez-Izquierdo, A., Direk, K., *et al.*, Data-driven identification of ageing-related diseases from electronic health records. *Sci. Rep.*, 11, 1, 2938, 2021.
122. Skolariki, K., Terrera, G.M., Danso, S.O., Predictive models for mild cognitive impairment to Alzheimer's disease conversion. *Neural Regener. Res.*, 16, 9, 1766–7, 2021.
123. Danso, S.O., Zeng, Z., Muniz-Terrera, G., Ritchie, C.W., Developing an Explainable Machine Learning-Based Personalised Dementia Risk Prediction Model: A Transfer Learning Approach With Ensemble Learning Algorithms. *Front. Big Data*, 4, 613047, 2021.
124. Bougea, A., Efthymiopoulou, E., Spanou, I., Zikos, P., A Novel Machine Learning Algorithm Predicts Dementia With Lewy Bodies Versus Parkinson's Disease Dementia Based on Clinical and Neuropsychological Scores. *J. Geriatr. Psychiatry Neurol.*, 35, 3, 317–20, 2022.
125. Dayan, P. and Abbott, L.F., *Theoretical neuroscience: computational and mathematical modeling of neural systems*, p. 480, MIT press, Cambridge, MA, 2005.
126. Baker, R.E., Peña, J.M., Jayamohan, J., Jérusalem, A., Mechanistic models versus machine learning, a fight worth fighting for the biological community? *Biol. Lett.*, 14, 5, 1–4, 2018.
127. Cassidy, C.M., Balsam, P.D., Weinstein, J.J., Rosengard, R.J., Slifstein, M., Daw, N.D., *et al.*, A Perceptual Inference Mechanism for Hallucinations Linked to Striatal Dopamine. *Curr. Biol. : CB*, 28, 4, 503–14.e4, 2018.
128. Kafadar, E., Fisher, V.L., Quagan, B., Hammer, A., Jaeger, H., Mourgues, C., *et al.*, Conditioned Hallucinations and Prior Overweighting Are State-Sensitive Markers of Hallucination Susceptibility. *Biol. Psychiatry*, 92, 10, 772–80, 2022.
129. Grujic, N., Brus, J., Burdakov, D., Polania, R., Rational inattention in mice. *Sci. Adv.*, 8, 9, eabj8935, 2022.
130. Reed, E.J., Uddenberg, S., Suthaharan, P., Mathys, C.D., Taylor, J.R., Groman, S.M., *et al.*, Paranoia as a deficit in non-social belief updating. *eLife*, 9, e56345, 1–39, 2020.

131. Bansal, S., Bae, G.Y., Robinson, B.M., Hahn, B., Waltz, J., Erickson, M., *et al.*, Association Between Failures in Perceptual Updating and the Severity of Psychosis in Schizophrenia. *JAMA Psychiatry*, 79, 2, 169–77, 2022.
132. Dawson, D.R. and Marcotte, T.D., Special issue on ecological validity and cognitive assessment. *Neuropsychol. Rehabil.*, 27, 5, 599–602, 2017.
133. Chaytor, N. and Schmitter-Edgecombe, M., The ecological validity of neuropsychological tests: a review of the literature on everyday cognitive skills. *Neuropsychol. Rev.*, 13, 4, 181–97, 2003.
134. Spooner, D.M. and Pachana, N.A., Ecological validity in neuropsychological assessment: a case for greater consideration in research with neurologically intact populations. *Arch. Clin. Neuropsychol.: Off. J. Natl. Acad. Neuropsychologists*, 21, 4, 327–37, 2006.
135. Smith, R., Taylor, S., Stewart, J.L., Guinjoan, S.M., Ironside, M., Kirlic, N., *et al.*, Slower learning rates from negative outcomes in substance use disorder over a 1-year period and their potential predictive utility, medRxiv, 2021–10, 2021.
136. Smith, R., Schwartenbeck, P., Stewart, J.L., Kuplicki, R., Ekhtiari, H., Paulus, M.P., Imprecise action selection in substance use disorder: Evidence for active learning impairments when solving the explore-exploit dilemma. *Drug Alcohol Depend.*, 215, 108208, 2020.
137. Smith, R., Kuplicki, R., Feinstein, J., Forthman, K.L., Stewart, J.L., Paulus, M.P., *et al.*, A Bayesian computational model reveals a failure to adapt interoceptive precision estimates across depression, anxiety, eating, and substance use disorders. *PLoS Comput. Biol.*, 16, 12, e1008484, 2020.
138. Smith, R., Kirlic, N., Stewart, J.L., Touthang, J., Kuplicki, R., Khalsa, S.S., *et al.*, Greater decision uncertainty characterizes a transdiagnostic patient sample during approach-avoidance conflict: a computational modelling approach. *J. Psychiatry Neurosci. : JPN*, 46, 1, E74–e87, 2021.
139. Smith, R., Kirlic, N., Stewart, J.L., Touthang, J., Kuplicki, R., McDermott, T.J., *et al.*, Long-term stability of computational parameters during approach-avoidance conflict in a transdiagnostic psychiatric patient sample. *Sci. Rep.*, 11, 1, 11783, 2021.
140. Wang, L.-H., Chen, B., Hwang, G.-J., Guan, J.-Q., Wang, Y.-Q., Effects of digital game-based STEM education on students' learning achievement: a meta-analysis. *Int. J. STEM Educ.*, 9, 1, 1–13, 2022.
141. Huys, Q.J.M., Browning, M., Paulus, M.P., Frank, M.J., Advances in the computational understanding of mental illness. *Neuropsychopharmacol.: Off. Publ. Am. Coll. Neuropsychopharmacol.*, 46, 1, 3–19, 2021.
142. Hauser, T.U., Skvortsova, V., De Choudhury, M., Koutsouleris, N., The promise of a model-based psychiatry: building computational models of mental ill health. *Lancet Digital Health*, 4, 11, e816–e28, 2022.
143. Winkler, R., Apple is working on iPhone features to help detect depression, cognitive decline. *Wall Street J.*, 4, 11, e816–e828, 2021.

144. Montague, P.R., Dolan, R.J., Friston, K.J., Dayan, P., Computational psychiatry. *Trends Cognit. Sci.*, 16, 1, 72–80, 2012.
145. Friston, K.J., Redish, A.D., Gordon, J.A., Computational nosology and precision psychiatry. *Comput. Psychiatry (Cambridge, Mass)*, 1, 2, 2017.
146. Stephan, K.E. and Mathys, C., Computational approaches to psychiatry. *Curr. Opin. Neurobiol.*, 25, 85–92, 2014.
147. Ziegler, G., Hauser, T.U., Moutoussis, M., Bullmore, E.T., Goodyer, I.M., Fonagy, P., *et al.*, Compulsivity and impulsivity traits linked to attenuated developmental frontostriatal myelination trajectories. *Nat. Neurosci.*, 22, 6, 992–9, 2019.
148. Koutsouleris, N., Kambeitz-Ilankovic, L., Ruhrmann, S., Rosen, M., Ruef, A., Dwyer, D.B., *et al.*, Prediction Models of Functional Outcomes for Individuals in the Clinical High-Risk State for Psychosis or With Recent-Onset Depression: A Multimodal, Multisite Machine Learning Analysis. *JAMA Psychiatry*, 75, 11, 1156–72, 2018.
149. Kennedy, R., Clifford, S., Burleigh, T., Waggoner, P.D., Jewell, R., Winter, N.J.G., The shape of and solutions to the MTurk quality crisis. *Polit. Sci. Res. Meth.*, 8, 4, 614–29, 2020.
150. Marek, S., Tervo-Clemmens, B., Calabro, F.J., Montez, D.F., Kay, B.P., Hatoum, A.S., *et al.*, Towards reproducible brain-wide association studies, BioRxiv, 2020–08, 2020.
151. Baker, M., 1,500 scientists lift the lid on reproducibility. *Nature*, 533, 7604, 452–4, 2016.
152. Gillan, C.M. and Rutledge, R.B., Smartphones and the Neuroscience of Mental Health. *Annu. Rev. Neurosci.*, 44, 129–51, 2021.
153. Hitchcock, P.F., Fried, E.I., Frank, M.J., Computational Psychiatry Needs Time and Context. *Annu. Rev. Psychol.*, 73, 243–70, 2022.
154. Thompson, P.M., Jahanshad, N., Ching, C.R.K., Salminen, L.E., Thomopoulos, S.I., Bright, J., *et al.*, ENIGMA and global neuroscience: A decade of large-scale studies of the brain in health and disease across more than 40 countries. *Transl. Psychiatry*, 10, 1, 100, 2020.
155. Koshiyama, D., Miura, K., Nemoto, K., Okada, N., Matsumoto, J., Fukunaga, M., *et al.*, Neuroimaging studies within Cognitive Genetics Collaborative Research Organization aiming to replicate and extend works of ENIGMA. *Hum. Brain Mapp.*, 43, 1, 182–93, 2022.
156. Durstewitz, D., Koppe, G., Meyer-Lindenberg, A., Deep neural networks in psychiatry. *Mol. Psychiatry*, 24, 11, 1583–98, 2019.
157. Mechelli, A. and Vieira, S., From models to tools: clinical translation of machine learning studies in psychosis. *npj Schizophr.*, 6, 1, 4, 2020.
158. Maia, T.V., Introduction to the series on computational psychiatry. *Clin. Psychol. Sci.*, 3, 3, 374–7, 2015.
159. Wang, X.J. and Krystal, J.H., Computational psychiatry. *Neuron*, 84, 3, 638–54, 2014.

160. Stephan, K.E., Iglesias, S., Heinzle, J., Diaconescu, A.O., Translational Perspectives for Computational Neuroimaging. *Neuron*, 87, 4, 716–32, 2015.
161. Freedman, R., Lewis, D.A., Michels, R., Pine, D.S., Schultz, S.K., Tamminga, C.A., *et al.*, The initial field trials of DSM-5: new blooms and old thorns. *Am. J. Psychiatry*, 170, 1, 1–5, 2013.
162. Gilpin, N.W., Herman, M.A., Roberto, M., The central amygdala as an integrative hub for anxiety and alcohol use disorders. *Biol. Psychiatry*, 77, 10, 859–69, 2015.
163. Insel, T., Cuthbert, B., Garvey, M., Heinssen, R., Pine, D.S., Quinn, K., *et al.*, Research domain criteria (RDoC): toward a new classification framework for research on mental disorders. *Am. J. Psychiatry*, 167, 7, 748–51, 2010.
164. Dwyer, D.B., Falkai, P., Koutsouleris, N., Machine Learning Approaches for Clinical Psychology and Psychiatry. *Annu. Rev. Clin. Psychol.*, 14, 91–118, 2018.
165. Paulus, M.P., Huys, Q.J., Maia, T.V., A Roadmap for the Development of Applied Computational Psychiatry. *Biol. Psychiatry Cognit. Neurosci. Neuroimaging*, 1, 5, 386–92, 2016.

7

Computational Neuropharmacology in Psychiatry

Amol D. Gholap[1], Pankaj R. Khuspe[2], Deepak K. Bharati[3], Sagar R. Pardeshi[1], Mohammad Dabeer Ahmad[4], ABM Sharif Hossain[5], Bhupendra G. Prajapati[6,7] and Md. Faiyazuddin[8*]

[1]*Department of Pharmaceutics, St. John Institute of Pharmacy and Research, Palghar, Maharashtra, India*
[2]*Department of Pharmaceutics, Shriram Shikshan Santhas, College of Pharmacy, Paniv, Solapur, Maharashtra, India*
[3]*Department of Pharmacology, St. John Institute of Pharmacy and Research, Palghar, Maharashtra, India*
[4]*College of Pharmacy, Jahangirabad Institute of Technology, Jahangirabad Fort, Jahangirabad, Barabanki, Uttar Pradesh, India*
[5]*Department of Biology, College of Science, Imam Mohammed Ibn Saud Islamic University, Riyadh, Kingdom of Saudi Arabia*
[6]*Department of Pharmaceutics and Pharmaceutical Technology, Shree S.K. Patel College of Pharmaceutical Education & Research, Faculty of Pharmacy, Ganpat University, Gujarat, India*
[7]*Department of Industrial Pharmacy, Faculty of Pharmacy, Silpakorn University, Nakhon Pathom, Thailand*
[8]*School of Pharmacy, Al-Karim University, Katihar, Bihar, India*

Abstract
The complexity of brain function and the interaction of neurochemical mechanisms involved in the same has induced a greater challenge regarding the treatment of mental disorders and illness. There is a requirement for a specific treatment strategy over the conventional route and dosage form to address the same. Many advancements in neuroscience have given greater benefits to the same category of patients across the globe. But dealing with vast complexities required powerful techniques for data and treatment analysis, interpretation, and future direction set up for the same. For more effective treatment selection and better prediction, the computational psychiatry

**Corresponding author*: md.faiyazuddin@gmail.com

Bhupendra Prajapati, Alok Tripathi, Rishabha Malviya and Lucy Mohapatra (eds.) Computational Neuropharmacology: Fundamentals and Clinical Aspects, (207–244) © 2025 Scrivener Publishing LLC

tool is implemented. There are a few approaches used in the same mainly data-driven, and theory-driven used for the same. Machine learning is mainly used for the data drive method which further deals with the improvisation of disease classification and selection of the effective treatment. A better prediction of the treatment outcomes can be done by using a data-driven method. The theory-driven approach is a model-driven technique used for better analysis and disease data abstraction. The impact of symptoms' contribution to the disease state is also explored in the data-driven method. On the other hand, algorithmic reinforcement learning models, biophysically realistic neural network models along with Bayesian models are used in the theory-driven approaches for the same.

***Keywords*:** Computational neuropharmacology, data, machine learning, network analysis, neuroscience, psychiatry

List of Abbreviations

fMRI	Functional magnetic resonance imaging
PTSD	Post-traumatic stress disorder
MCMC	Markov Chain Monte Carlo

7.1 Introduction

In the realm of neuroscience and psychiatry, understanding the intricate interplay between neural circuits, neurotransmitters, and complex behaviors has always been a challenge [1]. Mental illnesses such as depression, schizophrenia, bipolar disorder, and anxiety disorders have profound effects on the lives of millions worldwide. The traditional approaches to treating these disorders have often fallen short due to the complexity of the brain's functioning and the heterogeneity of their underlying causes [2, 3]. This has spurred the emergence of computational neuropharmacology techniques as a promising avenue for advancing our understanding of psychiatric disorders and improving the efficacy of their treatment [4, 5].

The human brain, a network of billions of neurons, communicates through the release and reception of neurotransmitters. These chemical messengers transmit signals across synapses, forming the basis of cognitive processes, emotions, and behaviors. Disruptions in these intricate signaling pathways are believed to contribute to the development of psychiatric disorders. However, the sheer complexity of neural networks and the nonlinear interactions among various neurotransmitters make it challenging to decipher the underlying mechanisms of these disorders through traditional experimental methods alone [6, 7]. This is where computational neuropharmacology comes into play. Combining principles from neuroscience, pharmacology, and computer science, this multidisciplinary field aims to unravel the complexities of neural signaling and pharmacological interventions using computational models [8]. These models simulate the behavior of neurons, neurotransmitters, and their receptors *in silico*, providing researchers and clinicians with invaluable insights into the effects of different drugs and potential therapeutic strategies. Through these simulations, computational neuropharmacology bridges the gap between experimental findings and clinical applications, offering a deeper understanding of how drugs impact neural circuits and behavior [9, 10].

7.2 Need for Computational Neuropharmacology in Psychiatry

The pressing need for computational neuropharmacology techniques in psychiatry. Below the description highlights the several key points and illustrated in Figure 7.1:

1. Complex Neural Circuitry: The brain's neural circuits are highly complex, with interconnected regions governing various cognitive and emotional processes [11, 12]. Traditional reductionist approaches often overlook these intricate connections, hindering our understanding of how different brain regions contribute to psychiatric disorders. Computational neuropharmacology models can simulate the activity of multiple interconnected brain regions simultaneously, shedding light on their coordinated functioning in health and disease [13, 14].
2. Heterogeneity of Disorders: Unlike many physical illnesses with well-defined biomarkers, psychiatric disorders exhibit significant heterogeneity. For instance, two individuals diagnosed with depression may present with entirely different symptoms and responses to treatment [15, 16]. This variability underscores the need for personalized approaches to treatment. Computational neuropharmacology allows researchers to model individual variability in neural circuits, aiding in the identification of personalized treatment strategies [17, 18].
3. Drug Development and Testing: Developing new drugs for psychiatric disorders is a lengthy and expensive process [18, 19]. Many potential compounds fail during

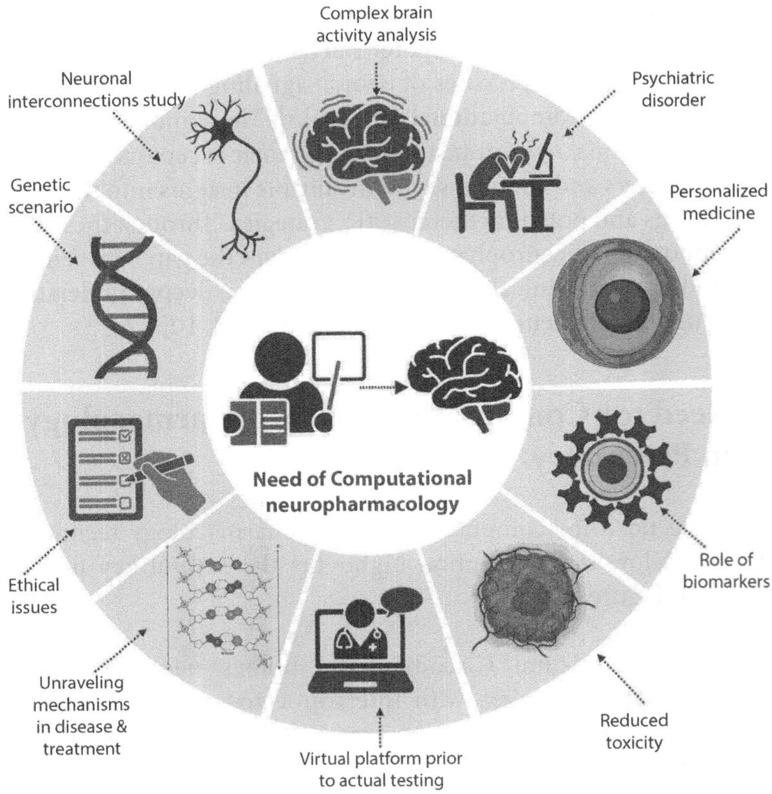

Figure 7.1 Summary of need for computational neuropharmacology in psychiatry (created by using Biorender.com).

clinical trials due to unforeseen side effects or lack of efficacy. Computational neuropharmacology offers a cost-effective platform for testing potential drugs in virtual environments before advancing to clinical trials. This predictive capability not only accelerates drug development but also reduces the risks associated with adverse effects [20–22].

4. Unraveling Mechanisms: The mechanisms underlying psychiatric disorders are often elusive, involving intricate interactions between genetics, environment, and neural activity [23]. Computational models can integrate data from various sources, helping researchers elucidate the complex relationships between genetic predisposition, neural dysfunction, and symptom manifestation. Such insights are crucial for identifying novel therapeutic targets [2, 24].

5. Ethical Considerations: Conducting experiments directly on human brains to understand psychiatric disorders is fraught with ethical challenges [25]. Computational neuropharmacology offers an ethical alternative by allowing researchers to explore various scenarios *in silico* without the need for invasive procedures. This ensures patient safety while advancing our understanding of these disorders [26, 27].

The integration of computational neuropharmacology techniques into the field of psychiatry has the potential to revolutionize our approach to understanding and treating psychiatric disorders. The complexity of neural circuits, the heterogeneity of disorders, challenges in drug development, and ethical considerations necessitate innovative solutions [28, 29]. Computational models provide a powerful tool to explore these intricacies, offering insights that traditional methods cannot. As we delve deeper into the complexities of the human brain, computational neuropharmacology stands as a beacon of hope, guiding us toward more effective treatments and a comprehensive understanding of psychiatric illnesses [30].

7.3 Data-Driven Computational Approaches in Psychiatry

In the realm of modern healthcare, the integration of cutting-edge technologies with medical practices has brought forth revolutionary advancements [31–33]. Among these, data-driven computational approaches have emerged as a transformative force, reshaping the landscape of diagnostics, treatment, and research across various medical disciplines. In particular, these approaches have ushered in a new era in the field of psychiatry, offering unprecedented insights into the complexities of mental health disorders and redefining the way clinicians approach diagnosis, treatment, and patient care [34]. The human mind, with its intricate neural networks and interwoven psychological processes, has long captivated scientists and medical professionals [35–37]. Mental health disorders, ranging from depression and anxiety to schizophrenia and bipolar disorder, represent a profound challenge due to their multifaceted nature and often enigmatic etiology. Traditional psychiatric methodologies, while valuable, often fall short in capturing the nuances and heterogeneity inherent to these disorders [38]. This is where data-driven computational approaches emerge as a game-changing paradigm, unlocking the potential to unravel the mysteries of the mind and provide personalized care on an unprecedented scale. At its core, data-driven computational approaches harness the power of advanced technologies, including machine learning, artificial intelligence, and data analytics, to analyze vast and complex datasets. These datasets encompass an array of information, ranging from genomic and neuroimaging data to clinical records and patient-reported outcomes. By processing and extracting meaningful patterns from these datasets, computational models can reveal insights that might otherwise remain hidden within the sheer volume of information [16, 39].

In the context of psychiatry, these approaches are poised to address some of the most pressing challenges that have traditionally hindered progress. The accurate diagnosis of mental health disorders has often been a delicate and subjective task, heavily reliant on clinical expertise and observable symptoms. However, data-driven computational approaches introduce objectivity and precision into the diagnostic process by identifying biomarkers, genetic factors, and neural signatures that distinguish between different disorders [40]. This not only enhances the accuracy of diagnosis but also contributes to the development of personalized treatment strategies. Furthermore, predicting treatment responses for psychiatric patients has historically been a complex and uncertain endeavor. The trial-and-error approach to finding the most effective intervention can lead to prolonged suffering and delayed recovery. In contrast, data-driven computational models leverage historical treatment response data to predict how an individual might react to specific therapies. By analyzing a patient's unique characteristics, such as genetic makeup, neural activity, and clinical history, these models provide a data-driven rationale for treatment decisions, minimizing the guesswork and optimizing the chances of successful outcomes. In the following sections, this paper delves into the multifaceted roles of data-driven computational approaches in psychiatry [41]. From refining diagnostic accuracy and predicting treatment responses to guiding treatment selection and unraveling the impact of symptoms, these approaches stand as a testament to the synergy between technological innovation and the complexities of human mental health. As we navigate the intricate pathways of the mind, data-driven computational approaches illuminate the way forward, promising a future where mental health care is not only more effective but also more personalized and empathetic [42].

7.4 Role of Diagnostic Classification

In the intricate landscape of mental health disorders, the accurate classification and diagnosis of conditions such as depression, schizophrenia, and anxiety disorders have long posed significant challenges [42, 43]. These disorders often manifest with a diverse array of symptoms and present substantial heterogeneity, making their precise identification and differentiation a complex task. Enter data-driven computational approaches, which have emerged as a powerful tool to revolutionize the role of diagnostic classification in psychiatry [44–46]. Diagnostic classification forms the foundation of effective psychiatric treatment, influencing the development of tailored interventions and patient-centered care plans. Traditionally, diagnosis in psychiatry has relied on symptomatology, clinical interviews, and expert judgment. However, this approach can be limited by subjectivity and the variability of individual experiences. Data-driven computational approaches offer a transformative shift by incorporating multidimensional data sources, such as brain imaging scans, genetic markers, and behavioral assessments [47, 48].

7.5 Machine Learning and Diagnostic Precision

One of the cornerstones of data-driven diagnostic classification is machine learning, a subset of artificial intelligence that excels at pattern recognition within large datasets [32, 49]. Machine learning algorithms are designed to detect subtle correlations and intricate relationships within complex data, often revealing insights beyond the scope of human cognition. In psychiatry, these algorithms can analyze vast amounts of diverse information to extract meaningful patterns that contribute to accurate diagnostic classification [50]. For example, consider the use of functional magnetic resonance imaging (fMRI) data in schizophrenia diagnosis. Machine learning algorithms can process and analyze brain connectivity patterns, identifying distinct neural signatures associated with different subtypes of the disorder. By recognizing these unique patterns, computational models contribute to the differentiation of subgroups that might otherwise be challenging to distinguish using traditional methods [51].

Data-driven computational approaches also play a pivotal role in identifying biomarkers, which are biological indicators that can provide crucial insights into the underlying mechanisms of mental health disorders. These biomarkers may include genetic variations, neural activity patterns, or even behavioral responses to specific stimuli. By integrating these biomarkers into diagnostic classification, data-driven models enhance the precision of identifying specific disorders and their subtypes. For instance, the identification of biomarkers associated with treatment-resistant depression has the potential to transform diagnosis and intervention strategies. Data-driven analyses of genetic data, neuroimaging findings, and clinical records can uncover distinctive patterns linked to treatment resistance. Armed with this information, clinicians can tailor treatments to target the underlying mechanisms contributing to resistance, thus increasing the likelihood of successful outcomes [47, 52, 53].

The integration of data-driven approaches into diagnostic classification holds the promise of ushering in an era of personalized psychiatry. By considering an individual's unique genetic makeup, neural architecture, and clinical history, these approaches move beyond the one-size-fits-all approach to treatment [43]. Instead, they empower clinicians with a comprehensive understanding of a patient's condition, enabling the customization of interventions based on data-derived insights. Furthermore, data-driven diagnostic classification contributes to a more objective and evidence-based practice. The inherent transparency of computational models allows clinicians to trace the reasoning behind a diagnosis, enhancing communication and trust between patients and healthcare providers [54]. Additionally, these approaches can also aid in early detection and intervention, potentially preventing the exacerbation of symptoms and promoting more favorable treatment outcomes. In the evolving landscape of psychiatry, data-driven computational approaches have emerged as a game-changer in the realm of diagnostic classification. By harnessing the power of machine learning, genetic analysis, and neuroimaging, these approaches transcend the limitations of traditional diagnostic methods. They offer a data-driven lens through which to view the complexities of mental health disorders, enabling more accurate and personalized classifications that hold the potential to reshape the future of psychiatric care. As technology continues to advance, the integration of data-driven insights into diagnostic classification is poised to lead us towards a new era of precision and efficacy in mental health treatment [45, 52].

In the realm of mental health treatment, the ability to accurately predict how an individual will respond to a particular intervention is a pivotal challenge. The intricate interplay of genetic predisposition, neural circuitry, and environmental factors makes treatment outcomes highly variable and often unpredictable. Enter data-driven computational approaches, which are poised to revolutionize the prediction of treatment response in psychiatry, offering a new paradigm for personalized and effective therapeutic strategies [55–57].

7.6 The Challenges of Treatment Response Prediction

The field of psychiatry has historically faced difficulties in predicting treatment responses for various mental health disorders. The trial-and-error approach to finding the right treatment can result in prolonged suffering, delayed recovery, and even adverse effects [58]. Traditional clinical methods rely heavily on subjective assessment and generalized treatment guidelines, which may not capture the nuances of an individual's unique profile [59].

Data-driven computational approaches, fueled by machine learning and advanced analytics, have the potential to transform treatment response prediction. By integrating diverse datasets encompassing genetics, neuroimaging, clinical history, and even lifestyle factors, these approaches leverage the wealth of available information to develop predictive models [60, 61]. Machine learning algorithms excel in processing large and complex datasets to identify patterns, trends, and correlations that might not be apparent through traditional analysis. In the context of psychiatry, these algorithms analyze the intricate relationships between various variables to create predictive models capable of estimating how a patient is likely to respond to a specific treatment [53, 57].

One of the strengths of data-driven computational approaches lies in their ability to personalize predictions. Each individual's response to treatment is influenced by a unique combination of factors, including genetic variations, neural circuitry, and past treatment history. Data-driven models take this complexity into account, generating predictions that are tailored to an individual's distinct characteristics [53, 62]. For instance, in the realm of antidepressant treatment, data-driven models can consider an individual's genetic makeup, neuroimaging findings related to brain activity, and biomarkers associated with treatment response. By analyzing historical treatment outcomes from a diverse patient population, these models can identify patterns that correlate with positive responses. The result is a predictive tool that empowers clinicians with valuable insights into the likelihood of success for different treatment options [45, 63].

Data-driven approaches not only provide predictions but also equip clinicians with evidence-based information to inform their decision-making process. Instead of relying solely on trial and error, clinicians can consult predictive models to weigh the potential benefits and risks of different treatment strategies [64]. This data-driven rationale enhances treatment planning, leading to more informed choices that increase the chances of positive outcomes. Moreover, the integration of data-driven insights into treatment response prediction fosters a collaborative approach between patients and healthcare providers. Patients can gain a deeper understanding of their treatment options, potential outcomes, and the rationale behind recommended interventions. This shared decision-making process contributes to a higher level of patient engagement, satisfaction, and treatment adherence [65, 66].

7.7 Future Implications and Ethical Considerations

As data-driven computational approaches continue to evolve, their impact on treatment response prediction extends beyond the clinical setting. These approaches hold the potential to transform clinical trials, drug development processes, and regulatory practices within psychiatry. By identifying biomarkers associated with treatment response, these models can expedite the identification of effective interventions and reduce the time and resources required for new treatment development [67]. However, the adoption of data-driven approaches also raises ethical considerations. The responsible use of patient data, ensuring transparency in model development, and addressing potential biases are paramount to maintaining the integrity of these predictive tools. Striking a balance between technological advancement and ethical practice is crucial to the successful integration of data-driven treatment response prediction in psychiatric care. Data-driven computational approaches are poised to redefine the landscape of treatment response prediction in psychiatry. By leveraging advanced analytics, machine learning, and personalized data, these approaches offer clinicians a powerful tool to enhance treatment planning, minimize the trial-and-error approach, and tailor interventions to individual patient profiles. As technology continues to advance and our understanding of mental health deepens, the integration of data-driven predictions has the potential to unlock new levels of precision, efficacy, and patient-centered care in the realm of psychiatric treatment [68, 69].

Selecting the most suitable treatment strategy for individuals grappling with mental health disorders is a multifaceted challenge [70–72]. The intricate interplay of diverse factors, including genetic makeup, neural functioning, symptom severity, and lifestyle, makes the process of treatment selection complex and nuanced [73]. In this dynamic landscape, data-driven computational approaches emerge as a beacon of hope, offering a transformative way to guide the selection of treatments that are tailored to each individual's unique profile [74]. The field of psychiatry often confronts the complexities of identifying the most effective treatment option for a given individual. Mental health disorders can manifest in a multitude of ways, and no two individuals experience these conditions in precisely the same manner. As a result, treatment responses vary significantly, and the traditional one-size-fits-all approach may fall short in optimizing outcomes [75, 76].

Data-driven computational approaches rise to the challenge by integrating an array of data sources to inform treatment selection. These sources encompass a wide spectrum, including genetic information, neuroimaging data, clinical history, biomarkers, and even real-time patient-reported outcomes [77, 78]. By processing and analyzing these multidimensional datasets, computational models can unearth patterns and relationships that inform the decision-making process [79].

7.8 Machine Learning for Informed Decisions

At the heart of data-driven treatment selection lies machine learning, a branch of artificial intelligence capable of identifying intricate patterns within large datasets. Machine learning algorithms excel at recognizing nonlinear correlations and hidden trends that might elude conventional analysis. In the realm of psychiatry, these algorithms scrutinize the relationships between variables to create predictive models that guide the selection of treatments. For example, consider an individual diagnosed with anxiety disorder [34]. By analyzing a dataset that includes genetic markers, neuroimaging findings, and clinical assessments, a machine learning algorithm can identify patterns that correlate with positive treatment outcomes for specific interventions. These patterns provide clinicians with a data-driven rationale for recommending treatments that align with the individual's profile [80]. One of the remarkable strengths of data-driven computational approaches is their capacity for personalization [81–84].

Recognizing that treatment effectiveness is influenced by a diverse array of individual factors, these approaches tailor treatment recommendations to each person's unique attributes. By factoring in genetics, neural activity, lifestyle, and more, computational models generate insights that foster personalized interventions [85, 86]. For instance, the integration of genetic data into treatment selection is a promising avenue. Genetic markers can offer clues about an individual's potential response to medications and therapies. By analyzing a patient's genetic profile, computational models can predict how well a specific drug might work, minimizing the risk of adverse effects and maximizing therapeutic benefit [87].

Data-driven approaches to treatment selection not only empower clinicians but also foster collaborative decision-making between healthcare providers and patients. Patients become active participants in their treatment journeys, armed with comprehensive information about the potential benefits and risks of different interventions. This shared decision-making process enhances patient engagement, satisfaction, and adherence to treatment plans. Furthermore, the transparency of data-driven models allows clinicians to explain the rationale behind treatment recommendations, bridging the communication gap and building trust with patients. By understanding the data-derived basis for treatment choices, patients can make informed decisions aligned with their preferences and values [88, 89].

While data-driven computational approaches hold immense promise, their integration into treatment selection is not without challenges. Ethical considerations surrounding data privacy, potential biases in model development, and the need for rigorous validation are paramount. Striking a balance between technological advancement and ethical practice remains crucial to ensure the responsible and effective implementation of these approaches. As the field of psychiatry continues to evolve, data-driven treatment selection has the potential to revolutionize clinical practice. By leveraging the power of machine learning, genetics, neuroimaging, and patient data, clinicians can transcend the limitations of the traditional trial-and-error approach, offering patients treatments that are optimized for their unique needs. The fusion of data-driven insights with compassionate care opens doors to a future where mental health interventions are not only more effective but also more tailored and empathetic [34, 90].

In the intricate tapestry of mental health disorders, the profound impact of symptoms on an individual's well-being and functioning cannot be overstated. Symptoms, ranging from mood fluctuations and cognitive impairments to behavioral changes, serve as the hallmark indicators of these disorders [91–94]. However, comprehending the intricate relationships and cascading effects of these symptoms is a complex endeavor. Enter data-driven computational approaches, which illuminate the multifaceted impact of symptoms and offer a novel lens through which to understand the dynamics of mental health [40, 95, 96]. In the realm of mental health, symptoms are not isolated entities but rather interconnected manifestations that often exert influence on one another. Traditionally, clinical understanding of symptoms has been rooted in their individual presentation. However, this approach may overlook the complex interactions and dependencies that contribute to the overall experience of a disorder [97, 98].

7.9 Network Analysis: Unraveling Symptom Dynamics

Data-driven computational approaches introduce the concept of network analysis to the study of symptoms. Network analysis is a technique that visualizes and quantifies the relationships between symptoms, depicting them as interconnected nodes in a network. By mapping these relationships, data-driven models provide insights into the central symptoms that hold pivotal roles within the network [99, 100]. For instance, in depression, network analysis might reveal that disturbances in sleep, loss of interest, and feelings of hopelessness are central symptoms that drive the manifestation of other symptoms. This information is invaluable in treatment planning, as targeting these central symptoms may disrupt the network and lead to symptom alleviation [98]. Data-driven approaches extend beyond static representations of symptom networks by incorporating longitudinal data analysis. Understanding how symptoms evolve over time offers critical insights into the course of a disorder, the effectiveness of interventions, and the identification of potential triggers or exacerbating factors [101]. For example, by analyzing data collected at multiple time points, computational models can track the trajectory of symptoms in schizophrenia. Longitudinal analysis may reveal patterns of symptom exacerbation before relapses or shed light on the impact of medication changes on symptom severity. This knowledge empowers clinicians to make more informed decisions and adapt treatment strategies based on real-time data [102, 103].

The understanding of symptom networks and their dynamics has direct implications for treatment planning [104–106]. Rather than solely targeting individual symptoms, data-driven approaches allow clinicians to identify interventions that disrupt the central nodes within the symptom network. By doing so, treatments may have a cascading effect, alleviating multiple symptoms and promoting holistic well-being [81, 107, 108]. Consider an individual with post-traumatic stress disorder (PTSD) who experiences hyperarousal, nightmares, and emotional numbness [97, 109]. Data-driven models might reveal that emotional numbness serves as a central node, influencing the intensity of other symptoms. Tailoring interventions to specifically address emotional numbness could lead to a ripple effect, diminishing other symptoms and ultimately improving the individual's quality of life. While data-driven computational approaches hold immense promise for understanding the impact of symptoms, ethical considerations are paramount. Protecting patient privacy, ensuring transparent model development, and addressing potential biases are essential to maintain the integrity and responsible use of these approaches. Additionally, clinical integration requires collaboration between data scientists, researchers, and clinicians. Data-driven insights complement rather than replace clinical expertise, emphasizing the importance of a multidisciplinary approach in translating these insights into meaningful patient care.

Data-driven computational approaches have the potential to transform our understanding of the impact of symptoms in mental health disorders. By applying network analysis and longitudinal data examination, these approaches reveal the intricate relationships between symptoms and their cascading effects. Through this enhanced understanding, clinicians can devise personalized interventions that address central symptoms, leading to comprehensive symptom alleviation and improved overall well-being. As the synergy between technology and mental health deepens, data-driven insights offer a promising avenue toward more effective, patient-centered psychiatric care [98, 99, 102]. The applications of data driven, computational tools techniques for computational neuropharmacology are presented in Figure 7.2.

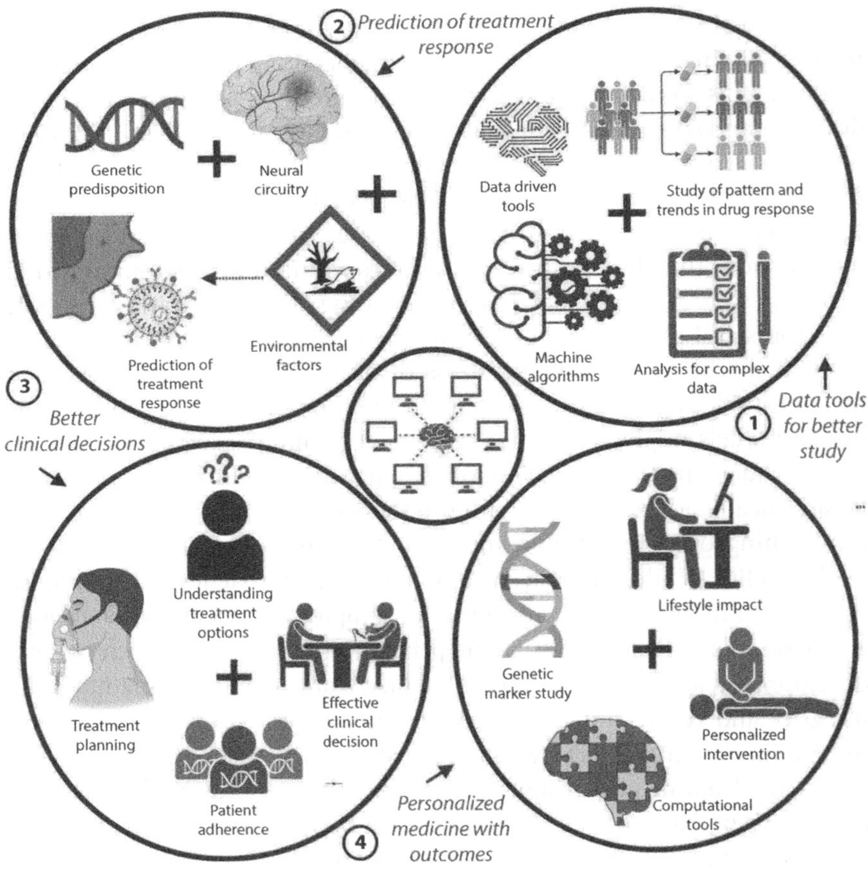

Figure 7.2 Applications of data driven technology and computational tools for computational neuropharmacology. It used for 1. Better study of selected case studies 2. Predication of treatment response 3. Effective clinical decision 4. Personalized medicines with anticipated outcomes (created by using Biorender.com).

7.10 Theory-Driven Computational Approaches: Integrating Knowledge and Data

In the dynamic realm of scientific inquiry, the interplay between theory and data has long been a cornerstone of knowledge advancement. This synergy is particularly pronounced in theory-driven computational approaches, where established theoretical frameworks guide the exploration of complex phenomena through the lens of data analysis. In diverse fields, from physics to biology, theory-driven approaches provide a structured pathway for interpreting data, generating insights, and expanding the boundaries of understanding. In this narrative, we delve into the essence of theory-driven computational approaches and their significance across various domains [110]. At the heart of theory-driven computational approaches lies the notion that theories serve as conceptual frameworks that organize and explain phenomena. These theories are often grounded in existing knowledge and observations. When combined with computational methodologies, theories gain the power to be tested, validated, and expanded using vast datasets. This integration bridges the gap between conceptual understanding and empirical evidence, facilitating deeper insights into complex systems [111–113].

Theory-driven computational approaches provide a roadmap for data exploration. Instead of treating data as raw inputs, these approaches guide the analysis by focusing on specific variables, relationships, and patterns derived from existing theories. This guidance is especially valuable in dealing with large datasets, where the risk of data overload can lead to confusion or spurious correlations. For instance, in ecology, theoretical models might predict the impact of certain environmental variables on species distribution. By incorporating these theories into computational models, ecologists can sift through ecological data more effectively, revealing patterns that align with the underlying theoretical framework [114–116]. The integration of theories into computational analyses allows for rigorous hypothesis testing and validation. Theories provide testable predictions that can be evaluated against real-world data. The congruence or divergence between theoretical predictions and empirical findings informs the refinement and advancement of theories. In the context of genetics, for example, a theoretical model might predict the distribution of genetic variations within populations. By comparing these predictions with genetic data from diverse populations, researchers can assess the validity of the theory and refine its assumptions to better fit observed patterns [102, 113, 114].

Despite their advantages, theory-driven computational approaches come with challenges. Theoretical models might oversimplify complex systems, and assumptions embedded within theories could introduce bias into the analysis. Additionally, not all phenomena are well-described by existing theories, which can limit the applicability of this approach in some cases. Moreover, striking the right balance between theory and data can be delicate. Over-reliance on theories might hinder the discovery of unexpected patterns, while excessive data-driven analysis could lead to data dredging without meaningful interpretation. Theory-driven computational approaches epitomize the marriage of knowledge and data, offering a structured framework for data analysis, hypothesis testing, and theory advancement. These approaches empower researchers to navigate the complexities of diverse fields, from natural sciences to social sciences, by integrating established theories with the power of computational methodologies. As technology advances and our understanding of complex systems deepen, the integration of theory-driven computational approaches holds the promise of enriching our knowledge landscape and driving innovation in scientific exploration [117, 118].

7.11 Biophysically Realistic Neural Network Models: Bridging the Gap Between Biology and Computation

In the quest to unravel the mysteries of the brain's intricate workings, biophysically realistic neural network models stand as a bridge between the biological complexity of neurons and the computational power of modern technology [119]. These models are designed to capture the physiological and anatomical intricacies of neural systems, enabling researchers to simulate and understand the behaviors of neural networks with a level of detail that approaches the elegance of nature itself. In this exploration, we delve into the essence of biophysically realistic neural network models, their significance, and their potential to transform our understanding of brain function [120].

The brain, composed of billions of neurons connected by trillions of synapses, exhibits a level of complexity that has fascinated scientists for centuries [121, 122]. While traditional artificial neural network models capture certain aspects of neural behavior, they often fall short in emulating the biological intricacies of neurons, ion channels, and dendritic processes. Biophysically realistic models, on the other hand, aim to replicate the physiological behavior of neurons as closely as possible, embracing the principles of biophysics, electrophysiology, and neuroanatomy [123, 124].

At the core of biophysically realistic neural network models lies the incorporation of neuronal dynamics. These models simulate the dynamic interplay of ion channels, neurotransmitter release, and electrical potentials within neurons. By integrating Hodgkin-Huxley type equations that describe the biophysical properties of ion channels, these models can replicate the firing patterns, refractory periods, and adaptation mechanisms observed in real neurons [125, 126]. For instance, a biophysically realistic model of a cortical neuron might incorporate equations representing the opening and closing of ion channels responsible for action potentials. This allows researchers to investigate how changes in channel conductance or synaptic inputs impact the neuron's firing behavior [127, 128]. The beauty of biophysically realistic neural network models lies in their ability to illuminate emergent properties that arise from the interaction of individual neurons within a network. These models can simulate the propagation of action potentials, the synchronization of neural firing, and the emergence of oscillatory patterns, providing insights into how neural networks process information and generate complex behaviors [129, 130]. In the study of epilepsy, for example, biophysically realistic models can simulate the spread of abnormal electrical activity across a network of neurons. By manipulating parameters such as synaptic strength or connectivity, researchers can explore how seizures might initiate and propagate within the brain [131].

The impact of biophysically realistic neural network models extends beyond the laboratory, offering potential applications in clinical research and therapeutic interventions. These models provide a platform for investigating the underlying mechanisms of neurological disorders and testing potential treatments [132]. For instance, in Parkinson's disease, biophysically realistic models can simulate the activity of the basal ganglia and explore how disruptions in the balance of excitatory and inhibitory inputs lead to motor symptoms. These models could aid in optimizing deep brain stimulation parameters or developing novel interventions [133, 134].

While biophysically realistic neural network models hold immense promise, they are not without challenges. Constructing these models requires a deep understanding of biophysics, as well as the availability of detailed experimental data for model parameterization. The computational complexity of these models can also be demanding, necessitating high-performance computing resources. As technology evolves, efforts are being made to refine these models by incorporating even more biological detail, such as the influence of neuromodulators or dendritic computations. Moreover, the integration of these models with experimental techniques like optogenetics and *in vivo* recordings holds the potential to provide a more holistic view of neural network function [135].

Biophysically realistic neural network models embody the intersection of biology and computation, offering a tantalizing glimpse into the underlying principles of neural systems. By capturing the nuanced behaviors of neurons and networks, these models empower researchers to explore the intricate dance of brain function, from basic mechanisms to complex behaviors. As our understanding of the brain deepens and technology advances, the integration of biophysically realistic models promises to illuminate the enigmatic workings of the brain and revolutionize our approach to neurological research and clinical interventions [122, 136].

In the realm of artificial intelligence, the marriage of theory and computation has given rise to a powerful paradigm: theory-driven computational approaches. Within this framework, algorithmic reinforcement learning models stand out as a prime example [137, 138]. These models, grounded in the foundations of reinforcement learning theory, harness the computational prowess of modern technology to develop intelligent agents capable of learning and decision-making. At the heart of algorithmic reinforcement learning models lies a fundamental theory of learning and decision-making. Reinforcement learning is rooted in the idea that an agent interacts with an environment, takes actions, and learns from feedback in the form of rewards or penalties. This theoretical framework provides a blueprint for learning optimal strategies through trial and error. In this exploration, we delve into the essence of algorithmic reinforcement learning models, their significance, and their potential to revolutionize diverse applications. Algorithmic reinforcement learning models leverage the computational capabilities of modern hardware to implement and execute the theoretical principles of reinforcement learning [138]. These models employ algorithms that enable agents to learn from data, optimize decision-making, and adapt their strategies to changing environments. By combining the theoretical understanding of reinforcement learning with computational efficiency, these models offer a dynamic approach to problem-solving [139, 140].

In algorithmic reinforcement learning, agents interact with an environment, taking actions and receiving feedback. The agent's goal is to learn a policy—a strategy that maximizes cumulative rewards over time. This learning process involves exploration (trying new actions) and exploitation (choosing known actions with high expected rewards). Algorithms guide this balance, ensuring that agents explore sufficiently to discover optimal strategies while also exploiting learned knowledge to make efficient decisions [137, 141, 142].

Algorithmic reinforcement learning models find applications across a wide spectrum of domains. From robotics to finance, healthcare to gaming, these models are employed to tackle complex problems where decision-making and learning from data is paramount [138, 143]. For instance, in robotics, reinforcement learning algorithms can train robotic agents to perform tasks such as grasping objects or walking. Through interactions with the environment, the robot learns motor control policies that optimize task performance [144]. Algorithmic reinforcement learning is not without challenges. The exploration-exploitation trade-off can be delicate, and learning optimal policies in high-dimensional state and action spaces can be computationally demanding. Theoretical understanding guides the development of algorithms that address these challenges, but there's a constant need to strike a balance between the complexity of the problem and the resources available. Recent advances in deep reinforcement learning, where neural networks are used to approximate policies and value functions, have shown remarkable success in complex tasks. These advances demonstrate the potential of algorithmic reinforcement learning to handle intricate scenarios with real-world implications [145].

7.12 Bayesian Models

In the realm of probabilistic reasoning and inference, Bayesian models shine as a prominent example of theory-driven computational approaches. These models elegantly integrate Bayesian probability theory with the power of computational methods, offering a framework to quantitatively represent uncertainty, update beliefs, and make informed decisions. Within this exploration, we delve into the essence of Bayesian models, their significance, and their transformative potential across various domains. At the core of Bayesian models lies the Bayesian probability theory, a principled approach to reasoning under uncertainty. This theory centers on the concept of updating beliefs based on new evidence, encapsulated by Bayes' theorem. It provides a formal mechanism to quantify uncertainty, combining prior knowledge (prior probabilities) with observed data (likelihood) to yield updated beliefs (posterior probabilities) [146].

Bayesian models leverage the computational capabilities of modern technology to implement the principles of Bayesian probability theory. These models encompass a wide range of techniques, from simple ones like Bayes' theorem to complex methods like Markov Chain Monte Carlo (MCMC) and Variational Inference. By combining theoretical foundations with computational prowess, Bayesian models offer a versatile approach to decision-making and inference [147, 148].

Bayesian models find applications across diverse domains, from finance and economics to healthcare, natural language processing, and more.

Their ability to handle uncertainty, incorporate prior knowledge, and provide probabilistic outputs makes them versatile tools for decision-making and prediction. For instance, in finance, Bayesian models can be used to estimate the uncertainty in stock prices, incorporate expert opinions, and generate probabilistic forecasts that guide investment strategies. Bayesian models are not without challenges. They can be computationally intensive, particularly when dealing with high-dimensional data or complex probabilistic structures. Balancing the trade-off between model complexity and computational efficiency is an ongoing concern [149]. Advances in computational methods, such as scalable variational inference and probabilistic programming languages, are addressing these challenges and expanding the scope of problems that Bayesian models can tackle. Bayesian models epitomize the fusion of theory and computation, offering a robust framework for probabilistic reasoning, inference, and decision-making. By embracing the principles of Bayesian probability theory and harnessing computational advancements, these models empower researchers and practitioners to tackle uncertainty and make informed choices across diverse domains. As technology continues to evolve and our understanding of uncertainty deepens, Bayesian models hold the promise of unlocking new dimensions of insights and transforming the way we approach complex problems [146, 150].

7.13 Combining Data-Driven and Theory-Driven Computational Approaches

In the ever-evolving landscape of scientific exploration and problem-solving, two powerful paradigms have emerged: data-driven and theory-driven computational approaches [151]. These approaches, while distinct in their origins and methodologies, hold the potential for a transformative synergy when combined. The fusion of data-driven insights and theoretical frameworks can unlock a new realm of understanding, innovation, and informed decision-making across a spectrum of domains. In this exploration, we delve into the significance of combining these approaches and the unique strengths they bring to the table [152].

Data-driven and theory-driven computational approaches stem from different philosophical roots, each offering distinct perspectives on how knowledge is acquired and applied. Data-driven approaches, fueled by the proliferation of big data and advanced analytics, focus on extracting patterns, correlations, and trends from vast datasets [153, 154]. These approaches often lack preconceived notions and are driven solely by the information contained within the data itself. On the other hand, theory-driven approaches are grounded in existing knowledge, often in the form of well-established theories and principles. These approaches seek to explain and predict phenomena based on foundational understanding, which guides the formulation of hypotheses and predictions [155].

Data-driven approaches are adept at uncovering hidden patterns and relationships within complex datasets, often revealing insights that may not be apparent through conventional analysis. These approaches excel in scenarios where the underlying mechanisms are not well understood or when the system's behavior is nonlinear and intricate. For example, in the field of genomics, data-driven approaches can identify genetic markers associated with diseases by analyzing large-scale genetic datasets. These markers might not have been initially hypothesized but can lead to novel insights into disease susceptibility [149, 152, 156]. Theory-driven approaches, while rooted in established principles, can provide valuable context and structure to data analysis. They guide the formulation of hypotheses and help interpret data within a broader framework. These approaches are particularly advantageous when dealing with sparse data or when the available data needs to be interpreted in light of existing knowledge. In the realm of climate science, for instance, theoretical models can provide a framework to understand observed climate patterns. Data-driven approaches can then validate and refine these models by comparing their predictions with real-world data [154–156].

The convergence of data-driven and theory-driven approaches capitalizes on the strengths of both paradigms. Data-driven insights can inspire new hypotheses or challenge existing theories, prompting researchers to explore new avenues of inquiry. Theory-driven frameworks can guide the interpretation of data, providing context and meaning to complex patterns. Consider neuroscience, where the combination of theoretical models of neural behavior with real-world brain imaging data can enhance our understanding of cognitive processes. Theoretical models might predict neural responses to certain stimuli, and data-driven analysis can validate or refine these predictions using observed brain activity [49, 157].

The combination of these approaches offers a holistic approach to discovery, enabling researchers to move beyond the limitations of each paradigm individually. This synergy not only enhances the robustness of findings but also fosters creativity, innovation, and interdisciplinary collaboration. However, challenges exist, such as reconciling conflicting insights from data and theory or integrating disparate data sources into theoretical frameworks [158]. Striking the right balance between data exploration and theoretical grounding requires a multidisciplinary perspective and careful consideration. The marriage of data-driven and theory-driven computational approaches marks a pivotal shift in our approach to understanding complex phenomena and making informed decisions [159]. By capitalizing on the strengths of both paradigms, researchers and practitioners can harness the power of data to inform and refine theoretical frameworks, while also using theory to guide the interpretation of data. This symbiotic relationship holds the promise of accelerating innovation, deepening insights, and ultimately advancing our understanding of the intricate world around us. As technology continues to evolve, the synergy of data and theory stands as a beacon of progress and discovery across a multitude of disciplines [160].

7.14 Conclusion

In the ever-evolving landscape of scientific inquiry and problem-solving, the convergence of diverse computational approaches has unveiled new dimensions of understanding, innovation, and progress. Whether in the realms of neural networks, Bayesian models, or the fusion of data-driven and theory-driven paradigms, each approach offers a unique lens through which to explore complex phenomena. The power of computation, harnessed by these methodologies, has illuminated pathways to discovery that were previously inaccessible. Biophysically realistic neural network models bridge the gap between biological intricacies and computational capabilities, enabling us to simulate and comprehend the complexities of neural systems with unprecedented detail. These models hold the key to unraveling the mysteries of brain function, informing clinical interventions, and fostering advancements across domains. Bayesian models, firmly rooted in probability theory and driven by computational prowess, have redefined probabilistic reasoning, offering a structured framework to quantify uncertainty, make informed decisions, and learn from data. These models empower us to tackle problems across diverse fields, from finance to healthcare, with a sophisticated understanding of uncertainty. The integration of data-driven and theory-driven computational approaches exemplifies the power of synergy. By combining the insights gleaned from vast datasets with the guiding principles of established theories, we usher in a new era of discovery. This synergy fuels innovation, enriches understanding, and propels us toward solutions that bridge the gap between theory and practice. The computational approaches stand as beacons of progress, illuminating the intricate tapestry of our world. The marriage of theory and computation offers a pathway to innovation that is greater than the sum of its parts. As technology evolves and our understanding deepens, these approaches continue to push boundaries, empowering us to unravel complex mysteries, make informed decisions, and shape a future illuminated by knowledge and discovery.

References

1. Duman, R.S. and Aghajanian, G.K., Synaptic Dysfunction in Depression: Potential Therapeutic Targets. *Science (80-.), 338*, 6103, 68–72, 2012. Available from: https://www.science.org/doi/10.1126/science.1222939.
2. Insel, T.R., Rethinking Schizophrenia. *Nature, 468*, 7321, 187–193, 2010. Available from: https://www.nature.com/articles/nature09552.
3. Le-Niculescu, H., Patel, S.D., Bhat, M., Kuczenski, R., Faraone, S.V., Tsuang, M.T., McMahon, F.J., Schork, N.J., Nurnberger, J.I., Niculescu, A.B., Convergent Functional Genomics of Genome-Wide Association Data for Bipolar Disorder: Comprehensive Identification of Candidate Genes, Pathways and Mechanisms. *Am. J. Med. Genet. Part B Neuropsychiatr. Genet., 150B*, 2, 155–181, 2009. Available from: https://onlinelibrary.wiley.com/doi/10.1002/ajmg.b.30887.
4. Reddy, D.S., Neurosteroids, in: *Progress in Brain Research*, I. Savic (ed.), vol. 186, pp. 113–137, 2020. [Internet]. Available from: https://linkinghub.elsevier.com/retrieve/pii/B9780444536303000087.
5. Belmaker, R.H. and Bersudsky, Y., Mechanism of atypicality of antipsychotic drugs. *Prog. Neuropsychopharmacol. Biol. Psychiatry*, 27, 7, 1067–1069, 2003 Oct. [Internet]. Available from: https://linkinghub.elsevier.com/retrieve/pii/S027858460300215X.
6. Roth, B.L., DREADDs for Neuroscientists. *Neuron, 89*, 4, 683–694, 2016. Available from: https://linkinghub.elsevier.com/retrieve/pii/S0896627316000659.
7. Durstewitz, D., Vittoz, N.M., Floresco, S.B., Seamans, J.K., Abrupt Transitions between Prefrontal Neural Ensemble States Accompany Behavioral Transitions during Rule Learning. *Neuron, 66*, 3, 438–448, 2010. Available from: https://linkinghub.elsevier.com/retrieve/pii/S0896627310002321.
8. Gandolfi, D., Boiani, G.M., Bigiani, A., Mapelli, J., Modeling Neurotransmission: Computational Tools to Investigate Neurological Disorders. *Int. J. Mol. Sci.*, 22, 9, 4565, 2021 Apr 27. [Internet]. Available from: https://www.mdpi.com/1422-0067/22/9/4565.
9. Okano, H., Sasaki, E., Yamamori, T., Iriki, A., Shimogori, T., Yamaguchi, Y., *et al.*, Brain/MINDS: A Japanese National Brain Project for Marmoset Neuroscience. *Neuron*, 92, 3, 582–590, 2016 Nov. [Internet]. Available from: https://linkinghub.elsevier.com/retrieve/pii/S089662731630719X.
10. Clark, L., Chamberlain, S.R., Sahakian, B.J., Neurocognitive Mechanisms in Depression: Implications for Treatment. *Annu. Rev. Neurosci., 32*, 1, 57–74,

2009. Available from: https://www.annualreviews.org/doi/10.1146/annurev.neuro.31.060407.125618.

11. Friston, K.J., Functional and Effective Connectivity: A Review. *Brain Connect.*, *1*, 1, 13–36, 2011. Available from: http://www.liebertpub.com/doi/10.1089/brain.2011.0008.
12. Lipton, J.O. and Sahin, M., The Neurology of MTOR. *Neuron*, *84*, 2, 275–291, 2014. Available from: https://linkinghub.elsevier.com/retrieve/pii/S0896627314008927.
13. Tavares, R.M., Mendelsohn, A., Grossman, Y., Williams, C.H., Shapiro, M., Trope, Y., Schiller, D.A., Map for Social Navigation in the Human Brain. *Neuron*, *87*, 1, 231–243, 2015. Available from: https://linkinghub.elsevier.com/retrieve/pii/S0896627315005243.
14. Honey, C.J., Sporns, O., Cammoun, L., Gigandet, X., Thiran, J.P., Meuli, R., Hagmann, P., Predicting Human Resting-State Functional Connectivity from Structural Connectivity. *Proc. Natl. Acad. Sci.*, *106*, 6, 2035–2040, 2009. Available from: https://pnas.org/doi/full/10.1073/pnas.0811168106.
15. Kessler, R.C. and Bromet, E.J., The Epidemiology of Depression Across Cultures. *Annu. Rev. Public Health*, *34*, 1, 119–138, 2013. Available from: https://www.annualreviews.org/doi/10.1146/annurev-publhealth-031912-114409.
16. Wang, A.M., Pradhan, S., Coughlin, J.M., Trivedi, A., DuBois, S.L., Crawford, J.L., Sedlak, T.W., Nucifora, F.C., Nestadt, G., Nucifora, L.G., *et al.*, Assessing Brain Metabolism With 7-T Proton Magnetic Resonance Spectroscopy in Patients With First-Episode Psychosis. *JAMA Psychiatry*, *76*, 3, 314, 2019. Available from: http://archpsyc.jamanetwork.com/article.aspx?doi=10.1001/jamapsychiatry.2018.3637.
17. Kim, S.W., Grant, J.E., Adson, D.E., Shin, Y.C., Zaninelli, R.A., Double-Blind Placebo-Controlled Study of the Efficacy and Safety of Paroxetine in the Treatment of Pathological Gambling. *J. Clin. Psychiatry*, *63*, 6, 501–507, 2002. Available from: http://article.psychiatrist.com/?ContentType=START&ID=10001511.
18. Plans, L., Nieto, E., Benabarre, A., Vieta, E., Completed Suicide in Bipolar Disorder Patients: A Cohort Study after First Hospitalization. *J. Affect. Disord.*, *257*, 340–344, 2019. Available from: https://linkinghub.elsevier.com/retrieve/pii/S0165032719304094.
19. Leeson, P.D. and Springthorpe, B., The Influence of Drug-like Concepts on Decision-Making in Medicinal Chemistry. *Nat. Rev. Drug Discov.*, *6*, 11, 881–890, 2007. Available from: https://www.nature.com/articles/nrd2445.
20. Giralt, M., Albaladejo, R., Tarro, L., Moriña, D., Arija, V., Solà, R., A Primary-School-Based Study to Reduce Prevalence of Childhood Obesity in Catalunya (Spain) - EDAL-Educació En Alimentació: Study Protocol for a Randomised Controlled Trial. *Trials*, *12*, 1, 54, 2011. Available from: https://trialsjournal.biomedcentral.com/articles/10.1186/1745-6215-12-54.
21. Sillitoe, I., Cuff, A.L., Dessailly, B.H., Dawson, N.L., Furnham, N., Lee, D., Lees, J.G., Lewis, T.E., Studer, R.A., Rentzsch, R., *et al.*, New Functional

Families (FunFams) in CATH to Improve the Mapping of Conserved Functional Sites to 3D Structures. *Nucleic Acids Res.*, *41*, D490–D498, 2012, doi: 10.1093/nar/gks1211.
22. Kola, I. and Landis, J., Can the Pharmaceutical Industry Reduce Attrition Rates? *Nat. Rev. Drug Discov.*, *3*, 8, 711–716, 2004. Available from: https://www.nature.com/articles/nrd1470.
23. Geschwind, D.H. and Flint, J., Genetics and Genomics of Psychiatric Disease. *Science (80-.)*, *349*, 6255, 1489–1494, 2015. Available from: https://www.science.org/doi/10.1126/science.aaa8954.
24. Bullmore, E. and Sporns, O., The Economy of Brain Network Organization. *Nat. Rev. Neurosci.*, *13*, 5, 336–349, 2012. Available from: https://www.nature.com/articles/nrn3214.
25. Nixon, R.A., Autophagy in Neurodegenerative Disease: Friend, Foe or Turncoat? *Trends Neurosci.*, *29*, 9, 528–535, 2006. Available from: https://linkinghub.elsevier.com/retrieve/pii/S0166223606001469.
26. Csermely, P., London, A., Wu, L.-Y., Uzzi, B., Structure and Dynamics of Core/Periphery Networks. *J. Complex Netw.*, *1*, 2, 93–123, 2013. Available from: https://academic.oup.com/comnet/article-lookup/doi/10.1093/comnet/cnt016.
27. De Schutter, E., Why Are Computational Neuroscience and Systems Biology So Separate? *PLoS Comput. Biol.*, *4*, 5, e1000078, 2008. Available from: https://dx.plos.org/10.1371/journal.pcbi.1000078.
28. Csermely, P., Agoston, V., Pongor, S., The Efficiency of Multi-Target Drugs: The Network Approach Might Help Drug Design. *Trends Pharmacol. Sci.*, *26*, 4, 178–182, 2005. Available from: https://linkinghub.elsevier.com/retrieve/pii/S0165614705000556.
29. Insel, T.R., Psychiatry as a Clinical Neuroscience Discipline. *JAMA*, *294*, 17, 2221, 2005. Available from: http://jama.jamanetwork.com/article.aspx?doi=10.1001/jama.294.17.2221.
30. Bullmore, E.T. and Bassett, D.S., Brain Graphs: Graphical Models of the Human Brain Connectome. *Annu. Rev. Clin. Psychol.*, *7*, 1, 113–140, 2011. Available from: https://www.annualreviews.org/doi/10.1146/annurev-clinpsy-040510-143934.
31. Gholap, A., Pardeshi, S., Giram, P., Preparation, Antibacterial and Antiviral Activity Measurements and Detection Methods, in: *Antibacterial and Antiviral Functional Materials,* [Internet]. p. 33–64, ACS Publications, 2023. Available from: https://pubs.acs.org/doi/abs/10.1021/bk-2023-1458.ch002.
32. Gholap, A.D., Hatvate, N.T., Khuspe, P.R., Mandhare, T.A., Kashid, P., Gaikwad, V.D., Proteomics in Oncology: Retrospect and Prospects, in: *Novel Technologies in Biosystems, Biomedical & Drug Delivery.* [Internet]. pp. 243–69, Springer Nature Singapore, Singapore, 2023. Available from: https://link.springer.com/10.1007/978-981-99-5281-6_10.
33. Gholap, A.D., Hatvate, N.T., Sawant, S.D., Khemka, P.N., Vora, L.K., Unlocking the potential of nanobiohybrids to combat environmental

pollution, in: *Nanobiohybrids for Advanced Wastewater Treatment and Energy Recovery.* [Internet]. pp. 169–90, IWA Publishing, 2023. Available from: https://iwaponline.com/ebooks/book/886/chapter/3516052/Unlocking-the-potential-of-nanobiohybrids-to.
34. Dwyer, D.B., Falkai, P., Koutsouleris, N., Machine Learning Approaches for Clinical Psychology and Psychiatry. *Annu. Rev. Clin. Psychol.*, *14*, 1, 91–118, 2018. Available from: https://www.annualreviews.org/doi/10.1146/annurev-clinpsy-032816-045037.
35. Bothiraja, C., Gholap, A.D., Shaikh, K.S., Pawar, A.P., Investigation of ethyl cellulose microsponge gel for topical delivery of eberconazole nitrate for fungal therapy. *Ther. Deliv.* [Internet]. 5, 7, 781–94, 2014 Jul. Available from: https://www.future-science.com/doi/10.4155/tde.14.43.
36. Puri, A.V. and Gholap, A.D., Comparative Evaluation of Papaya and Banana Starch as Disintegrant in Diclofenac Sodium Tablet Formulation. *Int. J. Pharm. Biol. Sci.*, 8, 3, 512–522, 2018.
37. Singh, T., Ravi Kumar, V., Gholap, A.D., Sanap, G., Anjaiah, B., Roja, K., *In vitro* antipyrial activity of psidium guajava leaf extract. *Eur. J. Pharm. Med. Res.* 3, 1, 312–5, 2016.
38. Leucht, S., Cipriani, A., Spineli, L., Mavridis, D., Örey, D., Richter, F., Samara, M., Barbui, C., Engel, R.R., Geddes, J.R., *et al.*, Comparative Efficacy and Tolerability of 15 Antipsychotic Drugs in Schizophrenia: A Multiple-Treatments Meta-Analysis. *Lancet*, *382*, 9896, 951–962, 2013. Available from: https://linkinghub.elsevier.com/retrieve/pii/S0140673613607333.
39. Nomura, M., Ohira, H., Haneda, K., Iidaka, T., Sadato, N., Okada, T., Yonekura, Y., Functional Association of the Amygdala and Ventral Prefrontal Cortex during Cognitive Evaluation of Facial Expressions Primed by Masked Angry Faces: An Event-Related FMRI Study. *Neuroimage*, *21*, 1, 352–363. Available from: https://linkinghub.elsevier.com/retrieve/pii/S1053811903005706.
40. Cuthbert, B.N. and Insel, T.R., Toward the Future of Psychiatric Diagnosis: The Seven Pillars of RDoC. *BMC Med.*, *11*, 1, 126, 2013. Available from: http://bmcmedicine.biomedcentral.com/articles/10.1186/1741-7015-11-126.
41. Anticevic, A., Cole, M.W., Murray, J.D., Corlett, P.R., Wang, X.-J., Krystal, J.H., The Role of Default Network Deactivation in Cognition and Disease. *Trends Cognit. Sci.*, *16*, 12, 584–592, 2012. Available from: https://linkinghub.elsevier.com/retrieve/pii/S1364661312002446.
42. O'Reilly, R.C. and Frank, M.J., Making Working Memory Work: A Computational Model of Learning in the Prefrontal Cortex and Basal Ganglia. *Neural Comput.*, *18*, 2, 283–328, 2006. Available from: https://direct.mit.edu/neco/article/18/2/283-328/7028.
43. Insel, T., Cuthbert, B., Garvey, M., Heinssen, R., Pine, D.S., Quinn, K., Sanislow, C., Wang, P., Research Domain Criteria (RDoC): Toward a New Classification Framework for Research on Mental Disorders. *Am. J. Psychiatry*, *167*, 7, 748–751, 2010. Available from: http://psychiatryonline.org/doi/abs/10.1176/appi.ajp.2010.09091379.

44. Rodriguez-Carranza, C.E., Mukherjee, P., Vigneron, D., Barkovich, J., Studholme, C., A Framework for *in Vivo* Quantification of Regional Brain Folding in Premature Neonates. *Neuroimage, 41*, 2, 462–478, 2008. Available from: https://linkinghub.elsevier.com/retrieve/pii/S1053811908000311.
45. Drysdale, A.T., Grosenick, L., Downar, J., Dunlop, K., Mansouri, F., Meng, Y., Fetcho, R.N., Zebley, B., Oathes, D.J., Etkin, A., *et al.*, Resting-State Connectivity Biomarkers Define Neurophysiological Subtypes of Depression. *Nat. Med., 23*, 1, 28–38, 2017. Available from: https://www.nature.com/articles/nm.4246.
46. Hyman, S.E., Can Neuroscience Be Integrated into the DSM-V? *Nat. Rev. Neurosci., 8*, 9, 725–732, 2007. Available from: https://www.nature.com/articles/nrn2218.
47. Fornito, A., Zalesky, A., Breakspear, M., The Connectomics of Brain Disorders. *Nat. Rev. Neurosci., 16*, 3, 159–172, 2015, doi: 10.1038/nrn3901.
48. Vowles, K.E., McEntee, M.L., Siyahhan Julnes, P., Frohe, T., Ney, J.P., van der Goes, D.N., On the Importance of Clear Comparisons and a Methodologically Rigorous Empirical Literature in Evaluating Opioid Use in Chronic Pain. *Pain, 156*, 8, 1577–1578, 2015. Available from: https://www.nature.com/articles/nrn3901.
49. Vora, L.K., Gholap, A.D., Jetha, K., Thakur, R.R.S., Solanki, H.K., Chavda, V.P., Artificial Intelligence in Pharmaceutical Technology and Drug Delivery Design. *Pharmaceutics, 15*, 7, 1916, 2023. Available from: https://www.mdpi.com/1999-4923/15/7/1916.
50. Soelkner, J., Very Many Variables and Limited Numbers of Observations; The P<<n Problem in Current Statistical Applications, in: *Proceedings of the Proceedings of the ITI 2012 34th International Conference on Information Technology Interfaces*, University Computing Centre - SRCE, Zagreb, pp. 13–14, 2012. Available from: http://iti.srce.unizg.hr/index.php/iti2012a/index/pages/view/proc2012_01.html.
51. Craddock, R.C., Holtzheimer, P.E., Hu, X.P., Mayberg, H.S., Disease State Prediction from Resting State Functional Connectivity. *Magn. Reson. Med., 62*, 6, 1619–1628, 2009. Available from: https://onlinelibrary.wiley.com/doi/10.1002/mrm.22159.
52. Menon, V., Large-Scale Brain Networks and Psychopathology: A Unifying Triple Network Model. *Trends Cognit. Sci., 15*, 10, 483–506, 2011. Available from: https://linkinghub.elsevier.com/retrieve/pii/S1364661311001719.
53. Arbabshirani, M.R., Plis, S., Sui, J., Calhoun, V.D., Single Subject Prediction of Brain Disorders in Neuroimaging: Promises and Pitfalls. *Neuroimage, 145*, 137–165, 2017. Available from: https://linkinghub.elsevier.com/retrieve/pii/S105381191600210X.
54. Zhang, B., Gaiteri, C., Bodea, L.-G., Wang, Z., McElwee, J., Podtelezhnikov, A.A., Zhang, C., Xie, T., Tran, L., Dobrin, R., *et al.*, Integrated Systems Approach Identifies Genetic Nodes and Networks in Late-Onset Alzheimer's

Disease. *Cell*, *153*, 707–720, 2013. Available from: https://linkinghub.elsevier.com/retrieve/pii/S0092867413003875.
55. Nutter, L.M.J., Heaney, J.D., Lloyd, K.C.K., Murray, S.A., Seavitt, J.R., Skarnes, W.C., Teboul, L., Brown, S.D.M., Moore, M., Response to "Unexpected Mutations after CRISPR–Cas9 Editing *in Vivo*." *Nat. Methods*, *15*, 4, 235–236, 2018. Available from: https://www.nature.com/articles/nmeth.4559.
56. Kwak, S., Park, S., Kim, J., Park, S., Lee, J.-Y., Multivariate Neuroanatomical Correlates of Behavioral and Psychological Symptoms in Dementia and the Moderating Role of Education. *NeuroImage Clin.*, *28*, 102452, 2020. Available from: https://linkinghub.elsevier.com/retrieve/pii/S2213158220302898.
57. Sale, M.V., Mattingley, J.B., Zalesky, A., Cocchi, L., Imaging Human Brain Networks to Improve the Clinical Efficacy of Non-Invasive Brain Stimulation. *Neurosci. Biobehav. Rev.*, *57*, 187–198, 2015. Available from: https://linkinghub.elsevier.com/retrieve/pii/S0149763415002444.
58. Hollon, S.D., Evidence-based Assessment: A Commentary. *Clin. Psychol. Sci. Pract.*, *24*, 4, 364–366, 2017. Available from: http://doi.apa.org/getdoi.cfm?doi=10.1111/cpsp.12208.
59. Chen, J. and MacCarthy, T., The Preferred Nucleotide Contexts of the AID/APOBEC Cytidine Deaminases Have Differential Effects When Mutating Retrotransposon and Virus Sequences Compared to Host Genes. *PLoS Comput. Biol.*, *13*, 3, e1005471, 2017. Available from: https://dx.plos.org/10.1371/journal.pcbi.1005471.
60. Redlich, R., Opel, N., Grotegerd, D., Dohm, K., Zaremba, D., Bürger, C., Münker, S., Mühlmann, L., Wahl, P., Heindel, W., *et al.*, Prediction of Individual Response to Electroconvulsive Therapy via Machine Learning on Structural Magnetic Resonance Imaging Data. *JAMA Psychiatry*, *73*, 6, 557, 2016. Available from: http://archpsyc.jamanetwork.com/article.aspx?doi=10.1001/jamapsychiatry.2016.0316.
61. Chekroud, A.M., Zotti, R.J., Shehzad, Z., Gueorguieva, R., Johnson, M.K., Trivedi, M.H., Cannon, T.D., Krystal, J.H., Corlett, P.R., Cross-Trial Prediction of Treatment Outcome in Depression: A Machine Learning Approach. *Lancet Psychiatry*, *3*, 3, 243–250, 2016. Available from: https://linkinghub.elsevier.com/retrieve/pii/S221503661500471X.
62. Horne, C.M., Vanes, L.D., Verneuil, T., Mouchlianitis, E., Szentgyorgyi, T., Averbeck, B., Leech, R., Moran, R.J., Shergill, S.S., Cognitive Control Network Connectivity Differentially Disrupted in Treatment Resistant Schizophrenia. *NeuroImage Clin.*, *30*, 102631, 2021. Available from: https://linkinghub.elsevier.com/retrieve/pii/S2213158221000759.
63. Shen, X., Finn, E.S., Scheinost, D., Rosenberg, M.D., Chun, M.M., Papademetris, X., Constable, R.T., Using Connectome-Based Predictive Modeling to Predict Individual Behavior from Brain Connectivity. *Nat. Protoc.*, *12*, 3, 506–518, 2017. Available from: https://www.nature.com/articles/nprot.2016.178.

64. Acar, I. and Butt, S.E., Modeling Nurse-Patient Assignments Considering Patient Acuity and Travel Distance Metrics. *J. Biomed. Inform.*, *64*, 192–206, 2016. Available from: https://linkinghub.elsevier.com/retrieve/pii/S1532046416301393.
65. Weng, S.F., Reps, J., Kai, J., Garibaldi, J.M., Qureshi, N., Can Machine-Learning Improve Cardiovascular Risk Prediction Using Routine Clinical Data? *PLoS One*, *12*, 4, e0174944, 2017. Available from: https://dx.plos.org/10.1371/journal.pone.0174944.
66. Cardonatto, M.C. and Melchor, R.N., Large Mammal Burrows in Late Miocene Calcic Paleosols from Central Argentina: Paleoenvironment, Taphonomy and Producers. *PeerJ*, *6*, e4787, 2018. Available from: https://peerj.com/articles/4787.
67. Obermeyer, Z. and Emanuel, E.J., Predicting the Future — Big Data, Machine Learning, and Clinical Medicine. *N. Engl. J. Med.*, *375*, 13, 1216–1219, 2016. Available from: http://www.nejm.org/doi/10.1056/NEJMp1606181.
68. Robinson, O.J. and Chase, H.W., Learning and Choice in Mood Disorders: Searching for the Computational Parameters of Anhedonia. *Comput. Psychiatry*, 1, 208, 2017 Dec 1. [Internet]. Available from: https://cpsyjournal.org/article/10.1162/CPSY_a_00009/.
69. Chen, J.H. and Asch, S.M., Machine Learning and Prediction in Medicine — Beyond the Peak of Inflated Expectations. *N. Engl. J. Med.*, 376, 26, 2507–2509, 2017 Jun 29. [Internet]. Available from: http://www.nejm.org/doi/10.1056/NEJMp1702071.
70. Pardeshi, S., Gholap, A., More, M., Togre, N., Rebello, N., Giram, P., Dendrimers Based Antibacterial and Antiviral Materials, in: *Antibacterial and Antiviral Functional Materials*, pp. 139–169, ACS Publications, 2023. [Internet]. Available from: https://pubs.acs.org/doi/abs/10.1021/bk-2023-1458.ch005.
71. Gholap, A.D., Sayyad, S.F., Hatvate, N.T., Dhumal, V.V., Pardeshi, S.R., Chavda, V.P., *et al.*, Drug Delivery Strategies for Avobenzone: A Case Study of Photostabilization. *Pharmaceutics*, 15, 3, 1008, 2023 Mar 21. [Internet]. Available from: https://www.mdpi.com/1999-4923/15/3/1008.
72. Gholap, A.D., Sawant, S.D., Sayyad, S.F., Hatvate, N.T., Chavan, M., Rojekar, S. *et al.*, Food Biopackaging for Human Benefits: Status and Perspectivesm in: *Biobased Packaging Materials* [Internet]. pp. 1–35, Springer Nature Singapore, Singapore, 2023. Available from: https://link.springer.com/10.1007/978-981-99-6050-7_1.
73. Newman, A.B., Sanders, J.L., Kizer, J.R., Boudreau, R.M., Odden, M.C., Al Hazzouri, A.Z., *et al.*, Trajectories of function and biomarkers with age: the CHS All Stars Study. *Int. J. Epidemiol.*, 45, 4, 1135–1145, 2016 Jun 6. [Internet]. Available from: https://academic.oup.com/ije/article-lookup/doi/10.1093/ije/dyw092.
74. Kennedy, S.H., Lam, R.W., McIntyre, R.S., Tourjman, S.V., Bhat, V., Blier, P., Hasnain, M., Jollant, F., Levitt, A.J., MacQueen, G.M., *et al.*, Canadian

Network for Mood and Anxiety Treatments (CANMAT) 2016 Clinical Guidelines for the Management of Adults with Major Depressive Disorder. *Can. J. Psychiatry*, *61*, 9, 540–560, 2016. Available from: http://journals.sagepub.com/doi/10.1177/0706743716659417.

75. Daley, D., van der Oord, S., Ferrin, M., Danckaerts, M., Doepfner, M., Cortese, S., Sonuga-Barke, E.J.S., Behavioral Interventions in Attention-Deficit/Hyperactivity Disorder: A Meta-Analysis of Randomized Controlled Trials Across Multiple Outcome Domains. *J. Am. Acad. Child Adolesc. Psychiatry*, *53*, 8, 835–847.e5, 2014. Available from: https://linkinghub.elsevier.com/retrieve/pii/S0890856714004080.
76. Harmer, C.J., O'Sullivan, U., Favaron, E., Massey-Chase, R., Ayres, R., Reinecke, A., Goodwin, G.M., Cowen, P.J., Effect of Acute Antidepressant Administration on Negative Affective Bias in Depressed Patients. *Am. J. Psychiatry*, *166*, 10, 1178–1184, 2009. Available from: http://psychiatryonline.org/doi/abs/10.1176/appi.ajp.2009.09020149.
77. Fox, A.S., Oler, J.A., Tromp, D.P.M., Fudge, J.L., Kalin, N.H., Extending the Amygdala in Theories of Threat Processing. *Trends Neurosci.*, *38*, 5, 319–329, 2015. Available from: https://linkinghub.elsevier.com/retrieve/pii/S0166223615000612.
78. Woo, C.-W., Chang, L.J., Lindquist, M.A., Wager, T.D., Building Better Biomarkers: Brain Models in Translational Neuroimaging. *Nat. Neurosci.*, *20*, 365–377, 2017. Available from: https://www.nature.com/articles/nn.4478.
79. Keshavan, A., Paul, F., Beyer, M.K., Zhu, A.H., Papinutto, N., Shinohara, R.T., Stern, W., Amann, M., Bakshi, R., Bischof, A., *et al.*, Power Estimation for Non-Standardized Multisite Studies. *Neuroimage*, *134*, 281–294, 2016. Available from: https://linkinghub.elsevier.com/retrieve/pii/S1053811916002561.
80. Hariri, A.R., Drabant, E.M., Weinberger, D.R., Imaging Genetics: Perspectives from Studies of Genetically Driven Variation in Serotonin Function and Corticolimbic Affective Processing. *Biol. Psychiatry*, *59*, 10, 888–897, 2006. Available from: https://linkinghub.elsevier.com/retrieve/pii/S0006322305013910.
81. Rojekar, S., Gholap, A.D., Togre, N., Bhoj, P., Haeck, C., Hatvate, N., *et al.*, Current status of mannose receptor-targeted drug delivery for improved anti-HIV therapy. *J. Control. Release*, [Internet], 372, 494–521, 2024 Aug. Available from: https://linkinghub.elsevier.com/retrieve/pii/S0168365924003444.
82. Gholap, A.D., Said, R.P., Pawar, R.D., Ambore, G.S., Hatvate, N.T., Chapter Five - Importance of carbohydrate-drug conjugates in vaccine development: A detailed review, in: *Bioconjugated Materials Part 2 Applications in Drug Delivery, Vaccine Formulations and Important Conjugates for Cancer Therapy*. S.K. Verma, A.K. Das (eds.). [Internet]. Comprehensive Analytical Chemistry, vol. 103, p. 191–256, Elsevier, 2023. Available from: https://www.sciencedirect.com/science/article/pii/S0166526X23000296.
83. Gholap, A.D., Bhowmik, D.D., Deshmukh, A.Y., Hatvate, N.T., Chapter Six - Quintessential impact of dendrimer bioconjugates in targeted drug delivery,

in: *Bioconjugated Materials Part 2 Applications in Drug Delivery, Vaccine Formulations and Important Conjugates for Cancer Therapy,* S.K. Verma, A.K. Das (eds.) [Internet]. Comprehensive Analytical Chemistry, vol. 103, p. 257–302, Elsevier, 2023. Available from: https://www.sciencedirect.com/science/article/pii/S0166526X23000272.

84. Gholap, A.D., Gupta, J.S., Kamandar, P.A., Banchhod, G.V., Hatvate, N.T., Chapter Four - Antibody-drug conjugates for cancer therapy: An up-to-date review on the chemistry and pharmacology, in: *Bioconjugated Materials Part 2 Applications in Drug Delivery, Vaccine Formulations and Important Conjugates for Cancer Therapy.* S.K. Verma, A.K. Das (eds.). [Internet]. Comprehensive Analytical Chemistry, vol. 103, pp. 105–90, Elsevier, 2023. Available from: https://www.sciencedirect.com/science/article/pii/S0166526X23000259.
85. Cao, H. and Cannon, T.D., New Evidence Supporting a Role of Hippocampus in the Development of Psychosis. *Biol. Psychiatry, 87,* 3, 200–201, 2020. Available from: https://linkinghub.elsevier.com/retrieve/pii/S0006322319318153.
86. Yao, L., Kusakunniran, W., Wu, Q., Zhang, J., Tang, Z., Yang, W., Robust Gait Recognition Using Hybrid Descriptors Based on Skeleton Gait Energy Image. *Pattern Recognit. Lett., 150,* 289–296, 2021. Available from: https://linkinghub.elsevier.com/retrieve/pii/S0167865519301618.
87. Torkamani, A., Wineinger, N.E., Topol, E.J., The Personal and Clinical Utility of Polygenic Risk Scores. *Nat. Rev. Genet., 19,* 9, 581–590, 2018. Available from: https://www.nature.com/articles/s41576-018-0018-x.
88. Elwyn, G., Frosch, D., Thomson, R., Joseph-Williams, N., Lloyd, A., Kinnersley, P., Cording, E., Tomson, D., Dodd, C., Rollnick, S., *et al.*, Shared Decision Making: A Model for Clinical Practice. *J. Gen. Intern. Med., 27,* 10, 1361–1367, 2012. Available from: http://link.springer.com/10.1007/s11606-012-2077-6.
89. Légaré, F., Ratté, S., Stacey, D., Kryworuchko, J., Gravel, K., Graham, I.D., Turcotte, S., Interventions for Improving the Adoption of Shared Decision Making by Healthcare Professionals, in: *Cochrane Database of Systematic Reviews,* F. Légaré (Ed.), John Wiley & Sons, Ltd, Chichester, UK, 2010. Available from: https://doi.wiley.com/10.1002/14651858.CD006732.pub2.
90. Bruwer, Z., Futter, M., Ramesar, R., Communicating Cancer Risk within an African Context: Experiences, Disclosure Patterns and Uptake Rates Following Genetic Testing for Lynch Syndrome. *Patient Educ. Couns., 92,* 1, 53–60, 2013. Available from: https://linkinghub.elsevier.com/retrieve/pii/S0738399113000529.
91. Gholap, A.D., Pardeshi, S.R., Hatvate, N.T., Dhorkule, N., Sayyad, S.F., Faiyazuddin, M., et al., Environmental implications and nanotechnological advances in octocrylene-enriched sunscreen formulations: A comprehensive review. *Chemosphere,* [Internet]. 358, 142235, 2024 Jun. Available from: https://linkinghub.elsevier.com/retrieve/pii/S0045653524011287.

92. Gholap, A.D., Kapare, H.S., Pagar, S., Kamandar, P., Bhowmik, D., Vishwakarma, N., *et al.*, Exploring modified chitosan-based gene delivery technologies for therapeutic advancements. *Int. J. Biol. Macromol.*, 260, 129581, 2024 Mar. [Internet]. Available from: https://linkinghub.elsevier.com/retrieve/pii/S0141813024003842.
93. Gholap, A.D., Gupta, J., Kamandar, P., Bhowmik, D.D., Rojekar, S., Faiyazuddin, M., *et al.*, Harnessing Nanovaccines for Effective Immunization — A Special Concern on COVID-19: Facts, Fidelity, and Future Prospective. *ACS Biomater. Sci. Eng.*, [Internet]. 10, 1, 271–97, 2024 Jan 8. Available from: https://pubs.acs.org/doi/10.1021/acsbiomaterials.3c01247
94. Gholap, A.D., Khuspe, P.R., Pardeshi, S.R., Uddin, M.J., Das, U., Hatvate, N.T., *et al.*, Achieving Optimal Health With Host-Directed Therapies (HDTs) in Infectious Diseases—A New Horizon. *Adv. Ther.*, 2024 Sep 18. [Internet]. Available from: https://onlinelibrary.wiley.com/doi/10.1002/adtp.202400169.
95. Fournier, J.C., DeRubeis, R.J., Hollon, S.D., Dimidjian, S., Amsterdam, J.D., Shelton, R.C., *et al.*, Antidepressant Drug Effects and Depression Severity. *JAMA*, 303, 1, 47, 2010 Jan 6. [Internet]. Available from: http://jama.jamanetwork.com/article.aspx?doi=10.1001/jama.2009.1943.
96. Pirone, A., Alexander, J., Lau, L.A., Hampton, D., Zayachkivsky, A., Yee, A., *et al.*, APC conditional knock-out mouse is a model of infantile spasms with elevated neuronal β-catenin levels, neonatal spasms, and chronic seizures. *Neurobiol. Dis.*, 98, 149–157, 2017 Feb. [Internet]. Available from: https://linkinghub.elsevier.com/retrieve/pii/S0969996116302534.
97. Epskamp, S., Borsboom, D., Fried, E.I., Estimating psychological networks and their accuracy: A tutorial paper. *Behav. Res. Methods*, 50, 1, 195–212, 2018 Feb 24. [Internet]. Available from: http://link.springer.com/10.3758/s13428-017-0862-1.
98. Delgadillo, J., Moreea, O., Lutz, W., Different people respond differently to therapy: A demonstration using patient profiling and risk stratification. *Behav. Res. Ther.*, 79, 15–22, 2016 Apr. [Internet]. Available from: https://linkinghub.elsevier.com/retrieve/pii/S0005796716300249.
99. Fried, E.I. and Cramer, A.O.J., Moving Forward: Challenges and Directions for Psychopathological Network Theory and Methodology. *Perspect. Psychol. Sci.*, 12, 6, 999–1020, 2017 Nov 5. [Internet]. Available from: http://journals.sagepub.com/doi/10.1177/1745691617705892.
100. Zavos, H.M.S., Eley, T.C., McGuire, P., Plomin, R., Cardno, A.G., Freeman, D., *et al.*, Shared Etiology of Psychotic Experiences and Depressive Symptoms in Adolescence: A Longitudinal Twin Study. *Schizophr. Bull.*, 42, 5, 1197–1206, 2016 Sep. [Internet]. Available from: https://academic.oup.com/schizophreniabulletin/article-lookup/doi/10.1093/schbul/sbw021.
101. Borsboom, D. and Cramer, A.O.J., Network Analysis: An Integrative Approach to the Structure of Psychopathology. *Annu. Rev. Clin. Psychol.*, 9, 1, 91–121, 2013. Available from: https://www.annualreviews.org/doi/10.1146/annurev-clinpsy-050212-185608.

102. Steinley, D., Hoffman, M., Brusco, M.J., Sher, K.J., A Method for Making Inferences in Network Analysis: Comment on Forbes, Wright, Markon, and Krueger (2017). *J. Abnorm. Psychol.*, *126*, 7, 1000–1010, 2017. Available from: http://doi.apa.org/getdoi.cfm?doi=10.1037/abn0000308.
103. He, Y. and Evans, A., Graph Theoretical Modeling of Brain Connectivity. *Curr. Opin. Neurol.*, *23*, 4, 341–350, 2010. Available from: https://journals.lww.com/00019052-201008000-00002.
104. Parit, S., Manchare, A., Gholap, A.D., Mundhe, P., Hatvate, N., Rojekar, S., *et al.*, Antibody-Drug Conjugates: A promising breakthrough in cancer therapy. *Int. J. Pharm.*, [Internet]. 659, 124211, 2024 Jun. Available from: https://linkinghub.elsevier.com/retrieve/pii/S0378517324004459.
105. Shaikh, M.S., Faiyazuddin, M., Khan, M.S., Pathan, S.K., Syed, I.J., Gholap, A.D., *et al.* Chikungunya virus vaccine: a decade of progress solving epidemiological dilemma, emerging concepts, and immunological interventions. *Front. Microbiol.*, [Internet]. 15, 2024 Jul 22. Available from: https://www.frontiersin.org/articles/10.3389/fmicb.2024.1413250/full.
106. Gholap, A.D., Rojekar, S., Kapare, H.S., Vishwakarma, N., Raikwar, S., Garkal, A., *et al.* Chitosan scaffolds: Expanding horizons in biomedical applications. *Carbohydr. Polym.*, [Internet]. 323, 121394, 2024 Jan. Available from: https://linkinghub.elsevier.com/retrieve/pii/S0144861723008597.
107. Gholap, A.D., Hatvate, N.T., Khuspe, P.R., Mandhare, T.A., Kashid, P., Vora, L.K., Decoding of Nanoneutraceuticals and Their Delivery, in: *Sustainable Green Nanotechnology*, [Internet]. p. 284–309, CRC Press, New York, 2024. Available from: https://www.taylorfrancis.com/books/9781003389408/chapters/10.1201/9781003389408-16.
108. Mesut, B., Al-Mohaya, M., Gholap, A.D., Yeşilkaya, E., Das, U., Akhtar, M.S., *et al.*, Demystifying the potential of lipid-based nanocarriers in targeting brain malignancies. *Naunyn Schmiedebergs Arch. Pharmacol.*, [Internet]. 2024 Jul 4. Available from: https://link.springer.com/10.1007/s00210-024-03212-6.
109. Bringmann, L.F., Vissers, N., Wichers, M., Geschwind, N., Kuppens, P., Peeters, F., Borsboom, D., Tuerlinckx, F., A Network Approach to Psychopathology: New Insights into Clinical Longitudinal Data. *PLoS One*, *8*, e60188, 2013, doi: 10.1371/journal.pone.0060188.
110. Wang, X.-J. and Krystal, J.H., Computational Psychiatry. *Neuron*, *84*, 3, 638–654, 2014. Available from: https://linkinghub.elsevier.com/retrieve/pii/S0896627314009167.
111. Gueguen, M.C., Schweitzer, E.M., Konova, A.B., Computational Theory-Driven Studies of Reinforcement Learning and Decision-Making in Addiction: What Have We Learned? *Curr. Opin. Behav. Sci.*, *38*, 40–48, 2021. Available from: https://linkinghub.elsevier.com/retrieve/pii/S2352154620301261.
112. Khaleghi, A., Mohammadi, M.R., Shahi, K., Nasrabadi, A.M., Computational Neuroscience Approach to Psychiatry: A Review on Theory-Driven Approaches. *Clin. Psychopharmacol. Neurosci.*, *20*, 1, 26–36, 2022.

Available from: http://www.cpn.or.kr/journal/view.html?doi=10.9758/cpn.2022.20.1.26.

113. Baden, C., Kligler-Vilenchik, N., Yarchi, M., Hybrid Content Analysis: Toward a Strategy for the Theory-Driven, Computer-Assisted Classification of Large Text Corpora. *Commun. Methods Meas., 14*, 3, 165–183, 2020. Available from: https://www.tandfonline.com/doi/full/10.1080/19312458.2020.1803247.
114. Tenenbaum, J.B., Kemp, C., Griffiths, T.L., Goodman, N.D., How to Grow a Mind: Statistics, Structure, and Abstraction. *Science (80-.), 331*, 6022, 1279–1285, 2011. Available from: https://www.science.org/doi/10.1126/science.1192788.
115. Pearl, J., Causality [Internet]. Cambridge University Press, 2009. Available from: https://www.cambridge.org/core/product/identifier/9780511803161/type/book.
116. Greene, J.D., Why are VMPFC patients more utilitarian? A dual-process theory of moral judgment explains. *Trends Cogn. Sci.*, 11, 8, 322–323, 2007 Aug. [Internet]. Available from: https://linkinghub.elsevier.com/retrieve/pii/S1364661307001544.
117. Erdem, T., Srinivasan, K., Amaldoss, W., Bajari, P., Che, H., Ho, T., Hutchinson, W., Katz, M., Keane, M., Meyer, R., *et al.*, Theory-Driven Choice Models. *Mark. Lett.*, *16*, 225–237, 2005, doi: 10.1007/s11002-005-5887-z.
118. Franke, M. and Bergen, L., Theory-Driven Statistical Modeling for Semantics and Pragmatics: A Case Study on Grammatically Generated Implicature Readings. *Language (Baltim)*, *96*, e77–e96, 2020, doi: 10.1353/lan.2020.0034.
119. Markram, H., Muller, E., Ramaswamy, S., Reimann, M.W., Abdellah, M., Sanchez, C.A., *et al.,* Reconstruction and Simulation of Neocortical Microcircuitry. *Cell,* [Internet]. 163, 2, 456–92, 2015 Oct. Available from: https://linkinghub.elsevier.com/retrieve/pii/S0092867415011915.
120. Gerstner, W. and Kistler, W.M., *Spiking Neuron Models.* Cambridge University Press, United Kingdom, 2002. [Internet]. Available from: https://www.cambridge.org/core/product/identifier/9780511815706/type/book.
121. Poirazi, P., Brannon, T., Mel, B.W., Pyramidal Neuron as Two-Layer Neural Network. *Neuron*, 37, 6, 989–999, 2003 Mar. [Internet]. Available from: https://linkinghub.elsevier.com/retrieve/pii/S0896627303001491.
122. Brette, R. and Gerstner, W., Adaptive Exponential Integrate-and-Fire Model as an Effective Description of Neuronal Activity. *J. Neurophysiol.*, *94*, 5, 3637–3642, 2005. Available from: https://www.physiology.org/doi/10.1152/jn.00686.2005.
123. Izhikevich, E.M., Simple Model of Spiking Neurons. *IEEE Trans. Neural Netw.*, *14*, 6, 1569–1572, 2003. Available from: http://ieeexplore.ieee.org/document/1257420/.
124. Pospischil, M., Toledo-Rodriguez, M., Monier, C., Piwkowska, Z., Bal, T., Frégnac, Y., Markram, H., Destexhe, A., Minimal Hodgkin–Huxley Type

Models for Different Classes of Cortical and Thalamic Neurons. *Biol. Cybern.*, *99*, 4–5, 427–441, 2008. Available from: http://link.springer.com/10.1007/s00422-008-0263-8.
125. Guryanova, O.A., Shank, K., Spitzer, B., Luciani, L., Koche, R.P., Garrett-Bakelman, F.E., Ganzel, C., Durham, B.H., Mohanty, A., Hoermann, G., *et al.*, DNMT3A Mutations Promote Anthracycline Resistance in Acute Myeloid Leukemia via Impaired Nucleosome Remodeling. *Nat. Med.*, *22*, 12, 1488–1495, 2016. Available from: https://www.nature.com/articles/nm.4210.
126. Connor, S.A., Ammendrup-Johnsen, I., Chan, A.W., Kishimoto, Y., Murayama, C., Kurihara, N., Tada, A., Ge, Y., Lu, H., Yan, R., *et al.*, Altered Cortical Dynamics and Cognitive Function upon Haploinsufficiency of the Autism-Linked Excitatory Synaptic Suppressor MDGA2. *Neuron*, *91*, 5, 1052–1068, 2016. Available from: https://linkinghub.elsevier.com/retrieve/pii/S0896627316305086.
127. Marder, E. and Goaillard, J.-M., Variability, Compensation and Homeostasis in Neuron and Network Function. *Nat. Rev. Neurosci.*, *7*, 7, 563–574, 2006. Available from: https://www.nature.com/articles/nrn1949.
128. Carreiras, M., Armstrong, B.C., Perea, M., Frost, R., The What, When, Where, and How of Visual Word Recognition. *Trends Cognit. Sci.*, *18*, 2, 90–98, 2014. Available from: https://linkinghub.elsevier.com/retrieve/pii/S1364661313002696.
129. Deco, G. and Jirsa, V.K., Ongoing Cortical Activity at Rest: Criticality, Multistability, and Ghost Attractors. *J. Neurosci.*, *32*, 10, 3366–3375, 2012. Available from: https://www.jneurosci.org/lookup/doi/10.1523/JNEUROSCI.2523-11.2012.
130. Marella, S. and Ermentrout, G.B., Class-II Neurons Display a Higher Degree of Stochastic Synchronization than Class-I Neurons. *Phys. Rev. E*, *77*, 4, 041918, 2008. Available from: https://link.aps.org/doi/10.1103/PhysRevE.77.041918.
131. Buzsáki, G. and Draguhn, A., Neuronal Oscillations in Cortical Networks. *Science (80-.)*, *304*, 5679, 1926–1929, 2004. Available from: https://www.science.org/doi/10.1126/science.1099745.
132. Hothorn, T., Brandl, R., Müller, J., Large-Scale Model-Based Assessment of Deer-Vehicle Collision Risk. *PLoS One*, *7*, 2, e29510, 2012. Available from: https://dx.plos.org/10.1371/journal.pone.0029510.
133. McIntyre, C.C. and Hahn, P.J., Network Perspectives on the Mechanisms of Deep Brain Stimulation. *Neurobiol. Dis.*, *38*, 3, 329–337, 2010. Available from: https://linkinghub.elsevier.com/retrieve/pii/S0969996109002745.
134. Hedrich, U.B.S., Smarandache, C.R., Stein, W., Differential Activation of Projection Neurons by Two Sensory Pathways Contributes to Motor Pattern Selection. *J. Neurophysiol.*, *102*, 2866–2879, 2009, doi: 10.1152/jn.00618.2009.
135. Brunton, B.W., Beyeler, M., Data-driven models in human neuroscience and neuroengineering. *Curr. Opin. Neurobiol.*, [Internet], 58, 21–9,

2019 Oct. Available from: https://linkinghub.elsevier.com/retrieve/pii/S0959438818302502.

136. Hay, E., Hill, S., Schürmann, F., Markram, H., Segev, I., Models of Neocortical Layer 5b Pyramidal Cells Capturing a Wide Range of Dendritic and Perisomatic Active Properties. *PLoS Comput. Biol.*, *7*, 7, e1002107, 2011. Available from: https://dx.plos.org/10.1371/journal.pcbi.1002107.
137. Kaelbling, L.P., Littman, M.L., Moore, A.W., Reinforcement Learning: A Survey. *J. Artif. Intell. Res.*, *4*, 237–285, 1996. Available from: https://www.jair.org/index.php/jair/article/view/10166.
138. Mnih, V., Kavukcuoglu, K., Silver, D., Rusu, A.A., Veness, J., Bellemare, M.G., Graves, A., Riedmiller, M., Fidjeland, A.K., Ostrovski, G., *et al.*, Human-Level Control through Deep Reinforcement Learning. *Nature*, *518*, 7540, 529–533, 2015. Available from: https://www.nature.com/articles/nature14236.
139. Silver, D., Huang, A., Maddison, C.J., Guez, A., Sifre, L., van den Driessche, G., Schrittwieser, J., Antonoglou, I., Panneershelvam, V., Lanctot, M., *et al.*, Mastering the Game of Go with Deep Neural Networks and Tree Search. *Nature*, *529*, 7587, 484–489, 2016. Available from: https://www.nature.com/articles/nature16961.
140. Sutton, R.S., Precup, D., Singh, S., Between MDPs and Semi-MDPs: A Framework for Temporal Abstraction in Reinforcement Learning. *Artif. Intell.*, *112*, 1–2, 181–211, 1999. Available from: https://linkinghub.elsevier.com/retrieve/pii/S0004370299000521.
141. Cesa-Bianchi, N. and Lugosi, G., *Prediction, Learning, and Games* [Internet]. Cambridge University Press, United Kingdom, 2006. Available from: https://www.cambridge.org/core/product/identifier/9780511546921/type/book.
142. Wang, C., Ha, S., Wang, Y., Xuan, J., Hoffman, E., Computational Analysis of Muscular Dystrophy Sub-types Using a Novel Integrative Scheme, in: *2010 Ninth International Conference on Machine Learning and Applications*, [Internet], pp. 287–92, IEEE, 2010. Available from: http://ieeexplore.ieee.org/document/5708846/.
143. Levine, S., Pastor, P., Krizhevsky, A., Ibarz, J., Quillen, D., Learning Hand-Eye Coordination for Robotic Grasping with Deep Learning and Large-Scale Data Collection. *Int. J. Rob. Res.*, *37*, 421–436, 2018. Available from: http://journals.sagepub.com/doi/10.1177/0278364917710318.
144. Silver, D., Hubert, T., Schrittwieser, J., Antonoglou, I., Lai, M., Guez, A., *et al.*, A general reinforcement learning algorithm that masters chess, shogi, and Go through self-play. *Science (80-)*, 362, 6419, 1140–1144, 2018 Dec 7. [Internet]. Available from: https://www.science.org/doi/10.1126/science.aar6404.
145. Silver, D., Schrittwieser. J., Simonyan. K., Antonoglou. I., Huang. A., Guez. A., *et al.*, Mastering the game of Go without human knowledge. *Nature*, [Internet], 550, 7676, 354–9, 2017 Oct 19. Available from: https://www.nature.com/articles/nature24270.

146. Knill, D.C. and Pouget, A., The Bayesian Brain: The Role of Uncertainty in Neural Coding and Computation. *Trends Neurosci.*, *27*, 12, 712–719, 2004. Available from: https://linkinghub.elsevier.com/retrieve/pii/S0166223604003352.
147. Xie, J. and Kim, N.-K., Bayesian Models and Markov Chain Monte Carlo Methods for Protein Motifs with the Secondary Characteristics. *J. Comput. Biol.*, 12, 7, 952–970, 2005 Sep. [Internet]. Available from: http://www.liebertpub.com/doi/10.1089/cmb.2005.12.952.
148. Lukens, J.M. and Passian, A., Bayesian Inference for Plasmonic Nanometrology. *Phys. Rev. A*, *104*, 5, 053501, 2021. Available from: https://link.aps.org/doi/10.1103/PhysRevA.104.053501.
149. Sarker, I.H., AI-Based Modeling: Techniques, Applications and Research Issues Towards Automation, Intelligent and Smart Systems. *SN Comput. Sci.*, *3*, 158, 2022. Available from: https://link.springer.com/10.1007/s42979-022-01043-x.
150. Magris, M. and Iosifidis, A., Bayesian Learning for Neural Networks: An Algorithmic Survey. *Artif. Intell. Rev.*, *56*, 10, 11773–11823, 2023. Available from: https://link.springer.com/10.1007/s10462-023-10443-1.
151. Aldoseri, A., Al-Khalifa, K.N., Hamouda, A.M., Re-Thinking Data Strategy and Integration for Artificial Intelligence: Concepts, Opportunities, and Challenges. *Appl. Sci.*, *13*, 12, 7082, 2023. Available from: https://www.mdpi.com/2076-3417/13/12/7082.
152. Ali, S., Abuhmed, T., El-Sappagh, S., Muhammad, K., Alonso-Moral, J.M., Confalonieri, R., Guidotti, R., Del Ser, J., Díaz-Rodríguez, N., Herrera, F., Explainable Artificial Intelligence (XAI): What We Know and What Is Left to Attain Trustworthy Artificial Intelligence. *Inf. Fusion*, *99*, 101805, 2023. Available from: https://linkinghub.elsevier.com/retrieve/pii/S1566253523001148.
153. Pechmann, S. and Frydman, J., Interplay between Chaperones and Protein Disorder Promotes the Evolution of Protein Networks. *PLoS Comput. Biol.*, *10*, e1003674, 2014. Available from: https://dx.plos.org/10.1371/journal.pcbi.1003674.
154. Kawamae, N., Topic Chronicle Forest for Topic Discovery and Tracking, in: *Proceedings of the Proceedings of the Eleventh ACM International Conference on Web Search and Data Mining*, February 2 2018, ACM, New York, NY, USA, pp. 315–323. Available from: https://dl.acm.org/doi/10.1145/3159652.3159653.
155. Bernard, P., Romain, A.-J., Caudroit, J., Chevance, G., Carayol, M., Gourlan, M., Needham Dancause, K., Moullec, G., Cognitive Behavior Therapy Combined with Exercise for Adults with Chronic Diseases: Systematic Review and Meta-Analysis. *Heal. Psychol.*, *37*, 5, 433–450, 2018. Available from: http://doi.apa.org/getdoi.cfm?doi=10.1037/hea0000578.
156. Steinfeldt, T., Könen-Waisman, S., Tong, L., Pawlowski, N., Lamkemeyer, T., Sibley, L.D., Hunn, J.P., Howard, J.C., Correction: Phosphorylation of Mouse Immunity-Related GTPase (IRG) Resistance Proteins Is an Evasion Strategy

for Virulent Toxoplasma Gondii. *PLoS Biol.*, *13*, 7, e1002199, 2015. Available from: https://dx.plos.org/10.1371/journal.pbio.1002199.

157. Gholap, A.D., Uddin, M.J., Faiyazuddin, M., Omri, A., Gowri, S., Khalid, M., Advances in artificial intelligence for drug delivery and development: A comprehensive review. *Comput. Biol. Med.*, [Internet]. 178, 108702, 2024 Aug. Available from: https://linkinghub.elsevier.com/retrieve/pii/S001048252400787X.
158. Bibri, S.E., Huang, J., Jagatheesaperumal, S.K., Krogstie, J., The synergistic interplay of artificial intelligence and digital twin in environmentally planning sustainable smart cities: A comprehensive systematic review. *Environ. Sci. Ecotechnol.*, [Internet], 20, 100433, 2024 Jul. Available from: https://linkinghub.elsevier.com/retrieve/pii/S2666498424000474.
159. Zamiri, M. and Esmaeili, A., Methods and Technologies for Supporting Knowledge Sharing within Learning Communities: A Systematic Literature Review. *Adm. Sci.*, [Internet], 14, 1, 17, 2024 Jan 20. Available from: https://www.mdpi.com/2076-3387/14/1/17.
160. How, M.-L., Cheah, S.-M., Forging the Future: Strategic Approaches to Quantum AI Integration for Industry Transformation. *AI*, [Internet], 5, 1, 290–323, 2024 Jan 29. Available from: https://www.mdpi.com/2673-2688/5/1/15.

Part 2

CLINICAL ASPECTS OF COMPUTATIONAL NEUROPHARMACOLOGY

8

Computational Attitudes to Drug Discovery in Neurohumoral Transmission and Signal Transduction

Lucy Mohapatra[1], Alok S. Tripathi[2*], Deepak Mishra[1], Alka[1,3], Sambit Kumar Parida[4] and Bhupendra Gopalbhai Prajapati[5,6]

[1]Amity Institute of Pharmacy, Lucknow, Amity University Uttar Pradesh, Noida, Uttar Pradesh, India
[2]Era College of Pharmacy, Era University, Lucknow, Uttar Pradesh, India
[3]Faculty of Pharmaceutical Sciences, Rama University, Mandhana, Kanpur (Uttar Pradesh), India
[4]Amity Institute of Pharmacy, Amity University Rajasthan, RIICO Kant Kalwar Industrial Area, Jaipur-Delhi Highway (Main Road), Jaipur, Rajasthan, India
[5]Department of Pharmaceutics, Shree S.K.Patel College of Pharmaceutical Education and Research, Ganpat University, Gujarat, India
[6]Department of Industrial Pharmacy, Faculty of Pharmacy, Silpakorn University, Nakhon Pathom, Thailand

Abstract

Neurohumoral transmission is the term utilized to describe the release of humoral (chemical) substances at the synapse and neuro-effector junction during the transmission of an impulse. Neurotransmitters (NT) shape the anatomy and physiology of the nervous system for entire lifespan. Synaptic transmission is the process by which NT generally enable their effector cells to respond, though this is not always the condition. It is relatively recent to recognize that the primary means of communication between neurons is chemical transmission across the synapse. Synapses are specialized structures that control neurotransmission, the process by which neurons communicate with one another. Although the groundwork for synaptic biophysics was laid a century ago, the development of advanced single cell recording methods, like calcium imaging and patch clamping, has made it possible to characterize certain characteristics of several synapses in the mammalian brain, usually with rodents serving as the experimental model. Here, we demonstrate

**Corresponding author*: shloksk@gmail.com

Bhupendra Prajapati, Alok Tripathi, Rishabha Malviya and Lucy Mohapatra (eds.) Computational Neuropharmacology: Fundamentals and Clinical Aspects, (247–270) © 2025 Scrivener Publishing LLC

how cerebellar synapses' neurotransmission characteristics are recapitulated by data-driven computational models. Microcircuit modeling simulations are beginning to shed light on the various ways that different synaptic pathways influence the spatiotemporal patterns of computation and circuit activity.

Keywords: Computational model, neurotransmitters, synapse

Abbreviations

5-HT	5-hydroxytryptamine
Ach	Acetylcholine
ADD	Attention deficit disorder
ADHD	Attention deficit hyperactivity disorder
ALS	Amyotrophic lateral sclerosis
AMPA	α-amino-3-hydroxy-5-methyl-4-isoxazolepropionic acid
CNS	Central nervous system
DA	Dopamine
EEG	Electroencephalography
fMRI	Magnetic resonance imaging
GABA	Gamma-aminobutyric acid.
K+	Potassium
MEG	Magnetoencephalography
NA	Noradrenaline
Na+2	Sodium
QSAR	Quantitative Structure-Activity Relationships
QSP	Quantitative systems pharmacology
REM	Rapid-eye-movement
SBD	Sleep behavior disorder

8.1 Introduction

The advancement of computational and experimental methods together has benefited the discipline of neuroscience by enabling the simulation and reconstruction of intricate models of neurons and synapses through neurohumoral transmission and signal transduction. Neurohumoral transmission refers to the process where nerves communicate across synapses and neuroeffector junctions through the release of chemical messengers. Typically, nerve signal transmission occurs through electrical or chemical means [1]. Electrical transmission involves the propagation of a rapid membrane depolarization, known as an action potential, either continuously or through saltatory conduction.

This type of transmission is primarily confined to individual nerve cells, where depolarization initiates, often near the soma, and travels down the axon towards the nerve ending, frequently forming a synapse with another nerve or a target cell like a smooth muscle or glandular cell. However, the depolarization itself cannot cross the synaptic cleft. At this juncture, neurohumoral transmission becomes essential, involving the release of NTs from the presynaptic neuron and their recognition by receptors on the postsynaptic neuron or target cell. Understanding the dynamics of agonist/antagonist mechanisms and signal transduction via various receptors is crucial for drug discovery and the development of therapeutic strategies [2, 3].

Recent years have seen significant progress with the determination of high-resolution crystal structures of receptors related to NTs from different subfamilies, providing a solid foundation for both experimental and computational studies of NT-related diseases. Presynaptic vesicle cycling, NT diffusion, and postsynaptic receptor activation belong to the procedures that define chemical connections [4]. These steps lead to changes in postsynaptic currents and alterations to the membrane's potential. Within the cerebellar cortex's network, these processes have been successfully modelled for a variety of connections and transmitter subtypes like AMPA, NMDA, and GABA), allowing for the simulation of their effects on computational processes. Short-term synaptic plasticity is particularly significant due to the fact that it affects the spatiotemporal filtering in local microcircuits, manages burst transmission, and directs information flow in the network that result in correcting the various pathological conditions like Depression, Parkinson's disease (PD), attention deficit hyperactivity disorder (ADHD), multiple sclerosis (MS), Huntington's disease (HD) and Tourette syndrome (TS) [5, 6].

The cerebellum junctions' neural communication properties can now be reproduced by data-driven computer models. The impacts of various synaptic processes to the spatiotemporal dynamics of circuit activity and computation are starting to become clearer with the help of these models. Through the mathematical translation of subcellular and synaptic mechanics. By translating cellular and synaptic biophysics into mathematical equations, models of neurons and networks can be generated, offering a new avenue for computational exploration of neurotransmission [7, 8]. Cerebellar circuits are renowned for exhibiting a diversity of postsynaptic flows, depolarization of the membrane kinetics, and immediate memory patterns due to an array of synaptic attributes, including postsynaptic receptor mechanisms and presynaptic discharge. This review sheds perspective on the influence of synaptic heterogeneity on microcircuit computation by comparing neurotransmission mechanisms at various cerebellar synapses and investigating the relevant computational models [9, 10].

8.2 Neurohumoral Transmission and Signal Transduction

Neurohumoral transmission involves the transmission of messages between nerves, synapses and in between effector and neuronal region through the release of chemical messengers. Generally, nerve signals can be transmitted in two ways: electrically or chemically. Electrical transmission occurs when membrane depolarization, known as an action potential, propagates continuously or via saltatory conduction along individual nerve cells [11]. This type of transmission is typically limited to individual neurons, where depolarization begins near the soma and travels down the axon towards the nerve ending. However, the depolarization itself cannot cross the synaptic cleft. In such cases, neurohumoral transmission becomes essential. There are requirements that must be fulfilled for a substance to be recognized as a postjunctional neurohumoral transmitter. This process leads to the signal transduction of the neurological information. First of all, it must exist in the presynaptic neuron and be released into the medium in response to nerve activation. Applying this material should also cause reactions that are the same as those brought on by nerve stimulation. Finally, chemicals that either antagonize or potentiate the effects of nerve stimulation should also have an influence on its activity [12].

8.2.1 Basic Steps Involved in Neurohumoral Transmission

8.2.1.1 Impulse Conduction

The resting membrane potential of neurons, which is around seventy negative inside, and is established by the higher availability to potassium ions (K+) and the high concentration of K+ inside the axoplasm, combined with low permeability to sodium ions (Na+) and active extrusion of Na+ from the neuron. An electrical shock or stimulus induces a neuron's sodium admittance rise unexpectedly, causing the cell to lose polarization and exceed its membrane potential, attaining about plus twenty mV inside [13]. Repolarization arises as a consequence of potassium ions departing the cells along the gradient of concentration. During the reaction phase, the sodium potassium pumps attempt to return the ionic distribution to its typical state. At the next excitable region of the membrane (such as the subsequent node of Ranvier in myelinated fibers), the next action potential generates local circuit flows that open ionic channels, allowing the action potential to propagate without deterioration. When neuron terminals' sodium admittance expands, it is more frequently inhibited by aflatro-do-toxin discovered in puffer fish alongside shellfish consist of shalitoxin, that limits pulse propagation [14, 15].

8.2.1.2 Transmitter Release

"Synaptic vesicles" are the structures in which the transmitter either excitatory or inhibitory is kept in prejunctiional nerve terminals as shown in Figure 8.1. In fluidizing membranes by Ca2+ entry, nerve impulses facilitate the fusion of vesicular and axonal membranes. Exocytosis, or the extrusion of all of the vesicle's contents, including the transmitter, enzymes, and other proteins, occurs at the junctional cleft. In order to modify the expulsion process, the messenger and other chemicals may stimulate particular receptors on the prejunctional membrane. For example, dopamine (DA), which adenosine, the cytokines peptides called, and NA itself all minimize the release of noradrenaline, while the AT1 and α2 receptors strengthen the discharge of NA. A similar inhibition of acetylcholine (ACh) release in autonomic neuroeffector locations is observed with α2 and muscarinic agonists (but not in ganglia and skeletal muscles) [16].

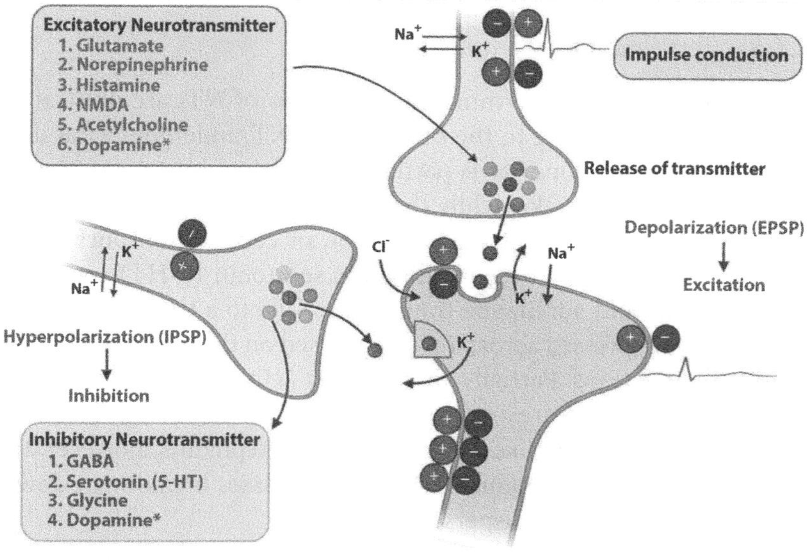

Figure 8.1 Diagrammatic representation of steps involved in excitatory and inhibitory neurohumoral transmission.

8.2.1.3 Transmitter Action on Postjunctional Membrane

NTs are released and, depending on their kind, bind to particular receptors on the postsynaptic membrane to cause either excitatory or inhibitory postsynaptic potentials (EPSPs or IPSPs). Depolarization and potassium efflux are the outcomes of EPSPs, which are characterized by an increase in permeability to cations like calcium and sodium [17]. Along their concentration gradients, these ionic motions take place passively. On the other hand, IPSPs result from an increase in permeability to anions, causing chloride ions to move into the neuron (since the axonal concentration of chloride goes lower than its extracellular concentration), resulting in membrane hyperpolarization. Additionally, membrane stabilization or hyperpolarization can occur due to a selective increase in permeability to potassium ions, which move out of the neuron carrying positive charges [18].

8.2.1.4 Postjunctional Activity

A suprathreshold EPSP triggers a propagated postsynaptic action potential (AP) in the neuron, leading to the generation of a nerve impulse. In muscles, this impulse results in contraction, while in glands, it leads to secretion. Conversely, IPSP stabilizes the postsynaptic membrane and opposes activated stimuli [19].

8.2.1.5 Termination of Transmitter Action

Various mechanisms for terminating the action of NTs are illustrated in Figure 8.1. Upon binding to the receptor, the NT undergoes local degradation (e.g., acetylcholine) or is partially reabsorbed into the presynaptic neuron via active reuptake, while the remainder diffuses away [20]. This process is made easier by the proliferation of certain transport proteins within the neuronal barrier, including the serotonin (5-HT) transporter, DA transporter, and adrenaline transporter. One to a thousand reactions per sec may be expressed across a synapse based on the speed at which NTs conclude their actions. Partially absorbed via active transport into nearby glial cells and synapses are amino acid messengers such as glutamate and GABA. For peptide NTs like VIP, NPY, and enkephalins, however, there is no active reuptake mechanism. Rather, peptidases break them down at different locations as they disperse away [21].

(Steps in neurohumoral transmission are shown in the figure in which excitatory neurotransmitter such as glutamate, norepinephrine, histamine,

NMDA, Acetylcholine leads to impulse conduction and Depolarization of cells takes place with intake of Na+ ions at EPSP which is countered with the Hyperpolarization caused by inhibitory neurotransmitter like GABA, serotonin and glycine at IPSP by K+ ions).

Abbreviations: EPSP- Excitatory postsynaptic potential; IPSP-Inhibitory postsynaptic potential.

8.2.2 Excitatory and Inhibitory Neurotransmitter

Early human development is significantly influenced by NTs that play a role in an assortment of functions including neural communication, neuronal growth, differentiation, and neuronal network development. Diverse stages of development may see the development of these transmitters; monoamines, such as NA are known to exist before neuron distinctions, specifically in early developmental structures like a notochord. Another factor in morphogenesis is serotonin. Late in the process of ontogenesis excitatory amino acid sequences become more common, and as novel synapses form, they increase their concentrations [22, 23]. After arriving during the neonatal period, some NTs, like glutamate, consolidate. The development of neural networks can be compromised by aspects like hypoxia and drug exposure, which can have unfavorable consequences over the long term. Acetylcholine, glutamate, gamma amino butyric acid (GABA), glycine, DA, norepinephrine, 5-HT, and histamine are just a few of the NTs that the body uses for diverse purposes. Primarily acting as an excitatory NT, glutamate serves as vital for nervous system remodeling and can also help in retaining memories. On the contrary hand, GABA and glycine constitute significant inhibitory neurotransmitters [24, 25]. Part of the brain's inhibitory action is associated with GABA alone, while glycine is primarily found in the spinal cord. A critical transmitter that is additionally involved in learning, encouragement, feelings, control of movement, and executive functions within the brain is DA [26]. Psychological and mental disorders are linked to the instability of DA. A lot of treatments for neurological and mental illnesses target 5-HT because it impacts several kinds of neurological functions and neuronal activity. Furthermore, serotonin affects the movements of the colon, the control of the bladder, and activity of the circulatory system. Synthesized in the sympathetic and central nervous systems, NA is a monoamine that modulates stress, sleep, focus, attention, inflammation, and autonomic nervous system responses. Another NT that decides feeding behavior, promotes wakefulness, modulates motivational behavior, and regulates homeostatic processes is histamine [27, 28].

8.2.3 Excitatory and Inhibitory Neurotransmitters in Various Pathological Conditions

GABA is the major NT performing inhibitory functions in the central nervous system that acts as a target for treating various pathological conditions such as anxiety disorders, insomnia, and epilepsy [29]. Drugs aimed at altering GABAergic function act on GABA-A&B receptors. In addition to being associated with many physiological processes, catecholamine mainly DA is also linked to the pathogenesis of cognitive and behavioral disorders. A variety of conditions are linked to abnormalities in DA neural communication: Tourette syndrome, schizophrenia, psychosis, depression, PD, MS, HD, and ADHD. According to research, PD is exacerbated by DAergic neurons in the substantia nigra pars compacta deterioration [30, 31].

Depression is linked to the pathogenesis of 5-HT, a chemical found in neurons that regulates various neuropsychiatric functions. According to research, individuals with endogenous anxiety have low plasma levels of tryptophan, which is precursor to 5-HT. Reduced 5-HT levels in the brain, observed postmortem, are associated with suicide among depressed individuals. As a result, pharmaceuticals that target 5-HT are created to treat depression. TCAs are one such treatment that does this by raising 5-HT levels in the synapses [32]. Because NA dysregulation, changed norepinephrine availability, modification of synaptic receptors, and alterations in locus coeruleus firing all contribute to the pathophysiology of neuropsychiatric diseases, norepinephrine is implicated in these disorders. Posttraumatic stress disorder, attention-deficit hyperactivity disorder, anxiety disorders, mood disorders, and Alzheimer's disease are among the illnesses linked to norepinephrine malfunction [33]. Neural circuit malfunction involving norepinephrine is the direct cause of many symptoms associated with these illnesses. Since glutamate is the primary excitatory NT in the brain, it may have an impact on memory storage as well as NS plasticity. Histamine, a NT produced in mast cells, contributes to the pathogenesis of IgE-mediated diseases. It binds to specific histamine receptors, influencing features like bronchospasm and mucosal edema in asthma. Histamine also plays a role in multiple sclerosis pathogenesis, affecting blood-brain barrier permeability and facilitating neuro-inflammation through increased central nervous system cell infiltration [34, 35]. Table 8.1 enlists the different types of neurotransmitters and their role in various neurological disorders.

8.2.4 Cotransmission

It is now clear that the traditional "one neuron—one transmitter" concept is very simplistic. Upon activation, the majority of central and peripheral

Table 8.1 Neurotransmitters and their role in various neurological disorders.

Neurotransmitters	Area of location	Nature of neurotransmitter	Role in pathological conditions	Target receptors	References
GABA	Spinal cord	Inhibitory	Anxiety disorders, insomnia, and epilepsy	GABA-A and GABA-B	[36]
DA	Substantia nigra and ventral tegmental	Both excitatory and inhibitory	Parkinson's disease, schizophrenia, psychosis, depression, Tourette syndrome, attention deficit hyperactivity disorder, multiple sclerosis, and Huntington's disease	D1 and D5	[37]
Glutamate	Synaptic vesicles of Glial cells	Excitatory	Anxiety, Parkinson's Disease, Alzheimer's Disease, Huntington's disease, OCD and Schizophrenia	NMDA and AMPA	[38]
5-HT	Raphe nuclei of Brainstem	Inhibitory	Depression, Migraine	5-HT 1A	[39]
Norepinephrine	Locus coeruleus of Brain stem	Excitatory	Depressive disorders, Schizophrenia, ADHD, Sleeping disorder.	α and β	[40]

(*Continued*)

Table 8.1 Neurotransmitters and their role in various neurological disorders. (*Continued*)

Neurotransmitters	Area of location	Nature of neurotransmitter	Role in pathological conditions	Target receptors	References
Histamine	Neurons of hypothalamus and Mast cells	Excitatory	Neuroinflammation	H1- H4	[41]
Glycine	Spinal cord	Inhibitory	Spinal nociceptive control	Strychnine-sensitive glycine receptors and NMDA	[42]
NMDA	Near Astrocyte, glial cells and also within the hippocampal region	Excitatory	Epilepsy, Parkinson's Disease, Alzheimer's Disease and Huntington's Disease	NMDA-R1	[43]
Acetylcholine	Ventral tegmental area hippocampus, prefrontal cortex and amygdala region	Excitatory	Parkinson's disease, Alzheimer's Disease	$M_1 M_4 M_5$	[44]

neurones have been demonstrated to produce several active substances. Apart from the major transmitters, ACh and NA, neurones have also been shown to produce co-transmitters in the ANS, including peptides (such substance P, enkephalins, somatostatin, and vasoactive intestinal peptide, or VIP), purines (ATP, adenosine), nitric oxide (NO), and prostaglandins. Most autonomic cholinergic neurones have ATP linked to both ACh and NA, although VIP is associated to Ach [45].

Some parasympathetic sites known as nitrergic nerves transmit Nitric Oxide as a transmitter. NPY is a substance found in vascular adrenergic neurons that results in persistent vasoconstriction. The cotransmitter is kept in different synaptic vesicles or sites inside the same neurone. On the other hand, NA and ATP are kept in the same vesicle [46]. The neuromodulator job of the contransmitter is to control the postsynaptic sensitivity to the main transmitter and/or its presynaptic release upon release by a nerve impulse. In addition, the cotransmitter may act as a separate alternative transmitter and/or have a trophic effect on the synaptic structures [47].

Nonadrenergic, noncholinergic (NANC) transmission has been shown to play a role in the autonomic input of the urinary path, glands that produce saliva, stomach, vas deferens, and some blood vessels. Even when total adrenergic and cholinergic inhibition is present, nerve stimulation can still cause restricted responses in certain areas. One example is the biphasic contractile response that results from stimulating the sympathetic nerve to the guinea pig vas deferens. ATP mediates the first phase of this response through P2 receptors, whereas NA mediates the second phase through $\alpha 1$ receptors. The primary transmitter and the cotransmitter often have distinct time courses of activity [48].

The other cotransmitter (VIP) of the parasympathetic cells stirs a more gradual and broadened action, while NO works in a transitional time-course among VIP and ACh. The cotransmitter NPY acts more slowly in sympathetic neurones than NA, but ATP acts more quickly. Cotransmitters such as NO, VIP, and NPY can also influence receptors some distance from the site of release by diffusing to a larger region [49].

8.3 Computational Approach in Creating Neurohumoral and Synaptic Models

Neurotransmission is the mechanism by which synapses, which are designed specifically, control communication among nerve cells. The tenets of synaptic biological physics have been developed in the course of the last century, but the creation of sophisticated one cell capturing techniques, like imaging with calcium and patch clenching that have primarily been tested on animals as an experimental platform, has recently opened the door for the examination of particular aspects of several synapses in the nervous system of mammals [50, 51].

The cerebellum circuit is a pertinent example, as seen by the many features displayed by its synapses. A wide range of postsynaptic currents, membrane depolarization kinetics, and short-term plasticity patterns are produced by these, including presynaptic releases and postsynaptic receptor processes [52]. The combining of this enormous collection of experimental data, computational models can be used to reconstitute a wide range of synaptic receptor kinetics (AMPA, NMDA, and GABA) and neurotransmission properties [53, 54].

8.3.1 Mechanisms and Models of Synaptic Transmission

8.3.1.1 Electrical Synapses

A vast network of electrical connections forms the structure of the mammalian neural network, encompassing the cerebral cortex, gap junctions (GJs) that are formed out of two hemichannels called connexons and mediate electrical synaptic transmission. These are generated by integral membrane proteins known as connexins [55, 56]. According to new research, electrical synapses are intricate, multifunctional, and highly mutable structures that control when spikes occur and how rhythmic activity synchronizes in connected neural networks [57, 58].

8.3.1.2 Chemical Synapses

It signifies inter axonal and dendritic connectivity that must be nearby for a typical chemical synapse to occur [59, 60]. The two areas, referred to as the pre & postsynaptic, have no contact, providing an empty area referred to as the cleft of synaptic neurons. The procedure of compartments fusing with the presynaptic membrane leads to the release of NT. Eventually at the postsynaptic receptors, this NT diffuses and can mix with G proteins along with additional kinases to generate either ionotropic or metabotropic second messenger receptors. So, neurotransmission modeling requires to incorporate the vesicle cycle, NT propagation, and stimulation to receptors [61]. Figure 8.2 shows the different elements and working of chemical synapse.

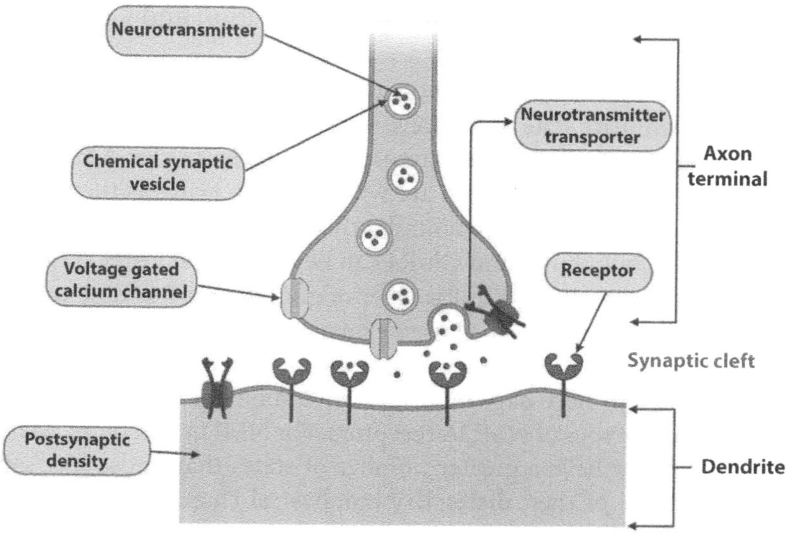

Figure 8.2 Diagram of the chemical synapse.

(The postsynaptic element of the effector cell, such as the dendrite, and the presynaptic component, the axon terminal, are present in typical chemical synapses. NTs housed in synaptic vesicles can be more easily exocytosed when synapses have voltage-dependent calcium channels. The actions of neurotransmitters are mediated by postsynaptic neurotransmitter receptors, which become active after the neurotransmitter is released. Synapses have transporters to retake the released neurotransmitter in order to stop the transmitter's function).

8.3.1.2.1 Presynaptic Site—Vesicle Cycle

Neurotransmitter-containing vesicles are produced, stored, and released at the presynaptic location by sophisticated biochemical machinery.

8.3.1.2.2 Synaptic Cleft—Neurotransmitter Diffusion

Complex mechanisms, including diffusion, receptor binding, and glial cell uptake, are involved in maintaining neurotransmitter homeostasis in and around a synaptic cleft [62, 63]. Numerous studies using neurotransmitter diffusion-based modeling have examined the activation of synaptic receptors and assessed the neurotransmitters' access to synaptic and extra-synaptic locations while taking glial cell sequestration into account [64].

8.3.1.2.3 Postsynaptic Site—Receptor Activation and Current Generation

The most basic Markov model, known as the Markov chain, is frequently employed to simulate state transitions, such as those between closed, open, and inactivated states [65]. The heterotetramer receptors for AMPA make up hetero pairs of dimers, which are with GluR2 serving as the principal subunit. GluR1 and GluR2 are abundantly expressed in every level of the cerebral cortex, but GluR3 and GluR4 can be detected at lower concentrations on Bergmann glia and in the layer of granules on Purkinje cell (PC) dendrites that are required to cause permanent potentiation. The cerebellum has four subunits at numerous locations. The hetero-tetramers that comprised the two necessary NR1 and NR2 subunits (A–D) give rise to four different subtypes of NMDA receptors. For NMDA receptors, Markov models frequently have a greater number of states than those for AMPA receptors because of their distinctive biophysical characteristics given by the wide range of the NR2 subunit [66, 67]. Three closed and two desensitized states, as well as two open states, were identified by kinetic study of postsynaptic currents produced by glutamate after extended exposure to glycine. According to one of the study the most prevalent stoichiometry in the mammalian central nervous system is two a, two b, and one g subunit (s). Macroscopic currents with quick activation, bi-exponential deterioration, and desensitized kinetics that match protracted closed regimes occur when brief GABA bursts were delivered to external macro sections obtained from synapses [68].

8.3.1.3 The Granular Layer

The granular layer is the input layer of the cerebellar cortex. It receives mossy fibers, one of the major inputs to the cerebellum, originating from multiple brainstem nuclei, and is composed of different types of cells [69]. The most numerous are small and closely packed granule cells (GrCs), which emit four short dendrites on average and a single axon ascending vertically into the molecular layer, where they bifurcate to form parallel fibers. A few types of interneurons, including excitatory unipolar brush cells (UBCs), inhibitory Golgi cells (GoCs), mossy fiber-granule cell (Mf-GrC) and Lugaro cells, are located in this layer. GoC axons are profusely arborized and terminate on GrC dendrites, together with mossy fibers, in a peculiar synaptic structure called a "glomerulus" [70, 71].

8.3.1.4 *The Molecular Layer*

The outermost area of the cerebellar cortex is called the molecular layer. This layer comprises massive parallel fibers and parasagittal PC dendritic trees, which separate at the end of the GrC ascending axon. At least 400 PCs' spines have glutamatergic synapses formed on them thanks to thousands of parallel fibers penetrating their dendritic trees. As some of the most complicated in the central nervous system, these neurons have several voltage-dependent ionic channels as well as a large number of proteins that buffer calcium entry [72, 73].

In addition, the molecular layer includes the two main types of regulatory interneurons likely stellate cells (SCs) and basket cells (BCs). Though BCs are often found in the bottom third of the molecular surface and establish GABAergic synapses on PC entities, SCs are usually considered to reside in the top two-thirds of the molecular layer and form GABAergic synapses on PC dendrites. SC aspects include pacemaker activity, virtually gradually raising spike regularity in response to electric current strength, slumping inward correcting in reactions to injection of hyperpolarizing current, and rebound excitation and pause following the current depolarization step after the hyperpolarization [74]. Additionally, during repeated parallel fiber transmission, SCs demonstrated notable short-term facilitation. Simulators, interestingly, suggested that SCs might play a new function in molecular layer low-pass and band-pass filtering, recoding parallel fiber bursts and controlling PC gain in a frequency-dependent way [75].

8.4 Primitive Computational Models

8.4.1 Metabotropic Receptor

Metabotropic receptors do not cause postsynaptic currents; instead, they influence ionic channels more slowly because of the way different enzymes translate the receptor signal via cytoplasmic cascades and second messengers. According to Yamasaki *et al.* (2021) this multi-step procedure is essential for regulating delayed responses, development, and cellular survival [76]. All mGluR subunits—aside from mGluR3 and mGluR6—are expressed in the cerebellum. The most prevalent is mGluR1, which is expressed in other cerebellar neurons in addition to PCs. While mGluR4 is limited to parallel fibers, some subtypes of GoC possess mGluR2 and mGluR5 [76]. As per Wu and Kapfhammer (2022), this receptor controls delayed electrical responses via TRPC channels, which are essential for

PCs' dendritic structure formation and restoration [77, 78]. Particularly in PC spines, dendritic shafts, and parallel fibers, the GABA (B) receptor is present in certain regions of the cerebellum. According to Deskexhe *et al.* (1994), a computational model of the ionic channels Kir3 and GABA (B) has been developed to mimic physiological reactions [79].

8.4.2 Electrical Synapses and Ephaptic Coupling

Electrical connections between GoCs are established through GJs, which enhance low-frequency oscillatory synchronization and resonance in GoC networks. When synchronized GoCs receive asynchronous excitatory inputs, desynchronization occurs quickly and significantly [80]. A better-validated computational model involving GoCs via synapses linked in the granular layer corroborated previous research showing that GJs between GoCs amplified the strength of oscillations in the cerebellar cortex, which had been synaptically induced by the feedback loop among Mf-GoCs and GrCs [81].

Compared to chemical synapses, GJs are electrical synapses that offer instantaneous transmission without any failures [82]. They can send signals in both ways; however, they only send signals in one direction using synaptic rectification [82]. The many reactions that PCs generate can be used to track the physiological effects of this arrangement. A delay lasting up to 60–70 ms is produced by synapses generated by SCs within the vast dendritic tree of PCs [83]. Table 8.2 provides a list of different properties of cerebral circuits in NMDA and GABA-A neuronal synapse evaluated in local microcircuit.

Table 8.2 Properties of cerebral circuit in NMDA and GABA-A neuronal synapse evaluated in local microcircuit.

Synapses	Gmax NMDA	Subtype	GABA-A	Subtype	References
GrC-BC	Above 2000	GluR2	No action	-	[87]
SC-PC	No action	-	Above 2500	α-1	[87]
SC-SC	No action	-	Above 1500	α-1	[87]
BC-BC	No action	-	Above 1500	α-1	[87]
Mf-GoC	Above 10000	NR2B	No action	-	[88]
GrC-GoC aa	Above 10000	NR2B	No action	-	[88]
Mf-GrC	Above 18000	NR2A-NR2C	No action	-	[89]

It was originally found in mouse spinal nerve roots between spontaneously activated fibers that haptic coupling, which permits neurons to exchange electrical impulses across extracellular space, occurs. Shortly afterward, it was shown that ephaptic signaling controlled hippocampal cell synchronization and time-spiking [84].

The tactile connection facilitated by BC synapses onto PCs is not the same as that between PCs. Phaptic transmission has an inhibiting effect in the latter scenario. The BC pinceau generates an extracellular field that inhibits PC firing quickly due to dual current components: via voltage-activated potassium conductance, a capacitive current and an ionic current [85]. On the other hand, sodium current flow dominates the extracellular field of PC AISs, resulting in a strong excitatory signal that in turn encourages the synchronization of neighboring PCs [86].

8.5 Conclusion

Based on the inherent electro-responsiveness and the transmission of signal characteristics of the brain shown in experiments, precise biophysical models are being created. The cellular and subcellular architecture of neurons and synapses and their function can be understood from a different angle thanks to these models. The initiation of intricate computational models can offer novel avenues for experimental research as well as clues towards comprehending neuro-pathologies, even though an overwhelming amount of molecular features still require system integration and theoretical justification. Cerebellar illnesses that impact the cerebellum include paroxysmal dyskinesia, dystonia, ataxia, and autism. Investigation into modifications to structure, receptors, and metabolic processes would be necessary to translate these disorders into models since they impact synaptic behavior. This means expanding the models to incorporate the process of long-term synaptic plasticity that is dependent on calcium and neuromodulation that is dependent on metabotropic receptors. The challenge is threefold: first, it involves optimizing single-neuron and synaptic models. Initially, the candelabrum cells, unipolar brush, Lugaro, and other surviving neurons and synapses of the cerebellar network adjust the activity of the cerebellar circuit using recurrent inhibitory loops. Second, a revised version of the cerebellar cortical model incorporates these additional cellular models and synapses. We hope that in the end, this roadmap will clarify the intrinsic algorithms of the cerebellar circuit and provide fresh insight into the computational consequences of neurohumoral transmission and synaptic mechanisms.

References

1. Hoehne, A., McFadden, M.H., DiGregorio, D.A., Feed-forward recruitment of electrical synapses enhances synchronous spiking in the mouse cerebellar cortex. *Elife*, 9, e57344, 2020 Sep 29.
2. Kita, K., Albergaria, C., Machado, A.S., Carey, M.R., Müller, M., Delvendahl, I., GluA4 facilitates cerebellar expansion coding and enables associative memory formation. *Elife*, 10, e65152, 2021 Jul 5.
3. Fryer, A.D., Christopoulos, A., Nathanson, N.M., editors. *Muscarinic Receptors*. Handbook of Experimental Pharmacology. Springer Berlin Heidelberg; 2012.
4. Kleinle, J., Vogt, K., Lüscher, H.R., Müller, L., Senn, W., Wyler, K., Streit, J., Transmitter concentration profiles in the synaptic cleft: an analytical model of release and diffusion. *Biophys. J.*, 71, 5, 2413–26, 1996 Nov 1.
5. Beemer, G.H. and Rozental, P., Postoperative neuromuscular function. *Anaesth. Intensive Care*, 14, 1, 41–5, 1986 Feb.
6. Masoli, S., Tognolina, M., Laforenza, U., Moccia, F., D'Angelo, E., Parameter tuning differentiates granule cell subtypes enriching transmission properties at the cerebellum input stage. *Commun. Biol.*, 3, 1, 222, 2020 May 8.
7. Hughes, J., Isolation of an endogenous compound from the brain with pharmacological properties similar to morphine. *Brain Res.*, 88, 2, 295–308, 1975 May 2.
8. Hughes, J.T., Smith, T.W., Kosterlitz, H.W., Fothergill, L.A., Morgan, B.A., Morris, H.R., Identification of two related pentapeptides from the brain with potent opiate agonist activity. *Nature*, 258, 5536, 577–9, 1975 Dec 18.
9. Masoli, S., Ottaviani, A., Casali, S., D'Angelo, E., Cerebellar Golgi cell models predict dendritic processing and mechanisms of synaptic plasticity. *PLoS Comput. Biol.*, 16, 12, e1007937, 2020 Dec 30.
10. Avoli, M. and Krnjević, K., The long and winding road to gamma-amino-butyric acid as neurotransmitter. *Can. J. Neurol. Sci.*, 43, 2, 219–26, 2016 Mar.
11. Docherty, J.R., Pre-junctional factors influencing adrenergic transmission, University of Glasgow, United Kingdom, 1979.
12. Reddy, S., Fung, A., Manlhiot, C., Selamet Tierney, E.S., Chung, W.K., Blume, E., Kaufman, B.D., Goldmuntz, E., Colan, S., Mital, S., Adrenergic receptor genotype influences heart failure severity and β-blocker response in children with dilated cardiomyopathy. *Pediatr. Res.*, 77, 2, 363–9, 2015 Feb.
13. Begg, E., *Book Review: Clinical pharmacology essentials*. Australian Prescriber [Internet]. 24, 4, 97, 2001 Aug 1.
14. Burstein, E.S., Spalding, T.A., Brann, M.R., Pharmacology of muscarinic receptor subtypes constitutively activated by G proteins. *Mol. Pharmacol.*, 51, 2, 312–9, 1997 Feb 1.
15. Tripathi K.D., *Essentials of Medical Pharmacology*, JP Medical Ltd, 411–24, Sep 30, 2013.

16. Fillenz, M., *Noradrenergic neurons*. Cambridge University Press, Cambridge, New York, 1990 May 25.
17. Kerwin, R. and Pycock, C., Baclofen (β-p-chlorophenyl-γ-aminobutyric acid) enhances [3H] γ-aminobutyric acid (3H-GABA) release from rat globus pallidus *in vitro*. *J. Pharm. Pharmacol.*, 30, 1, 622–7, 1978 Sep.
18. Helmer, H., Frequently asked questions on tocolytics. *BJOG: Int. J. Obstet. Gynaecol.*, 112, 94–6, 2005 Mar.
19. Straughan, D.W. and Legge, K.F., The pharmacology of amygdaloid neurones. *J. Pharm. Pharmacol.*, 17, 10, 675–7, 1965 Oct.
20. Ruffolo Jr., R.R., Nichols, A.J., Stadel, J.M., Hieble, J.P., Pharmacologic and therapeutic applications of alpha2-adrenoceptor subtypes. *Annu. Rev. Pharmacol. Toxicol.*, 33, 1, 243–79, 1993 Apr.
21. Blum, B.P. and Mann, J.J., The GABAergic system in schizophrenia. *Int. J. Neuropsychopharmacol.*, 5, 2, 159–79, 2002 Jun 1.
22. Hideomi, F., Yoshihisa, K., Hideki, O., Effects of β-(p-chlorophenyl)-GABA (baclofen) on spinal synaptic activity. *Eur. J. Pharmacol.*, 44, 1, 17–24, 1977 Jul 1.
23. Curtis, D.R. and Crawford, J.M., Central synaptic transmission-microelectrophoretic studies. *Annu. Rev. Pharmacol.*, 9, 1, 209–40, 1969 Apr.
24. Gray, E.G., Axo-somatic and axo-dendritic synapses of the cerebral cortex: an electron microscope study. *J. Anat.*, 93, Pt 4, 420, 1959 Oct.
25. Todorov, L.D., Mihaylova-Todorova, S., Westfall, T.D., Sneddon, P., Kennedy, C., Bjur, R.A., Westfall, D.P., Neuronal release of soluble nucleotidases and their role in neurotransmitter inactivation. *Nature*, 387, 6628, 76–9, 1997 May 1.
26. Chanowitz, B. and Langer, E.J., Premature cognitive commitment. *J. Pers. Soc. Psychol.*, 41, 6, 1051, 1981 Dec.
27. Segev, I., What do dendrites and their synapses tell the neuron? *J. Neurophysiol.*, 95, 3, 1295–7, 2006 Mar.
28. Menticoglou, S.M., Morrison, I., Harman, C.R., Manning, F.A., Lange, I.R., Maximum possible impact of tocolytics in preventing preterm birth: a retrospective assessment. *Am. J. Perinatol.*, 9, 05/06, 394–7, 1992 Sep.
29. Olson, D.M. and Zakar, T., Intrauterine tissue prostaglandin synthesis: regulatory mechanisms. *Semin. Reprod. Endocrinol.*, 11, 03, 234–249, 1993 Aug.
30. Massey, S.C. and Redburn, D.A., A tonic gamma-aminobutyric acid-mediated inhibition of cholinergic amacrine cells in rabbit retina. *J. Neurosci.*, 2, 11, 1633–43, 1982 Nov 1.
31. Villalobos, N. and Magdaleno-Madrigal, V.M., Pallidal GABA B receptors: involvement in cortex beta dynamics and thalamic reticular nucleus activity. *J. Physiol. Sci.*, 73, 1, 14, 2023 Jun 16.
32. Kano, M., Ohno-Shosaku, T., Hashimotodani, Y., Uchigashima, M., Watanabe, M., Endocannabinoid-Mediated Control of Synaptic Transmission. *Physiological Reviews [Internet]*. 89, 1, 309–80, 2009 Jan.

33. Allen, N.J., Astrocyte regulation of synaptic behavior. *Annu. Rev. Cell Dev. Biol.*, 30, 439–63, 2014 Oct 6.
34. Corty, M.M. and Freeman, M.R., Architects in neural circuit design: Glia control neuron numbers and connectivity. *J. Cell Biol.*, 203, 3, 395–405, 2013 Nov 11.
35. Watanabe, M., Maemura, K., Kanbara, K., Tamayama, T., Hayasaki, H., GABA and GABA receptors in the central nervous system and other organs. *Int. Rev. Cytol.*, 213, 1–47, 2002 Jan 1.
36. Berke, J.D., What does dopamine mean? *Nat. Neurosci.*, 21, 6, 787–93, 2018 Jun.
37. Vandenberg, R.J. and Ryan, R.M., Mechanisms of glutamate transport. *Physiol. Rev.*, 93, 4, 1621–57, 2013 Oct.
38. Silberstei, S.D., Serotonin (5-HT) and migraine. *Headache: J. Head Face Pain*, 34, 7, 408–17, 1994 Jul.
39. Delgado, P.L. and Moreno, F.A., Role of norepinephrine in depression. *J. Clin. Psychiatry*, 61, 5–12, 2000 Jan 1.
40. Hu, W. and Chen, Z., The roles of histamine and its receptor ligands in central nervous system disorders: An update. *Pharmacol. Ther.*, 175, 116–32, 2017 Jul 1.
41. Salceda, R., Glycine neurotransmission: Its role in development. *Front. Neurosci.*, 16, 947563, 2022 Sep 16.
42. Hansen, K.B., Yi, F., Perszyk, R.E., Menniti, F.S., Traynelis, S.F., NMDA receptors in the central nervous system, in: *NMDA receptors: Methods and protocols*, pp. 1–80, 2017.
43. Singh, S., Agrawal, N., Goyal, A., Role of Alpha-7-Nicotinic Acetylcholine Receptor in Alzheimer's Disease. *CNS Neurol. Disord.-Drug Targets (Formerly Curr. Drug Targets-CNS Neurol. Disord.)*, 23, 3, 384–94, 2024 Mar 1.
44. Shoemaker, J.K. and Gros, R., A century of exercise physiology: key concepts in neural control of the circulation. *Eur. J. Appl. Physiol.*, 5, 1–4, 2024 Mar.
45. Han, Y., Ren, J., Liu, B., Li, Y., Ullah, R., Mao, Y., Wu, X., Chen, S., Wang, B., Wu, Y., Zhao, L., Highly secure non-orthogonal multiple access based on key accompanying transmission in training sequence. *Opt. Express*, 32, 2, 1979–97, 2024 Jan 15.
46. Harb, A., Tacke, C., Vafadari, B., Hülsmann, S., The Relative Contribution of Glycine–GABA Cotransmission in the Core of the Respiratory Network. *Int. J. Mol. Sci.*, 25, 6, 3128, 2024 Mar 8.
47. McLennan, H. and York, D.H., The action of dopamine on neurones of the caudate nucleus. *J. Physiol.*, 189, 3, 393–402, 1967 Apr 1.
48. Koch, C., Cable theory in neurons with active, linearized membranes. *Biol. Cybern.*, 50, 1, 15–33, 1984 Feb.
49. Brorson, J., *Synapses*. Perspectives in Biology and Medicine. 46, 4, 600-2, 2003.

50. Hickman, D.L., Johnson, J., Vemulapalli, T.H., Crisler, J.R., Shepherd, R., Commonly used animal models, in: *Principles of animal research for graduate and undergraduate students*, p. 117, 2017.
51. D'Angelo, E., Antonietti, A., Casali, S., Casellato, C., Garrido, J.A., Luque, N.R., Mapelli, L., Masoli, S., Pedrocchi, A., Prestori, F., Rizza, M.F., Modeling the cerebellar microcircuit: new strategies for a long-standing issue. *Front. Cell. Neurosci.*, 10, 176, 2016 Jul 8.
52. Schepper, R.D., Geminiani, A., Masoli, S., Rizza, M.F., Antonietti, A., Casellato, C., D'Angelo, E., Scaffold modelling captures the structure-function-dynamics relationship in brain microcircuits, BioRxiv, 2021-07, 2021 Aug 1.
53. D'Angelo, E. and Jirsa, V., The quest for multiscale brain modeling. *Trends Neurosci.*, 45, 10, 777–90, 2022 Oct 1.
54. Kumar, N.M. and Gilula, N.B., The gap junction communication channel. *Cell*, 84, 3, 381–8, 1996 Feb 9.
55. Evans, W.H. and Martin, P.E., Gap junctions: structure and function. *Mol. Membr. Biol.*, 19, 2, 121–36, 2002 Jan 1.
56. Nielsen, M.S., Nygaard Axelsen, L., Sorgen, P.L., Verma, V., Delmar, M., Holstein-Rathlou, N., Gap Junctions. *Comprehensive Physiology [Internet].* 1981–2035, 2012 Jul..
57. Destexhe, A., Contreras, D., Steriade, M., Mechanisms underlying the synchronizing action of corticothalamic feedback through inhibition of thalamic relay cells. *J. Neurophysiol.*, 79, 2, 999–1016, 1998 Feb 1.
58. Pham, T. and Haas, J.S., Electrical synapses between inhibitory neurons shape the responses of principal neurons to transient inputs in the thalamus: a modeling study. *Sci. Rep.*, 8, 1, 7763, 2018 May 17.
59. Andreae, L.C. and Burrone, J., The role of neuronal activity and transmitter release on synapse formation. *Curr. Opin. Neurobiol.*, 27, 47–52, 2014 Aug 1.
60. Clements, J.D., Lester, R.A., Tong, G., Jahr, C.E., Westbrook, G.L., The time course of glutamate in the synaptic cleft. *Science*, 258, 5087, 1498–501, 1992 Nov 27.
61. Franks, K.M., Bartol, T.M., Sejnowski, T.J., A Monte Carlo model reveals independent signaling at central glutamatergic synapses. *Biophys. J.*, 83, 5, 2333–48, 2002 Nov 1.
62. Barbour, B., An evaluation of synapse independence. *J. Neurosci.*, 21, 20, 7969–84, 2001 Oct 15.
63. Wässle, H., Puller, C., Müller, F., Haverkamp, S., Cone contacts, mosaics, and territories of bipolar cells in the mouse retina. *J. Neurosci.*, 29, 1, 106–17, 2009 Jan 7.
64. Yamada, W.M. and Zucker, R.S., Time course of transmitter release calculated from simulations of a calcium diffusion model. *Biophys. J.*, 61, 3, 671–82, 1992 Mar 1.
65. Jones, M.V. and Westbrook, G.L., Desensitized states prolong GABAA channel responses to brief agonist pulses. *Neuron*, 15, 1, 181–91, 1995 Jul 1.

66. Mugnaini, E., Di, M.R., Jaarsma, D., The unipolar brush cells of the mammalian cerebellum and cochlear nucleus: cytology and microcircuitry. *Prog. Brain Res.*, 114, 131–50, 1997 Jan 1.
67. Mugnaini, E., Sekerková, G., Martina, M., The unipolar brush cell: a remarkable neuron finally receiving deserved attention. *Brain Res. Rev.*, 66, 1-2, 220–45, 2011 Jan 7.
68. Voogd, J. and Glickstein, M., The anatomy of the cerebellum. *Trends Cognit. Sci.*, 2, 9, 307–13, 1998 Sep 1.
69. Akemann, W. and Knöpfel, T., Interaction of Kv3 potassium channels and resurgent sodium current influences the rate of spontaneous firing of Purkinje neurons. *J. Neurosci.*, 26, 17, 4602–12, 2006 Apr 26.
70. Rizza, M.F., Locatelli, F., Masoli, S., Sánchez-Ponce, D., Muñoz, A., Prestori, F., D'Angelo, E., Stellate cell computational modeling predicts signal filtering in the molecular layer circuit of cerebellum. *Sci. Rep.*, 11, 1, 3873, 2021 Feb 16.
71. Vikram Singh, A., Gharat, T., Batuwangala, M., Park, B.W., Endlein, T., Sitti, M., Three-dimensional patterning in biomedicine: Importance and applications in neuropharmacology. *J. Biomed. Mater. Res. B Appl. Biomater.*, 106, 3, 1369–82, 2018 Apr.
72. Masoli, S., Solinas, S., D'Angelo, E., Action potential processing in a detailed Purkinje cell model reveals a critical role for axonal compartmentalization. *Front. Cell. Neurosci.*, 9, 47, 2015 Feb 24.
73. Anwar, H., Hong, S., De Schutter, E., Controlling Ca 2+-activated K+ channels with models of Ca 2+ buffering in Purkinje cells. *Cerebellum*, 11, 681–93, 2012 Sep.
74. Sultan, F. and Bower, J.M., Quantitative Golgi study of the rat cerebellar molecular layer interneurons using principal component analysis. *J. Comp. Neurol.*, 393, 3, 353–73, 1998 Apr 13.
75. Yamasaki, M., Aiba, A., Kano, M., Watanabe, M., mGluR1 signaling in cerebellar Purkinje cells: Subcellular organization and involvement in cerebellar function and disease. *Neuropharmacology*, 194, 108629, 2021 Aug 15.
76. Knöpfel, T. and Grandes, P., Metabotropic glutamate receptors in the cerebellum with a focus on their function in Purkinje cells. *Cerebellum*, 1, 19–26, 2002 Mar.
77. Kulik, Á., Nakadate, K., Nyíri, G., Notomi, T., Malitschek, B., Bettler, B., Shigemoto, R., Distinct localization of GABAB receptors relative to synaptic sites in the rat cerebellum and ventrobasal thalamus. *Eur. J. Neurosci.*, 15, 2, 291–307, 2002 Jan.
78. Destexhe, A., Mainen, Z.F., Sejnowski, T.J., Synthesis of models for excitable membranes, synaptic transmission and neuromodulation using a common kinetic formalism. *J. Comput. Neurosci.*, 1, 195–230, 1994 Aug.
79. Dugué, G.P., Brunel, N., Hakim, V., Schwartz, E., Chat, M., Lévesque, M., Courtemanche, R., Léna, C., Dieudonné, S., Electrical coupling mediates

tunable low-frequency oscillations and resonance in the cerebellar Golgi cell network. *Neuron*, 61, 1, 126–39, 2009 Jan 15.

80. Shen, W., Plotkin, J.L., Francardo, V., Ko, W.K., Xie, Z., Li, Q., Fieblinger, T., Wess, J., Neubig, R.R., Lindsley, C.W., Conn, P.J., M4 muscarinic receptor signaling ameliorates striatal plasticity deficits in models of L-DOPA-induced dyskinesia. *Neuron*, 88, 4, 762–73, 2015 Nov 18.
81. Simões de Souza, F.M. and De Schutter, E., Robustness effect of gap junctions between Golgi cells on cerebellar cortex oscillations. *Neural Syst. Circuits*, 1, 1–9, 2011 Dec.
82. De Schutter, E., editor. *Computational neuroscience: realistic modeling for experimentalists*. CRC press, Taylor and Francis, New York, 2000 Nov 22.
83. Masoli, S., Rizza, M.F., Tognolina, M., Prestori, F., D'Angelo, E., Computational models of neurotransmission at cerebellar synapses unveil the impact on network computation. *Front. Comput. Neurosci.*, 16, 1006989, 2022 Oct 28.
84. Snipas, M., Rimkute, L., Kraujalis, T., Maciunas, K., Bukauskas, F.F., Functional asymmetry and plasticity of electrical synapses interconnecting neurons through a 36-state model of gap junction channel gating. *PLoS Comput. Biol.*, 13, 4, e1005464, 2017 Apr 6.
85. Steuber, V., Mittmann, W., Hoebeek, F.E., Silver, R.A., De Zeeuw, C.I., Häusser, M., De Schutter, E., Cerebellar LTD and pattern recognition by Purkinje cells. *Neuron*, 54, 1, 121–36, 2007 Apr 5.
86. Bobik, M., Ellisman, M.H., Rudy, B., Martone, M.E., Potassium channel subunit Kv3.2 and the water channel aquaporin-4 are selectively localized to cerebellar pinceau. *Brain Res.*, 1026, 168–178, 2004, doi: 10.1016/j.brainres.2004.07.088.
87. Bartos, M., Vida, I., Frotscher, M., Geiger, J.R., Jonas, P., Rapid signaling at inhibitory synapses in a dentate gyrus interneuron network. *J. Neurosci.*, 21, 8, 2687–98, 2001 Apr 15.
88. Cesana, E., Dieudonné, S., Isope, P., Bidoret, C., D'Angelo, E., Forti, L., Novel granule cell-Golgi cell excitatory input in the cerebellar granular layer. *InFENS Abstr*, 5, 052.7, 2010.
89. Nieus, T., Sola, E., Mapelli, J., Saftenku, E., Rossi, P., D'Angelo, E., LTP regulates burst initiation and frequency at mossy fiber–granule cell synapses of rat cerebellum: experimental observations and theoretical predictions. *J. Neurophysiol.*, 95, 2, 686–99, 2006 Feb.

9

Computational Attitude to Drug Discovery in Parkinson's Disease

Chitra Vellapandian[1], Ankul Singh S.[1], Swathi Suresh[1] and Bhupendra Prajapati[2,3]*

[1]*Department of Pharmacology, SRM College of Pharmacy, SRM Institute of Science and Technology, Kattankulathur, Chengalpattu, Tamil Nadu, India*
[2]*Shree S. K. Patel College of Pharmaceutical Education and Research, Ganpat University, Kherva, Gujarat, India*
[3]*Department of Industrial Pharmacy, Faculty of Pharmacy, Silpakorn University, Nakhon Pathom, Thailand*

Abstract

Blood sample analysis, leveraging data-driven techniques, holds promise for monitoring crucial neurological traits and disorders. This potential stems from identifying genes that regulate expression both in blood cells and brain tissue. Utilizing gene ontology (GO) and KEGG enrichment analysis aids in pinpointing differentially expressed genes (DEGs) crucial for disease understanding. By identifying hub genes based on their significance, it facilitates selecting precise targets for specific diseases and pathways, enhancing ligand targeting. Parkinson's disease (PD), among diseases not naturally occurring in animals, poses challenges in replication due to hazardous chemical requirements. Current research emphasizes α-synuclein's pivotal role in PD etiology, driving exploration for disease-modifying medications. Mutations in genes encoding DJ1, LRRK2, Parkin, PTEN-induced kinase 1, SNCA, and VPS35 lead to PD and disrupt calcium homeostasis upon apoptotic pathway activation. Gene ontology studies predict disease progression and biomarker identification. Analyses revealed 2078 DEGs in PD genes, offering insight into vital pathways. Protein-protein interaction networks were constructed

**Corresponding author*: bhupen27@gmail.com; ORCID: https://orcid.org/0000-0001-8242-4541
Chitra Vellapandian: ORCID: https://orcid.org/0000-0002-3052-9517
Ankul Singh S.: ORCID: https://orcid.org/0000-0001-7927-0983
Swathi Suresh: ORCID: https://orcid.org/0000-0002-6669-7379

Bhupendra Prajapati, Alok Tripathi, Rishabha Malviya and Lucy Mohapatra (eds.) Computational Neuropharmacology: Fundamentals and Clinical Aspects, (271–312) © 2025 Scrivener Publishing LLC

via the STRING tool to observe interactions. This comprehensive approach showcases the potential of blood sample analysis in unraveling neurological disorders' molecular mechanisms and therapeutic targets.

***Keywords*:** Computational approach, parkinson's, gene ontology, KEGG analysis, STRING

List of Abbreviations

AS	Alpha Synuclein
ATG	Autophagy related proteins
ATP13A2	ATPase cation transporting 13A2
BST1	Bone marrow stromal cell antigen 1
CNS	Central nervous system
CSF	Cerebro spinal fluid
DA	Dopaminergic
DBS	Deep brain stimulation
DDC	Dopa decarboxylase
DEGs	Differentially expressed genes
DisGeNET	Database of gene-disease associations
DJ1	Parkinson's disease protein 7
DRD1	Dopamine receptor D1
DRD2	Dopamine receptor D2
ER-stress	Endoplasmic reticulum stress
G51D	α-Synuclein gene
GAK	Cyclin G associated kinase
GBA	Glucosylceramidase beta
GO	Gene Ontology
GTP	Guanosine triphosphate
IGF1R	Insulin like growth factor 1 receptor
KEGG	Kyoto Encyclopedia of Genes and Genomes
LB	Lewy bodies
lincRNA-p21	p53-induced long noncoding RNA
lncRNAs	Long non-coding RNAs
LRRK2	Leucine-rich repeated kinase 2
MALAT1	Metastasis associated lung adenocarcinoma transcript 1
MAOB	Monoamine oxidase B
MAPK	Mitogen-activated protein kinase
MAPT	Microtubule associated protein tau

MiRNAs	MicroRNA
MPTP	1-methyl-4-phenyl-1,2,3,6-tetrahydropyridine
MTK458	Selective and brain penetrant PINK1 activator in mitophagy
ncRNAs	non-coding RNAs
NfL	Neurofilament light chain
NF-κB	Nuclear Factor-Kappa B
NMS	Neuroleptic Malignant Syndrome
NURR1	Nuclear receptor 4A2
PARK2	Parkin 2
PARK7	Parkin 7
PD	Parkinson's disease
PEDF	Pigment epithelium derived factor
PET	Positron emission tomography
PI3K/AKT	Phosphoinositide 3-kinase
PINK1	PTEN induced kinase 1
PINK1	PTEN-induced kinase 1
PINK1/TOM complex's	PTEN induced kinase 1/Translocase of the outer membrane
piRNA	PIWI-interacting RNAs
PPI	Protein-Protein Interaction
PRKN	Parkin RBR E3 ubiquitin protein ligase
SLC18A2	Solute carrier family 18 member A2
SN	Substantia nigra
SNCA	Synuclein alpha
SNGH1	Long noncoding RNA small molecule RNA host gene 1
SPECT	Single-photon emission computed tomography
STRING	Search Tool for the Retrieval of Interacting Genes
TH	Tyrosine hydroxylase
TncRNA	Transfer RNA-derived non-coding RNA
UCH-L1	Ubiquitin C-terminal hydrolase-L1
VPS35	Vacuolar protein sorting 35

9.1 Introduction

The time and expense of developing new drugs have been significantly reduced due to various computational technologies that have substantially helped drug screening and design. Virtual screening can swiftly explore massive chemical databases for lead compounds. Integrating computational technologies can speed up the development of novel medications with potential therapeutic benefits for use in complex biological systems [1]. Studies have shown that the use of evolutionary algorithms to evaluate human movement can be used to distinguish Parkinson disease (PD) patients from healthy controls and to categorize the degree of dyskinesia in PD patients. The path to an earlier and more accurate diagnosis and successful therapeutic approaches may be paved by identifying and analyzing the combinatorial impact of several PD-associated biochemical alterations [2, 3]. Machine learning has been used to analyze and interpret huge, extremely complicated genomic and transcriptome data to acquire a better knowledge of PD. Machine learning is a sophisticated computer approach that enables individuals to find complex pattern insights [4]. It is also clear that computational techniques have laid a foundation stone for the discovery of drugs by setting priorities and criteria to establish a network and make a profound research outcome. Neurotherapeutic and Inhibitory properties of Leucine-rich repeated kinase 2 domain are carried out via molecular modelling and computational simulation methods. It has shown astounding results in progression towards the pharmacological and pharmacokinetic optimization in treating PD [5]. Preclinical and clinical studies are increasingly using *in silico* analysis to investigate the neurotransmission mechanism in PD. A set of equations representing changes in brain functions that have been parametrized based on experimental data is needed for computer models. For computational models, a set of equations describing changes in brain functions parametrized on experimental data is required. It enables the creation and design of experimental methods, while also enabling the modification of parameter settings and the exploration of conditions that are typically difficult to test empirically [6]. Assessment and diagnosis of a disease is a cumbersome process once its pathology is multifactorial and thus it remains elusive. Various methodologies have been adopted in assessing PD risk where the feasibility of smartphone cognitive assessment in people with PD offers circumstantial support for the impact of context on cognitive variability in PD [7]. The discovery of genes whose expression regulation is functionally similar in brain tissue and blood cells may, in theory, enable the data-driven analysis of blood samples to monitor important neurological features and illnesses [8]. Docking approaches are also emerging as a vital source for identification of ligand binding along with receptors but again it is a computational approach which we cannot rely upon and thus such computational techniques can be employed to minimize the time taken for research failure and can work systematically to understand the core concept of the disease with advanced tools available. A study also carried out molecular docking by using Autodock Vina and found out two phytochemicals out of 11 energy minimized structure by using cluster analysis and Confusion Matrix for further studies on the PD [9]. To design and assess closed-loop adaptive DBS systems, modelling frameworks have frequently been combined with model-based control techniques. However, additional research is still needed to prevent unwanted interactions with brain functions that are not impacted by PD [10]. A recent study showed that structural and ligand-based screening methods could be kept as basis for associating identified ligands to risk factor of disease. Thus, it was revealed that the identified ligands were promising candidates for NURR1 receptor as in for PD; yet, needs more *in vitro* and *in vivo* studies to be done for proper validation [11]. To better understand the disease, gene ontology (GO) and KEGG enrichment analysis (KEGG) are used to identify differentially expressed genes (DEGs). It may be useful in identifying hub genes based on their strength, and it may provide a reason for selecting a specific target for a specific disease with a specific pathway to target the ligand [12]. Significant PD-related KEGG pathways screening from the Disgenet Database can be identified and correlated with other receptors interaction to identify potential candidate genes for PD progression [13]. In this regard, a fundamental computational method that reveals the identification of key genes necessary for the progression of PD is well established and researched.

9.2 PD and Drug Development

Dopaminergic neurons in the substantia nigra pars compacta, a part of the midbrain, die as a result of Lewy bodies, cytoplasmic inclusions containing insoluble alpha-synuclein clumps. Lewy bodies are associated with PD. However, more severe disease in other brain regions, such as non-dopaminergic neurons, distinguishes PD from other forms of the condition [14]. PD affects 2 in 1000 people, and its frequency rises with age, affecting 1% of people over 60 [15]. Incidence rises five to ten times between the sixth and ninth decades of life. After receiving a diagnosis of PD, mortality does not increase in comparison to the general population for the first ten years, but it does in the years that follow, suggesting that the patient may still be saved if appropriate treatment options are put into place. Strong evidence from epidemiological studies shows no hereditary origins for disease and that behavioral and environmental variables exerts a significant role in the progression and development [16, 17]. There is strong evidence for preventing α synuclein aggregating by vagotomy which reduces the risk of PD in animals and humans. It describes the need to study the dysautonomic symptoms in preventing PD-associated symptoms [18]. Men are twice as likely as women to have PD, but women die from it more frequently and it progresses faster in women. Furthermore, men and women differ in terms of illness risk factors, treatment response, and motor and nonmotor symptoms [19]. PD has been connected to environmental pollutants, ageing, family history, pesticide exposure, and synthetic heroin usage. Resting tremors, postural instability, expressionlessness, lead pipe rigidity, and less frequently cognitive impairment are all clinical markers of PD [20]. For the diagnosis of PD, molecular imaging methods including positron emission tomography and single-photon emission computed tomography have been developed [21]. However, PD diagnosis remains challenging, and the absence of validated biomarkers makes it difficult to identify the disease's early warning signs [22]. The clinical development of trial phases is a critical rate-limiting stage in the provision of innovative medications to patients with PD because disease-modifying treatment is only available in phase 2 to phase 3 trials [23]. The failure rate for drug development in the setting of PD is considerable. Ineffective trial data can be turned into knowledge by MID3 to improve strategies in current or forthcoming drug development programmes [24].

9.3 Animal Models and Translational Discovery

There are currently few animal models suitable for medication research and delivery for these previously incurable illnesses like PD. Furthermore, it is recognised that the most recent α-Synuclein model closely mimics important clinical and morphological features but falls short of accurately simulating the whole spectrum of human illnesses [25]. One of the many human diseases that do not appear to have developed naturally in animals is Parkinson's disease. Replicating the condition with a variety of dangerous compounds is difficult [26]. The first neurological condition to receive palliative neurotransmitter replacement therapy was PD. One of the numerous human diseases that do not seem to have arisen naturally in animals is PD. It is challenging to reproduce the state with several dangerous chemicals. A detailed neurobiological characterization of the connection between motor symptoms and cognitive abnormalities in PD will also be essential for developing future treatments that may effectively treat all facets of the illness [27]. The primary factor in the aetiology of PD has so far been identified as α-synuclein, and it has been the focus of a state-of-the-art research study in the search for brand-new medications with disease-modifying capabilities. It mimicked the progression of neurodegeneration and the spread of pathogenic α-synuclein through the usage of viral vector-induced α-synuclein expression and LB seeding. It is nevertheless a significant step towards establishing a translational platform for successful therapeutic results [28]. The goal of parkin knockout models and A30P synuclein transgenic mice is to mimic human familial Parkinson's disease. A30P animals showed significantly higher carbonylation than controls in the oxidative modification of lactate dehydrogenase 2, carbonic anhydrase 2, and α-enolase, which resulted in function loss [29]. Higher levels of β-enolase and β-actin, which were found to be oxidatively changed in the striatum and SN brain areas of hemiparkinsonian rats compared to controls, have been linked to AD in previous studies. Drosophila that produce the human A30P mutant α-synuclein provide more evidence that pathways related to the cytoskeleton and mitochondria are involved in the aetiology of Parkinson's disease (PD). The aetiology of PD is thought to involve processes related to oxidative stress, mitochondrial function, energy metabolism, and protein folding and degradation, as evidenced by the changed protein expression levels in A53T transgenic α-synuclein Drosophila [30, 31].

In the development and evaluation of innovative treatments, the use of monkey models of PD plays a distinctive role. When it comes to the neuroanatomy of lost neurotransmitters, the MPTP-based model in marmosets matches the disease [32]. Mammalian animal models may not be the best choice for studying endogenous dopaminergic neuroregeneration since the adult mammalian CNS is unable to adapt for severe neuronal loss. Mammalian models have the most genetic resemblance to humans, allowing them to accurately imitate PD symptoms and have the greatest translatability to humans [33]. A single model appears unlikely to accurately depict the complexity of this human disease, just as a single medicine appears unlikely to totally cure PD's motor and NMS symptoms. To research the gradual neurodegeneration associated with PD, future models should include neurotoxins and genetic animal models. They should also make it simple to test medicines for symptoms other than motor ones [34]. Understanding the pathophysiological significance in disease alleviation and the therapeutic potential of pathways disturbed in PD with a focus on synaptic transmission, vesicle trafficking, and mitochondrial function is crucial [35]. To understand the aetiology of neurodegenerative illnesses better and to create new therapy methods, nonhuman models of PD are essential. PD progression is often researched using both genetic and toxic models. Typically, PD-related motor, non-motor, and neurodegenerative symptoms, as well as adverse effects, are treated with medications that are first investigated in toxic models [36]. Molecular genetics, protein biochemistry, pharmacology, physiology, and comparative anatomy, in particular, are very active research areas that are in constant communication with other research disciplines. These fields offer opportunities to recreate and study nearly all significant aspects of PD pathogenesis in laboratory animals [37]. Cellular models are useful for studying protein aggregation connected to neuronal death and testing medicines for therapeutic applications. However, only animal models may be utilised to examine pathogenic processes involving interactions across different cell types and tissues [38]. The model organism Drosophila melanogaster offers many benefits for biological research. The potential to find conserved genetic pathways related to human disease has therefore been demonstrated in fly research. Additionally, Multiple subclasses of DA neurons regulate DA-dependent behaviours in flies, and their distinct functions and connectivity are still being investigated [39]. Even if there are more and more complementing rodent models of PD (both genetic and non-genetic models), the applicability of their findings to people is constrained by several problems. Only a few features of human PD are accurately replicated in animal model.

9.4 Pathophysiology

One of the pathological characteristics of PD is α- synuclein buildup. Due to the complexity of PD, different patient subgroups may respond differently to VPS35-targeted therapies which is a critical element of mitochondrial homeostasis that regulates the accumulation of synuclein [40]. A significant part of PD aetiology involves increased alpha-synuclein build-up, mitochondrial dysfunction, oxidative stress, excitotoxicity, and neuroinflammation. The pathogenic insult caused by protein misfolding, aggregation, and accumulation results in neuronal apoptosis [41]. The molecular interaction between the transcription factor XBP1s and the kinase PINK1 is important for the mitochondrial homeostasis that is disturbed in PD and the ER-stress response and mitophagy flux at endogenous levels [42]. It is believed that α-Synuclein build up, mitochondrial dysfunction, oxidative stress, and excitotoxicity are key factors in the pathogenesis of PD. Mutations in Synuclein alpha (SNCA), and LRRK2 can impair mitochondrial fusion and fission, which can lead to mitochondrial malfunction and the emergence of PD [43]. Stress-induced p38 MAPK activation reduced the expression of SIRT1, which in turn led to an increase in the acetylation of many ATG proteins in the autophagic complex and, ultimately, to deficiencies in autophagy [44]. Drooling is more likely to occur in PD patients who have poor sleep habits, whereas excellent sleep hygiene is linked to less morning motor symptoms. Drooling has been connected to orthostatic hypotension, diverse gastrointestinal symptoms, and dysautonomias (urinary problems, sexual dysfunction, and obstipation) in PD [45]. Nuclear Factor-Kappa B (NF-κB), a crucial regulator of a wide range of cellular processes, is implicated in these PD-related mechanisms, spreading apoptosis via a variety of pathways, as shown by the buildup of alpha-synuclein [46]. To learn more about how the miR-185 and SEPT5 genes affect the pathogenesis of PD, researchers researched these genes. Leucine-rich repeat kinase 2 (LRRK2) gene augmentation was found to have protective benefits [47]. Eventually, mutations in the genes encoding DJ1, LRRK2, Parkin, PTEN-induced kinase 1, SNCA, and VPS35 cause PD and disturb calcium homeostasis by activating apoptotic pathways. Furthermore, miR-7 can regulate the PI3K/AKT signalling pathway to regulate the synthesis of α-synuclein [48].

Changes in microglial histamine H4R function and disruptions of histaminergic neurotransmission may hold the key to understanding a number of neurological disorders with perplexing pathogenesis that are yet difficult to treat [49]. LRRK2 mutations can enhance kinase activity and GTP binding activity, decrease the rate of guanosine triphosphate hydrolysis, and ultimately cause cell death. They share the same histological and clinical characteristics as sporadic PD [50]. In the animal model of PD induced by 1-methyl-4-phenyl-1, 2, 3, 6-tetrahydropyridine (MPTP), MALAT1 has been found to be up-regulated. Among the up-regulated lncRNAs were TncRNA, SNGH1, MALAT1, and lincRNA-p21 [51]. MiRNAs are becoming important molecular tools for the development of novel, state-of-the-art therapeutic and diagnostic techniques. While miR-21, miR-34a, miR-34c, and miR-155 are not changed, the expression levels of miR-146a, miR-335-3p, and miR-335-5p are significantly lower in PD patients compared to controls [52]. Neuroinflammation, pathogenic damage to SN neurons, mitochondrial dysfunction, and oxidative damage can all be brought on by activated microglia and astrocytes. Because mitophagy eliminates dysfunctional mitochondria from SN neurons, deficiencies in autophagy of defective proteins produced by mutations might result in dangerous protein build up in the cell [53]. Numerous studies have shown that long non-coding RNAs and microRNAs are dysregulated in the brain tissues, plasma exosomes, and leukocytes of PD patients and animal models [54].

9.5 Validated Biomarkers

Proteins including α-synuclein, tau, and DJ-1 (PARK7) have all been thoroughly researched for their potential to serve as PD diagnostic biomarkers. Nonetheless, a-synuclein have never been discovered to be differently regulated in the CSF by proteomic investigations. Proteins including serpinA1, PEDF, and netrinG1 (α-synuclein) have already undergone the subsequent phase of confirmation in a targeted proteomic approach. All have also been successfully quantified in CSF, indicating that they can be used as surrogate indicators [55]. Genetic studies have uncovered mutations and genetic variations that can be used as trait biomarkers to identify illness susceptibility but do not always represent the presence of the disease [56]. The compounds derived from diabetes and insulin resistance mechanisms stand out as interesting candidates for PD and neurodegeneration with a novel pathogenetic basis, both therapeutically and for the creation of pertinent sensitive and precise biomarkers [57]. The pathophysiology of PD, imaging, and biochemical markers have all been identified to identify the disease in its early or preclinical phases. Using these potential indicators in combination, we may be able to make more accurate diagnoses and implement more suitable management regimens for these life-threatening conditions [58]. The majority of biomarkers have already been identified, and the research is limited to validation, prospective evaluation, and refinement as discussed in Table 9.1 and Table 9.2, respectively [59]. The most accurate diagnoses were made using a combination of CSF-synuclein species, conventional Alzheimer's disease biomarkers, and NfL [60]. MiRNAs may serve as PD biomarkers and therapeutic targets since they control genes linked to the development of PD, according to a number of studies [61, 62]. Dysregulation of ncRNAs like miRNAs may be a source of these circulating

Table 9.1 Biomarkers that predict PD progression in morphology and clinical level.

Level	Biomarker	Condition	Reference
Molecular	oxidized DJ-1, PET, SPECT, miRNA	Oxidative stress	[79–81]
Cellular	PARK7	Cellular damage	[82]
Tissue	DJ-1 (PARK7); neuronal and peripheral tissues	Mitochondrial function and protein homeostasis with PD related proteins	[83]
Clinical effects	α-synuclein, CSF lactate, 42-amyloid-β peptide, total and 181-phosphorylated tau	Neurodegenerative disorders	[84, 85]

Table 9.2 Biomarkers that predict PD progression in homeostasis level.

Level	Biomarker	Condition	Reference
Pharmacological	miRNA, Aptamers	Nucleic acid-based therapeutics	[86]
Biochemical	MiRNA, PIWI-interacting RNAs, Valosin-containing human protein	Personalized medicine for PD patients	[61, 87]
Haematological	Neurofilament light chain, Urate, alpha/beta-mannosidase	Malignant PD phenotype, purine pathway metabolite upstream marker	[88–90]
Immunological	α-Synuclein	Distinguish PD from multiple system atrophy	[91]
Genetic	Matrix metalloproteinase 3	Neural stress and inflammation responses	[92]

biomarkers, resulting in overlapping symptoms and making disease diagnosis more difficult. Circulating biomarkers may be crucial for the differential diagnosis of neurological disorders [63, 64]. The autophagy-lysosomal route is one of the major pathways that is altered during PD, according to genetic and functional research. There is proof that peripheral blood samples can be used as possible biomarkers to identify autophagy-lysosomal pathway disruption, which is the aetiology of PD [65].

Previous studies have discovered that the expression profile of long non-coding ribonucleic acids (lncRNAs) is disrupted in human illnesses such as PD, and that lncRNAs are of interest to be employed as possible tools for PD diagnosis, therapy, and prognosis [66, 67]. Studies reporting the changes in AS levels (presence/absence) have been reported so far in PD patients along with phosphorylated AS and total AS which leads to dilemma in the discovery of essential validated quantitative biomarkers [68–71]. Several forms of α-synuclein (both soluble and aggregated), neurofilament light chain protein, brain-derived neurotrophic factor, insulin-like growth factor-1, protein DJ-1, orexin, ubiquitin C-terminal hydrolase-L1 (UCH-L1),

-glucocerebrosidase, and 8-hydroxyguanine are possible biochemical markers for PD [72]. For instance, there is no radioligand for the pathogenic protein associated with PD known as alpha-synuclein, which may make detection more challenging due to its intracellular location [73]. While other indicators need further verification, total oligomeric, Ser129 phosphorylated α-synuclein (blood plasma and cerebrospinal fluid), α-synuclein oligomers (Red blood cells), genetic polymorphism in α-synuclein locus (genome), and tau/-synuclein (cerebrospinal fluid) are regarded credible [74]. MiRNAs appear to be one of the easily accessible instruments for assisting in the differential diagnosis of PD and associated illnesses, according to recent study, and they play a significant role in the pathobiology of various neurodegenerative disorders, including Parkinsonism. Interest in miRNA mimics and antagonists as potential biomarkers and therapeutic approaches is growing [75]. So far, a comprehensive public repository for quantitative evaluation of autoantigen-based serological biomarkers has been established in order to establish additional validated biomarkers and distinguish between various diseases. It will encourage protein annotation and the study of disease immunological processes [76]. According to one study, the PD2 biomarker binds significantly more IgG3 antibodies in PD patients and can be 84% reliable in identifying de novo PD individuals [77]. Both parkinsonism and atypical parkinsonism syndromes can be confused and there is a need to necessitate the novel biomarkers in differentiating diagnosis between both. Integrated plasma biomarkers approach has been developed for differential diagnosis and found to be suitable in its improvement [78].

9.6 Computational Drug Discovery

Clinical diagnosis of PD also has a low level of specificity. One of the most popular computational tools for tagging genes and their offspring with functional characteristics is the Gene Ontology (GO). However, GO annotation standards are based on low-throughput experimental setups that allow results to be evaluated in context, taking into consideration prior knowledge of the gene, the experimental premise, the physiological relevance of the assay, and other factors [93]. For GO annotations, which are relationships between genes and GO words that are either directly or indirectly supported by the scientific literature, a GO "evidence code" and either a published citation or a description of the process used to construct the annotation are required [94]. Serum miR-221 has been identified as a potential biomarker for PD. Other research has found that inhibiting miR-34b and miR-34c improves

PD symptoms by boosting -synuclein expression. The results of the study cited above suggest that miR plays a part in the onset and development of PD [95, 96]. Gene ontology studies that are currently publicly available have been thoroughly analysed for their applicability in predicting the course of disease and identifying potential biomarkers. The roles of discovered differentially expressed genes (DEGs) are predicted using protein–protein interaction (PPI) networks, Kyoto Encyclopaedia of Genes and Genomes pathway analysis (KEGG), and Gene Ontology enrichment [97]. Between the PD genes studied, 2078 DEGs were found. Following an integrated study of protein-protein interaction in PD expression datasets, with 0.5 to 1.0 strength, datasets were filtered with 17 genes to obtain the expression analysis of affinity-based interaction.

Therefore, PD-related pathways and genes are predicted using gene expression profiling. It demonstrates the utility of computational approaches in the study of diseases and their pathways, and the PD ontology (PDON) approach is a tool that represents modelling of the PD knowledge domain. In order to organise, integrate, and standardise the knowledge domains particular to diseases, disease ontologies offer an innovative approach. PD Ontology provides the quantum of knowledge in the domain of PD with a condensed, computer-readable form, allowing the scientific community to update and enrich it as well as build, depict, and automatically extend computable models of PD [98]. GO is frequently used to study high-throughput datasets, particularly for enrichment analysis that look for GO concepts in a gene-set of interest. Using the Search Tool for the Retrieval of Interacting Genes (STRING) v11.0, a PPI link between PD and periodontitis is established, and an interaction with a combined score is selected for further study. In order to identify important candidate genes, a PPI network was created in PD and tested utilising a number of bioinformatics methods for gene expression profiling [99]. The biological processes, molecular functions, and cellular components that are associated to genes were investigated using GO enrichment analysis. However, further studies are always warranted to predict the reliability of a disease but can minimize the usage of animals and narrow down the study [100]. Throughout development and adulthood, the brain expresses all the PARK genes. Since various interrelated cell and synaptic networks are affected throughout a person's lifespan, including and after the onset of neuronal degeneration, PD-causing gene mutations have the ability to fundamentally alter these networks. There is evidence that several PARK gene mutations change brain networks and have long-term effects on non-motor behaviour. This is most likely true for all PARK gene mutations, and it is similar to other brain disorders that have both genetic and environmental causes [101].

9.6.1 Screening of Active Genes

The DisGeNET App utilises the most recent version 7.0 of the DisGeNET database, which includes 369,554 VDAs between 194,515 variations and 14,155 diseases and 1,134,942 GDAs between 21,671 genes and 30,170 disorders. The DisGeNET Cytoscape App uses Cytoscape features, including the prioritisation metrics and characteristics offered in the most recent DisGeNET (v7.0), to examine genotype-phenotype connections supplied by DisGeNET [102].

9.6.2 Protein-Protein Interaction (PPI) Network Establishment

Protein-protein interaction (PPI) analysis is essential for understanding biological processes and breaking down intricate cellular mechanisms. The PPI network was built using the STRING tool. To identify the PPI network relationship, the intersecting targets were put into STRING, the species was set to "homo sapiens," and the interaction score was set to 0.4. To build the PPI network diagram, CytoScape_v3.9.1 was used. The accuracy of the value was evaluated using the network analyzer feature. The primary goals could be defined as the top 10 goals with the highest degree scores.

9.6.3 Gene Ontology Functional Enrichment and the Kyoto Encyclopedia of Genes and Genomes Pathway Analysis

The bioinformatics enrichment analysis on potential targets for Parkinson's disease (PD) was conducted using the STRING database and the GO and KEGG databases [103]. It is noteworthy that a significant 0.05 p-value cut limit was assigned to the parameter. Three categories were created out of the GO word: biological process, cellular component, and molecular function. The GO results, arranged from small to large according to the p-value parameter.

9.7 Outcomes From Gene Ontology and KEGG Analysis

9.7.1 PD-Related Target Screening Results

A total of 2078 PD-related targets were identified through DisGeNET and were screened for Score_gda with range 0.5 to 1.0 and observed 17 unique genes which have highest score and could be possible relevant to pathway. Targeting of these genes could be of interest to focus on its pathway to mitigate the progression of PD.

9.7.2 PPI Network Construction

A total of 2078 targets were detected in the DisGeNet database based on the gda score of 0.5 to 1.0, and 17 targets were found to obtain PPI network information. The PPI network was constructed with 18 nodes and 75 edges, where nodes represent proteins and edges represent PPI (Figure 9.1). Cytoscape 3.7.2 was used to load protein interaction data from the imported PPI network for visual analysis (Figure 9.2). After gathering the PPI network, the degree value was determined using the network analyzer tool. With degree values of 13, 13, 12, 12, 11, 11, 10, and 10, the top nine targets by degree value were SNCA, TH, LRRK2, PARK7, MAOB, PINK1, PARK2, MAPT, and SLC18A2 as discussed in Table 9.3. These targets are crucial nodes in the PPI network that link other nodes together.

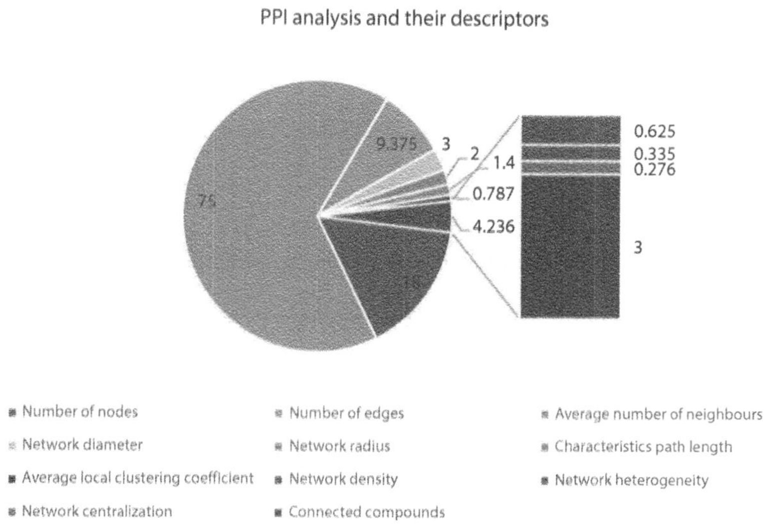

Figure 9.1 PPI analysis and their descriptors.

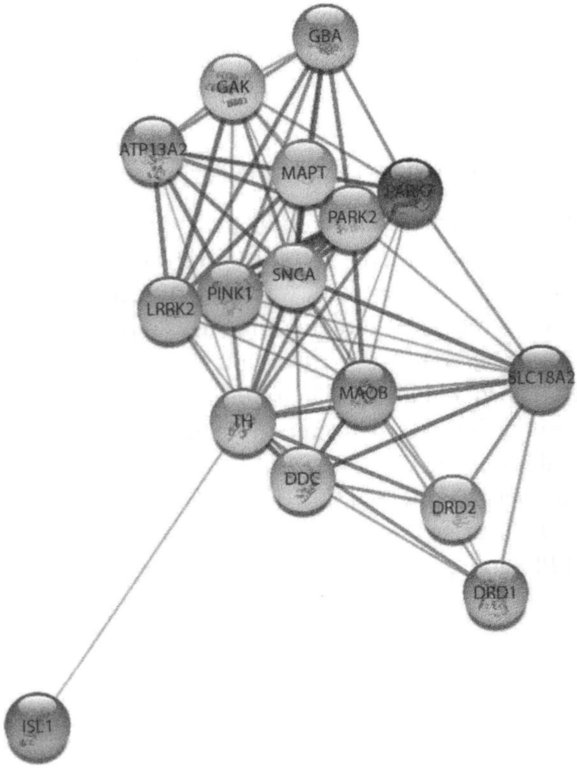

Figure 9.2 PPI network of selected genes in Parkinson disease.

Table 9.3 DEGs and their protein classification to predict PD progression based on score_gda.

Gene	Gene full name	UniProt ID	Protein class	Score_gda
SNCA	Synuclein alpha	P37840	Transporter	0.7
PARK7	Parkinsonism associated deglycase	Q99497	Enzyme	0.6
DDC	Dopa decarboxylase	P20711	-	0.6

(*Continued*)

Table 9.3 DEGs and their protein classification to predict PD progression based on score_gda. (*Continued*)

Gene	Gene full name	UniProt ID	Protein class	Score_gda
DRD2	Dopamine receptor D2	P14416	G-protein coupled receptor	0.6
ATP13A2	ATPase cation transporting 13A2	Q9NQ11	-	0.6
MAOB	Monoamine oxidase B	P27338	-	0.6
PRKN	Parkin RBR E3 ubiquitin protein ligase	O60260	Enzyme	0.6
PINK1	PTEN induced kinase 1	Q9BXM7	Kinase	0.6
SLC18A2	Solute carrier family 18 member A2	Q05940	Transporter	0.6
TH	Tyrosine hydroxylase	P07101	-	0.6
DRD1	Dopamine receptor D1	P21728	G-protein coupled receptor	0.57
IGF1R	Insulin like growth factor 1 receptor	P08069	Kinase	0.51
LRRK2	Leucine rich repeat kinase 2	Q5S007	Kinase	0.5
GAK	Cyclin G associated kinase	O14976	Kinase	0.5
GBA	Glucosylceramidase beta	P04062	-	0.5
MAPT	Microtubule associated protein tau	P10636	-	0.5
BST1	Bone marrow stromal cell antigen 1	Q10588	Enzyme	0.5

9.7.3 Protein–Protein Interaction Enrichment Analysis

With the help of the GO Consortium's annotation extensions, GO annotations can now include more context data. Extensions depict spatial and temporal features of a process, such as cell type, tissue, and developmental stage, as well as links between effectors and targets, such as dependence on localization, substrates for protein modifiers, and targets for signalling pathways and transcription factor regulatory factors [104]. Betweenness centrality of the selected score and text mining was carried out with R value of 0.875 and showed better regression curve and is illustrated in Figure 9.3 and Table 9.4, respectively.

Gene ontology IDs related to molecular function were taken into consideration because they more precisely identify a particular gene than those referring to a gene's participation in a biological process or the cellular component of the corresponding protein. A score between 0 and 1 was used to evaluate a gene's functional similarity to every other gene. Finding potential PD genes by combining semantic similarity and gene expression is successful. This study also introduces a novel "three-way comparative approach." The variety of genes and the associated pathways serve as an illustration of the complexity of the PD condition [105].

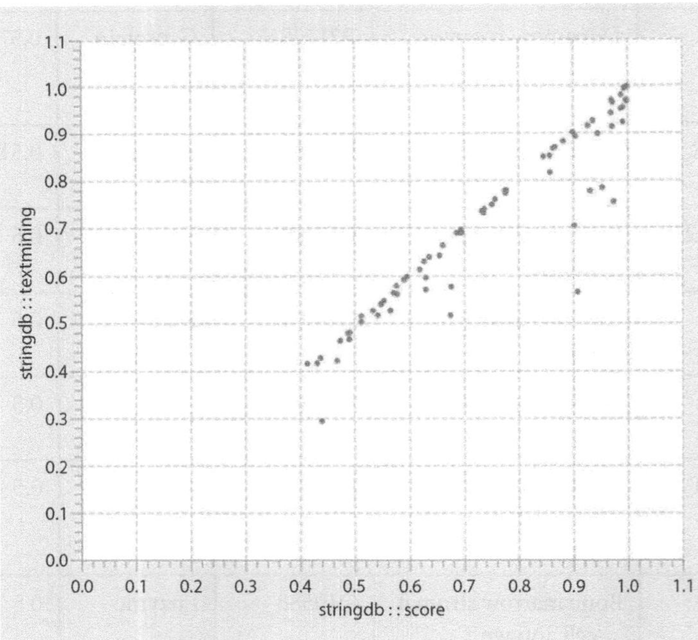

Figure 9.3 Linear regression betweenness centrality analysis of screened targets.

Table 9.4 Some common PD related genes and its degree analysis.

Common name	Degree	Closeness centrality	Between centrality
SNCA	13	0.882352941	0.059972888
TH	13	0.882352941	0.191640047
LRRK2	12	0.833333333	0.028765158
PARK7	12	0.833333333	0.028765158
MAOB	12	0.833333333	0.05949719
PINK1	11	0.789473684	0.015335207
PARK2	11	0.789473684	0.015335207
MAPT	10	0.75	0.008481219
SLC18A2	10	0.75	0.026832298

9.7.4 GO Functional Enrichment and KEGG Pathway Analysis

On 17 potential PD targets, GO and KEGG pathway enrichment analysis was carried out using the string platform. 469 GO enrichment analysis outcomes were produced as a result of the investigation, comprising 414 biological process analysis findings, 37 results from cellular component analysis, and 18 results from molecular function analysis. The main biological processes involved in the visualisation were Phytoalexin metabolic process, Negative regulation of hydrogen peroxide-induced neuron intrinsic apoptotic signalling pathway, Aminergic neurotransmitter loading into synaptic vesicle, Isoquinoline alkaloid metabolic process. Lewy body, dopaminergic synapse, inclusion body, glial cell projection, terminal bouton, and multivascular body were the most important cellular elements and is discussed in brief in Table 9.5, respectively. Cupric ion binding, Dopamine neurotransmitter receptor activity, Dopamine binding, Dopamine binding, and Ubiquitin-specific protease binding were the primary molecular processes involved. KEGG analysis yielded a total of ten findings. Parkinson disease, dopaminergic synapse, tyrosine metabolism, phenylalanine metabolism, serotonergic synapse, and tryptophan metabolism were the most common KEGG pathways.

Table 9.5 Top 20 biological process from identified 414 genes.

GO-ID	Descriptors	Strength	Matching proteins
GO:0052314	Phytoalexin metabolic process	3.06	TH, DDC
GO:1903384	Negative regulation of hydrogen peroxide-induced neuron intrinsic apoptotic signaling pathway	3.06	PINK1, PARK7
GO:0015842	Aminergic neurotransmitter loading into synaptic vesicle	2.94	SLC18A2, TH, DDC
GO:0033076	Isoquinoline alkaloid metabolic process	2.88	TH, DDC
GO:0099074	Mitochondrion to lysosome transport	2.88	PARK2, PINK1
GO:1902958	Positive regulation of mitochondrial electron transport, NADH to ubiquinone	2.88	PINK1, PARK7
GO:2000468	Regulation of peroxidase activity	2.88	LRRK2, SNCA
GO:0051583	Dopamine uptake involved in synaptic transmission	2.84	SNCA, PARK2, PARK7
GO:1902956	Regulation of mitochondrial electron transport, NADH to ubiquinone	2.84	SNCA, PINK1, PARK7
GO:1903377	Negative regulation of oxidative stress-induced neuron intrinsic apoptotic signaling pathway	2.84	PARK2, PINK1, PARK7
GO:0046684	Response to pyrethroid	2.76	TH, DDC
GO:0090258	Negative regulation of mitochondrial fission	2.76	MAPT, PINK1
GO:1902803	Regulation of synaptic vesicle transport	2.69	LRRK2, PARK2, PINK1

(Continued)

Table 9.5 Top 20 biological process from identified 414 genes. (*Continued*)

GO-ID	Descriptors	Strength	Matching proteins
GO:0051582	Positive regulation of neurotransmitter uptake	2.66	DRD2, PARK2
GO:1905279	Regulation of retrograde transport, endosome to Golgi	2.66	LRRK2, PARK2
GO:2000152	Regulation of ubiquitin-specific protease activity	2.66	ATP13A2, PARK7
GO:0015872	Dopamine transport	2.65	SLC18A2, SNCA, PARK2, DRD1, PARK7
GO:0051584	Regulation of dopamine uptake involved in synaptic transmission	2.64	SNCA, DRD2, DRD1
GO:0001963	Synaptic transmission, dopaminergic	2.61	SNCA, DRD2, PARK2, TH, DRD1, PARK7
GO:0071287	Cellular response to manganese ion	2.58	LRRK2, ATP13A2, PARK2, TH

9.7.5 Molecular Function

Molecular functions carried out by gene products such as "catalysis" and "transport" are used to define molecular level processes. In contrast to the entities (molecules or complexes) that carry out the actions, GO molecular function phrases refer to activities rather than the specific times, places, or circumstances in which the action occurs. The majority of molecular functions can be carried out by individual gene products, but some functions are carried out by molecular complexes made up of several gene products whose matching proteins along with strength and descriptors is mentioned in Table 9.6, respectively.

Table 9.6 Molecular function and its strength for 18 filtered genes.

GO-ID	Descriptors	Strength	Matching proteins
GO:1903135	Cupric ion binding	2.76	ATP13A2, PARK7
GO:0004952	Dopamine neurotransmitter receptor activity	2.66	DRD2, DRD1
GO:0035240	Dopamine binding	2.64	DRD2, TH, DRD1
GO:1903136	Cuprous ion binding	2.58	SNCA, PARK7
GO:1990381	Ubiquitin-specific protease binding	2.04	PARK2, PARK7
GO:0008198	Ferrous iron binding	1.96	SNCA, TH
GO:0005507	Copper ion binding	1.78	ATP13A2, SNCA, PARK7
GO:0051087	Chaperone binding	1.52	GAK, MAPT, PARK2
GO:0031072	Heat shock protein binding	1.46	SNCA, MAPT, PARK2
GO:0002020	Protease binding	1.4	PARK2, PINK1, PARK7
GO:0015631	Tubulin binding	1.11	LRRK2, SNCA, MAPT, PARK2
GO:0019904	Protein domain specific binding	0.91	MAPT, PARK2, PINK1, TH, DDC
GO:0044877	Protein-containing complex binding	0.88	IGF1R, LRRK2, SNCA, MAPT, DRD2, PARK2, PINK1, PARK7
GO:0042802	Identical protein binding	0.69	IGF1R, LRRK2, SNCA, MAPT, DRD2, PARK2, MAOB, PARK7
GO:0043168	Anion binding	0.61	IGF1R, LRRK2, GAK, ATP13A2, SNCA, MAPT, PINK1, MAOB, TH, DDC

(Continued)

Table 9.6 Molecular function and its strength for 18 filtered genes. (*Continued*)

GO-ID	Descriptors	Strength	Matching proteins
GO:0097159	Organic cyclic compound binding	0.44	IGF1R, LRRK2, GAK, ATP13A2, SNCA, MAPT, DRD2, PARK2, PINK1, MAOB, TH, DRD1, DDC, PARK7
GO:0003824	Catalytic activity	0.44	BST1, IGF1R, LRRK2, GAK, GBA, ATP13A2, SNCA, PARK2, PINK1, MAOB, TH, DDC, PARK7
GO:0043167	Ion binding	0.42	IGF1R, LRRK2, GAK, ATP13A2, SNCA, MAPT, DRD2, PARK2, PINK1, MAOB, TH, DRD1, DDC, PARK7

9.7.6 Cellular Component Analysis

The places in relation to cellular structures where a gene product functions, such as cellular compartments (mitochondria) or stable macromolecular complexes (ribosome) of which they are constituents. Cellular component classes, in contrast to other GO components, refer to a cellular anatomy rather than processes. Various cellular component analysis identified from screened genes have been depicted in Table 9.7, respectively.

Table 9.7 Top 20 cellular component analysis descriptors from 37 identified genes.

GO-ID	Descriptors	Strength	Matching proteins
GO:0097413	Lewy body	2.52	PARK2, PINK1
GO:0098691	Dopaminergic synapse	2.25	SLC18A2, DRD2
GO:0016234	Inclusion body	1.86	LRRK2, SNCA, MAPT, PARK2, PINK1

(*Continued*)

Table 9.7 Top 20 cellular component analysis descriptors from 37 identified genes. (*Continued*)

GO-ID	Descriptors	Strength	Matching proteins
GO:0097386	Glial cell projection	1.83	MAPT, PINK1
GO:0043195	Terminal bouton	1.8	SLC18A2, LRRK2, TH
GO:0005771	Multivesicular body	1.57	LRRK2, ATP13A2
GO:0043679	Axon terminus	1.56	SLC18A2, LRRK2, DRD2, TH
GO:0030672	Synaptic vesicle membrane	1.52	SLC18A2, LRRK2, DRD2
GO:0008021	Synaptic vesicle	1.45	SLC18A2, LRRK2, DRD2, TH, DDC
GO:0150034	Distal axon	1.41	SLC18A2, LRRK2, SNCA, MAPT, DRD2, PINK1, TH
GO:0030426	Growth cone	1.39	LRRK2, SNCA, MAPT, PINK1
GO:0098793	Presynapse	1.34	SLC18A2, LRRK2, GAK, SNCA, DRD2, PARK2, TH, DRD1, DDC, PARK7
GO:0043204	Perikaryon	1.34	LRRK2, DRD2, TH
GO:0043197	Dendritic spine	1.3	MAPT, DRD2, DRD1
GO:0030424	Axon	1.25	IGF1R, SLC18A2, LRRK2, SNCA, MAPT, DRD2, PINK1, TH, DDC, PARK7
GO:0005741	Mitochondrial outer membrane	1.25	LRRK2, PINK1, MAOB
GO:0030133	Transport vesicle	1.23	SLC18A2, LRRK2, ATP13A2, DRD2, TH, DDC

(*Continued*)

Table 9.7 Top 20 cellular component analysis descriptors from 37 identified genes. (*Continued*)

GO-ID	Descriptors	Strength	Matching proteins
GO:0044297	Cell body	1.19	LRRK2, ATP13A2, MAPT, DRD2, PINK1, TH, DDC, PARK7
GO:0043025	Neuronal cell body	1.12	LRRK2, ATP13A2, MAPT, DRD2, TH, DDC
GO:0043005	Neuron projection	1.04	IGF1R, SLC18A2, LRRK2, ATP13A2, SNCA, MAPT, DRD2, PARK2, PINK1, TH, DRD1, DDC, PARK7

9.7.7 KEGG Analysis

KEGG is a database resource that helps users understand high-level biological system functions and utility from molecular-level data, especially from massive molecular datasets generated by high-throughput experimental technologies like genome sequencing. Biological systems studied include the cell, the organism, and the ecosystem. KEGG analysis of the matched proteins and genes are provided in Table 9.8, respectively.

Based on available omics data, a static network of molecular interactions can anticipate the potential interaction pairings by transmitting information through its nodes and edges [106]. The limitations of this new paradigm, whose operability is strongly related to the non-living elements of their surroundings, are being learned by humans [107]. In order to identify PD states that vary in their motor and non-motor symptoms and progression, researchers used the Parkinson's Progression Markers Initiative data to apply a contrastive latent variable model followed by a distinctive hidden Markov model variation [108]. The fundamental drawback of computational models is that their outputs can only be informed by knowledge and input data that already exist. As a result, it is essential to feed and update models with new, emerging, human-relevant data [109]. The idea that PD is a group of illnesses, each affected by a particular confluence of environmental, genetic, and possibly coexisting factors, is one that is increasingly accepted. Given that PD is a composite illness, there is no one biomarker or treatment for the onset or progression of the condition [110]. PD treatments that are efficient and secure are still in high demand. Levodopa is a

Table 9.8 KEGG analysis of the matched proteins and genes to identify PD targets.

GO-ID	Descriptors	Strength	Matching proteins
hsa05012	Parkinson disease	1.72	SLC18A2, LRRK2, SNCA, MAPT, DRD2, PARK2, PINK1, MAOB, TH, DRD1, PARK7
hsa05030	Cocaine addiction	2.15	SLC18A2, DRD2, MAOB, TH, DRD1, DDC
hsa04728	Dopaminergic synapse	1.73	SLC18A2, DRD2, MAOB, TH, DRD1, DDC
hsa05034	Alcoholism	1.68	SLC18A2, DRD2, MAOB, TH, DRD1, DDC
hsa05031	Amphetamine addiction	1.94	SLC18A2, MAOB, TH, DRD1, DDC
hsa00350	Tyrosine metabolism	1.99	MAOB, TH, DDC
hsa00360	Phenylalanine metabolism	2.13	MAOB, DDC
hsa04726	Serotonergic synapse	1.5	SLC18A2, MAOB, DDC
hsa00380	Tryptophan metabolism	1.75	MAOB, DDC
hsa04137	Mitophagy - animal	1.56	PARK2, PINK1

medicine that is very effective but can also have serious side effects. Many suffer from severe movement problems and need surgical procedures like deep brain stimulation [111]. PD may be impacted by immune system dysfunction, according to a number of lines of evidence. The lines of evidence include inadequate cellular and humoral immune responses in PD, imaging evidence of activated inflammatory cells, evidence of immunological dysregulation in PD experimental models, clinical and genetic linkages between autoimmune illness and PD, and more [112]. Clinical studies have shown that Lewy pathology in individuals with specific leucine-rich repeat kinase 2 mutations is not a reliable indicator of PD cellular dysfunction and that the site of synuclein accumulation in association with PD's early stages does not make sense [113].

PD-based co-expression networks are helpful for the analysis of expressed genes and identification of immune signatures. The conclusion that the expression of module genes was higher in healthy controls suggests that genes related to the immune system are significantly correlated with PD and may have significant functions in the condition [114]. Using network pharmacology and transcriptome sequencing, it has offered a promising method for identifying the bioactive substances [115]. Utilising network pharmacology and molecular docking technologies, the main chemical components of extracts/decoctions used in the treatment of PD and its molecular mechanism can be investigated [116]. Weighted gene co-expression network analysis has also been researched in order to efficiently identify the significant modules associated with PD from the dataset [117]. Using bioinformatics methods, network analysis discovered a putative miRNA-mRNA regulation network associated with the aetiology of PD. Blood samples taken from PD patients then confirmed this network [118]. The study of differentially expressed genes discovered using the GO and KEGG methodologies, as well as the discovery of hub genes associated with PD disease using algorithms, are two methods for tackling disease [100].

Several Disgenet targets were preliminarily screened out for PD pathology utilising the string network in this investigation. The KEGG study findings suggest that Disgenet pathways in Parkinson disease screening are related to Parkinson disease, Cocaine addiction, Dopaminergic synapse, Alcoholism, Amphetamine addiction, Tyrosine metabolism, Phenylalanine metabolism, Serotonergic synapse, Tryptophan metabolism, and Mitophagy. The crucial target among these pathways include Parkinson disease, cocaine abuse [119], Dopaminergic system [120], Amphetamine abuse [121], serum amino acid metabolism [122, 123], and mitophagy [124]. Among these pathways, cocaine addiction and phenylalanine metabolism, tyrosine metabolism, and amphetamine addiction showed highest strength ranging from 1.9-2.5. Astrocytes are important in the regulation of the ventral tegmental region local circuit and cocaine reward. Cocaine exposure enhances the astrocytic release of tonic GABA through volume-regulated anion channels as well as the tonic inhibition of VTA GABA neurons, which lowers their activity and disinhibits DA neurons that transmit to the nucleus accumbens [125]. Due to levodopa's metabolism by catechol-O-methyltransferase, levodopa treatment for PD has a tendency to cause circulating homocysteine levels to increase even further [126]. It is yet unknown how APP and α-synuclein regulate themselves physiologically on mitochondria, however these two proteins share regulatory mechanisms that cause mitochondrial malfunction, which contributes to the onset of neurodegenerative disorders [127]. According to a study, the G51D SNCA mutation specifically results in the production of a slowly proliferating α-synuclein strain that more closely matches the -synuclein aggregates connected to PD than Multiple System Atrophy [128]. Neuromelanin, a dark pigment associated with tyrosine hydroxylase, is abundant in the cell bodies of norepinephrine neurons linked to PD and dopamine neurons in the substantia nigra [129]. In the early stages, serum NFL rather than all PD patients or LRRK2-PD from healthy individuals may be a more reliable biomarker for differentiating idiopathic PD from healthy participants and neurodegeneration may be similar to that of healthy patients and lower than that of idiopathic PD, although it may diminish in the later stages in the early stages of LRRK2-PD individuals [130, 131]. The LRRK2 p.F1700L variant shows that replacement is one such variant with prominent kinase activity among all LRRK2 variants examined till date, impairs protein function more severely, and represents the most common source of clinically typical monogenic PD [132, 133].

Autosomal recessive early-onset PD is caused by the protein DJ-1 encoded by PARK7 losing its function. Mutations in PARK7 result in autosomal recessive inheritance of early-onset Parkinson disease (PD) and complete inactivation or absence of the DJ-1 protein. In general, PARK7/DJ-1 dysfunctions have a substantial impact on microglia's immunological responses and neuroinflammatory processes [134]. Monoamine oxidase B inhibitors are known to be used to treat symptoms by lowering the amount of dopamine that is degraded by this enzyme, maintaining functioning levels of dopamine. However, there is little proof of their effectiveness and their part in PD models, which calls for more research [135]. In the PARK7/DJ-1 gene, deletions or mutations are associated with early-onset familial PD. The DJ-1 protein protects against neurodegeneration by controlling oxidative stress, mitochondrial function, transcription, and signal transduction [136]. By preventing MAO-B regulation in mediating DA breakdown functions such as ROS production, and mitochondrial dysfunction, DJ-1 has exerted neuroprotective benefits [137]. The use of inhibitors of monoamine oxidase B and catechol-O-methyltransferase prolongs and increases the effect of levodopa on striatal dopaminergic neurotransmission in individuals with PD [138]. The second-most prevalent gene linked to autosomal recessive PD is PINK1. Early symptom onset is linked to homozygous mutations in this gene. All affected family members had homozygous mutations that co-segregated, and the Sudanese community's huge effective population size raises the likelihood of novel discoveries in monogenic and polygenic disorders there [139]. Preclinical PD models indicate that pharmacological stimulation of PINK1 is a promising therapeutic approach for PD disease modification and warrants further investigation. MTK458 preferentially activates PINK1 via encouraging the PINK1/TOM complex's dimerization and stability [140]. Analyses of the metabolome and transcriptome in PINK1-mutant flies and neural progenitor cells produced from human induced pluripotent stem cells with a PINK1 mutation revealed changes in cysteine metabolism in both the fly and human PD models. Changes in a number of metabolites connected to elevated glutathione levels and cysteine production were brought on by mitochondrial malfunction in the neural precursor cells [141]. There have been many documented PINK1 gene mutations, although not all of these variants are pathogenic, which are the causal agents of autosomal recessive Parkinson disease, which is PD with an early onset [142]. The neuroprotective gene programme was suppressed during the pro-inflammatory transition, and PARK2-deficient glia had less TNFα reactivity than glial cells from healthy donors [143]. As a result, PARK2 copy number variations may be more harmful; this is particularly true for patients who have exonic duplications [144]. It was inferred that the microtubule-associated protein tau gene H1 haplotype was not related with an elevated risk or age at onset of PD despite the importance of MAPT in the causal pathway of neurodegenerative illnesses, including PD [145].

9.8 Conclusion

The present work demonstrated that the utilization of Gene ontology and KEGG enrichment analysis as a computational strategy could play a vital role in identification of hub genes and identify the novel biomarkers for diagnosing PD. Nevertheless, a worth investigation of those genes is required in further studies to understand its clear role in identification of genes. We believe that analyzing the genes through this approach can be time saving and additional studies can be focussed to apprehend molecular investigation which will have a profound significance. Herein, we contend that a computational method should take the important markers of PD pathogenesis into account. By acting on various targets, such active components may influence calcium signalling pathways, neuroactive ligand-receptor interaction, and signalling pathways in PD. With the help of this method, it can be assumed to pinpoint a potential mechanism for treating PD that could serve as a guide for clinical treatments as well as provide some theoretical justification and evidence for the creation of new medications to treat the illness.

Acknowledgments

The authors would like to thank the management of SRM College of Pharmacy, SRM Institute of Science and Technology for extending the support in completion of this work. Bhupendra Prajapati would like to thank Shree S. K. Patel College of Pharmaceutical Education and Research, Ganpat University for the valuable support in this work.

References

1. Lin, X., Li, X., Lin, X., A Review on Applications of Computational Methods in Drug Screening and Design. *Molecules*, 25, 6, 1375, 2020.
2. Smith, S.L., Lones, M.A., Bedder, M., Alty, J.E., Cosgrove, J., Maguire, R.J., Pownall, M.E., Ivanoiu, D., Lyle, C., Cording, A., Elliott, C.J.H., Computational Approaches for Understanding the Diagnosis and Treatment of Parkinson's Disease. *IET Syst. Biol.*, 9, 6, 226–233, 2015.
3. Glaab, E., Computational Systems Biology Approaches for Parkinson's Disease. *Cell Tissue Res.*, 373, 1, 91–109, 2018.
4. Su, C., Tong, J., Wang, F., Mining Genetic and Transcriptomic Data Using Machine Learning Approaches in Parkinson's Disease. *npj Parkinson's Dis.*, 6, 1, 24, 2020.
5. Subair, T.I., Soremekun, O.S., Olotu, F.A., Soliman, M.E.S., From a Computational Perspective: Elucidating the Neurotherapeutic and Inhibitory Properties of LRRK2 Kinase Domain by a Benzothiazole-Based Compound. *Curr. Pharm. Biotechnol.*, 23, 2022.
6. Gandolfi, D., Boiani, G.M., Bigiani, A., Mapelli, J., Modeling Neurotransmission: Computational Tools to Investigate Neurological Disorders. *Int. J. Mol. Sci.*, 22, 9, 4565, 2021.
7. Weizenbaum, E.L., Fulford, D., Torous, J., Pinsky, E., Kolachalama, V.B., Cronin-Golomb, A., Smartphone-Based Neuropsychological Assessment in Parkinson's Disease: Feasibility, Validity, and Contextually Driven Variability in Cognition. *J. Int. Neuropsychol. Soc.*, 28, 4, 401–413, 2022.
8. Moni, M.A., Rana, H.K., Islam, M.B., Ahmed, M.B., Xu, H., Hasan, M.A.M., Lei, Y., Quinn, J.M.W., A Computational Approach to Identify Blood Cell-Expressed Parkinson's Disease Biomarkers That Are Coordinately Expressed in Brain Tissue. *Comput. Biol. Med.*, 113, 103385, 2019.
9. Jayadeepa, R.M. and Niveditha, M.S., Computational Approaches to Screen Candidate Ligands with Anti- Parkinson's Activity Using R Programming. *Curr. Top. Med. Chem.*, 12, 16, 1807–14, 2012.
10. Santaniello, S., Gale, J.T., Sarma, S.V., Systems Approaches to Optimizing Deep Brain Stimulation Therapies in Parkinson's Disease. *WIREs Syst. Biol. Med.*, 10, 5, e1421, 2018.

11. John Marshal, J., Kuriakose, B.B., Alhazmi, A.H., Muthusamy, K., Mechanistic Insights into the Role of Vitamin D and Computational Identification of Potential Lead Compounds for Parkinson's Disease. *J. Cell. Biochem.*, 124, 3, 434–445, 2023.
12. Chen, X., Cao, W., Zhuang, Y., Chen, S., Li, X., Integrative Analysis of Potential Biomarkers and Immune Cell Infiltration in Parkinson's Disease. *Brain Res. Bull.*, 177, 53–63, 2021.
13. Yang, M., Wu, X.-Q., Ding, C.-B., Zhang, G.-F., Li, M., Lv, L.-N., Li, Y.-H., Sun, D.-W., Zhao, J.-J., Weighted Gene Co-Expression Network Analysis Identifies Specific Modules and Hub Genes Related to Parkinson's Disease. *Neuroreport*, 32, 13, 1073–1081, 2021.
14. Simon, D.K., Tanner, C.M., Brundin, P., Parkinson Disease Epidemiology, Pathology, Genetics, and Pathophysiology. *Clin. Geriatr. Med.*, 36, 1, 1–12, 2020.
15. Tysnes, O.-B. and Storstein, A., Epidemiology of Parkinson's Disease. *J. Neural Transm.*, 124, 8, 901–905, 2017.
16. Ascherio, A. and Schwarzschild, M.A., The Epidemiology of Parkinson's Disease: Risk Factors and Prevention. *Lancet Neurol.*, 15, 12, 1257–1272, 2016.
17. Day, J.O. and Mullin, S., The Genetics of Parkinson's Disease and Implications for Clinical Practice. *Genes (Basel)*, 12, 7, 1006, 2021.
18. Chen, Z., Li, G., Liu, J., Autonomic Dysfunction in Parkinson's Disease: Implications for Pathophysiology, Diagnosis, and Treatment. *Neurobiol. Dis.*, 134, 104700, 2020.
19. Cerri, S., Mus, L., Blandini, F., Parkinson's Disease in Women and Men: What's the Difference? *J. Parkinson's Dis.*, 9, 3, 501–515, 2019.
20. Khan, A.U., Akram, M., Daniyal, M., Zainab, R., Awareness and Current Knowledge of Parkinson's Disease: A Neurodegenerative Disorder. *Int. J. Neurosci.*, 129, 1, 55–93, 2019.
21. Haider, A., Elghazawy, N.H., Dawoud, A., Gebhard, C., Wichmann, T., Sippl, W., Hoener, M., Arenas, E., Liang, S.H., Translational Molecular Imaging and Drug Development in Parkinson's Disease. *Mol. Neurodegener.*, 18, 1, 11, 2023.
22. Tolosa, E., Garrido, A., Scholz, S.W., Poewe, W., Challenges in the Diagnosis of Parkinson's Disease. *Lancet Neurol.*, 20, 5, 385–397, 2021.
23. McFarthing, K., Rafaloff, G., Baptista, M., Mursaleen, L., Fuest, R., Wyse, R.K., Stott, S.R.W., Parkinson's Disease Drug Therapies in the Clinical Trial Pipeline: 2022 Update. *J. Parkinson's Dis.*, 12, 4, 1073–1082, 2022.
24. Conrado, D.J., Duvvuri, S., Geerts, H., Burton, J., Biesdorf, C., Ahamadi, M., Macha, S., Hather, G., Francisco Morales, J., Podichetty, J., Nicholas, T., Stephenson, D., Trame, M., Romero, K., Corrigan, B., Challenges in Alzheimer's Disease Drug Discovery and Development: The Role of Modeling, Simulation, and Open Data. *Clin. Pharmacol. Ther.*, 107, 4, 796–805, 2020.

25. Jellinger, K.A., Animal Models of Synucleinopathies and How They Could Impact Future Drug Discovery and Delivery Efforts. *Expert Opin. Drug Discov.*, 14, 10, 969–982, 2019.
26. Gerlach, M. and Riederer, P., Animal Models of Parkinson's Disease: An Empirical Comparison with the Phenomenology of the Disease in Man. *J. Neural Transm.*, 103, 8–9, 987–1041, 1996.
27. Solari, N., Bonito-Oliva, A., Fisone, G., Brambilla, R., Understanding Cognitive Deficits in Parkinson's Disease: Lessons from Preclinical Animal Models. *Learn. Mem.*, 20, 10, 592–600, 2013.
28. Ko, W.K.D. and Bezard, E., Experimental Animal Models of Parkinson's Disease: A Transition from Assessing Symptomatology to α-Synuclein Targeted Disease Modification. *Exp. Neurol.*, 298, 172–179, 2017.
29. Fai Poon, H., Frasier, M., Shreve, N., Calabrese, V., Wolozin, B., Butterfield, D.A., Mitochondrial Associated Metabolic Proteins Are Selectively Oxidized in A30P α-Synuclein Transgenic Mice—a Model of Familial Parkinson's Disease. *Neurobiol. Dis.*, 18, 3, 492–498, 2005.
30. Sowell, R.A., Owen, J.B., Allan Butterfield, D., Proteomics in Animal Models of Alzheimer's and Parkinson's Diseases. *Ageing Res. Rev.*, 8, 1, 1–17, 2009.
31. Marmion, D.J. and Kordower, J.H., α-Synuclein Nonhuman Primate Models of Parkinson's Disease. *J. Neural Transm.*, 125, 3, 385–400, 2018.
32. Eslamboli, A., Marmoset Monkey Models of Parkinson's Disease: Which Model, When and Why? *Brain Res. Bull.*, 68, 3, 140–149, 2005.
33. Vijayanathan, Y., Lim, S.M., Tan, M.P., Lim, F.T., Majeed, A.B.A., Ramasamy, K., Adult Endogenous Dopaminergic Neuroregeneration Against Parkinson's Disease: Ideal Animal Models? *Neurotox. Res.*, 39, 2, 504–532, 2021.
34. Trigo-Damas, I., del Rey, N.L.-G., Blesa, J., Novel Models for Parkinson's Disease and Their Impact on Future Drug Discovery. *Expert Opin. Drug Discov.*, 13, 3, 229–239, 2018.
35. Valadas, J.S., Vos, M., Verstreken, P., Therapeutic Strategies in Parkinson's Disease: What We Have Learned from Animal Models. *Ann. N. Y. Acad. Sci.*, 1338, 1, 16–37, 2015.
36. Le, W., Sayana, P., Jankovic, J., Animal Models of Parkinson's Disease: A Gateway to Therapeutics? *Neurotherapeutics*, 11, 1, 92–110, 2014.
37. Cenci, M.A. and Björklund, A., Animal Models for Preclinical Parkinson's Research: An Update and Critical Appraisal. *Prog. Brain Res.*, 252, 27–59, 2020.
38. Falkenburger, B.H., Saridaki, T., Dinter, E., Cellular Models for Parkinson's Disease. *J. Neurochem.*, 139, 121–130, 2016.
39. Nagoshi, E., Drosophila Models of Sporadic Parkinson's Disease. *Int. J. Mol. Sci.*, 19, 11, 3343, 2018.
40. Cutillo, G., Simon, D.K., Eleuteri, S., VPS35 and the Mitochondria: Connecting the Dots in Parkinson's Disease Pathophysiology. *Neurobiol. Dis.*, 145, 105056, 2020.

41. Kaur, R., Mehan, S., Singh, S., Understanding Multifactorial Architecture of Parkinson's Disease: Pathophysiology to Management. *Neurol. Sci.*, 40, 1, 13–23, 2019.
42. El Manaa, W., Duplan, E., Goiran, T., Lauritzen, I., Vaillant Beuchot, L., Lacas-Gervais, S., Morais, V.A., You, H., Qi, L., Salazar, M., Ozcan, U., Chami, M., Checler, F., Alves da Costa, C., Transcription- and Phosphorylation-Dependent Control of a Functional Interplay between XBP1s and PINK1 Governs Mitophagy and Potentially Impacts Parkinson Disease Pathophysiology. *Autophagy*, 17, 12, 4363–4385, 2021.
43. Pan, L., Meng, L., He, M., Zhang, Z., Tau in the Pathophysiology of Parkinson's Disease. *J. Mol. Neurosci.*, 71, 11, 2179–2191, 2021.
44. Zhu, L., Sun, C., Ren, J., Wang, G., Ma, R., Sun, L., Yang, D., Gao, S., Ning, K., Wang, Z., Chen, X., Chen, S., Zhu, H., Gao, Z., Xu, J., Stress-Induced Precocious Aging in PD-Patient IPSC-Derived NSCs May Underlie the Pathophysiology of Parkinson's Disease. *Cell Death Dis.*, 10, 2, 105, 2019.
45. Polychronis, S., Nasios, G., Dardiotis, E., Messinis, L., Pagano, G., Pathophysiology and Symptomatology of Drooling in Parkinson's Disease. *Healthcare*, 10, 3, 516, 2022.
46. Dolatshahi, M., Ranjbar Hameghavandi, M.H., Sabahi, M., Rostamkhani, S., Nuclear Factor-kappa B (NF-κB) in Pathophysiology of Parkinson Disease: Diverse Patterns and Mechanisms Contributing to Neurodegeneration. *Eur. J. Neurosci.*, 54, 1, 4101–4123, 2021.
47. Rahimmi, A., Peluso, I., Rajabi, A., Hassanzadeh, K., MiR-185 and SEPT5 Genes May Contribute to Parkinson's Disease Pathophysiology. *Oxid. Med. Cell. Longev.*, 2019, 1–8, 2019.
48. Khezri, M.R., Yousefi, K., Zolbanin, N.M., Ghasemnejad-Berenji, M., MicroRNAs in the Pathophysiology of Alzheimer's Disease and Parkinson's Disease: An Overview. *Mol. Neurobiol.*, 59, 3, 1589–1603, 2022.
49. Schneider, E.H., Microglial Histamine H4R in the Pathophysiology of Parkinson's Disease—a New Actor on the Stage? *Naunyn. Schmiedebergs. Arch. Pharmacol.*, 392, 6, 641–645, 2019.
50. Chen, M.-L. and Wu, R.-M., LRRK 2 Gene Mutations in the Pathophysiology of the ROCO Domain and Therapeutic Targets for Parkinson's Disease: A Review. *J. Biomed. Sci.*, 25, 1, 52, 2018.
51. Abrishamdar, M., Jalali, M.S., Rashno, M., MALAT1 LncRNA and Parkinson's Disease: The Role in the Pathophysiology and Significance for Diagnostic and Therapeutic Approaches. *Mol. Neurobiol.*, 59, 9, 5253–5262, 2022.
52. Oliveira, S.R., Dionísio, P.A., Correia Guedes, L., Gonçalves, N., Coelho, M., Rosa, M.M., Amaral, J.D., Ferreira, J.J., Rodrigues, C.M.P., Circulating Inflammatory MiRNAs Associated with Parkinson's Disease Pathophysiology. *Biomolecules*, 10, 6, 945, 2020.
53. Zaman, V., Shields, D.C., Shams, R., Drasites, K.P., Matzelle, D., Haque, A., Banik, N.L., Cellular and Molecular Pathophysiology in the Progression of Parkinson's Disease. *Metab. Brain Dis.*, 36, 5, 815–827, 2021.

54. Rezaei, O., Nateghinia, S., Estiar, M.A., Taheri, M., Ghafouri-Fard, S., Assessment of the Role of Non-Coding RNAs in the Pathophysiology of Parkinson's Disease. *Eur. J. Pharmacol.*, 896, 173914, 2021.
55. Halbgebauer, S., Öckl, P., Wirth, K., Steinacker, P., Otto, M., Protein Biomarkers in Parkinson's Disease: Focus on Cerebrospinal Fluid Markers and Synaptic Proteins. *Mov. Disord.*, 31, 6, 848–860, 2016.
56. Gasser, T., Genomic and Proteomic Biomarkers for Parkinson Disease. *Neurology*, 72, Issue 7, Supplement 2, S27–S31, 2009.
57. Ho, G., Takamatsu, Y., Waragai, M., Wada, R., Sugama, S., Takenouchi, T., Fujita, M., Ali, A., Hsieh, M.H.-I., Hashimoto, M., Current and Future Clinical Utilities of Parkinson's Disease and Dementia Biomarkers: Can They Help Us Conquer the Disease? *Expert Rev. Neurother.*, 19, 11, 1149–1161, 2019.
58. Waragai, M., Sekiyama, K., Fujita, M., Tokuda, T., Hashimoto, M., Biomarkers for the Diagnosis and Management of Parkinson's Disease. *Expert Opin. Med. Diagn.*, 7, 1, 71–83, 2013.
59. Xie, F., Gao, X., Yang, W., Chang, Z., Yang, X., Wei, X., Huang, Z., Xie, H., Yue, Z., Zhou, F., Wang, Q., Advances in the Research of Risk Factors and Prodromal Biomarkers of Parkinson's Disease. *ACS Chem. Neurosci.*, 10, 2, 973–990, 2019.
60. Parnetti, L., Gaetani, L., Eusebi, P., Paciotti, S., Hansson, O., El-Agnaf, O., Mollenhauer, B., Blennow, K., Calabresi, P., CSF and Blood Biomarkers for Parkinson's Disease. *Lancet Neurol.*, 18, 6, 573–586, 2019.
61. Fonseca Cabral, G., Schaan, A.P., Cavalcante, G.C., Sena-dos-Santos, C., de Souza, T.P., Souza Port's, N.M., dos Santos Pinheiro, J.A., Ribeiro-dos-Santos, Â., Vidal, A.F., Nuclear and Mitochondrial Genome, Epigenome and Gut Microbiome: Emerging Molecular Biomarkers for Parkinson's Disease. *Int. J. Mol. Sci.*, 22, 18, 9839, 2021.
62. Angelopoulou, E., Paudel, Y.N., Piperi, C., MiR-124 and Parkinson's Disease: A Biomarker with Therapeutic Potential. *Pharmacol. Res.*, 150, 104515, 2019.
63. Arnerić, S.P., Kern, V.D., Stephenson, D.T., Regulatory-Accepted Drug Development Tools Are Needed to Accelerate Innovative CNS Disease Treatments. *Biochem. Pharmacol.*, 151, 291–306, 2018.
64. Roser, A.E., Caldi Gomes, L., Schünemann, J., Maass, F., Lingor, P., Circulating MiRNAs as Diagnostic Biomarkers for Parkinson's Disease. *Front. Neurosci.*, 12, 625, 2018.
65. Xicoy, H., Peñuelas, N., Vila, M., Laguna, A., Autophagic- and Lysosomal-Related Biomarkers for Parkinson's Disease: Lights and Shadows. *Cells*, 8, 11, 1317, 2019.
66. Taghizadeh, E., Gheibihayat, S.M., Taheri, F., Afshani, S.M., Farahani, N., Saberi, A., LncRNAs as Putative Biomarkers and Therapeutic Targets for Parkinson's Disease. *Neurol. Sci.*, 42, 10, 4007–4015, 2021.
67. Maiese, K., Biomarkers for Parkinson's Disease and Neurodegenerative Disorders: A Role for Non-Coding RNAs. *Curr. Neurovasc. Res.*, 19, 2, 127–130, 2022.

68. Majbour, N.K., Vaikath, N.N., Eusebi, P., Chiasserini, D., Ardah, M., Varghese, S., Haque, M.E., Tokuda, T., Auinger, P., Calabresi, P., Parnetti, L., El-Agnaf, O.M.A., Longitudinal Changes in CSF Alpha-Synuclein Species Reflect Parkinson's Disease Progression. *Mov. Disord.*, 31, 10, 1535–1542, 2016.
69. Öhrfelt, A., Grognet, P., Andreasen, N., Wallin, A., Vanmechelen, E., Blennow, K., Zetterberg, H., Cerebrospinal Fluid α-Synuclein in Neurodegenerative Disorders—A Marker of Synapse Loss? *Neurosci. Lett.*, 450, 3, 332–335, 2009.
70. Wang, Y., Shi, M., Chung, K.A., Zabetian, C.P., Leverenz, J.B., Berg, D., Srulijes, K., Trojanowski, J.Q., Lee, V.M.-Y., Siderowf, A.D., Hurtig, H., Litvan, I., Schiess, M.C., Peskind, E.R., Masuda, M., Hasegawa, M., Lin, X., Pan, C., Galasko, D., Goldstein, D.S., Jensen, P.H., Yang, H., Cain, K.C., Zhang, J., Phosphorylated α-Synuclein in Parkinson's Disease. *Sci. Transl. Med.*, 4, 121ra20, 2012.
71. Polissidis, A., Petropoulou-Vathi, L., Nakos-Bimpos, M., Rideout, H.J., The Future of Targeted Gene-Based Treatment Strategies and Biomarkers in Parkinson's Disease. *Biomolecules*, 10, 6, 912, 2020.
72. Khan, M.A., Haider, N., Singh, T., Bandopadhyay, R., Ghoneim, M.M., Alshehri, S., Taha, M., Ahmad, J., Mishra, A., Promising Biomarkers and Therapeutic Targets for the Management of Parkinson's Disease: Recent Advancements and Contemporary Research. *Metab. Brain Dis.*, 38, 3, 873–919, 2023.
73. Tropea, T.F. and Chen-Plotkin, A.S., Unlocking the Mystery of Biomarkers: A Brief Introduction, Challenges and Opportunities in Parkinson Disease. *Parkinsonism Relat. Disord.*, 46, S15–S18, 2018.
74. Khodadadian, A., Hemmati-Dinarvand, M., Kalantary-Charvadeh, A., Ghobadi, A., Mazaheri, M., Candidate Biomarkers for Parkinson's Disease. *Biomed. Pharmacother.*, 104, 699–704, 2018.
75. Ramaswamy, P., Christopher, R., Pal, P.K., Yadav, R., MicroRNAs to Differentiate Parkinsonian Disorders: Advances in Biomarkers and Therapeutics. *J. Neurol. Sci.*, 394, 26–37, 2018.
76. Pan, J., Liu, S., Zhu, H., Qian, J., AAgMarker 1.0: A Resource of Serological Autoantigen Biomarkers for Clinical Diagnosis and Prognosis of Various Human Diseases. *Nucleic Acids Res.*, 46, D1, D886–D893, 2018.
77. Feng, N., Simanski, S., Islam, K., Hynan, L.S., Kodadek, T., German, D.C., Antibody Biomarker for de Novo Parkinson Disease: Attempted Validation. *npj Parkinson's Dis.*, 4, 1, 28, 2018.
78. Lin, C.-H., Yang, S.-Y., Horng, H.-E., Yang, C.-C., Chieh, J.-J., Chen, H.-H., Liu, B.-H., Chiu, M.-J., Plasma Biomarkers Differentiate Parkinson's Disease From Atypical Parkinsonism Syndromes. *Front. Aging Neurosci.*, 10, 123, 2018.
79. Jang, J., Jeong, S., Lee, S.I., Seol, W., Seo, H., Son, I., Ho, D.H., Oxidized DJ-1 Levels in Urine Samples as a Putative Biomarker for Parkinson's Disease. *Parkinsons Dis.*, 2018, 1–9, 2018.

80. de Natale, E.R., Niccolini, F., Wilson, H., Politis, M., Molecular Imaging of the Dopaminergic System in Idiopathic Parkinson's Disease. *Int. Rev. Neurobiol.*, 141, 131–172, 2018.
81. Schilling, M. and Lill, C.M., MicroRNAs as Molecular Biomarkers for Parkinson's Disease Progression. *Mov. Disord.*, 36, 8, 1793–1793, 2021.
82. Heremans, I.P., Caligiore, F., Gerin, I., Bury, M., Lutz, M., Graff, J., Stroobant, V., Vertommen, D., Teleman, A.A., Van Schaftingen, E., Bommer, G.T., Parkinson's Disease Protein PARK7 Prevents Metabolite and Protein Damage Caused by a Glycolytic Metabolite. *Proc. Natl. Acad. Sci.*, 119, 4, e2111338119, 2022.
83. Hu, S., Tan, J., Qin, L., Lv, L., Yan, W., Zhang, H., Tang, B., Wang, C., Molecular Chaperones and Parkinson's Disease. *Neurobiol. Dis.*, 160, 105527, 2021.
84. Kluge, A., Bunk, J., Schaeffer, E., Drobny, A., Xiang, W., Knacke, H., Bub, S., Lückstädt, W., Arnold, P., Lucius, R., Berg, D., Zunke, F., Detection of Neuron-Derived Pathological α-Synuclein in Blood. *Brain*, 145, 9, 3058–3071, 2022.
85. Schirinzi, T., Di Lazzaro, G., Sancesario, G.M., Summa, S., Petrucci, S., Colona, V.L., Bernardini, S., Pierantozzi, M., Stefani, A., Mercuri, N.B., Pisani, A., Young-Onset and Late-Onset Parkinson's Disease Exhibit a Different Profile of Fluid Biomarkers and Clinical Features. *Neurobiol. Aging*, 90, 119–124, 2020.
86. Li, D., Mastaglia, F.L., Fletcher, S., Wilton, S.D., Progress in the Molecular Pathogenesis and Nucleic Acid Therapeutics for Parkinson's Disease in the Precision Medicine Era. *Med. Res. Rev.*, 40, 6, 2650–2681, 2020.
87. Alieva, A., Rudenok, M., Filatova, E., Karabanov, A., Doronina, O., Doronina, K., Kolacheva, A., Ugrumov, M., Illarioshkin, S., Slominsky, P., Shadrina, M., VCP Expression Decrease as a Biomarker of Preclinical and Early Clinical Stages of Parkinson's Disease. *Sci. Rep.*, 10, 1, 827, 2020.
88. Pilotto, A., Imarisio, A., Conforti, F., Scalvini, A., Masciocchi, S., Nocivelli, S., Turrone, R., Gipponi, S., Cottini, E., Borroni, B., Rizzetti, M.C., Pizzi, M., Bonanni, L., Sturchio, A., Espay, A.J., Zetterberg, H., Ashton, N.J., Hye, A., Padovani, A., Plasma NfL, Clinical Subtypes and Motor Progression in Parkinson's Disease. *Parkinsonism Relat. Disord.*, 87, 41–47, 2021.
89. Chen, X., Wu, G., Schwarzschild, M.A., Urate in Parkinson's Disease: More Than a Biomarker? *Curr. Neurol. Neurosci. Rep.*, 12, 4, 367–375, 2012.
90. Collins, L.M., Drouin-Ouellet, J., Kuan, W.-L., Cox, T., Barker, R.A., Dermal Fibroblasts from Patients with Parkinson's Disease Have Normal GCase Activity and Autophagy Compared to Patients with PD and GBA Mutations. *F1000Research*, 6, 1751, 2018.
91. Dutta, S., Hornung, S., Kruayatidee, A., Maina, K.N., del Rosario, I., Paul, K.C., Wong, D.Y., Duarte Folle, A., Markovic, D., Palma, J.-A., Serrano, G.E., Adler, C.H., Perlman, S.L., Poon, W.W., Kang, U.J., Alcalay, R.N., Sklerov, M., Gylys, K.H., Kaufmann, H., Fogel, B.L., Bronstein, J.M., Ritz, B., Bitan, G., α-Synuclein in Blood Exosomes Immunoprecipitated Using Neuronal and

Oligodendroglial Markers Distinguishes Parkinson's Disease from Multiple System Atrophy. *Acta Neuropathol.*, 142, 3, 495–511, 2021.

92. Li, R., Li, Y., Mu, M., Yang, B., Chen, X., Lee, W.Y.W., Ke, Y., Yung, W.H., Tang, B.Z., Bian, L., Multifunctional Nanoprobe for the Delivery of Therapeutic SiRNA and Real-Time Molecular Imaging of Parkinson's Disease Biomarkers. *ACS Appl. Mater. Interfaces*, 13, 10, 11609–11620, 2021.
93. Poux, S. and Gaudet, P., Best Practices in Manual Annotation with the Gene Ontology. *The Gene Ontology Handbook*, Dessimoz, C., Škunca, N., eds. Springer Protocol, Department of Genetics Evolution and Environment, University College of London, London, United Kingdom, Department of Computer Science, ETH Zurich, Zurich, Switzerland, pp. 41–54, 2017.
94. Expansion of the Gene Ontology Knowledgebase and Resources. *Nucleic Acids Res.*, 45, D1, D331–D338, 2017.
95. Ma, W., Li, Y., Wang, C., Xu, F., Wang, M., Liu, Y., Serum MiR-221 Serves as a Biomarker for Parkinson's Disease. *Cell Biochem. Funct.*, 34, 7, 511–515, 2016.
96. Kabaria, S., Choi, D.C., Chaudhuri, A.D., Mouradian, M.M., Junn, E., Inhibition of MiR-34b and MiR-34c Enhances α-Synuclein Expression in Parkinson's Disease. *FEBS Lett.*, 589, 3, 319–325, 2015.
97. Wang, J., Liu, Y., Chen, T., Identification of Key Genes and Pathways in Parkinson's Disease through Integrated Analysis. *Mol. Med. Rep.*, 16, 4, 3769–3776, 2017.
98. Younesi, E., Malhotra, A., Gündel, M., Scordis, P., Kodamullil, A.T., Page, M., Müller, B., Springstubbe, S., Wüllner, U., Scheller, D., Hofmann-Apitius, M., PDON: Parkinson's Disease Ontology for Representation and Modeling of the Parkinson's Disease Knowledge Domain. *Theor. Biol. Med. Model.*, 12, 1, 20, 2015.
99. Wang, X., Shi, N., Wu, B., Yuan, L., Chen, J., Ye, C., Hao, M., Bioinformatics Analysis of Gene Expression Profile and Functional Analysis in Periodontitis and Parkinson's Disease. *Front. Aging Neurosci.*, 14, 1029637, 2022.
100. Yang, Y., Wang, Y., Wang, C., Xu, X., Liu, C., Huang, X., Identification of Hub Genes of Parkinson's Disease through Bioinformatics Analysis. *Front. Neurosci.*, 16, 974838, 2022.
101. Benson, D.L. and Huntley, G.W., Are We Listening to Everything the PARK Genes Are Telling Us? *J. Comp. Neurol.*, 527, 9, 1527–1540, 2019.
102. Piñero, J., Saüch, J., Sanz, F., Furlong, L.I., The DisGeNET Cytoscape App: Exploring and Visualizing Disease Genomics Data. *Comput. Struct. Biotechnol. J.*, 19, 2960–2967, 2021.
103. Zhang, T., Wei, W., Chang, S., Liu, N., Li, H., Integrated Network Pharmacology and Comprehensive Bioinformatics Identifying the Mechanisms and Molecular Targets of Yizhiqingxin Formula for Treatment of Comorbidity With Alzheimer's Disease and Depression. *Front. Pharmacol.*, 13, 853375, 2022.

104. Huntley, R.P., Harris, M.A., Alam-Faruque, Y., Blake, J.A., Carbon, S., Dietze, H., Dimmer, E.C., Foulger, R.E., Hill, D.P., Khodiyar, V.K., Lock, A., Lomax, J., Lovering, R.C., Mutowo-Meullenet, P., Sawford, T., Van Auken, K., Wood, V., Mungall, C.J., A Method for Increasing Expressivity of Gene Ontology Annotations Using a Compositional Approach. *BMC Bioinf.*, 15, 1, 155, 2014.
105. George, G., Singh, S., Lokappa, S.B., Varkey, J., Gene Co-Expression Network Analysis for Identifying Genetic Markers in Parkinson's Disease- a Three-Way Comparative Approach. *Genomics*, 111, 4, 819–830, 2019.
106. Yue, R. and Dutta, A., Computational Systems Biology in Disease Modeling and Control, Review and Perspectives. *npj Syst. Biol. Appl.*, 8, 1, 37, 2022.
107. Barisoni, L., Lafata, K.J., Hewitt, S.M., Madabhushi, A., Balis, U.G.J., Digital Pathology and Computational Image Analysis in Nephropathology. *Nat. Rev. Nephrol.*, 16, 11, 669–685, 2020.
108. Severson, K.A., Chahine, L.M., Smolensky, L.A., Dhuliawala, M., Frasier, M., Ng, K., Ghosh, S., Hu, J., Discovery of Parkinson's Disease States and Disease Progression Modelling: A Longitudinal Data Study Using Machine Learning. *Lancet Digit. Health*, 3, 9, e555–e564, 2021.
109. Cassotta, M., The Future of Parkinson's Disease Research: A New Paradigm of Human-Specific Investigation Is Necessary… and Possible. *ALTEX*, 39, 694–709, 2022.
110. Farrow, S.L., Cooper, A.A., O'Sullivan, J.M., Redefining the Hypotheses Driving Parkinson's Diseases Research. *npj Parkinson's Dis.*, 8, 1, 45, 2022.
111. Salat, D. and Tolosa, E., Levodopa in the Treatment of Parkinson's Disease: Current Status and New Developments. *J. Parkinson's Dis.*, 3, 3, 255–269, 2013.
112. Tan, E.-K., Chao, Y.-X., West, A., Chan, L.-L., Poewe, W., Jankovic, J., Parkinson Disease and the Immune System — Associations, Mechanisms and Therapeutics. *Nat. Rev. Neurol.*, 16, 6, 303–318, 2020.
113. Rietdijk, C.D., Perez-Pardo, P., Garssen, J., van Wezel, R.J.A., Kraneveld, A.D., Exploring Braak's Hypothesis of Parkinson's Disease. *Front. Neurol.*, 8, 37, 2017.
114. Dong, X., Li, Y., Li, Q., Li, W., Wu, G., Identification of Immune Signatures in Parkinson's Disease Based on Co-Expression Networks. *Front. Genet.*, 14, 1090382, 2023.
115. Wu, Y., Bai, Y., Lu, Y., Zhang, Z., Zhao, Y., Huang, S., Tang, L., Liang, Y., Hu, Y., Xu, C., Transcriptome Sequencing and Network Pharmacology-Based Approach to Reveal the Effect and Mechanism of Ji Chuan Jian against Parkinson's Disease. *BMC Complement. Med. Ther.*, 23, 1, 182, 2023.
116. Ni, P., Zhao, B., Pang, Y., Pan, K., Mechanism of Tianma Gouteng Decoction in the Treatment of Parkinson's Disease Based on Network Pharmacology and Molecular Docking. *Am. J. Transl. Res.*, 15, 1, 596–611, 2023.
117. Xing, N., Dong, Z., Wu, Q., Zhang, Y., Kan, P., Han, Y., Cheng, X., Wang, Y., Zhang, B., Identification of Ferroptosis Related Biomarkers and Immune Infiltration in Parkinson's Disease by Integrated Bioinformatic Analysis. *BMC Med. Genomics*, 16, 1, 55, 2023.

118. Yin, X., Wang, M., Wang, W., Chen, T., Song, G., Niu, Y., Jiang, Z., Gao, Z., Wang, Z., Identification of Potential MiRNA-MRNA Regulatory Network Contributing to Parkinson's Disease. *Parkinsons Dis.*, 2022, 1–12, 2022.
119. Mash, D.C., Ouyang, Q., Pablo, J., Basile, M., Izenwasser, S., Lieberman, A., Perrin, R.J., Cocaine Abusers Have an Overexpression of α-Synuclein in Dopamine Neurons. *J. Neurosci.*, 23, 7, 2564–2571, 2003.
120. Mishra, A.K. and Dixit, A., Dopaminergic Axons: Key Recitalists in Parkinson's Disease. *Neurochem. Res.*, 47, 2, 234–248, 2022.
121. Curtin, K., Fleckenstein, A.E., Robison, R.J., Crookston, M.J., Smith, K.R., Hanson, G.R., Methamphetamine/Amphetamine Abuse and Risk of Parkinson's Disease in Utah: A Population-Based Assessment. *Drug Alcohol Depend.*, 146, 30–38, 2015.
122. Yan, J., Feng, X., Zhou, X., Zhao, M., Xiao, H., Li, R., Shen, H., Identification of Gut Metabolites Associated with Parkinson's Disease Using Bioinformatic Analyses. *Front. Aging Neurosci.*, 14, 927625, 2022.
123. Figura, M., Kuśmierska, K., Bucior, E., Szlufik, S., Koziorowski, D., Jamrozik, Z., Janik, P., Serum Amino Acid Profile in Patients with Parkinson's Disease. *PLoS One*, 13, 1, e0191670, 2018.
124. Clark, E.H., Vázquez de la Torre, A., Hoshikawa, T., Briston, T., Targeting Mitophagy in Parkinson's Disease. *J. Biol. Chem.*, 296, 100209, 2021.
125. Yang, J., Chen, J., Liu, Y., Chen, K.H., Baraban, J.M., Qiu, Z., Ventral Tegmental Area Astrocytes Modulate Cocaine Reward by Tonically Releasing GABA. *Neuron*, 111, 7, 1104–1117.e6, 2023.
126. Ahlskog, J.E., Levodopa, Homocysteine and Parkinson's Disease: What's the Problem? *Parkinsonism Relat. Disord.*, 109, 105357, 2023.
127. Wang, Y., Miao, Z., Xu, C., Cai, Y., Yang, Y., Hu, Y., Zhao, M., Shao, Y., Li, Z., Chen, J., Chen, S., Wang, L., Pathological Convergence of APP and SNCA Deficiency in Hippocampal Degeneration of Young Rats. *Cell Death Dis.*, 14, 5, 325, 2023.
128. Lau, H.H.C., Martinez-Valbuena, I., So, R.W.L., Mehra, S., Silver, N.R.G., Mao, A., Stuart, E., Schmitt-Ulms, C., Hyman, B.T., Ingelsson, M., Kovacs, G.G., Watts, J.C., The G51D SNCA Mutation Generates a Slowly Progressive α-Synuclein Strain in Early-Onset Parkinson's Disease. *Acta Neuropathol. Commun.*, 11, 1, 72, 2023.
129. Nagatsu, T., Nakashima, A., Watanabe, H., Ito, S., Wakamatsu, K., Zucca, F.A., Zecca, L., Youdim, M., Wulf, M., Riederer, P., Dijkstra, J.M., The Role of Tyrosine Hydroxylase as a Key Player in Neuromelanin Synthesis and the Association of Neuromelanin with Parkinson's Disease. *J. Neural Transm.*, 130, 5, 611–625, 2023.
130. Nabizadeh, F., Mohamadzadeh, O., Hosseini, H., Rasouli, K., Afyouni, N.E., Serum Neurofilament Light Chain in LRRK2 Related Parkinson's Disease: A Five Years Follow-Up. *J. Clin. Neurosci.*, 110, 12–18, 2023.
131. Vissers, M.F.J.M., Troyer, M.D., Thijssen, E., Heuberger, J.A.A.C., Groeneveld, G.J., Huntwork-Rodriguez, S., A leucine-rich Repeat Kinase 2 (LRRK2)

Pathway Biomarker Characterization Study in Patients with Parkinson's Disease with and without LRRK2 Mutations and Healthy Controls. *Clin. Transl. Sci.*, 16, 1408–1420, 2023.

132. Borsche, M., Pratuseviciute, N., Schaake, S., Hinrichs, F., Morel, G., Uter, J., Lohmann, K., Klein, C., Alessi, D.R., Hagenah, J., Sammler, E., The New p. F1700L LRRK2 Variant Causes Parkinson's Disease by Extensively Increasing Kinase Activity. *Mov. Disord.*, 38, 1105–1107, 2023.
133. Mancini, A., Stoops, E., Demeyer, L., Bellomo, G., Paolini Paoletti, F., Gaetani, L., Di Filippo, M., Parnetti, L., LRRK2 Quantification in Cerebrospinal Fluid of Patients with Parkinson's Disease and Atypical Parkinsonian Syndromes. *Mov. Disord.*, 38, 4, 682–688, 2023.
134. Lind-Holm Mogensen, F., Scafidi, A., Poli, A., Michelucci, A., PARK7/DJ-1 in Microglia: Implications in Parkinson's Disease and Relevance as a Therapeutic Target. *J. Neuroinflammation*, 20, 1, 95, 2023.
135. Chew, Z.X., Lim, C.L., Ng, K.Y., Chye, S.M., Ling, A.P.K., Koh, R.Y., The Role of Monoamine Oxidase B Inhibitors in the Treatment of Parkinson's Disease - An Update. *CNS Neurol. Disord. Drug Targets*, 22, 3, 329–352, 2023.
136. Kung, H.-C., Lin, K.-J., Kung, C.-T., Lin, T.-K., Oxidative Stress, Mitochondrial Dysfunction, and Neuroprotection of Polyphenols with Respect to Resveratrol in Parkinson's Disease. *Biomedicines*, 9, 8, 918, 2021.
137. Liu, L., Han, Y., Zhang, Z., Wang, Y., Hu, Y., Kaznacheyeva, E., Ding, J., Guo, D., Wang, G., Li, B., Ren, H., Loss of DJ-1 Function Contributes to Parkinson's Disease Pathogenesis in Mice via RACK1-Mediated PKC Activation and MAO-B Upregulation. *Acta Pharmacol. Sin.*, 44, 1948–1961, 2023.
138. Regensburger, M., Ip, C.W., Kohl, Z., Schrader, C., Urban, P.P., Kassubek, J., Jost, W.H., Clinical Benefit of MAO-B and COMT Inhibition in Parkinson's Disease: Practical Considerations. *J. Neural Transm.*, 130, 6, 847–861, 2023.
139. Bakhit, Y., Ibrahim, M.O., Tesson, C., Elhassan, A.A., Ahmed, M.A., Alebeed, M.A., Elrasheed, S.M., Omar, M.A., Abubaker, R., Eltom, K., Shaheen, M.T., Ibrahim, Y.A., Almak, M.E., Ali, H.A., Abugrain, A.A., Almahal, M.A., MohamedSharif, A.A., Tahir, M.Y., Malik, S.M., Eldirdiri Abdelrahman, H., Khidir, R.J., Mohamed, M.T., Abdalla, A., Elsayed, L.E.O., Lesage, S., Corvol, J.-C., Seidi, O., Wüllner, U., Intrafamilial and Interfamilial Heterogeneity of PINK1-Associated Parkinson's Disease in Sudan. *Parkinsonism Relat. Disord.*, 111, 105401, 2023.
140. Wang, Z., Chan, S.-W., Zhao, H., Miu, K.-K., Chan, W.-Y., Outlook of PINK1/Parkin Signaling in Molecular Etiology of Parkinson's Disease, with Insights into Pink1 Knockout Models. *Zool. Res.*, 44, 3, 559–576, 2023.
141. Travaglio, M., Michopoulos, F., Yu, Y., Popovic, R., Foster, E., Coen, M., Martins, L.M., Increased Cysteine Metabolism in PINK1 Models of Parkinson's Disease. *Dis. Model. Mech.*, 16, 1, 2023.
142. Biswas, S. and Bagchi, A., Analysis of the Structural Dynamics of the Mutations in the Kinase Domain of PINK1 Protein Associated with Parkinson's Disease. *Gene*, 857, 147183, 2023.

143. Gerasimova, T., Stepanenko, E., Novosadova, L., Arsenyeva, E., Shimchenko, D., Tarantul, V., Grivennikov, I., Nenasheva, V., Novosadova, E., Glial Cultures Differentiated from IPSCs of Patients with PARK2-Associated Parkinson's Disease Demonstrate a Pro-Inflammatory Shift and Reduced Response to TNFα Stimulation. *Int. J. Mol. Sci.*, 24, 3, 2000, 2023.

144. Ahmad, A., Nkosi, D., Iqbal, M.A., PARK2 Microdeletion or Duplications Have Been Implicated in Different Neurological Disorders Including Early Onset Parkinson Disease. *Genes (Basel)*, 14, 3, 600, 2023.

145. Fernandes, H.J.R., Kent, J.P., Bruntraeger, M., Bassett, A.R., Koulman, A., Metzakopian, E., Snowden, S.G., Mitochondrial and Endoplasmic Reticulum Stress Trigger Triglyceride Accumulation in Models of Parkinson's Disease Independent of Mutations in MAPT. *Metabolites*, 13, 1, 112, 2023.

10

Computational Attitudes to Drug Discovery in Epilepsy

Shama Mujawar[1*], Aarohi Deshpande[1], Avni Bhambure[1], Shreyash Kolhe[1] and Bhupendra Prajapati[2,3]

[1]MIT School of Bioengineering Sciences and Research, MIT-Art, Design and Technology University, Loni Kalbhor, Pune, India
[2]Shree S.K. Patel College of Pharmaceutical Education & Research, Faculty of Pharmacy, Ganpat University, Gujarat, India
[3]Department of Industrial Pharmacy, Faculty of Pharmacy, Silpakorn University, Nakhon Pathom, Thailand

Abstract

Epilepsy, a neurological condition that affects around 50 million people worldwide, is marked by recurrent seizures. Even with the abundance of antiepileptic drugs available, a significant number of patients lack access to adequate seizure management and treatment, discovering novel medications continues to be a serious obstacle. The drug candidates could be efficiently identified using remarkable computational technologies, such as molecular docking, virtual screening, machine learning, and artificial intelligence, which have grown in popularity in the field of drug development. This review paper gives a general overview of the current computational techniques to drug discovery in epilepsy while expressing the advantages and difficulties of these methods. More specifically, we go through how to employ computational methods to find new therapeutic targets, repurpose already-approved medications, and enhance drug efficacy and safety. We also examine the drawbacks of these methods and focus on the requirement for more precise predictive models based on artificial intelligence and the difficulties in locating appropriate biomarkers. Overall, the use of computational techniques for drug discovery substantially speeds up the development of innovative and successful epilepsy medications, which have been brought into light.

***Keywords*:** Epilepsy, artificial intelligence, electro-encephelography, computational models, artificial intelligence, seizures

**Corresponding author*: shamamujawar@gmail.com

Bhupendra Prajapati, Alok Tripathi, Rishabha Malviya and Lucy Mohapatra (eds.) Computational Neuropharmacology: Fundamentals and Clinical Aspects, (313–334) © 2025 Scrivener Publishing LLC

List of Abbreviations

AIML	Artificial intelligence and machine learning algorithms
BACE-1	Beta-secretase 1
CNN	Convolutional Neural Networks
CRP	C-Reactive Protein
DBS	Deep brain stimulation
EEG	Electroencephalogram
GABA	Gamma-Aminobutyric Acid
HTS	High Throughput Screening
NFL	Neurofilament Light Chain
QSP	Quantitative systems pharmacology
RNN	Recurrent Neural Networks
RNS	Responsive neurostimulation
SVM	Support Vector Machines
SVM	Support Vector Machines
VNS	Vagus Nerve Stimulator

10.1 Introduction

Millions of individuals throughout the world suffer with epilepsy, a condition of the nervous system that can lead to recurring seizures and significantly lower quality of life. Despite decades of research, the underlying causes of epileptic episodes are still unknown, and the current treatments are not always effective or have serious side effects. Because epilepsy is associated with a wide range of factors and seizure types, the negative effects of epilepsy on various persons may differ significantly. In certain individuals, epilepsy may coexist with other medical disorders or several different kinds of epileptic events, which might alter how severe the disorder is for them and how it affects their everyday lives [1]. Prolonged electrical impulses in a group of neurons trigger instances of convulsions. These electrical discharges may take place in multiple locations of the brain. Even minor muscle jerks or loss of focus can manifest as seizures, alongside serious, prolonged seizures. Seizures can also occur at different rates; they can occur up to once a year or on a daily basis [2]. As a result, computational approaches are needed to accurately diagnose and treat multiple types of epilepsy, as well as to prevent future occurrences [1]. High-efficiency computing systems and machine learning are two recent technological developments that have opened up new research directions for treating epilepsy

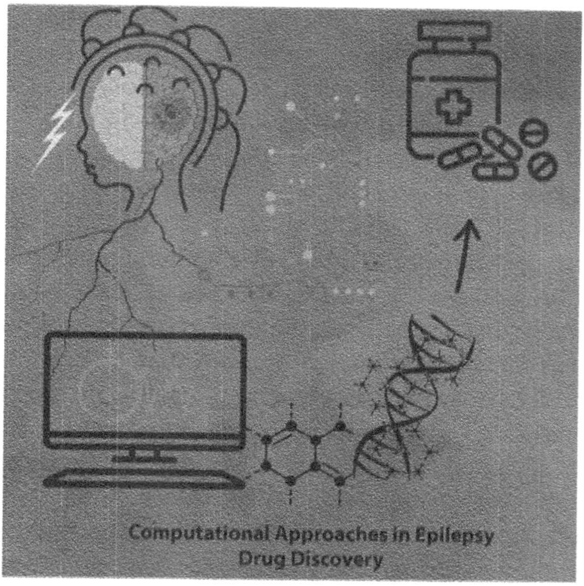

Figure 10.1 Bridging of EEG data in epilepsy drug discovery.

and may lead to more individualized and efficient therapies. In this study, we examine the intricate interplay of a complex neurodegenerative disease such as epilepsy, along with some drug discovery approach currently developed for the treatment of epilepsy. This is done to address the critical concerns of the underlying processes governing the episodes in epilepsy which can be eliminated using potential and rising drug discovery approaches as shown in Figure 10.1. Our goal is to come to terms with the mechanisms that lie beneath and propose novel solutions to these challenges.

10.2 Traditional Drug Discovery Approaches for Epilepsy

10.2.1 Animal Models

Seizure models were used to identify the potential therapeutic effect of the drugs. During the past three decades, several animal models exhibiting different levels of pharmacoresistance have been established. These include rats that develop spontaneous recurrent seizures following the inducing seizures after being stimulated chemically or electrically, the 6-Hz mouse system representing focal seizure, and the intrahippocampal kainate system in mice. The antiseizure action of all therapeutically used drugs was primarily established using the acute seizure models in rats utilizing the maximal electroshock, the rats stimulated electrically and subcutaneous PTZ seizure assessments [3]. Novel anti-seizure medications have been identified and differentiated using the 6Hz model, which has been applied extensively. But this model has proved to establish results only in mice and the corresponding assay has not yet been developed for rat [4]. In the lamotrigine-resistant amygdala seizure induction model, lamotrigine resistance is created by continuously administering a tiny dosage of the drug during the process. This resistance additionally applies to numerous anti-seizure medications. Sodium channel inhibitors, including rufinamide, lacosamide, lamotrigine, eslicarbazepine, carbamazepine, and phenytoin, had failed to work at the highest studied levels or caused side effects. Lamotrigine-resistant kindling rats tolerated GABA-acting substances better than naive rats. Therefore, it appears that this model is sensitive to GABAergic compounds and is resistant to sodium channel blockers [5].

It was thought that they would help to study the resistance of drug mechanisms and would also offer a way to identify a novel anti-epileptic drug. However, no potential compound has been validated by any of the models [3]. Although employment of a variety of models in the development of new approaches for the antiseizure medications is possible.

10.2.2 High Throughput Screening

Another traditional method for drug discovery is high throughput screening. It is employed in the discovery of drugs of many neurodegenerative diseases, epilepsy being one of them. This step is essential as it identifies potential candidates from chemical libraries which might be further developed through optimization of drug discovery [6]. Few hundred "hits" may be found after screening a large number of small molecules, which might give rise to the discovery of potential therapeutic compounds. The development of novel diagnostic techniques that allow for the temporal study of the neural region and integrity of cellular membrane has led to the creation of assays for monitoring the progression of neurodegeneration [7]. Some common types of HTS assays used in epilepsy are Ion channel assays, GABA and Glutamate receptor assays, Neuronal Excitability Assays, and Phenotypic Assays. HTS has proved to be an important method in drug discovery, however, it faces some limitations. It has a poor hit frequency which may be because of incongruent libraries and potential false results [8]. The magnitude of data collected is very huge, hence the data analytics and data handling can be overwhelming. Further, it does not provide any information for underlying mechanisms of action. The HTS library might not include all the natural compounds and potential targets may be overlooked. Along with it the cost for screening these large libraries is high [9]. Identification of novel drugs through HTS and assessment of drugs before experimental work through computer aided drug design can facilitate therapeutic medicine.

10.2.3 Computational Approaches in Epilepsy Drug Discovery

Epilepsy is a critical brain condition that affects millions of individuals worldwide, yet there are presently no medications at hand, readily accessible to cure or treat it. Even existing anti-epileptic medications fail to work for approximately one-third of individuals. For evaluating brain function, multiple novel methods for imaging are currently in use, such as PET scanning, magnetoencephalography, computed tomography and MRI. Despite these advances, EEG remains the most significant tool for pinpointing epileptic network activity [10]. Drug development for central nervous system illnesses such as epilepsy is difficult Because of inadequate validation of drug targets in the initial stages of discovery, which frequently results in high rates of disintegration [11]. Even after the success of pharmacological combination therapy in the diagnosis and treatment of many complicated diseases, discovering efficacious disease-altering medicines for brain disorders continues to be a significant problem. In spite of extensive research, there is a pressing need for innovative medications that might prevent epilepsy from progressing in high-risk individuals or alter how it develops [12].

10.2.4 Computational Methods Applied to Enhance Drug Discovery Processes in Epilepsy

The idea behind drug combination therapy is to enhance the effectiveness of treatment and overcome resistance by using drugs with diverse mechanisms of action, known activity, and minimal overlapping toxicities. This approach is made possible by scientific progress that has improved our understanding of the fundamental pathways contributing to complex disorders. In fact, a thorough comprehension of the origin and physiological mechanism of a disease is necessary to establish a logical biological rationale for implementing combination therapy. With this knowledge, drugs with differing mechanisms of action can be rationally combined to achieve better treatment outcomes [12].

With the advancement of high-performance computing technologies and the development of various mathematical algorithms, computational studies are being actively conducted for clinical application in epilepsy.

These studies aim to analyze large amounts of data to derive a proper solution for individual patients. Some studies have used scalp-EEG recordings labeled with seizure types from multiple patients to train computational models that can classify seizure types with high accuracy based on spatio-temporal features. To implement these models, classic machine learning algorithms such as SVM and k-NN have been used, as well as deep learning algorithms such as CNN [13].

Exploring the repositioning of approved drugs has become increasingly popular as a means of quickly identifying and developing new treatments for conditions with inadequate treatment options, including personalized medicine for rare genetic epileptic conditions. This approach Has become increasingly popular in recent times because of its potential to rapidly identify new therapies for complex diseases. Both the FDA and researchers are keenly interested in this field, with the goal of development of new therapies that can enhance patient clinical results [14].

To summarize, computational models have emerged as a significant tool in epilepsy investigations, gaining insights into epileptic causes and influencing the research and development of new therapeutics. Various drug discovery approaches have proven to be effective in producing a therapeutic effect for minimizing seizures.

10.2.5 Identifying Potential Antiepileptic Compounds for Drug Discovery

Computing approaches including virtual screening, pharmacophore modeling, and molecular docking are frequently employed to find promising anti-epileptic drugs. With the use of these techniques, scientists may examine huge databases of compounds and rank those with the best chances of being successful against epilepsy in the early phases of drug discovery. Here, we'll go into further depth about each technique:

10.2.5.1 Molecular Docking

A computer method for the prediction of binding interactions within a ligand and a receptor is known as molecular docking. Molecular docking has been widely used in the context of anti-epileptic drug development to locate substances that can interact with certain protein targets related to epilepsy. Practically, the interactions of numerous antiepileptic medications with the gamma-aminobutyric acid type A (GABA-A) receptor using molecular docking. The structural characteristics and binding interactions crucial to their antiepileptic efficacy were revealed by the docking study [15].

10.2.5.2 Virtual Screening

This computational approach is employed to explore through enormous databases of compounds and find probable molecules with required pharmacological characteristics. There are two basic methods for doing it: 1. structure-based screening 2. ligand based screening.

To find related molecules in a compound library, ligand-based virtual screening compares the chemical characteristics and structural attributes of known active compounds. This strategy is predicated on the idea that molecules with comparable structures would have related biological effects. An experiment used ligand-based virtual screening to find new beta-secretase 1 (BACE1) inhibitors for Alzheimer's. Potential anti-epileptic drugs were found throughout the screening procedure by comparing the structural characteristics of known BACE1 inhibitors with a chemical library [13].

Contrarily, structure-based virtual screening includes docking candidate compounds into the target protein's three-dimensional structure and assessing their binding affinity. This method depends on the availability of homology models or protein structures. Structure-based virtual screening has been used in the context of epilepsy medication development to find substances that interact with certain targets implicated in epileptogenesis.

10.2.5.3 Pharmacophore Modeling

The process of identifying and characterizing the crucial characteristics or chemical groups necessary for a ligand to interact with its target protein. Hydrophobic areas, aromatic rings, hydrogen bond acceptors, donors, and other particular chemical groups are a few examples of these characteristics. To screen chemical libraries and find compounds that fit the required properties, pharmacophore models are used as a template. Drug development for anti-epileptics has effectively used pharmacophore modeling to find substances with unique pharmacophoric properties that interact with epileptic targets. Pharmacophore modeling was used in research to find prospective anti-epileptic drugs that target certain proteins implicated in epilepsy development [16].

10.3 Computer Simulations in Understanding and Optimizing Drug Efficacy

The implementation of computer modeling to simulate the behavior of various drugs and estimate their potential efficacy in treating epilepsy is one technique. Researchers can uncover intriguing possibilities for further investigation more rapidly. Another method is to employ artificial intelligence and machine learning algorithms (AIML) to analyze huge amounts of data and detect trends that can be applied to guide drug discovery. These strategies can assist researchers in identifying novel therapeutic targets and developing improved epilepsy interventions.

Computational modeling techniques are additionally employed to simulate epileptic episodes and explore the mechanics behind epilepsy. For instance, neural network computational models have been used to mimic seizure activity and discover potential candidates for antiepileptic medicines. Furthermore, neurological simulations utilizing neuronal mass algorithms have been utilized to investigate the patterns of seizures associated with epilepsy and to develop biomarkers towards seizure projections. The interlocking mechanism structures involving positive and negative feedback loops can also be used to characterize dynamics. Numerous channels of feedback in the brain yield unexpected results. Some feedback loops are partially adequate and alleviate impairment, while others may aggravate the illness. Similarly, certain medications that help with one type of epilepsy can aggravate another. Benzodiazepines, for example, promote blocking and are utilized as an auxiliary or short-term therapy in a variety of epilepsies [17].

Using quantum mechanics, the behavior of molecules can be simulated, and their properties can be predicted, which can aid in the design of novel drugs [18]. Additionally, the process of discovering drugs can be accelerated by using quantum technology to complete complicated calculations that are not feasible for traditional computers [19]. The research and development of medications to treat seizures may be greatly aided by these developments in quantum physics and computation, leading to more precise and potent therapies [20].

These models can also be used to test the efficacy of different treatments and interventions in a virtual environment before being tested on real patients [21]. Furthermore, computer simulations can help researchers understand the underlying mechanisms of epileptic seizures and identify potential targets for drug development [22]. Additionally, computer simulations can help clinicians optimize treatment plans for individual patients by predicting how different medications will affect their seizures [18]. Overall, computer simulations provide a powerful tool for epilepsy research and have the potential to improve the lives of individuals with epilepsy.

10.4 Development of Computational Models

10.4.1 Models for Neural Networks

Brain activity and synaptic connections are simulated by neural network models. These simulations can capture the intricate dynamics of epileptic convulsions and provide understanding of the underlying causes. To characterize the electrical activity of neurons, synaptic connections, and network features, they use biophysical and mathematical equations. It has become possible to understand the mechanisms of seizure start, progression, and termination thanks to the use of such models. The model explains how action potentials in neurons are generated and spread. This model served as the cornerstone for comprehending the function of ion channels in epilepsy and has been widely utilized as a complex network model [23].

10.4.2 Seizure Models

Simulating the interactions of several neurons and their synaptic connections is the goal of neural circuit models. To reproduce the intricate dynamics seen in epileptic convulsions, our models include realistic connection patterns and synaptic dynamics. By including elements like synaptic plasticity, network remodeling, and excitation-inhibition imbalances, computational models of brain circuits may recreate seizure-like behavior. Epileptic seizures' patterns and traits are replicated by seizure models. These models mimic the aberrant synchronized firing of neurons seen during seizures by combining experimental data and mathematical formulae. They may include several elements including network connections, synaptic changes, and ion channel dysfunctions.

One such is the Izhikevich model, which was put up by Izhikevich in 2003 and offers a straightforward yet effective framework for replicating various neural firing patterns, including those seen during seizures. This model has been applied to research the dynamics of seizure formation as well as the impact of various pharmaceutical therapies [24].

10.4.3 Use of Computational Models of Brain Networks

10.4.3.1 Recognizing Epileptogenesis

Understanding the mechanics of epileptogenesis—the process by which a healthy brain becomes susceptible to seizures—has been made possible in large part by computational models. The development of hyperexcitability, variations in synaptic strength, and network rewiring are only a few examples of the long-term changes in neural networks that these models can represent. These models help to clarify the important variables and dynamics that play a role in the onset of epilepsy. A neural mass model that accurately depicted the change from epileptic activity to normal brain activity. The synaptic connection and excitability changes in this model led to the appearance of aberrant oscillations that resembled epileptic convulsions [25].

10.4.3.2 The Creation of Antiepileptic Medications

Computational models are essential to the creation and improvement of anti-epileptic medications. These models may replicate the pharmacodynamics (drug-target interactions and effects) and pharmacokinetics (absorption, distribution, metabolism, and excretion) of medications. Computational models offer insights into the effectiveness, safety, and ideal dosage regimens of anti-epileptic medications by incorporating experimental data on drug characteristics and their impact on brain networks.

For instance, the quantitative systems pharmacology (QSP) model included the pharmacokinetic and pharmacodynamic characteristics of anti-epileptic medications to forecast their effectiveness and optimize dose regimens. This model made it possible to forecast the best medication dosages and concentrations, which paved the way for the creation of individualized epilepsy treatment plans [26].

10.4.4 Planning for Epilepsy Surgery

The purpose of using computational models in epilepsy surgery planning is to identify the epileptic focus—the area of the brain from which seizures begin—and reduce the possibility of functional losses following surgery. To provide customized simulations, these models use patient-specific data such as neuroimaging, electrophysiological recordings, and anatomical details.One illustration is the localization of the epileptic focal using computational models based on electroencephalography (EEG) data. These simulations can help pinpoint the area of the brain that causes seizures by simulating the distribution of epileptic activity across the brain.

10.4.5 Connectivity and Network Topology

The connections and connectivity patterns between various brain areas are simulated by computational models of brain networks. These models offer insights into the mechanics of seizure propagation and the identification of crucial seizure-generating areas by modeling the spread of epileptic activity across brain regions.

To comprehend the structure of epileptic networks, for instance, graph theory-based models examine the network features of brain connection, such as node degree, clustering coefficient, and small-worldness. These models have shown the significance of certain brain areas and network connections in the formation and maintenance of epilepsy, revealing typical changes in network topology in epilepsy [27].

10.4.6 Seizure Initiation and Propagation

By including the dynamics of excitation and inhibition, computational models of brain networks may replicate the initiation and propagation of seizures. Such models use mathematical formulas to explain how neurons fire and how electrical impulses move between different parts of the brain. These models recreate the spatiotemporal patterns noticed during seizures by integrating experimental data and theoretical assumptions.

The use of brain mass models, which show populations of neurons as collective entities with streamlined dynamics, is one instance. Understanding the mechanics of seizure start and spread across many brain areas is possible thanks to neural mass models [28, 29].

10.4.7 Predictive Modeling and Personalized Medicine

Brain network computational models offer the potential to advance epilepsy personalized therapy. These models may recreate the unique dynamics of seizure activity by including patient-specific data, such as neuroimaging, electroencephalography (EEG), and clinical data. Because of this, seizure onset, propagation patterns, and medication response may all be predicted, making it possible to develop customized treatment plans. To help with seizure forecasting and preventative actions, patient-specific models built from EEG data have been utilized to predict the chance of seizure onset. These models can improve treatment regimens created for specific individuals [30, 31].

10.5 Computational Models for Predicting Effects on Seizure Activity

Finding appropriate pharmacological targets and predicting their effects on seizure activity are prerequisites for the creation of potent medications for the treatment of epilepsy. Here mathematical modeling and computer simulations come into play by giving us accurate representation of the resulting model. This section describes the use of computational models to predict the impact of drug candidates on seizure activity and identify possible therapeutic targets.

10.5.1 Pharmacophore Modeling

Predicting the impact of potential drug candidates on seizure activity is made possible by building pharmacophore models based on existing ligands. Pharmacophore algorithms consider important chemical relationships, such as hydrogen bonds, hydrophobic associations, and electrostatic forces in order to generate concepts concerning potential effects on medicinal product candidates. By comparing their fit to the pharmacophore model of well-known anti-epileptic medications, these models have been utilized to predict the action of new compounds [14].

10.5.2 Systems Pharmacology Methods

Computational modeling and experimental data are used in systems pharmacology to forecast how drug candidates may affect intricate biological systems. Systems pharmacology models evaluate the interactions between drug candidates and various parts of the epileptic system using techniques such dynamic network modeling, pharmacokinetic-pharmacodynamic modeling, and pathway analysis. These models have been used to forecast how medication candidates may affect seizure threshold, network activity, and other important biomarkers [32].

10.5.3 Implantable Devices

As the search for new drugs and surgical techniques continues, another epilepsy therapy option is gradually gaining traction: implantable devices meant to anticipate, recognise, minimize and abort episodes [30]. Types of implants can be divided into 3 types:

a) **Vagus nerve stimulator:** Developed by Houston, Texas-based Cyberonics, Inc., the Vagus Nerve Stimulator (VNS) is the initial approved FDA instrument for managing seizures. The vagus nerve, that lies within the neck, can be stimulated by a surgically placed device called vagus nerve stimulation (VNS) [33]. It is usually recommended to avoid stimulating the correct vagus nerve as this may promote the atrioventricular nodes of the cardiac muscle and result in bradyarrhythmias. Stimulation of the vagus nerve is able to avoid epilepsy and a number of individuals report that this treatment may induce epileptic attacks when personally initiated in response to a neurological aura, though the exact mechanism by which it changes epileptic attacks is still unclear [30].

b) **Deep brain stimulation:** The process known as "deep brain stimulation" (DBS) involves the placement of sensors in particular brain regions that undergo stimulation electrically. Drug-resistant individuals with epilepsy have shown benefit from DBS in terms of fewer seizures per episode [33].

c) **Responsive neurostimulation (RNS):** RNS is a device that examines neurological activity and gives electrical impulses to the nervous system in response to aberrant activity that may result in seizures [33].

10.6 Data Integration and Analysis in Epilepsy Research

Electroencephalography (EEG) data, proteomics data, and genomics data are just a few of the many forms of data that are combined and analyzed in epilepsy research. In addition to discussing the significance of computational tools in gaining insights and identifying new targets or biomarkers for the development of epilepsy drugs, this section analyzes the difficulties of integrating and analyzing a variety of data in epilepsy research.

10.6.1 Epilepsy Research: Challenges in Integrating and Analyzing Diverse Data Types

10.6.1.1 Genomics Data

Genomics data, such as gene expression profiles and DNA sequencing, offer important insights into the underlying genetics of epilepsy. Due to the enormous number of genetic variants and the intricate ways that gene regulation works, there are difficulties in integrating and analyzing these data. To find possible genetic targets and pathways linked to epilepsy, methods including variant calling, genomic variant annotation, and pathway analysis are used [34].

10.6.1.2 Proteomics Data

Proteomics data shed light on the post-translational changes and patterns of protein expression in epilepsy. However, the intricacy of protein interactions, the dynamic nature of protein expression patterns, and the requirement for precise quantification make it difficult to analyze proteomics data. Identification of dysregulated proteins and pathways in epilepsy is aided by computational techniques such functional enrichment analysis, protein-protein interaction networks, and mass spectrometry data analysis [35].

10.6.1.3 EEG Data

The electrical activity of the brain during epileptic seizures can be better understood using EEG data. However, the number of continuous recordings, the variation in seizure types, and the presence of artifacts make it difficult to analyze EEG data. Using computational methods including feature extraction, time-frequency analysis, and machine learning algorithms, seizure patterns may be recognized, seizure types can be categorized, and seizure occurrence can be predicted [35]. The potential to combine fmri data with eeg has emerged as a major area of interest for research in the study of epilepsy with the advancement of fmri technology.

The benefit of MRI scans is their ability to define the areas of the brain where blood flow and metabolic activity are concentrated [36].

10.6.2 Role of Computational Techniques in Extracting Insights From Large-Scale Datasets

10.6.2.1 Data Mining

Data mining techniques involve the extraction of patterns, correlations, and relationships from large datasets. In epilepsy research, data mining approaches are applied to identify potential associations between clinical features, genetic variations, and treatment outcomes. Association rule mining, clustering, and classification algorithms assist in identifying subgroups of patients, identifying risk factors, and predicting treatment responses [37].

10.6.2.2 Machine Learning

The analysis of massive datasets in epilepsy research relies heavily on machine learning technologies. These algorithms generate predictions or classifications without explicit programming by learning from data patterns. Machine learning approaches are used in epilepsy research for seizure detection and prediction, identifying disease subgroups, and forecasting treatment outcomes. Research on epilepsy frequently makes use of Deep Learning architectures including Convolutional Neural Networks (CNN) and Recurrent Neural Networks (RNN), Random Forests, and Support Vector Machines (SVM) [38].

10.6.2.3 Network Analysis

Network analysis entails the investigation of intricate linkages and interactions between elements within a network. Gene regulatory networks, protein-protein interaction networks, and brain connection networks obtained from EEG data are all constructed and analyzed in epilepsy research using network analysis techniques. Using network-based methods, it is possible to spot dysregulated pathways, identify important nodes or genes linked to epilepsy, and forecast prospective therapeutic targets [39].

10.6.3 Identification of New Targets or Biomarkers for the Development of Epilepsy Drugs

10.6.3.1 Targeted Pathways Identification

Computational analysis of genomes and transcriptomics data has assisted in the identification of certain pathways implicated in the pathophysiology of epilepsy. For instance, network analysis was used in a study to find dysregulated gene modules linked to epilepsy. This research identified novel pathways as possible therapeutic targets, including calcium signaling and synaptic transmission [40].

10.6.3.2 Biomarker Discovery for Treatment Responses

The development of biomarkers that can forecast treatment response in epilepsy has been made possible via computational analysis. Machine learning strategies have been applied to the examination of EEG as well as clinical records from individuals with epilepsy who were taking different antiepileptic pills. They uncovered possible biomarkers that might direct personalized medicine by detecting certain EEG patterns and clinical traits related with treatment response [41].

10.6.3.3 Biomarker Development for Seizure Prediction

Biomarkers for seizure prediction have been developed by computational analysis of EEG data. Using machine learning approaches the analysis of long-term EEG recordings from epileptic patients was done. They created predictive models that could foretell the development of seizures by recognizing certain patterns and traits that appeared before seizures, allowing for preventive interventions and individualized treatment plans Table 10.1 [42].

10.7 Challenges and Future Directions

Despite tremendous progress in antiepileptic drug therapy over the past decade, the percentage of people with epilepsy who do not receive therapies has stayed the same, indicating the need for new approaches to treatment [30]. Because of a number of obstacles, including the intricacy of neural systems as well as the challenging task of obtaining malfunctioning zones (which frequently necessitate surgical operations like craniotomies),

Table 10.1 Important epilepsy biomarkers and their role in detection of epilepsy.

Sr. no.	Epilepsy biomarkers	Role in epilepsy
1	S100B	Indication of glial cell activation and neuroinflammation.
2	Glutamate	Detection of post stroke seizures
3	Gamma-Aminobutyric Acid	Monitoring seizure frequency in MRI
4	Neuron-Specific Enolase	Detection of neuronal damage
5	C-Reactive Protein (CRP)	Prediction of subsequent cognitive decline
6	Neurofilament Light Chain (NFL)	Detection of neurodestruction in status epilepticus
7	Serum Prolactin	Differentiation of epileptic seizures from PNES
8	HMGB1	Detection of seizure relapse and response to therapy
9	Translocator protein	Detection of neuroinflammation

implantable cerebral tools haven't yet been widely employed in hospitals. As a result, the implementation of these devices in clinical settings lags behind their use in cardiology. Poor target recognition and validation in the beginning of the occurrence of episodes is one of the main factors hindering the succession of development of a target-based drug discovery approach. This might lead to employment of surgical procedures which cannot guarantee complete success. Since these activities are not only laborious and arduous, but recent studies have shown that even qualified medical professionals can struggle to discriminate against epilepsy types, automation systems based on standardized techniques are becoming increasingly relevant [11].

Data availability and quality of available data is the main challenge for computational drug discovery. The availability of well-curated dataset for epilepsy is very limited. Integration of various data types like clinical data, genomics and EEG data is crucial to develop a robust model. Epilepsy is a complex neurological disease and therefore the understanding of underlying molecular mechanisms is limited. The fact that all computational models are very context specific and cannot be extrapolated to other contexts is a common problem for model creation and model validation. Data privacy and patient data anonymity are legal and ethical concerns for computational modeling. In the event of a malfunction, problems of liability may also emerge. These questions can be challenging to answer and are a constant subject of debate [43]. The implementation of multi-target pharmaceutical therapy occurs without an in-depth understanding of the related consequences of a combination of drugs. These limitations reflect the difficulties in understanding the functioning of the brain [21].

Although initial results are promising, extensive development and controlled clinical trials are necessary before these treatments become part of our standard clinical care. With advances in computing power, engineering capabilities, and our understanding of neurophysiology, there are increasing opportunities to develop clinical neurophysiological devices that can leverage these new insights. As technology continues to progress, it is likely that implantable brain devices will become more accessible and effective for treating neurological disorders [30]. When compared to other neurological conditions, epilepsy offers a solid disease base for these recent advances. Diagnostic indicators are available; electroencephalogram (EEG) can be used to identify seizures; (2) numerous neurological disorders drugs currently on the market act on the voltage- or ligand-sensitive ion channels, thereby providing a predetermined set of properties to consider as pharmacological targets (3). The use of multiple medications and multitarget medications are widespread in epilepsy and give a level of additional complexity that is incomprehensible without explicit mathematical models. Seizures became among the initial illness manifestations to be depicted explicitly using multiscale modeling [17].

Devices should be made to fit a wide range of patients, and therapeutic intervention settings should be matched to the physiology of the patient to control seizures with fewer adverse effects.

There is a need for new intervention techniques. Electrical stimulation will undoubtedly be supplemented with additional therapies such as optogenetics, biofeedback, focused cooling, optical medication uncaging, and others to help control seizures [44].

10.8 Conclusion

This review paper has explored the significant role of computational approaches in the field of epilepsy drug discovery. The topics covered in this paper highlight the breadth and depth of computational attitudes towards advancing our understanding of epilepsy and identifying potential anti-epileptic compounds. By examining traditional drug discovery methods, the limitations they entail, and the subsequent advancements brought about by computational techniques, this paper underscores the significant impact of computational attitudes on advancing our understanding of epilepsy and identifying potential therapeutic options. The analysis of computational approaches in epilepsy drug discovery revealed a wide array of methods employed to enhance the drug development process which have been briefly discussed in this paper. These methods encompassed the identification of anti-epileptic compounds, the optimization of drug efficacy through computer simulations, and the development of sophisticated computational models. By leveraging these computational models, particularly those based on brain networks, researchers have been able to predict and understand the effects of drugs on seizure activity, thus facilitating targeted drug development efforts.

This review paper also examined the challenges associated with integrating and analyzing diverse data types in epilepsy research. It emphasized the crucial role of computational techniques in extracting meaningful insights from large-scale datasets and identifying novel targets or biomarkers for epilepsy drug development.The use of computers have made it possible to explore new therapeutic opportunities and expand the discipline by integrating and analyzing diverse data. The discussion also encompassed the translational applications and clinical implications of computational approaches in epilepsy drug discovery. Computational models aid in predicting drug responses, optimizing treatment approaches, and enabling more personalized care for individuals with epilepsy. These techniques not only can be incorporated for epileptic treatments, but also prove to be helpful in treatment strategies of other neurological disorders which lack a current and efficient treatment plan. We are optimistic that the aforementioned computational methodologies will reach their peak in neurological developments and serve as a solid foundation for the healthcare sector.

Acknowledgments

The authors would like to thank the Director and Head of School, MIT School of Bioengineering Sciences & Research for giving an opportunity to work independently and all their support with the best lab facilities.

References

1. Epilepsy [Internet], Centers for Disease Control and Prevention, [cited 2024 Jun 5]. Available from: https://www.cdc.gov/epilepsy/index.html.
2. Epilepsy [Internet], World Health Organization, [cited 2024 Jun 5]. Available from: https://www.who.int/news-room/fact-sheets/detail/epilepsy.
3. Löscher, W., Animal models of seizures and epilepsy: past, present, and future role for the discovery of antiseizure drugs. *Neurochem. Res.*, 42, 1873–88, 2017 Jul.
4. Metcalf, C.S., West, P.J., Thomson, K.E., Edwards, S.F., Smith, M.D., White, H.S., Wilcox, K.S., Development and pharmacologic characterization of the rat 6 Hz model of partial seizures. *Epilepsia*, 58, 6, 1073–84, 2017 Jun.
5. Metcalf, C.S., Huff, J., Thomson, K.E., Johnson, K., Edwards, S.F., Wilcox, K.S., Evaluation of antiseizure drug efficacy and tolerability in the rat lamotrigine-resistant amygdala kindling model. *Epilepsia Open*, 4, 3, 452–63, 2019 Sep.
6. Entzeroth, M., Flotow, H., Condron, P., Overview of high-throughput screening. *Curr. Protoc. Pharmacol.*, 44, 1, 9–4, 2009 Mar.
7. Aldewachi, H., Al-Zidan, R.N., Conner, M.T., Salman, M.M., High-throughput screening platforms in the discovery of novel drugs for neurodegenerative diseases. *Bioengineering*, 8, 2, 30, 2021 Feb 23.
8. High throughput screening. High Throughput Screening - an overview, ScienceDirect Topics, https://www.sciencedirect.com/topics/biochemistry-genetics-and-molecular-biology/high-throughput-screening.
9. Team, D., High throughput screening: Meeting the challenges of more cell-based assays and HCS, Drug Discovery World (DDW), 2020, October 13, https://www.ddw-online.com/high-throughput-screening-meeting-the-challenges-of-more-cell-based-assays-and-hcs-681-200512/.

10. Stacey, W.C. and Litt, B., Technology insight: Neuroengineering and epilepsy-designing devices for seizure control. *Nat. Clin. Pract. Neurol.*, 4, 190–201, 2008, April, https://www.ncbi.nlm.nih.gov/pmc/articles/PMC2904395/.
11. Timmins, G., Big Data Approach to drug discovery leads to New Epilepsy Target: Imperial News: Imperial College London. Imperial News, 2018, September 3, https://www.imperial.ac.uk/news/187827/big-data-approach-drug-discovery-leads/.
12. Löscher, W. and Klein, P., New approaches for developing multi-targeted drug combinations for disease modification of complex brain disorders. Does epilepsy prevention become a realistic goal? *Pharm. Ther.*, 229, 107934, 2022.
13. An, S., Kang, C., Lee, H.W., Artificial Intelligence and Computational Approaches for Epilepsy. *J. Epilepsy Res.*, 10, 1, 8–17, 2020, https://doi.org/10.14581/jer.20003.
14. Stefanescu, R.A., Shivakeshavan, R.G., Talathi, S.S., Computational models of epilepsy. *Seizure*, 21, 10, 748–759, 2012.
15. Mehta, P., Srivastava, S., Sharma, M., Singh, I., Malik, R., Identification of chemically diverse GABAA agonists as potential anti-epileptic agents using structure-guided virtual screening, ADMET, quantum mechanics and clinical validation through off-target analysis. *Int. J. Biol. Macromol.*, 119, 1113–1128, 2018.
16. Yang, S.Y., Pharmacophore modeling and applications in drug discovery: challenges and recent advances. *Drug Discov. Today*, 15, 11-12, 444–450, 2010.
17. Lytton, W.W., Computer modeling of epilepsy: Opportunities for drug discovery. *Drug Discov. Today Dis. Models*, 19, 27–30, 2016.
18. Arodola, O.A. and Soliman, M.E., Quantum mechanics implementation in drug-design workflows: does it really help? *Drug Des. Dev. Ther.*, 11, 2551–2564, 2017, https://doi.org/10.2147/DDDT.S126344.
19. Evers, M., Heid, A., Ostojic, I., Pharma's Digital RX: Quantum Computing in Drug Research and Development, McKinsey & Company, 2021, June 18, https://www.mckinsey.com/industries/life-sciences/our-insights/pharmas-digital-rx-quantum-computing-in-drug-research-and-development.
20. Kotev, M., Sarrat, L., Gonzalez, C.D., User-Friendly Quantum Mechanics: Applications for Drug Discovery. *Methods Mol. Biol. (Clifton, N.J.)*, 2114, 231–255, 2020, https://doi.org/10.1007/978-1-0716-0282-9_15.
21. Lytton, W.W., Computer modelling of epilepsy. *Nat. Rev. Neurosci.*, 9, 8, 626–637, 2008, https://doi.org/10.1038/nrn2416.
22. Case, M.J., Morgan, R.J., Schneider, C.J., Soltesz, I., Computer Modeling of Epilepsy, in: *Jasper's Basic Mechanisms of the Epilepsies*, 4th ed., J.L. Noebels, *et al.* (Eds.), National Center for Biotechnology Information (US), 2012.
23 Hodgkin, A.L. and Huxley, A.F., A quantitative description of membrane current and its application to conduction and excitation in nerve. *J. Physiol.*, 117, 4, 500, 1952.

24. Izhikevich, E.M., Simple model of spiking neurons. *IEEE Trans. Neural Networks*, 14, 6, 1569–1572, 2003.
25. Wendling, F., Bartolomei, F., Bellanger, J.J., Chauvel, P., Epileptic fast activity can be explained by a model of impaired GABAergic dendritic inhibition. *Eur. J. Neurosci.*, 15, 9, 1499–1508, 2002.
26. GHTS, T.C.P.S. and Ersto, O.M.N.R., Theme 1: Clinical trials: Methodology.
27. Wendling, F., Computational models of epilepsy. *Neurophysiol. Clin./Clin. Neurophysiol.*, 1, 42, 75–76, 2012.
28. Ellis, C.A., Miller, R.L., Calhoun, V.D., Towards greater neuroimaging classification transparency via the integration of explainability methods and confidence estimation approaches. *Inf. Med. Unlocked*, 37, 101176, 2023.
29. Buchlak, Q.D., Esmaili, N., Leveque, J.C., Farrokhi, F., Bennett, C., Piccardi, M., Sethi, R.K., Machine learning applications to clinical decision support in neurosurgery: an artificial intelligence augmented systematic review. *Neurosurg. Rev.*, 43, 1235–1253, 2020.
30. Xiao, C., Wang, S., Iasemidis, L., Wong, S., Chaovalitwongse, W.A., An adaptive pattern learning framework to personalize online seizure prediction. *IEEE Trans. Big Data*, 7, 5, 819–831, 2017.
31. Warren, C.P., Hu, S., Stead, M., Brinkmann, B.H., Bower, M.R., Worrell, G.A., Synchrony in normal and focal epileptic brain: the seizure onset zone is functionally disconnected. *J. Neurophysiol.*, 104, 6, 3530–3539, 2010.
32. Pitkänen, A., Löscher, W., Vezzani, A., Becker, A.J., Simonato, M., Lukasiuk, K., Beck, H., Advances in the development of biomarkers for epilepsy. *Lancet Neurol.*, 15, 8, 843–856, 2016.
33. Bryn Mawr Communications, Epilepsy essentials: Neuromodulation devices for epilepsy. *Pract. Neurol.*, 2018, https://practicalneurology.com/articles/2018-nov-dec/epilepsy-essentials-neuromodulation-devices-for-epilepsy.
34. León, A. and Pastor, O., Enhancing precision medicine: A big data-driven approach for the management of genomic data. *Big Data Res.*, 26, 100253, 2021.
35. Leitner, D.F., Lin, Z., Sawaged, Z., Kanshin, E., Friedman, D., Devore, S., Devinsky, O., Brain molecular mechanisms in Rasmussen encephalitis. *Epilepsia*, 64, 1, 218–230, 2023.
36. Independent component analysis for functional MRI. Appendix E in J. M. Dunn, *et al.* Understanding the abnormal brain activity in epilepsy as a potential predictor of the onset of an epileptic seizure. - Scientific Figure on ResearchGate, Available from: https://www.researchgate.net/figure/Electroencephalography-eeg-data-showing-onset-of-an-epileptic-event_fig1_249657525 [accessed 16 Jun, 2023].
37. Chien, J.H., Shiau, D.S., Sackellares, J.C., Halford, J.J., Kelly, K.M., Pardalos, P.M., Using Relative Power Asymmetry as a Biomarker for Classifying Psychogenic Nonepileptic Seizure and Complex Partial Seizure Patients, in: *Data Mining for Biomarker Discovery*, pp. 57–77, 2012.

38. Abbasi, B. and Goldenholz, D.M., Machine learning applications in epilepsy. *Epilepsia*, 60, 10, 2037–2047, 2019.
39. Acharya, U.R., Sree, S.V., Swapna, G., Martis, R.J., Suri, J.S., Automated EEG analysis of epilepsy: a review. *Knowledge-Based Syst.*, 45, 147–165, 2013.
40. Jin, Y., Zhao, C., Chen, L., Liu, X., Pan, S., Ju, D., Wei, B., Identification of novel gene and pathway targets for human epilepsy treatment. *Biol. Res.*, 49, 1, 1–9, 2016.
41. Yao, L., Cai, M., Chen, Y., Shen, C., Shi, L., Guo, Y., Prediction of antiepileptic drug treatment outcomes of patients with newly diagnosed epilepsy by machine learning. *Epilepsy Behav.*, 96, 92–97, 2019.
42. Nielsen, J.M., Rades, D., Kjaer, T.W., Wearable electroencephalography for ultra-long-term seizure monitoring: a systematic review and future prospects. *Expert Rev. Med. Devices*, 18, sup1, 57–67, 2021.
43. Ayyar, P. and Subramanian, U., Repurposing – second life for drugs. *Pharmacia*, 69, 1, 51–59, 2022, https://doi.org/10.3897/pharmacia.69.e72548.
44. Nagaraj, V., Lee, S.T., Krook-Magnuson, E., Soltesz, I., Benquet, P., Irazoqui, P.P., Netoff, T.I., Future of seizure prediction and intervention: closing the loop. *J. Clin. Neurophysiol.*, 32, 3, 194–206, 2015 Jun, doi: 10.1097/WNP.0000000000000139. PMID: 26035672; PMCID: PMC4455045.

11

Computational Attitudes to Drug Discovery in Alzheimer's Disease

Shubhrat Maheshwari[1], Aditya Singh[2*], Amita Verma[1], Juber Akhtar[2], Jigna B. Prajapati[3], Sudarshan Singh[4] and Bhupendra Prajapati[5,6†]

[1]Bioorganic and Medicinal Chemistry Research Laboratory, Department of Pharmaceutical Sciences, Sam Higginbottom University of Agriculture, Technology and Sciences, Prayagraj, India
[2]Department of Pharmacy, Faculty of Pharmacy, Integral University, Lucknow, Uttar Pradesh, India
[3]Acharya Motibhai Patel Institute of Computer Studies, Ganpat University, Gujarat, India
[4]Department of Pharmaceutical Sciences, Faculty of Pharmacy, Chiang Mai University, Chiang Mai, Thailand
[5]Shree S.K. Patel College of Pharmaceutical Education and Research, Ganpat University, Gujarat, India
[6]Department of Industrial Pharmacy, Faculty of Pharmacy, Silpakorn University, Nakhon Pathom, Thailand

Abstract
Alzheimer's disease (AD), neuronal death, and synaptic loss are all characterized by extracellular amyloid-(Aβ) plaques and neurofibrillary in the intracellular milieu, all of which lead to progressive cognitive impairment. The prevalence of AD exponentially rises after age 65 years, making age the most major risk factor for the disease. It is expected that during the next 20 years, the overall prevalence of AD will double as the average lifespan in emerging nation's increases. When a computer mimics cognitive conduct linked with the human brain when solving problems and learning, this behavior is often referred to as artificial intelligence (AI). To overcome the aforementioned limitations, AI can be a useful tool. Importantly, AI has the ability to impartially reveal trends within complex data and can also determine which gene target(s) deserves further study. The toxicity problems brought on by off-target interactions are also resolved by AI models.

*Corresponding author: simmu5adityasingh@gmail.com; ORCID: 0000-0002-0398-6130
†*Corresponding author*: bhupen27@gmail.com; ORCID: https://bit.ly/drbgprajapatiorcid

Bhupendra Prajapati, Alok Tripathi, Rishabha Malviya and Lucy Mohapatra (eds.) Computational Neuropharmacology: Fundamentals and Clinical Aspects, (335–348) © 2025 Scrivener Publishing LLC

To increase the precision of predicting and diagnosing cognitive decline and dementia, researchers are looking at the use of a multi-modal method. In order to create more precise predictive models, this method combines information from various sources, including digital tools, cognitive testing, and medical imaging focusing on the pathophysiology that underlies it. Finding the right targets (such as genes or proteins) implicated in disease pathophysiology is the first and most crucial phase in the drug development process in order to find effective medications that can interfere with these targets.

Keywords: Alzheimer's disease, AI models, cognitive impairment, pathophysiology, deep learning algorithms

List of Abbreviations

CNS	central nervous system
AI	Artificial intelligence
ML	Machine learning
AD	Alzheimer's disease
BBB	Blood-brain barrier
Aβ	Amyloid beta
AMBER	Assisted model building with energy refinement),
GROMACS	Groningen machine for chemical simulations.
NDDs	Neurodegenerative diseases
SBDD	Structure-based drug design
GPU	Graphics Processing Unit

11.1 Introduction

The creation of effective and sophisticated complex for the targeted administration of corrective substances with paramount effectiveness and minimal hazards has been a significant summons to botanist over the past two decades [1]. Another obstacle in the process of generating new medicinal agents was the expense of development and the amount of time it took. Researchers from all over the world turned to computational methods, also known as conventional methods, like molecular docking and virtual screening (VS), to reduce these difficulties and obstacles [2]. However, these methods also come with drawbacks like inefficiency and lack of accuracy. Pharmaceutical corporations and chemical expert's priorities research in the area of drug development and design [3].

Long and complex processes like target validation, therapeutic molecule optimization, conduct screening, clinical trial, design, and fabricate procedures are all part of the drug discovery and design process [4]. These all present a significant obstacle to finding a disease-specific medicine that works. Controlling the process's cost and speed is the main challenge facing pharmaceutical businesses. This approach required less time and money since AI provided clear, scientific answers to all of these concerns [4]. GPU acceleration results from huge data parallelism, which is created when equivalent independent operations are carried out on a large number of data elements [5]. The usage of a rotation matrix across coordinates to describe the positions of objects while a view is rotated in graphics is an illustration of a typical data parallel process. Data parallelism can be used to independently calculate the atomic potential energies in a molecular simulation. The development of GPU-centered molecular dynamics codes over the past ten years has resulted in a hundred-fold decrease in simulation computation costs when compared to CPU-based approaches [6]. The majority of molecular dynamics engines now offer GPU-accelerated versions, including AMBER (assisted model building with energy refinement), GROMACS (Groningen machine for chemical simulations), and NAMD (nanoscale molecular dynamics). GPUs not only scale well with system size using the spatial domain but are also well adapted to accelerate molecular dynamics simulations [7]. The term AI is frequently worn a computer imitates folded fibrils protein linked to the human brain during problem-solving and learning. AI is a tool that can be used to get around the aforementioned constraints. Importantly, AI has the capacity to identify which gene target(s) needs additional research in addition to being able to unbiasedly expose patterns inside complex data [1]. In other words, deep learning algorithms and artificial neural networks have modernized the field. Additionally, AI models solve the toxicity issues that result from off-target interactions [1].

Graphs and networks are mathematical constructs that represent a collection of data as a whole. Nodes (the network's constituent objects) and edges (the links separating them) make up networks. Biological networks can be visualized using free software like Cytoscape. In mice, their significance in the illness was validated, resulting in the creation of a brand-new IBD model [8]. These recently created technologies were successfully used to "treat" patient-derived IBD organoids, discovering new therapeutic targets and treatments for IBD. Fold change, p-values, used in the traditional method for finding genes that are differentially expressed. These approaches, however, are constrained and unlikely to offer the data required to draw firm conclusions regarding dysregulated genes that might be significant for AD [9]. First, it is fundamentally challenging to choose differentially expressed genes using traditional fold change cutoffs. The replication problem is likely exacerbated by the p-value, which is likewise known to be readily misconstrued and to only provide limited information about the data at hand [10].

The ELM makes it possible to foresee how a particular frame used to convey information about AD may result in more significant alterations that affect behavior and stimulate higher-order thinking. The communication argument must be pertinent to and consistent with the person's prior views and experiences in order to do this. Additionally, it is anticipated that the type of emotion the frame arouses will affect which arguments the person will think about and analyze in deeper detail [11]. Drug discovery is aided by comparison data regarding other bioactive agents and other targets, in addition to knowledge of medications and their therapeutic targets. The development of such tools necessitates the collection of data on weak binders and non-binders of a particular target [12]. The prodrug data already available may provide fresh perspectives on how to get around drug development issues that restrict formulation choices or produce undesirable biopharmaceutical/pharmacokinetic outcomes. Researchers working on drug discovery are therefore in dire need of the comparative data mentioned above. Additionally, the data on the 3D structure of the target is crucial information for drug discovery. In addition to the rise in experimentally resolved target crystal structures, improved AI methods, like AlphaFold, have made crystal structure prediction for targets possible [13].

Over 99.6% of AD clinical trials have failed thus far, and there are presently no effective therapies available [14]. Although previously limited

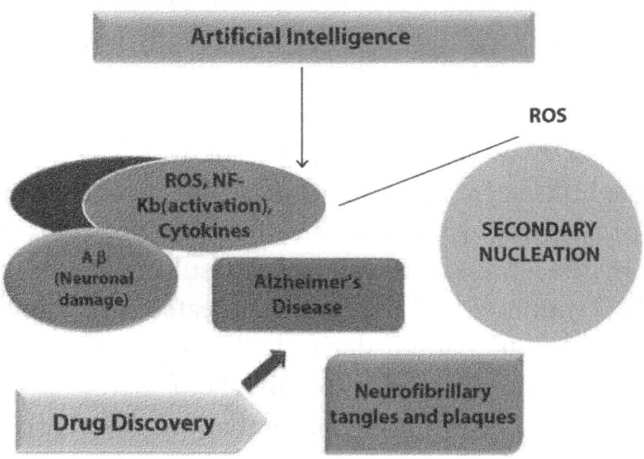

Figure 11.1 Computational attitudes of drug discovery for AD by different strategies.

to relatively modest sample sizes, using permits quantification of a large number of mRNA. To be more precise, the dataset is pruned using random fold changes and p-values that call for repeated corrections, both of which run the risk of missing potentially significant genes. Additionally, rather than employing microarray as a strictly experimental tool, there is a propensity to confirm a priori beliefs using microarray data, which may bias reported outcomes. Recent meta-analyses of publically accessible microarray data have shown to be a valuable source for discovering [15]. However, only a small number of research have employed this strategy to far to look at potential genetic causes of AD. As shown in Figure 11.1. One such study used comparison to identify computational attitudes of drug discovery for AD by different strategies [16].

11.2 Alzheimer's Disease

AD is extremely common as per Figure 11.2. In fact, it accounts for about 70% of all instances of dementia and is the most common neurodegenerative illness [17]. AD places a significant economic and societal burden. The difficulty of preventing, delaying the start of, or slowing the progression of AD is largely to blame for its devastating side effects [18]. Only four medications have been approved to treat AD in the last 20 years: memantine, donepezil, rivastigmine, and galantamine, NMDA receptor (rivastigmine) [19]. The cognitive and neuropsychiatric symptoms of the condition are only somewhat temporarily relieved by these medications. Phase III

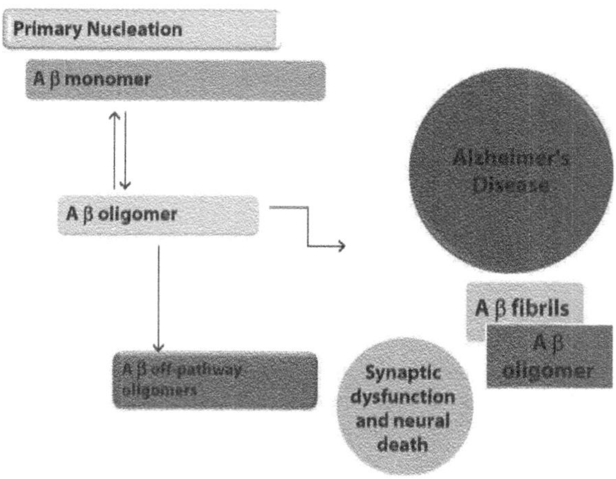

Figure 11.2 Molecular pathway of dementia and cognitive impairment (AD).

Table 11.1 Drugs being developed for AD (clinicaltrials.gov, accessed January 25, 2022).

Candidate	Report	Action	CT.gov ID	Start date-estimated end date sponsor by
Atuzaginstat	Neuroprotection	Hippocampal degeneration and neuroinflammation [24].	(NCT03823404)	Mar 2019- Dec 2022 Cortexyme
Aducanumab	Amyloid	A monoclonal antibody that targets A oligomers and plaques [25].	(NCT04241068)	Mar 2020- Oct 2023 Biogen
Donanemab & Aducanumab	Amyloid	Donanemab, a monoclonal antibody that targets the pyroglutamate form of A, and aducanumab, a monoclonal antibody that targets plaques and oligomers, were administered in different study arms [26].	(NCT05108922)	Nov 2021- Jun 2023 Eli Lilly
Brexpiprazole	Neurotransmitter receptors	Atypical antipsychotic, partial agonist of the D2 receptor, and serotonin-dopamine modulator [27].	(NCT03620981)	Aug 2018- Aug 2022 Otsuka
Caffeine	Neurotransmitter receptors	Antagonist of adenosine; non-specific inhibitor of phosphodiesterase [28].	(NCT04570085)	Nov 2021- Nov 2024 University Hospital, Lille
Hydralazine	Oxidative stress	Free radical scavenger [29].	(NCT04842552)	Jun 2021- Dec 2023 Shahid Sadoughi University, Iran
Omega-3 (DHA+EPA)	Oxidative stress	Antioxidant [30].	(NCT03691519)	Apr 2018-Dec 2023 University Hospital, Toulouse
Tricaprilin	Metabolism and bioenergetics	Caprylic triglyceride: promotes mitochondrial and neural health, ketosis [31].	(NCT04187547)	Jun 2022- Feb 2024 Cerecin

clinical trials for this medication are currently being conducted in the USA and Europe in order to obtain a global regulatory approval. Additionally, Biogen's -amyloid-directed monoclonal antibody aducanumab was licensed in the USA in June 2021 following much debate regarding its real clinical efficacy [20]. Acuranumab's maker, Biogen, had in fact reported that the drug had failed in pivotal Phase III clinical trials, but it later changed course based on a new analysis of the clinical data. In this situation, there is an urgent need for the development of effective medications that can stop or slow the progression of neurodegeneration [12].

While the majority of efforts should go towards creating medications that can alter the course of the disease, these efforts should also include looking for complementary therapies that can more effectively treat patients with AD [21]. The majority of the 2173 clinical trials for AD conducted prior to 2019 involved drug candidates intended to target one target that appears to be important in AD pathogenesis, including neurovascular mechanisms (8%) [22]. The expanding number of AD sufferers and the growing public health concern that the disease poses have made therapy for AD—which aims to prevent, postpone the onset, reduce the development, and ameliorate symptoms—imperative as per Table 11.1. Acuranumab, the first DMT for AD that was approved and put on the market in 2021 for those with moderate AD dementia and MCI owing to AD, represents a significant advance in treatment. An anti-amyloid monoclonal antibody is called aducanumab. In the near future, further fresh therapies for AD might be made available. 47 Phase 3 studies including 31 agents are underway. DMTs make up 21 (67.8%) of the medicines in Phase 3 trials, including 16 (51.6% of the agents) small molecules and five (16.1%) biologics. Putative cognitive enhancers make up five Phase 3 agents (16.1%) while medications that address behavioral symptoms make up five (16.1%) of the total [23].

11.3 Computational Attitudes to Drug Discovery

In order to discover effective drugs that can interfere with these targets, the first and most important step in the drug pioneering citation is to identify the proper targets (such as genes or proteins) involved in disease pathophysiology. Furthermore, the development of AI has construct awful lot of information much simpler since a variety of ML techniques are now readily available that may assist in uproot important facet, patterns, and design attending in these substantial biological data sets [31]. The most popular search phrase in September 2015, following the launch of ML, was AI, according to Google search trends. ML is sometimes referred to as the main application of AI, while other times it is referred to as a subset of AI as per Table 11.2 [31].

The term "artificial intelligence" (AI) refers to the ability of computer programs to think and act like people. In contrast, "machine learning" (ML) goes beyond this, allowing computers to learn on their own without being explicitly programmed [31]. Examples of ML algorithms include Naive Bayes, decision trees, hidden Markov models, and hidden Markov models (HMM) [31]. ChemSpider, ZINC, and BindingDB are crucial facet for conceal during the drug development and disclosure process. Instead of relying on quantum mechanical calculations to determine properties of compounds contained in the aforementioned databases was carefully selected and screened for pharmacological and physicochemical characteristics that were essential to the drug discovery process [31, 34]. The ChEMBL database is a large open-access database that contains information on a variety of bioactive substances with drug-like characteristics. The ChEMBL database also includes details on these substances' toxicity characteristics, absorption, distribution,

Table 11.2 The role of AI in the development of personalized treatment plans.

Role of AI	Functions
Brain imaging analysis	AI algorithms can be trained to analyze brain imaging [31].
Early detection and prevention	Analyzing data from wearable devices and other sources to identify changes in behavior and cognition that may be early signs of dementia. This can lead to early interventions that may help slow or prevent cognitive decline [31].
Patient monitoring and care	AI-powered monitoring systems can be used to track patient behavior and cognition in real-time, alerting caregivers and healthcare providers to changes that may require intervention [31].
Drug discovery	AI can be used for dementia and cognitive impairment [31, 32].
Improved clinical trials	AI-powered clinical trials can be used to test new treatments for dementia and cognitive impairment more efficiently by identifying eligible participants, monitoring patient outcomes, and predicting treatment responses [31, 33].

metabolism, and excretion (ADME), as well as target interactions [35]. DrugBank is a different open-access pharmaceutical data source that includes information on a variety of medications, their targets, and mechanisms. Another collection providing data on the L1000CDS2 contains information on medications. In addition, the protein data bank (PDB) is a different openly available online database that holds information about the three-dimensional structures of proteins, DNA, and RNA [36]. Reinforcement learning can be applied throughout the entire process. For QSAR research, matched molecular pairs (MMP) have been employed frequently. One alteration in a therapeutic candidate linked to MMP affects the compound's bioactivity in additional ways [37].

11.4 Applications of Computational Attitudes to Drug Development Process

In recent years, scientists have harnessed AI to their advantage to find novel peptides. For instance, the CNN-driven program Deep-AmPEP30, created by 2020, predicts short anti-microbial peptides (AMPs) using DNA sequence data. It is essential to establish a leery dose for therapeutic reasons because giving a patient the wrong dosage of any medication might result in unfavorable and even fatal side effects. Over time, it has proven difficult to determine the ideal dosage of a medication that will produce the crave efficacy with the fewest hazardous side effects [38]. Recently, a number of tools discovered to analyze medication of particular bioactive substances as a carrier [39]. The computational analysis is later validated using benchmark data sets. The most appropriate pharmacophore for such evaluations is one based on chemical characteristics. These models generate sizable 3D data sets from internal compound collections or *in silico* studies [40]. VS are one of the key CADD techniques used in drug design and discovery. Identification of a tiny chemical molecule that binds to a therapeutic target is referred to as VS. VS are a useful technique for selecting the most promising medicinal chemical from a collection of compounds. In high-throughput screening, when high costs and low accuracy rate are a problem, it therefore becomes a crucial instrument [41]. It is then synthesized in a lab and tested *in vitro* or *in vivo* after the most promising molecule has been chosen. Regression and classification models are two categories into which QSAR models can be generally divided. A reliable and effective technique for QSAR modelling is the use of Gaussian processes (GPs), a sort of QSAR constructing regression model [42]. GP techniques can sort through a vast number of descriptors to find the most important ones. Two classification models have recently been proven utilizing GP, one of which combines GP regression approach and probit analysis with intrinsic GP classification methods. The technique also works well for modelling nonlinear relationships and does not call for arbitrary model parameter determination. QSAR modelling has a lot of opportunities thanks to recent improvements and expanding uses of ML techniques including neural networks, DL, and SVM. For QSAR modelling, a number of web-based tools and algorithms have been created, including the VEGA platform, QSAR-Co, and FL-QSAR [5]. Several targets may be involved in various diseases, repositioning medications may be beneficial [4]. Another thing to consider is that the development of sizable data sets from studies on genomes, proteomics, and pharmacology in *in silico* and *in vitro* opens up a lot of possibilities for repositioning drugs [44]. The recent development of AI-based drug discovery tools and algorithms offers a foundation for further study. ML algorithms replace the traditional system biology methods based on chemical similarity and molecular docking with novel methods that can assess medication effects. As a result, various AI-based algorithms and web-based tools, including DrugNet, DPDR-CPI, PHARMGKB, PROMISCUOUS 2.0, and DRRS, have been developed recently. ML models to forecast potential anti-fibrosis medication candidates, specifically the structural profile prediction model and biological profile prediction model. The findings showed that the training set's and testing set's areas under the receiver intervene attribute curve, respectively, were 0.814 and 0.874. According to the findings, natural compounds have anti-fibrosis properties and could be used as fibrosis medication targets. Recently, COVID-19 becomes an international epidemic, and experts from all over the world began looking for promising therapeutics. AI-based medication repositioning is quite important in this area. For instance, using a multimodal DL method [39]. SNF-CVAE based on drug similitude network fusion revealed auspicious treatment drugs for juvenile rheumatoid arthritis and Alzheimer's disease (AD). Other machine learning (ML)-based methods include PhenoPredict and SDTNBI, which are used to forecast medication-target interactions and find disease respectively [45]. DL model predict medication response and synergy. The findings showed that response predictions stratify clinical outcomes for patients with ER-positive breast cancer and predicted combos increase progression-free survival. Another example of AI being applied in drug repurposing is found in the study 2020 conducted using bipartite graph convolution networks. The authors developed the BiFusion model. The negative impact a chemical molecule has on an organism or any component of the body is referred to as drug toxicity. The wide reach of

AI has captivated the scientists involved in the drug development process. Drug development time, cost, attrition rates, and human resources have all been decreased because to AI. For this, various web-based tools including LimTox, pkCSM, admetSAR, and Toxtree have been developed [46].

11.5 Conclusion

Alzheimer's disease (AD), a neurological ailment that gets worse over time and impairs memory, cognition, and behavior. There is currently no known cure for AD, and treatments only treat the symptoms. However, a number of novel formulations that target the underlying pathophysiology of AD are currently under investigation. Utilizing AI can be a helpful tool for overcoming the aforementioned restrictions. Importantly, AI can identify which gene targets merit additional research and can objectively uncover trends inside complex data. In other words, deep learning algorithms and artificial neural networks have modernized the field. AI models also address the toxicity issues brought on by off-target interactions. AI and ML can be used to diagnose cognitive impairment through the analysis of medical imaging data, including positron emission tomography (PET) and magnetic resonance imaging (MRI) scans. Researchers are considering using a multi-modal approach to diagnose and predict cognitive decline and dementia with greater accuracy. It's a gift that we now have access to many archives of biomedical data since they can aid us in this process. A number of ML techniques are now widely available and may aid with the extraction of important characteristics, patterns, and structures contained in these massive biological data sets, which has further simplified big data analytics due to the advent of AI.

References

1. Cascianelli, S., Scialpi, M., Amici, S., Forini, N., Minestrini, M., Luca Fravolini, M., *et al.*, Role of artificial intelligence techniques (automatic classifiers) in molecular imaging modalities in neurodegenerative diseases. *Curr. Alzheimer Res.*, 14, 2, 198–207, 2017.
2. Hauser, A.S., Attwood, M.M., Rask-Andersen, M., Schiöth, H.B., Gloriam, D.E., Trends in GPCR drug discovery: new agents, targets and indications. *Nat. Rev. Drug Discov.*, 16, 12, 829–42, 2017.
3. Hindi, S.S., Sabir, J.S., Dawoud, U.M., Ismail, I.M., Asiry, K.A., Mirdad, Z.M., *et al.*, Nanocellulose-Based Passivated-Carbon Quantum Dots (P-CQDs) for Antimicrobial Applications: A Practical Review. *Polymers*, 15, 12, 2660, 2023.

4. Carracedo-Reboredo, P., Liñares-Blanco, J., Rodríguez-Fernández, N., Cedrón, F., Novoa, F.J., Carballal, A., *et al.*, A review on machine learning approaches and trends in drug discovery. *Comput. Struct. Biotechnol. J.*, 19, 4538–58, 2021.
5. Batool, M., Ahmad, B., Choi, S., A structure-based drug discovery paradigm. *Int. J. Mol. Sci.*, 20, 11, 2783, 2019.
6. Basith, S., Cui, M., Macalino, S.J., Park, J., Clavio, N.A., Kang, S., *et al.*, Exploring G protein-coupled receptors (GPCRs) ligand space via chemin-formatics approaches: impact on rational drug design. *Front. Pharmacol.*, 9, 338587, 2018.
7. Bakkar, N., Kovalik, T., Lorenzini, I., Spangler, S., Lacoste, A., Sponaugle, K., *et al.*, Artificial intelligence in neurodegenerative disease research: use of IBM Watson to identify additional RNA-binding proteins altered in amyo-trophic lateral sclerosis. *Acta Neuropathol.*, 135, 227–47, 2018.
8. Arega, B., Tilahun, K., Minda, A., Agunie, A., Mengistu, G., Prevalence rate of undiagnosed tuberculosis in the community in Ethiopia from 2001 to 2014: systematic review and meta-analysis. *Arch. Public Health*, 77, 33, 2019.
9. Alfei, S., Schito, A.M., Zuccari, G., Considerable improvement of ursolic acid water solubility by its encapsulation in dendrimer nanoparticles: Design, synthesis and physicochemical characterization. *Nanomaterials*, 11, 9, 2196, 2021.
10. Hampel, H., Hardy, J., Blennow, K., Chen, C., Perry, G., Kim, S.H., *et al.*, The Amyloid-β Pathway in Alzheimer's Disease. *Mol. Psychiatry*, 26, 10, 5481–503, 2021.
11. Abitbol, A., Mallard, B., Tiralongo, E., Tiralongo, J., Mushroom natural products in neurodegenerative disease drug discovery. *Cells*, 11, 23, 3938, 2022 Dec 6.
12. Lei, P., Ayton, S., Bush, A.I., The essential elements of Alzheimer's disease. *J. Biol. Chem.*, 296, 100105, 2021.
13. Ferrari, C. and Sorbi, S., The complexity of Alzheimer's disease: an evolving puzzle. *Physiol. Rev.*, 101, 3, 1047–81, 2021.
14. De Simone, A., Naldi, M., Tedesco, D., Bartolini, M., Davani, L., Andrisano, V., Advanced analytical methodologies in Alzheimer's disease drug discovery. *J. Pharm. Biomed. Anal.*, 178, 112899, 2020.
15. Ismail, Z., Creese, B., Aarsland, D., Kales, H.C., Lyketsos, C.G., Sweet, R.A., *et al.*, Psychosis in Alzheimer disease - mechanisms, genetics and therapeutic opportunities. *Nat. Rev. Neurol.*, 18, 3, 131–44, 2022.
16. Mungenast, A.E., Siegert, S., Tsai, L.H., Modeling Alzheimer's disease with human induced pluripotent stem (iPS) cells. *Mol. Cell. Neurosci.*, 73, 13–31, 2016.
17. Scheltens, P., De Strooper, B., Kivipelto, M., Holstege, H., Chételat, G., Teunissen, C.E., *et al.*, Alzheimer's disease. *Lancet*, 397, 10284, 1577–90, 2021.

18. Scheltens, P., Blennow, K., Breteler, M.M., de Strooper, B., Frisoni, G.B., Salloway, S., *et al.*, Alzheimer's disease. *Lancet*, 388, 10043, 505–17, 2016.
19. Briggs, R., Kennelly, S.P., O'Neill, D., Drug treatments in Alzheimer's disease. *Clin. Med. (Lond)*, 16, 3, 247–53, 2016.
20. Mantzavinos, V. and Alexiou, A., Biomarkers for Alzheimer's Disease Diagnosis. *Curr. Alzheimer Res.*, 14, 11, 1149–54, 2017.
21. Jack, C.R., Jr., Bennett, D.A., Blennow, K., Carrillo, M.C., Dunn, B., Haeberlein, S.B., *et al.*, NIA-AA Research Framework: Toward a biological definition of Alzheimer's disease. *Alzheimers Dement.*, 14, 4, 535–62, 2018.
22. Almasoudi, H.H., Mashraqi, M.M., Alshamrani, S.A., Alharthi, A.A., Alsalmi, O., Nahari, M.H., *et al.*, Structure-Based *In Silico* Approaches Reveal IRESSA as a Multitargeted Breast Cancer Regulatory, Signalling, and Receptor Protein Inhibitor. *Pharmaceuticals*, 17, 2, 208, 2024.
23. Pavlopoulos, G.A. and Wegener, A.-L., Schneider R. A survey of visualization tools for biological network analysis. *Biodata Min.*, 1, 1–11, 2008.
24. Sanders, O.D., Virus-Like Cytosolic and Cell-Free Oxidatively Damaged Nucleic Acids Likely Drive Inflammation, Synapse Degeneration, and Neuron Death in Alzheimer's Disease. *J. Alzheimers Dis. Rep.*, 7, 1, 1–19, 2023.
25. Vitek, G.E., Decourt, B., Sabbagh, M.N., Lecanemab (BAN2401): an anti-beta-amyloid monoclonal antibody for the treatment of Alzheimer disease. *Expert Opin. Investig. Drugs*, 32, 2, 89–94, 2023.
26. Withington, C.G. and Turner, R.S., Amyloid-Related Imaging Abnormalities With Anti-amyloid Antibodies for the Treatment of Dementia Due to Alzheimer's Disease. *Front. Neurol.*, 13, 862369, 2022.
27. Ecevitoglu, A., Edelstein, G.A., Presby, R.E., Rotolo, R.A., Yang, J.H., Quiles, T., *et al.*, Effects of the atypical antipsychotic and D3/D2 dopamine partial agonist cariprazine on effort-based choice behavior: implications for modeling avolition. *Psychopharmacology (Berl)*, 240, 8, 1747–57, 2023.
28. Carli, M., Kolachalam, S., Longoni, B., Pintaudi, A., Baldini, M., Aringhieri, S., Fasciani, I., Annibale, P., Maggio, R., Scarselli, M., Atypical antipsychotics and metabolic syndrome: from molecular mechanisms to clinical differences. *Pharmaceuticals,* 14, 3, 238, 2021 Mar 8.
29. Hydralazine, *LiverTox: Clinical and Research Information on Drug-Induced Liver Injury*, National Institute of Diabetes and Digestive and Kidney Diseases, Bethesda (MD), 2012.
30. Kousparou, C., Fyrilla, M., Stephanou, A., Patrikios, I., DHA/EPA (Omega-3) and LA/GLA (Omega-6) as bioactive molecules in neurodegenerative diseases. *Int. J. Mol. Sci.*, 24, 13, 10717, 2023 Jun 27.
31. Lin, T.Y., Liu, H.W., Hung, T.M., The Ketogenic Effect of Medium-Chain Triacylglycerides. *Front. Nutr.*, 8, 747284, 2021.
32. Costamagna, G., Comi, G.P., Corti, S., Advancing drug discovery for neurological disorders using iPSC-derived neural organoids. *Int. J. Mol. Sci.*, 22, 5, 2659, 2021 Mar 6.

33. Wojcicki, J.M. and Heyman, M.B., Maternal omega-3 fatty acid supplementation and risk for perinatal maternal depression. *J. Matern. Fetal Neonatal Med.*, 24, 5, 680–6, 2011.
34. Jiang, F., Jiang, Y., Zhi, H., Dong, Y., Li, H., Ma, S., *et al.*, Artificial intelligence in healthcare: past, present and future. *Stroke Vasc. Neurol.*, 2, 4, 230–43, 2017.
35. Salama, S., Khan, M., Pardo, S., Izbudak, I., Levy, M., MOG antibody-associated encephalomyelitis/encephalitis. *Mult. Scler.*, 25, 11, 1427–33, 2019.
36. Philips, T. and Rothstein, J.D., Rodent Models of Amyotrophic Lateral Sclerosis. *Curr. Protoc. Pharmacol.*, 69, 5.67.1–5..21, 2015.
37. Treiman, D.M., GABAergic mechanisms in epilepsy. *Epilepsia*, 42 Suppl 3, 8–12, 2001.
38. Cummings, J., Zhou, Y., Lee, G., Zhong, K., Fonseca, J., Cheng, F., Alzheimer's disease drug development pipeline: 2023. *Alzheimers Dement. (N Y)*, 9, 2, e12385, 2023.
39. Hasan, M.R., Alsaiari, A.A., Fakhurji, B.Z., Molla, M.H., Asseri, A.H., Sumon, M.A., Park, M.N., Ahammad, F., Kim, B., Application of mathematical modeling and computational tools in the modern drug design and development process. *Molecules*. 27, 13, 4169, 2022 Jun 29.
40. Gupta, R., Srivastava, D., Sahu, M., Tiwari, S., Ambasta, R.K., Kumar, P., Artificial intelligence to deep learning: machine intelligence approach for drug discovery. *Mol. Divers.*, 25, 3, 1315–60, 2021.
41. Ferreira, L.G., Dos Santos, R.N., Oliva, G., Andricopulo, A.D., Molecular docking and structure-based drug design strategies. *Molecules*, 20, 7, 13384–421, 2015.
42. Ding, Y., Shi, Y., Yang, S., Molecular Regulation of Plant Responses to Environmental Temperatures. *Mol. Plant*, 13, 4, 544–64, 2020.
43. Andrade, C.H., Pasqualoto, K.F., Ferreira, E.I., Hopfinger, A.J., 4D-QSAR: perspectives in drug design. *Molecules*, 15, 5, 3281–94, 2010.
44. Leelananda, S.P. and Lindert, S., Computational methods in drug discovery. *Beilstein J. Org. Chem.*, 12, 2694–718, 2016.
45. Lin, X., Li, X., Lin, X., A Review on Applications of Computational Methods in Drug Screening and Design. *Molecules*, 25, 6, 2020.
46. Lim, K.G., Varatharajan, R., Muthuraman, A., The attenuating effect of Beta-carotene on Streptozotocin Induced Diabetic Vascular dementia symptoms in rats. *Molecules*. 27, 13, 4293, 2022 Jul 4.

12

The Integration of Molecular Docking and Machine Learning in Drug Discovery for Neurological Disorders

Aditya Singh[1]*, Shubhrat Maheshwari[2], Jigna B. Prajapati[3], Juber Akhtar[1], Syed Misbahul Hasan[1], Amita Verma[2], Sudarshan Singh[4] and Bhupendra Prajapati[5,6†]

[1]*Department of Pharmacy, Faculty of Pharmacy, Integral University, Lucknow, Uttar Pradesh, India*
[2]*Bioorganic and Medicinal Chemistry Research Laboratory, Department of Pharmaceutical Sciences, Sam Higginbottom University of Agriculture, Technology and Sciences, Prayagraj, India*
[3]*Acharya Motibhai Patel Institute of Computer Studies, Ganpat University, Gujarat, India*
[4]*Department of Pharmaceutical Sciences, Faculty of Pharmacy, Chiang Mai University, Chiang Mai, Thailand*
[5]*Shree S.K. Patel College of Pharmaceutical Education and Research, Ganpat University, Gujarat, India*
[6]*Department of Industrial Pharmacy, Faculty of Pharmacy, Silpakorn University, Nakhon Pathom, Thailand*

Abstract
In other therapeutic fields, neurological disorders are much more prevalent than diseases. Despite the lengthy timetables and high attrition rates, finding medications for disorders of the central nervous system (CNS) continues to be the most difficult area of drug development. Artificial intelligence (AI) and machine learning (ML) have become indispensable tools to derive relevant insights and enhance decision making in drug development as a result of the enormous growth of biomedical data made possible by cutting-edge experimental technologies [1]. Due to improvements in AI and ML algorithms, solutions based on these technologies currently have an unmatched potential to hasten the development of

**Corresponding author*: simmu5adityasingh@gmail.com; ORCID: 0000-0002-0398-6130
†*Corresponding author*: bhupen27@gmail.com; ORCID: https://bit.ly/drbgprajapatiorcid

Bhupendra Prajapati, Alok Tripathi, Rishabha Malviya and Lucy Mohapatra (eds.) Computational Neuropharmacology: Fundamentals and Clinical Aspects, (349–374) © 2025 Scrivener Publishing LLC

CNS drugs with a higher success rate. This chapter offers a thorough summary. Extracellular amyloid-(A) plaques and neurofibrillary in the intracellular milieu are the hallmarks of Alzheimer's disease (AD), neuronal death, and synaptic loss, all of which cause progressive cognitive impairment. After age 65, the prevalence of AD increases exponentially, making age the main risk factor for the condition. The overall prevalence of AD is anticipated to double over the next 20 years as the average lifespan in developing countries rises. Pharmaceutical research fueled by AI/ML and its applications in the CNS. After outlining the applications of AI/ML technologies to several crucial steps in drug discovery, such as target identification, compound screening, hit/lead generation and optimization, drug response and synergy prediction, de novo drug design, and drug repurposing, we introduce the AI/ML models as well as the conceptualization and data preparation. A well-known *in silico* structure-based technique for drug discovery is molecular docking. Without knowing in advance the chemical makeup of other target modulators, docking makes it possible to identify novel therapeutic compounds, anticipate ligand-target interactions at the molecular level, or define structure-activity correlations (SAR). Although docking was initially created to aid in the understanding of the mechanics of molecular recognition between tiny and large molecules, its usage and applications in drug development have undergone significant modification in recent years. This chapter explains how molecular docking was initially used to aid in drug discovery tasks.

Keywords: Artificial intelligence, machine learning (ML), structure-activity correlations, *in silico*, molecula

Abbreviations

CNS	central nervous system
AI	Artificial intelligence
ML	Machine learning
AD	Alzheimer's disease
BBB	Blood-brain barrier
Aβ	Amyloid beta
AMBER	Assisted model building with energy refinement),
GROMACS	Groningen machine for chemical simulations.
NDDs	Neurodegenerative diseases
SBDD	Structure-based drug design
GPU	Graphics Processing Unit

12.1 Introduction

Structure-based approaches rank databases of compounds based on the structural and electronic complementarity of ligands to a specific target and rely on the data gained from the knowledge of a target's 3D structure [2]. Molecular docking is one of the most well-known and productive structure-based *in silico* techniques that aid in predicting the interactions that take place between molecules and biological targets [3]. Multiple illness conditions with major economic and societal consequences are caused by disorders of the CNS. Significant advancements in our comprehension of the anatomy and physiology of the CNS, creating novel medications to treat CNS illnesses comes with its own set of difficulties [4]. Due to a number of factors, including inadequate target engagement, lack of efficacy in the early phases of research, the blood-brain barrier (BBB), and lack of understanding of the pathophysiology of complicated CNS disorders, CNS medications have a lower success rate than other therapeutic classes. These challenges have significantly increased the time required for CNS drugs to develop from discovery to regulatory approval; it currently takes 15 to 19 years on average [5]. However, these methods also come with drawbacks like inefficiency and lack of accuracy. Pharmaceutical corporations and chemical expert's priorities research in the area of drug development and design. ML, one of the AI ideas, has the ability to speed up pharmaceutical research by sifting through the enormous quantity of complicated data produced by the drug discovery process and identifying new and significant information. Recently, a variety of therapeutic domains have seen a wide use of AI/ML-based methodologies that have produced state-of-the-art results when tackling various issues in drug discovery [6]. The most difficult area of drug discovery, the creation of CNS medicines, has showed promise with such uses of AI/ML algorithms. But we have only just started to look into the possibilities of these technologies for finding new medicines and repurposing existing ones for CNS illnesses [7].

Docking has established itself as a crucial method for understanding how chemical compounds interact with their molecular targets as well as for drug discovery and development ever since it first appeared in the middle of the 1970s [8]. In reality, since its inception, there have been a significant number of publications describing the use of molecular docking to find structural factors required for effective ligand-receptor binding and the development of more precise docking approaches. The early 1980s study by Kuntz *et al.* on the application of docking in drug discovery and biology is one of the pioneering and most fascinating researches on the topic. The authors of this chapter presented a computational approach that enables the discovery of geometrically viable ligand-receptor alignments for the well-known heme-myoglobin/metmyoglobin and thyroxine/prealbumin structures [9]. In order to forecast probable conformations of molecular complexes, docking has been used before—not just in this study. The authors were also the first to think of the receptor as a solid, rigid body with "pockets" that make up the binding site. Recent advancements and uses for AI/ML-assisted drug discovery, notably for CNS illnesses [10]. AI algorithms and the molecular descriptors they use as input before summarizing AI/ML-based approaches in various stages of drug discovery, including target identification and characterization, virtual screening, lead discovery, and physicochemical pharmacokinetic property prediction. The development of GPU-centered molecular dynamics codes over the past ten years has resulted in a hundred-fold decrease in simulation computation costs when compared to CPU-based approaches [11]. The majority of molecular dynamics engines now offer GPU-accelerated versions, including AMBER (assisted model building with energy refinement), GROMACS (Groningen machine for chemical simulations) 10, and NAMD (nanoscale molecular dynamics). GPUs not only scale well with system size using the spatial domain but are also well adapted to accelerate molecular dynamics simulations [12].

Studies on DL methods have also been done to enhance docking outcomes. They specifically created a scoring mechanism that produces DL-based docking scores based on structural information defining a ligand-protein combination in the first place. The authors then carried out rigorous validation screenings on 40 different datasets presented in the DUD-e database to show that the proposed scoring function can produce better results [13]. The results showed that compared to normal scoring functions, the newly created "DeepVS" scoring function was able to deliver better screening performances. This chapter is notable since it was the first to describe the use of deep learning for rescoring docking postures without the need for human-defined parameters, making it a particularly suitable method for unsupervised, massive virtual screening campaigns [14]. To enhance docking scoring functions, ML techniques have also been applied. For instance, "RF-Score," one of the earliest ML-based scoring functions, was created by Ballester *et al.* and uses Random Forest to enhance predictions of protein-ligand binding affinity. To be more specific, the authors first created an RF-based scoring system utilizing several sets of ligand-protein complexes with known activity information published in the PDBbind database. Then they showed that "RF-Score" improved docking results in both virtual screening and lead optimization tasks, and that its performances are independent of the used training set, by comparing its performance with that of the other sixteen scoring functions implemented in currently available docking programs [15]. Additionally, they showed how using larger datasets for the RF-based scoring function correction improved the prediction performances of their scoring function. The term AI is frequently worn a computer imitates folded fibrils protein linked to the human brain during problem-solving and learning. AI is a tool that can be used to get around the aforementioned constraints. Importantly, AI has the capacity to identify which gene target(s) needs additional research in addition to being able to unbiasedly expose patterns inside complex data. Deep learning algorithms and artificial neural networks have modernized the field. Additionally, AI models solve the toxicity issues that result from off-target interactions [16]. Target identification, lead generation and optimization, and preclinical development are the three stages of the early drug discovery process where AI/ML has been applied. AI-based techniques have been utilized in target discovery to combine diverse data sets and uncover trends in order to comprehend the molecular pathways behind diseases and therapeutic actions. The scoring functions and QSAR models in virtual screening pipelines are improved by AI/ML algorithms for lead generation and optimization, and these algorithms also enable the automation and optimization of the de novo drug creation procedures. In preclinical research, AI/ML techniques are used to quickly evaluate vast amounts of chemical data to provide predictive models of physicochemical features and to further optimize absorption, distribution, and metabolism, and excretion— toxicity (ADME-T) profiles [16].

Graphs and networks are mathematical constructs that represent a collection of data as a whole. Nodes (the network's constituent objects) and edges (the links separating them) make up networks. Biological networks can be visualized using free software like Cytoscape. In mice, their significance in the illness was validated, resulting in the creation of a brand-new IBD model [17]. These recently created technologies were successfully used to "treat" patient-derived IBD organoids, discovering new therapeutic targets and treatments for IBD. Fold change, p-values, used in the traditional method for finding genes that are differentially expressed. These approaches, however, are constrained and unlikely to offer the data required to draw firm conclusions regarding dysregulated genes that might be significant for AD [18]. First, it is fundamentally challenging to choose differentially expressed genes using traditional fold change cutoffs. The replication problem is likely exacerbated by the p-value, which is likewise known to be readily misconstrued and to only provide limited information about the data at hand [19]. Along with knowledge of drugs and their therapeutic targets, comparing data regarding other bioactive compounds and other targets is helpful in the drug discovery process. The knowledge of poor binders or non-binders of individual targets is useful for developing drug discovery tools with improved performance. Information about prodrugs that facilitate drug design by improving pharmacokinetic/pharmacodynamic features, co-targets of therapeutic targets that facilitate the investigations of multi-target strategies, off-target & undesired effect, and other comparative data. Drug-likeness landscapes of the explored bioactive chemical space for therapeutic targets are provided by drug profiles of their drug-like properties, and significant pharmaceutical features like activity cliffs are revealed by collective structure-activity landscapes of drugs against individual targets [20]. The development of AI tools for the discovery of pharmaceuticals, including AI techniques for discovering bioactive chemicals, is a particularly relevant trend. The gathering of information on weak binders and non-binders of a certain target is necessary for the development of such instruments. Until then, the prodrug data already in existence might offer novel insights into how to circumvent problems with drug development that limit the options for formulation or lead to unfavorable biopharmaceutical/pharmacokinetic outcomes. Therefore, the above-mentioned comparative data are extremely important for researchers working on medication discovery. Additionally, knowledge regarding the target's 3D structure is essential for the development of new drugs. Improved AI techniques like AlphaFold have made crystal structure prediction possible in addition to the increase in empirically determined target crystal structures [21].

The most popular AI-based drug discovery algorithms to aid the reader in understanding AI/ML applications in CNS drug discovery [22]. In order

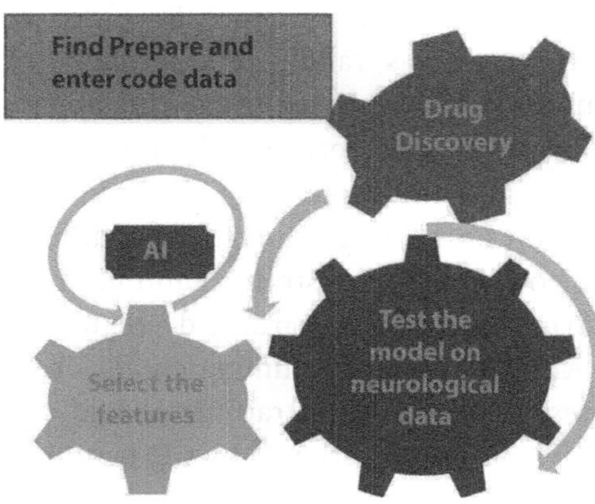

Figure 12.1 Utilization of AI in computational drug discovery.

to create intelligent systems that can be categorized, AI employs a wide range of models. Although ML algorithms and AI are frequently used interchangeably, they are not the same [23]. It would therefore be beneficial to first define both terms. The FDA's (US Food and Drug Administration) definition of AI is used in this chapter. They define ML as per Figure 12.1, "an artificial intelligence technique that can be used to design and train software algorithms to learn from and act on data," adding that all ML techniques are based on AI, which they define as "the science and engineering of making intelligent machines."

12.2 Neurodegenerative Disease

A decline in the population of specific neurons occurs in conjunction with neurodegenerative diseases (NDDs), which are chronic, progressive disorders that affect normal function, cause dementia, and reduce patient life expectancy. The difficulty of preventing, delaying the start of, or slowing the progression of AD is largely to blame for its devastating side effects [24]. The anatomical distribution, major clinical traits, and molecular anomalies are just a few of the criteria used to categorize NDDs and neurological diseases. Multiple sclerosis (MS), Parkinson's disease (PD), Alzheimer's disease (AD), amyotrophic lateral sclerosis (ALS), and Huntington's disease (HD) are examples of common NDDs/neurological disorders that result in the slow degeneration of neurons. Neurodegenerative diseases are on the rise, along with the ageing of the global population [25]. By 2050, there will be 36.7 million instances of dementia, up from 13.5 million in 2000. Extracellular vesicles (EVs), membrane-bound nano-microscale objects, are secreted by nearly all bacterial and eukaryotic cell types. They were initially found to be debris and "cell dust" in human platelet parasite cells and were assumed to be irrelevant in biology. Later, by conveying substances with physiological activity, their purpose was understood [26]. Previously, it was thought that intercellular communication was carried out through intracellular exchanges of soluble molecules such interleukins, cytokines, and soluble factors. It is understood that EVs are essential for long-distance communication between various cell types. When the plasma membrane is extensively infiltrated, an early sorting endosome—a saucer-shaped structure—is created. Proteins that have been absorbed, cytoplasmic genetic material, and cell membrane proteins are all present in this endosome as a result of membrane invagination [27]. AD is characterized as a prevalent neurological condition that has a direct correlation to the emergence of dementia. According to reports, AD predominates in the elderly population and has an impact on social interactions and quality of life. In addition, amyloid beta (Aβ) buildup, tau protein hyperphosphorylation, neuroinflammation, and oxidative stress are linked to AD pathogenesis [28]. The PI3K/Akt/mTOR axis is thought to be one of the most important signaling pathways in the pathophysiology of NDDs based on the prior research. Apoptosis, cell proliferation, metabolism, and angiogenesis are only a few examples of the biological processes in which this regulatory system plays a crucial role. It has been suggested by a number of lines of research that exosome treatment and PI3K/Akt/mTOR signaling are related to neuronal injury. Another study found that bone marrow mesenchymal stromal cells (BMSCs)-Exo carrying GDF-15 could reduce the damage that A-42 causes to SH-SY5Y cells by improving the Akt/glycogen synthase kinase-3 beta (GSK-3 beta)/-catenin signaling pathway. Notably, this therapeutic technique increased cell survival while decreasing apoptosis, TNF-, IL-6, IL-1, and IL-8 (inflammatory cytokines) in Aβ-42-induced SH-SY5Y cell damage [29].

While the majority of efforts should go towards creating medications that can alter the course of the disease, these efforts should also include looking for complementary therapies that can more effectively treat patients with AD [30]. The majority of the 2173 clinical trials for AD conducted prior to 2019 involved drug candidates intended to target one target that appears to be important in AD pathogenesis, including neurovascular mechanisms (8%), tau pathology (12%), mitochondrial dysfunction (17%), and -amyloid biology (22%), while 19% were to address neurotransmitter deficits. In January 2021, 126 candidates were undergoing clinical trials for preclinical, prodromal, and AD dementia populations (28 in Phase III, 74 in Phase II, and 24 in Phase I), in an effort to address the equally crucial needs of prevention, delaying the onset, slowing down the progression, and relieving the cognitive and neuropsychiatric symptoms. Among the 126 clinical candidates, 104 (82.5%) had the potential to be disease-modifying drugs, 13 (10.3%) to improve cognitive symptoms, and 9 (7.1%) to lessen neuropsychiatric symptoms. Candidates focusing on the diseases of "amyloid" (13.5%) or tau (8.7%), inflammatory/infectious/immune (15.1%), metabolism and bioenergetics (5.6%), or neurovascular processes (4.8%) continued to be among the most sought-after candidates. Indeed, numerous distinct signaling pathways that are involved in the overall development of AD have been discovered [18]. An anti-amyloid monoclonal antibody is called aducanumab. In the near future, further fresh therapies for AD might be made available. 47 Phase 3 studies including 31 agents are underway. DMTs make up 21 (67.8%) of the medicines in Phase 3 trials, including 16 (51.6% of the agents) small molecules and five (16.1%) biologics.

Putative cognitive enhancers make up five Phase 3 agents (16.1%) while medications that address behavioral symptoms make up five (16.1%) of the total [31].

12.3 Molecular Docking

When the 3D-structure of a biological target is known, the technique of structure-based drug design (SBDD), a subset of CADD, is used. The ability to screen compound libraries and specifically build compounds to meet the ligand binding site best is made possible by knowledge of the structure of a ligand binding site. The number of 3D-structures of significant pharmacological targets that are currently available is fast growing thanks to recent improvements in crystallization processes and cryogenic electron microscopy (cryo-EM) technology [32]. Furthermore, it is now possible to predict protein structures solely from sequence thanks to the Human Genome Project's completion and the application of artificial intelligence-based technologies like AlphaFold2 and RoseTTAFold. A technique called molecular docking anticipates the preferred orientation of a ligand against a receptor (a protein) to create a stable compound [31]. Favored orientation may be used to anticipate the strength of the bond or binding affinity between the ligand and the protein by making use of scoring systems. In order to forecast the affinity and activity of a drug, docking is frequently used to anticipate the binding orientation of drug candidates against protein targets. As a result, docking is essential to the process of developing new drugs [32]. In order to get an optimized conformation and reduce the free energy of the entire system, molecular docking's primary goals are to computationally simulate the molecular identification process and achieve these goals. The process of developing a new medicine is quite challenging. *In silico*-chemo-biological approaches are the mainstay of contemporary drug discovery [33]. The use of computer-aided approaches in the drug research and development process is swiftly growing in acceptance and use [34].

12.3.1 Pharmacophore Modeling

A pharmacophore model is a group of chemical characteristics that describes how a ligand interacts with a biological target to explain a physiological response. To determine whether a ligand would behave as an agonist or antagonist, pharmacophore modelling can be employed. This method is very helpful when there is a lack of specific target structure knowledge [35].

Pharmacophore modelling can also be used to infer oblique information about binding site characteristics based on the chemical characteristics of known active compounds. Prior to the 3D structure of this protein being solved, pharmacophore modelling was used to look for antagonists of the nicotinic acetylcholine receptor (nAChR) subtype 7. T761-0184, a new lead with great potency, was discovered and is now being used to treat CNS conditions. The chemical characteristics that enable the ligand to bind to the target receptor and trigger a pharmacological response are included in 3D-pharmacophore models [36]. Examples include steric and electrostatic properties. The pharmacophore identifies functional groups that are hydrogen bond donors or acceptors, charges, hydrophobic, and aromatic. Pharmacophoric modelling essentially creates a 3D map of the important chemical properties using a training set of at least two ligands, which is subsequently verified using a test set of additional active molecules and decoys [37]. The Database of Useful Decoys: Enhanced (DUD-E), a collection of biologically inert small compounds with related physicochemical features, is a source of decoys. This method is especially helpful as a pre-screening tool for ultra-large chemical libraries that need to be filtered before being further assessed using more computationally expensive techniques like molecular docking. Finding human dopamine transporter (hDAT) antagonists for novel Parkinson's disease treatments followed a similar process. Using 25 known active ligands and 50 decoys as the training set and 50 substrates and 2500 decoys as the validation set, a pharmacophore was also produced. Following pharmacophore screening on the top 1000 ligands from the 300,000 ligand docking experiments, six interesting compounds were found. In biological testing, three leads showed a statistically significant reduction of hDAT receptor uptake. Pharmacophore development must involve an iterative process where models are evaluated and improved until the best option is identified in order to create a solid model that is representative of all known active compounds. In general, overlay or root mean square derivative (RMSD) scoring can be used to score and generate pharmacophore models. The radii of chemical characteristics from the alignment of molecules in three-dimensional space are matched using overlay methods to construct and score a hypothesis for a pharmacophore [38]. A functional group on a query molecule's functional distance from the pharmacophore model is instead used by RMSD [39].

12.3.2 QSAR

The association between a molecule's physicochemical characteristics and topological features and the biological activity they exert on a target is the foundation of the QSAR concept. This connection is used in QSAR investigations to sort and order chemical libraries and forecast biological activity [40]. The application of statistical techniques to correlate molecular descriptors to biological data, such as binding affinity (KD) or functional potency (EC50 or IC50) values, allows for these predictions. Chemical databases are frequently utilized to get data for the training and testing of QSAR models, but data can also be obtained internally (for instance, utilizing proprietary data). By locating the regions relevant for biological activity, QSAR modelling is also appealing for lead optimization [41]. The lead optimization of PDE4 inhibitors that could be utilized to treat major depressive disorder was done using a 3D-QSAR model. Hydrophobic and hydrogen bonding groups that extended into additional pockets of the active site were added to improve prior leads. The novel compounds showed anti-inflammatory activities in microglial cells and had nanomolar IC50 values, according to *in vitro* analyses. In addition to an external validation set, the molecules utilized in model creation should ideally be divided into a set that is roughly 80% for training and 20% for testing [42]. To ensure the validity of predictions and avoid biases, it is crucial that the molecules in the training set are as chemically different from one another as feasible and that the bioactivity data of these chemicals is dispersed across the whole range of endpoints. It is crucial to specify an application domain (AD) because these molecules can't encompass the entirety of chemical space. An AD specifies the region of chemical space for which the model can produce reliable predictions. The Organisation for Economic Co-operation and Development's (OECD) recommendations for valid QSAR development include five guidelines, and this is one of them. After a dataset has been vetted and descriptors have been created, a regression algorithm must be chosen to complete the work. In general, linear and non-linear approaches can be employed to classify the algorithms utilized in QSAR. Partial least squares (PLSs), one of the more used linear regression algorithms PLSs convert vast, multidimensional data, like chemical descriptors, into predictions through linear solutions. Support vector machines (SVMs), random forests (RF), and k-nearest neighbors (k-NNs) are examples of non-linear approaches. The solutions to the QSAR regression issues are determined by k-NN and SVMs using the distances between parameters in the hyperplanes. RF, in comparison, assembles a number of decision trees into a single, reliable predictive model [43]. Validation tests must be carried out to make sure the QSAR model being utilized has strong predictive capabilities. This is yet another suggestion made by the OECD for the creation of a reliable QSAR model. The performance of a model should be evaluated using an external validation set that is roughly 15–20% the size of the full dataset. The two parameters that are most frequently used to evaluate the goodness of fit for QSAR models are R2 (1-(RSS/TSS), where RSS stands for residual sum of squares and TSS for total sum of squares. R2 is calculated using leave-one-out cross validation to provide the q2 value [44].

12.3.3 Homology Modeling

Selecting a target structure and locating the desired binding site are the initial steps in the structure-based drug design (SBDD) process. X-ray crystallography, cryo-EM, and nuclear magnetic resonance (NMR) are three experimental methods for determining the target structure. These structures have been determined experimentally, and they are stored in the Protein Data Bank (PDB) [45]. If the target of interest's structures is unavailable, they can also be computationally simulated using the structure of one or more proteins with evolutionary connections. Typically, it is advised that the template and target sequences share at least 30% of their similarities. At the time of writing, the PDB contains almost 200,000 experimentally determined structures, all of which can be utilized as templates. The Uniprot and Swissprot databases provide access to the sequences of the desired target and the template, which can then be aligned using tools like T-Coffee or BLAST. The homology modelling technique is carried out with tools like MODELLER and SWISS-MODEL or paid applications like Prime and MOE. In essence, predictions of protein folding are guided by the template structure by exploiting similarity areas [46]. Recently, more information regarding the steps involved in homology modelling has been evaluated. Most often, a variety of models based on several starting seeds are produced, and the most accurate model suited for further research is determined using scoring systems like the discrete optimized protein energy (DOPE) or qualitative model energy analysis (QMEAN). Both scoring functions are statistical evaluations that evaluate a homology model's local and global energy potentials to determine whether the structure is plausible. The human dopamine transporter (hDAT) homology models were a crucial first step in identifying potential treatments for Parkinson's disease. In the search for better treatments for Alzheimer's disease, homology modelling helped to identify dual-target ligands for the nicotinic acetylcholine receptor (nAChR) and acetylcholinesterase (AChE). A SBDD method that predicts a ligand's preferred orientation and conformation in a binding site is called molecular docking. In order to help in the discovery of potential hit molecules, it can be used to filter databases of ligands, offer scores for the estimated binding energy, and rank the ligands [47]. When binding into the protein, the ligand is allowed to be flexible in molecular docking studies.

This allows the lowest possible energy state of each ligand-protein complex to be established when scoring and ranking the ligands. Depending on the search technique used, conformers are either developed a priori, by methodically investigating all conceivable degrees of freedom with respect to ligand binding, or by randomly altering parameters, such as torsional angles [48].

12.4 Machine Learning in Drug Discovery

AI algorithms like ML to DL have become more prevalent in CADD due to technological developments and the creation of high-performance computers. The introduction of AI in drug discovery and development is not new, and neither is the goal of chemists to precisely predict chemical activity-structure connections [49]. But in the modern day, scientists' top priority is to use ML algorithms built on traditional chemistry activities to enhance the drug. This will motivate chemists to recognize the potential of AI techniques for addressing two important medicinal chemistry concerns, namely, "What should be the next compound?" and "What is the process of making a compound?" As a result, over the past 20 years, numerous methodologies and tools have been created for computational drug development, QSAR methods, and free-energy minimization strategies [50]. Use machine intelligence techniques, such as DT, CNN, SVM, LSTM networks, and gradient boosting machines, to, for instance, differentiate compound cell activity. Analyzing the compound's physical characteristics, bioactivity, and toxicity is a part of the secondary drug screening. Some of the physical characteristics that control a compound's bioavailability include melting point and partition coefficient, which are also necessary when creating new compounds [51]. Reinforcement learning can be applied throughout the entire process as per Figure 12.2. For QSAR research, matched molecular pairs (MMP) have been employed frequently. One alteration in a therapeutic candidate linked to MMP affects the compound's bioactivity in additional ways [52].

Finding appropriate and bioactive drug molecules in the enormous chemical space, which contains on the order of 1060 compounds, is the most difficult and discouraging stage in the drug discovery and development process [52]. Additionally, the process of medication research and development is thought to be time- and money-consuming. Implementing AI-based technologies and methodologies can overcome the aforementioned limitations of drug discovery and development [52]. In recent years, scientists have harnessed AI to their advantage to find novel peptides.

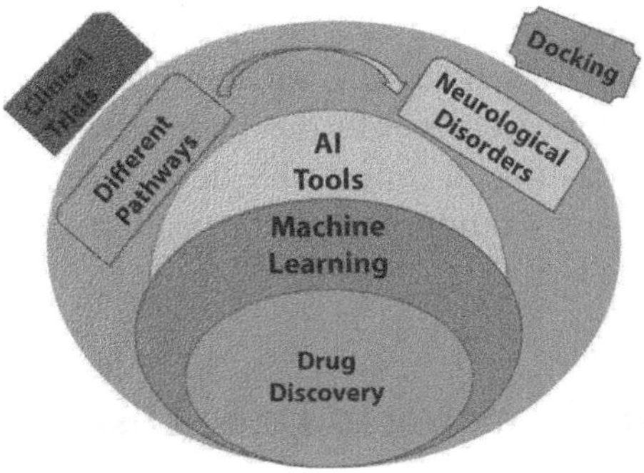

Figure 12.2 AI in drug discovery and treatment of neurological disorders.

For instance, the CNN-driven program Deep-AmPEP30, created by 2020, predicts short anti-microbial peptides (AMPs) using DNA sequence data. It is essential to establish a leery dose for therapeutic reasons because giving a patient the wrong dosage of any medication might result in unfavorable and even fatal side effects. Over time, it has proven difficult to determine the ideal dosage of a medication that will produce the crave efficacy with the fewest hazardous side effects [53].

However, the LBVS does not rely on the target protein's 3-D structure [54]. This technique is typically used to forecast interactions between active ligands and their associated targets as well as the amino acid residues implicated in drug-target binding. Developing the link between chemical structures, their physiochemical properties, and biological activities is essential for drug creation and discovery. QSAR modeling, then, is a computational method for developing quantitative mathematical models connecting chemical structure and biological activity. The key benefit of creating a mathematical model is discovering the many chemical structures in molecular databases that can be employed as therapeutic compounds against a disease target [55]. It is then synthesized in a lab and tested *in vitro* or *in vivo* after the most promising molecule has been chosen. Regression and classification models are two categories into which QSAR models can be generally divided. A reliable and effective technique for QSAR modelling is the use of Gaussian processes (GPs), a sort of QSAR constructing regression model. GP techniques can sort through a vast number of descriptors to find the most important ones. Two classification models have recently been proven utilizing GP, one of which combines GP regression approach and probit analysis with intrinsic GP classification methods. The technique also works well for modelling nonlinear relationships and does not call for arbitrary model parameter determination. QSAR modelling has a lot of opportunities thanks to recent improvements and expanding uses of ML techniques including neural networks, DL, and SVM. For QSAR modelling, a number of web-based tools and algorithms have been created, including the VEGA platform, QSAR-Co, and FL-QSAR [56].

The AI algorithm has been used in drug repurposing or drug repositioning techniques in addition to QSAR modelling. Drug repositioning in the context of drug developing and discovery refers to the examination of medications that have already been created for one diseased condition and their repositioning for additional diseases [56]. Because several targets may be involved in various diseases, repositioning medications may be beneficial. Another thing to consider is that the development of sizable data sets from studies on genomes, proteomics, and pharmacology in *in silico* and *in vitro* opens up a lot of possibilities for repositioning drugs [43]. The recent development of AI-based drug discovery tools and algorithms offers a foundation for further study. ML algorithms replace the traditional system biology methods based on chemical similarity and molecular docking with novel methods that can assess medication effects. As a result, various AI-based algorithms and web-based tools, including DrugNet, DPDR-CPI, PHARMGKB, PROMISCUOUS 2.0, and DRRS, have been developed recently. ML models to forecast potential anti-fibrosis medication candidates, specifically the structural profile prediction model and biological profile prediction model [50]. The findings showed that the training set's and testing set's areas under the receiver intervene attribute curve, respectively, were 0.814 and 0.874. According to the findings, natural compounds have anti-fibrosis properties and could be used as fibrosis medication targets [57]. Recently, COVID-19 becomes an international epidemic, and experts from all over the world began looking for promising therapeutics. AI-based medication repositioning is quite important in this area. ML is sometimes referred to as the main application of AI, while other times it is referred to as a subset of AI as per Table 12.1.

ChemSpider, ZINC, and BindingDB are crucial facet for conceal during the drug development and disclosure process. Instead of relying on quantum mechanical calculations to determine properties of compounds contained in the aforementioned databases was carefully selected and screened for pharmacological and physicochemical characteristics that were essential to the drug discovery process. The ChEMBL database is a large open-access database that contains information on a variety of bioactive substances with drug-like characteristics. The ChEMBL database also includes details on these substances' toxicity characteristics, absorption, distribution, metabolism, and excretion (ADME), as well as target interactions. Additionally, DrugBank is a different open-access pharmaceutical

Table 12.1 The role of AI in the development of personalized treatment plans.

Role of AI	Functions
Brain imaging analysis	AI algorithms can be trained to analyze brain imaging [58].
Early detection and prevention	Analyzing data from wearable devices and other sources to identify changes in behavior and cognition that may be early signs of dementia. This can lead to early interventions that may help slow or prevent cognitive decline [59].
Patient monitoring and care	AI-powered monitoring systems can be used to track patient behavior and cognition in real-time, alerting caregivers and healthcare providers to changes that may require intervention [59].
Drug discovery	AI can be used for dementia and cognitive impairment by analyzing large datasets of genetic and molecular information to identify potential drug targets [60].
Improved clinical trials	AI-powered clinical trials can be used to test new treatments for dementia and cognitive impairment more efficiently by identifying eligible participants, monitoring patient outcomes, and predicting treatment responses [60].

data source that includes information on a variety of medications, their targets, and mechanisms [61]. Another collection providing data on the L1000CDS2 contains information on medications. In addition, the protein data bank (PDB) is a different openly available online database that holds information about the three-dimensional structures of proteins, DNA, and RNA. In the pharmaceutical sector, AI has come to be seen as a potential answer to issues highlighted by traditional chemistry which impede medication discovery [62]. The wide reach of AI has captivated the scientists involved in the drug development process [63]. Drug development time, cost, attrition rates, and human resources have all been decreased because to AI as per Table 12.2. For this, various web-based tools including LimTox, pkCSM, admetSAR, and Toxtree have been developed.

Table 12.2 The use of AI in healthcare raises important ethical and privacy considerations that must be carefully addressed.

S. no	Name of AI	Functions
1	**MindMate**	An AI-powered app that helps individuals with dementia and their caregivers manage their daily activities, medications, and appointments [64].
2	**CogniFit**	An online platform that offers cognitive assessment and training programs that are personalized to individual users [65].
3	**Re-Mind**	An AI-powered platform that uses machine learning algorithms to personalize cognitive training programs for individuals with cognitive impairment [66].
4	**BrainHQ**	An online brain training platform that uses gamification and personalized training programs [67].
5	**MindMaze**	An AI-powered virtual reality platform that offers immersive cognitive therapy for individuals with dementia and other cognitive impairments [68].
6	**Virtual Assistants**	Amazon's Alexa or Google Assistant, can be used to provide reminders, facilitate communication with caregivers, and offer cognitive training exercises [69].
7	**Neurotrack**	An AI-powered cognitive assessment and monitoring tool that uses eye tracking technology to measure cognitive function [57].
8	**Altoida**	An AI-powered diagnostic tool that uses virtual reality and machine learning algorithms to assess cognitive function and predict the likelihood of developing dementia [70].
9	**Cognitive scale**	An AI-powered platform that provides personalized care plans for individuals with dementia and cognitive impairment by analyzing large datasets of patient information [71].
10	**AiCure**	An AI-powered diagnostic tool that uses virtual reality and machine learning algorithms to assess cognitive function and predict the likelihood of developing dementia [72].
11	**MemTrax**	An AI-based tool that measures cognitive function using a simple online test that is administered through a computer or mobile device [72].

12.5 Random Forest

As it clarifies by eliminating outliers and classifies and designates datasets based on relative characteristics defined for the specific algorithm, RF is a frequently used algorithm that is specifically built for huge datasets with various features. It is frequently trained using data gathered from numerous databases to account for big inputs, variables, and accessibility. It helps with a variety of tasks, including attributing missing data, dealing with outliers, and predicting classification-related attributes. Each tree in the ensemble of uncorrelated decision trees that forms the mathematical foundation of RF is in charge of making one prediction. Feature selection, classifiers, or regression are the three main uses of RFs in the drug development process [73]. RF techniques to virtual screen molecular descriptors based on a training data set for enzymes, such as ligands of kinases and nuclear hormone receptors, to improve affinity prediction between ligand and protein. The following are some crucial elements that RF must have while finding new drugs: It speeds up the training process, employs fewer parameters, fills in blanks, and leverages nonparametric data. Multivariate RF by including genomic sequencing information, which supported error and produced pharmacological responses based on genomic characterizations. Multivariate RFs are experts in reducing error by computing various error estimation methods within the system. Data incorporating genetic and epigenetic characterization combinations are fed into the computational framework, enabling the framework to forecast the mean and confidence interval of the medication reactions. Rahman *et al.* tried to improve the prediction based on the response profile by combining the modelling framework with functional RF. They made an effort to overcome the challenges people had finding the right chemicals for their unique tumors. The creation of the regression tree node and leaf nodes included RF. It collected the dose-response data points. The algorithms' leaf nodes are in charge of forecasting the dose-response profile while concurrently storing the functional data [74].

12.6 Naïve Bayesian

NB algorithms are a subset of supervised learning techniques that have evolved into a crucial tool for categorization in predictive modeling. Depending on the input qualities, factor correlation, and dimensionality of the data, standard NB algorithms can be one of the most effective methods for classifying dataset features. It is unknown how well works in conjunction with decision tree algorithms for text mining NB. These methods improve the accuracy of data sets that are retrieved, which typically come from huge, jumbled sources. The target finding subgroup of the drug development process, in particular, depends heavily on the classification of biomedical data. Recently, scientists have been able to apply NB approaches to many different aspects of the drug development process. NB models and other techniques as classifiers for active and inactive substances with potential action as estrogen receptor antagonists in breast cancer. The researchers made use of NB algorithms' capacity to process enormous amounts of data while possessing a special tolerance for noise. The method proved successful in gathering high-quality results when used in conjunction with other instruments like extended-connectivity fingerprint-6. Combinational approach of NB and support vector machine methods to suggest potential compounds that could be active against targets of human immunodeficiency virus type-1 and the hepatitis C virus [75].

12.7 Support Vector Machine

SVMs are supervised machine learning algorithms that divide classes of compounds depending on the feature selector by constructing a hyperplane. They are utilized in the drug development process. It creates an endless number of the hyperplanes by using the similarities between classes. When dealing with linear data, it trains by creating classes made up of compounds based on a few features and then putting those classes into chemical feature space. An ideal hyperplane is defined by a hyperplane, which is used to categorize data points by establishing decision boundaries in N-dimensional space (N is the number of features) [76]. It is obtained by maximizing the gap between classes. As it uses a query for datasets to predict, regression models are essential in understanding the link between the drug and ligand. SVM can be utilized in a variety of situations when numerous active drugs are screened against a single protein of interest. A subset binary class prediction in SVM classification could distinguish between molecules that are active and inactive. It might rank compounds from various databases for drug discovery based on their likelihood of being active for any computational screening. SVM can be extrapolated in a variety of ways to get results, with a particular focus on telling active molecules apart from inactive ones. By training the algorithm with different descriptors for feature selectors, such as 2D fingerprints and target proteins, the process might be controlled. It is specifically made to integrate ligands and proteins of interest information as a crucial part of SVM modelling for drug-target interaction [61]. In order to improve the prediction, Wang *et al.* looked into drug-target interactions and combined data from published research from diverse sources. In order to characterize the drug-target interactions, they employed kernel function to include data on drug pharmacological and therapeutic effects, drug chemical structures, and protein genetic information. The majority of the findings from the various sources were encouraging, with kernel function for the forecasting of pharmacological and therapeutic effects showing the most promise. SVM are widely used to forecast the potential bioactivities of pharmaceuticals [61].

12.8 Conclusion

Drug development is being revived by ML-based methods. These techniques are based on several applications in the discovery of targets, lead compounds, synthesis, protein-ligand interactions, etc. Algorithm-enhanced data query, analysis, and generation are being made possible by ML applications. One such instance is the application of ML to target identification, which mainly relies on the analysis and exploration of pre-existing omics and medical data. Viable targets can be determined utilizing data clustering, regression, and classification from enormous omics databases and sources through the integration of AI employing ML approaches. Currently, lead compound discovery is widely utilized to create reasonable molecular candidates based on training inputs, such as when employing QSAR. With the use of enormous amounts of acquired data, best-chance trees and NN-based retrosynthesis algorithms have also been used to speed up the synthesis of lead compounds. Alzheimer's disease (AD), a neurological ailment that gets worse over time and impairs memory, cognition, and behavior. There is currently no known cure for AD, and treatments only treat the symptoms. However, a number of novel formulations that target the underlying pathophysiology of AD are currently under investigation. Utilizing AI can be a helpful tool for overcoming the aforementioned restrictions. Importantly, AI can identify which gene targets merit additional research and can objectively uncover trends inside complex data. In other words, deep learning algorithms and artificial neural networks have modernized the field. AI models also address the toxicity issues brought on by off-target interactions. AI and ML can be used to diagnose cognitive impairment through the analysis of medical imaging data, including positron emission tomography (PET) and magnetic resonance imaging (MRI) scans. Researchers are considering using a multi-modal approach to diagnose and predict cognitive decline and dementia with greater accuracy. It's a gift that we now have access to many archives of biomedical data since they can aid us in this process. A number of ML techniques are now widely available and may aid with the extraction of important characteristics, patterns, and structures contained in these massive biological data sets, which has further simplified big data analytics due to the advent of AI.

References

1. Hauser, A.S., Attwood, M.M., Rask-Andersen, M., Schiöth, H.B., Gloriam, D.E., Trends in GPCR drug discovery: new agents, targets and indications. *Nat. Rev. Drug Discov.*, 16, 12, 829–42, 2017.
2. Basith, S., Cui, M., Macalino, S.J., Park, J., Clavio, N.A., Kang, S., *et al.*, Exploring G protein-coupled receptors (GPCRs) ligand space via cheminformatics approaches: impact on rational drug design. *Front. Pharmacol.*, 9, 338587, 2018.
3. Pinzi, L. and Rastelli, G., Molecular docking: shifting paradigms in drug discovery. *Int. J. Mol. Sci.*, 20, 18, 4331, 2019.
4. Olğaç, A., Orhan, I.E., Banoglu, E., The potential role of *in silico* approaches to identify novel bioactive molecules from natural resources. *Future Med. Chem.*, 9, 14, 1665–86, 2017.
5. Aamir, M., Singh, V.K., Dubey, M.K., Meena, M., Kashyap, S.P., Katari, S.K., *et al.*, *In silico* prediction, characterization, molecular docking, and dynamic studies on fungal SDRs as novel targets for searching potential fungicides against Fusarium wilt in tomato. *Front. Pharmacol.*, 9, 339383, 2018.
6. Hashemi, Z.S., Zarei, M., Fath, M.K., Ganji, M., Farahani, M.S., Afsharnouri, F., *et al.*, *In silico* approaches for the design and optimization of interfering peptides against protein–protein interactions. *Front. Mol. Biosci.*, 8, 669431, 2021.
7. Pandey, R.K., Kumbhar, B.V., Sundar, S., Kunwar, A., Prajapati, V.K., Structure-based virtual screening, molecular docking, ADMET and molecular simulations to develop benzoxaborole analogs as potential inhibitor against Leishmania donovani trypanothione reductase. *J. Recept. Signal Transduction*, 37, 1, 60–70, 2017.
8. Trosset, J.-Y. and Cavé, C., *In silico* drug–target profiling, in: *Target Identification and Validation in Drug Discovery: Methods and Protocols*, pp. 89–103, 2019.
9. Muhammed, M.T. and Aki-Yalcin, E., Molecular docking: principles, advances, and its applications in drug discovery. *Lett. Drug Des. Discov.*, 21, 3, 480–95, 2024.
10. Badhani, B. and Kakkar, R., *In silico* studies on potential MCF-7 inhibitors: a combination of pharmacophore and 3D-QSAR modeling, virtual screening,

molecular docking, and pharmacokinetic analysis. *J. Biomol. Struct. Dyn.*, 35, 9, 1950–67, 2017.
11. De Vita, S., Terracciano, S., Bruno, I., Chini, M.G., From natural compounds to bioactive molecules through NMR and *in silico* methodologies. *Eur. J. Org. Chem.*, 2020, 40, 6297–317, 2020.
12. Sirous, H., Campiani, G., Calderone, V., Brogi, S., Discovery of novel hit compounds as potential HDAC1 inhibitors: The case of ligand-and structure-based virtual screening. *Comput. Biol. Med.*, 137, 104808, 2021.
13. Moshawih, S., Lim, A.F., Ardianto, C., Goh, K.W., Kifli, N., Goh, H.P., *et al.*, Target-based small molecule drug discovery for colorectal cancer: a review of molecular pathways and *in silico* studies. *Biomolecules*, 12, 7, 878, 2022.
14. Sharma, S. and Bhatia, V., Appraisal of the role of *In silico* Methods in Pyrazole based drug design. *Mini-Rev. Med. Chem.*, 21, 2, 204–16, 2021.
15. Vidal-Limon, A., Aguilar-Toalá, J.E., Liceaga, A.M., Integration of molecular docking analysis and molecular dynamics simulations for studying food proteins and bioactive peptides. *J. Agric. Food. Chem.*, 70, 4, 934–43, 2022.
16. Almasoudi, H.H., Mashraqi, M.M., Alshamrani, S.A., Alharthi, A.A., Alsalmi, O., Nahari, M.H., *et al.*, Structure-Based *In Silico* Approaches Reveal IRESSA as a Multitargeted Breast Cancer Regulatory, Signalling, and Receptor Protein Inhibitor. *Pharmaceuticals*, 17, 2, 208, 2024.
17. Pavlopoulos, G.A. and Wegener, A.-L., Schneider R. A survey of visualization tools for biological network analysis. *Biodata Min.*, 1, 1–11, 2008.
18. Bader, G.D., Biological Networks and Pathways, in: *Bioinformatics*, p. 399, 2020.
19. Tosadori, G., Biological network analysis: from topological indexes to biological applications towards personalised medicine, 2017.
20. Zhou, Y., Zhang, Y., Lian, X., Li, F., Wang, C., Zhu, F., *et al.*, Therapeutic target database update 2022: facilitating drug discovery with enriched comparative data of targeted agents. *Nucleic Acids Res.*, 50, D1, D1398–D407, 2022.
21. González-Medina, M., Prieto-Martínez, F.D., Naveja, J.J., Méndez-Lucio, O., El-Elimat, T., Pearce, C.J., *et al.*, Chemoinformatic expedition of the chemical space of fungal products. *Future Med. Chem.*, 8, 12, 1399–412, 2016.
22. Batool, M., Ahmad, B., Choi, S., A structure-based drug discovery paradigm. *Int. J. Mol. Sci.*, 20, 11, 2783, 2019.
23. Paul, D., Sanap, G., Shenoy, S., Kalyane, D., Kalia, K., Tekade, R.K., Artificial intelligence in drug discovery and development. *Drug Discov. Today*, 26, 1, 80, 2021.
24. Tăuţan, A.-M., Ionescu, B., Santarnecchi, E., Artificial intelligence in neurodegenerative diseases: A review of available tools with a focus on machine learning techniques. *Artif. Intell. Med.*, 117, 102081, 2021.
25. Bakkar, N., Kovalik, T., Lorenzini, I., Spangler, S., Lacoste, A., Sponaugle, K., *et al.*, Artificial intelligence in neurodegenerative disease research: use of IBM Watson to identify additional RNA-binding proteins altered in amyotrophic lateral sclerosis. *Acta Neuropathol.*, 135, 227–47, 2018.

26. Myszczynska, M.A., Ojamies, P.N., Lacoste, A.M., Neil, D., Saffari, A., Mead, R., *et al.*, Applications of machine learning to diagnosis and treatment of neurodegenerative diseases. *Nat. Rev. Neurol.*, 16, 8, 440–56, 2020.
27. Izadpanah, M., Seddigh, A., Ebrahimi Barough, S., Fazeli, S.A.S., Ai, J., Potential of extracellular vesicles in neurodegenerative diseases: diagnostic and therapeutic indications. *J. Mol. Neurosci.*, 66, 172–9, 2018.
28. Loiotile, A.D., Dentamaro, V., Giglio, P., Impedovo, D. (Eds.), AI-Based Clinical Decision Support Tool on Mobile Devices for Neurodegenerative Diseases. *IFIP Conference on Human-Computer Interaction*, Springer, 2021.
29. Cascianelli, S., Scialpi, M., Amici, S., Forini, N., Minestrini, M., Luca Fravolini, M., *et al.*, Role of artificial intelligence techniques (automatic classifiers) in molecular imaging modalities in neurodegenerative diseases. *Curr. Alzheimer Res.*, 14, 2, 198–207, 2017.
30. Mishra, S. and Bhargavi, K. (Eds.), An AI Model for Neurodegenerative Diseases. *2021 International Conference on Computer Communication and Informatics (ICCCI)*, IEEE, 2021.
31. Jeong, S.Y., Suh, C.H., Park, H.Y., Heo, H., Shim, W.H., Kim, S.J., Brain MRI-based artificial intelligence software in patients with neurodegenerative diseases: current status. *Taehan Yongsang Uihakhoe Chi*, 83, 3, 473–85, 2022.
32. Singh, G. and Ramanathan, M., Repurposing artificial intelligence tools for disease modeling: case study of face recognition deficits in neurodegenerative diseases. *Clin. Pharmacol. Ther.*, 114, 4, 862–73, 2023.
33. Nichol, B.A., Hurlbert, A.C., Read, J.C., Predicting attitudes towards screening for neurodegenerative diseases using OCT and artificial intelligence: Findings from a literature review. *J. Public Health Res.*, 11, 4, 22799036221127627, 2022.
34. Olaniyan, O.T., Adetunji, C.O., Dare, A., Adeyomoye, O., Adeniyi, M.J., Enoch, A., Prevention and diagnosis of neurodegenerative diseases using machine learning models, in: *Artificial Intelligence for Neurological Disorders*, pp. 289–304, Academic Press, 2023 Jan 1.
35. Schaller, D., Šribar, D., Noonan, T., Deng, L., Nguyen, T.N., Pach, S., *et al.*, Next generation 3D pharmacophore modeling. *Wiley Interdiscip. Rev.: Comput. Mol. Sci.*, 10, 4, e1468, 2020.
36. Patil, A.D., Biousse, V., Newman, N.J., Artificial intelligence in ophthalmology: an insight into neurodegenerative disease. *Curr. Opin. Ophthalmol.*, 33, 5, 432–9, 2022.
37. Warszycki, D., Struski, Ł., Smieja, M., Kafel, R., Kurczab, R., Pharmacoprint: a combination of a pharmacophore fingerprint and artificial intelligence as a tool for Computer-aided drug design. *J. Chem. Inf. Model.*, 61, 10, 5054–65, 2021.
38. Langer, T. and Hoffmann, R.D., Pharmacophore modelling: applications in drug discovery. *Expert Opin. Drug Discov.*, 1, 3, 261–7, 2006.

39. Tyagi, R., Singh, A., Chaudhary, K.K., Yadav, M.K., Pharmacophore modeling and its applications, in: *Bioinformatics*, pp. 269–289, Academic Press, 2022 Jan 1.
40. Simeon, S. and Jongkon N., Construction of quantitative structure activity relationship (QSAR) Models to predict potency of structurally diversed janus kinase 2 inhibitors. *Molecules*, 24, 23, 4393, 2019 Dec 1.
41. Cortés-Ciriano, I., Škuta, C., Bender, A., Svozil, D., QSAR-derived affinity fingerprints (part 2): Modeling performance for potency prediction. *J. Cheminf.*, 12, 1, 41, 2020 Jun 5.
42. Muratov, E.N., Bajorath, J., Sheridan, R.P., Tetko, I.V., Filimonov, D., Poroikov, V., *et al.*, QSAR without borders. *Chem. Soc. Rev.*, 49, 11, 3525–64, 2020.
43. Li, J., Wang, C., Yue, L., Chen, F., Cao, X., Wang, Z., Nano-QSAR modeling for predicting the cytotoxicity of metallic and metal oxide nanoparticles: A review. *Ecotoxicol. Environ. Saf.*, 243, 113955, 2022.
44. Correa, M.F. and Dos Santos Fernandes, J.P., QSAR Modeling of Histamine H3R Antagonists/inverse Agonists as Future Drugs for Neurodegenerative Diseases. *Curr. Neuropharmacol.*, 16, 6, 749–57, 2018.
45. Louwette, S., Van Geet, C., Freson, K., Regulators of G protein signaling: role in hematopoiesis, megakaryopoiesis and platelet function. *J. Thromb. Haemost.*, 10, 11, 2215–22, 2012.
46. Gabella, C., Duvaud, S., Durinx, C., Managing the life cycle of a portfolio of open data resources at the SIB Swiss Institute of Bioinformatics. *Briefings Bioinf.*, 23, 1, bbab478, 2022 Jan.
47. Ferreira, L.G., Dos Santos, R.N., Oliva, G., Andricopulo, A.D., Molecular docking and structure-based drug design strategies. *Molecules*, 20, 7, 13384–421, 2015.
48. Haddad, Y., Adam, V., Heger, Z., Rotamer Dynamics: Analysis of Rotamers in Molecular Dynamics Simulations of Proteins. *Biophys. J.*, 116, 11, 2062–72, 2019.
49. Yasmin, F., Shah, S.M.I., Naeem, A., Shujauddin, S.M., Jabeen, A., Kazmi, S., *et al.*, Artificial intelligence in the diagnosis and detection of heart failure: the past, present, and future. *Rev. Cardiovasc. Med.*, 22, 4, 1095–113, 2021.
50. Carracedo-Reboredo, P., Liñares-Blanco, J., Rodríguez-Fernández, N., Cedrón, F., Novoa, F.J., Carballal, A., *et al.*, A review on machine learning approaches and trends in drug discovery. *Comput. Struct. Biotechnol. J.*, 19, 4538–58, 2021.
51. Kolluri, S., Lin, J., Liu, R., Zhang, Y., Zhang, W., Machine Learning and Artificial Intelligence in Pharmaceutical Research and Development: a Review. *AAPS J.*, 24, 1, 19, 2022.
52. Kim, H., Kim, E., Lee, I., Bae, B., Park, M., Nam, H., Artificial Intelligence in Drug Discovery: A Comprehensive Review of Data-driven and Machine Learning Approaches. *Biotechnol. Bioprocess Eng.*, 25, 6, 895–930, 2020.

53. Ching. T., Himmelstein, D.S., Beaulieu-Jones, B.K., Kalinin, A.A., Do, B.T., Way, G.P., Ferrero, E., Agapow, P.M., Zietz, M., Hoffman, M.M., Xie, W., Opportunities and obstacles for deep learning in biology and medicine. *J. R. Soc. Interface*, 15, 141, 20170387, 2018 Apr 30.
54. Shin, W.H., Zhu, X., Bures, M.G., Kihara, D., Three-dimensional compound comparison methods and their application in drug discovery. *Molecules*, 20, 7, 12841–62, 2015.
55. Lavecchia, A., Machine-learning approaches in drug discovery: methods and applications. *Drug Discov. Today*, 20, 3, 318–31, 2015.
56. Gupta, R., Srivastava, D., Sahu, M., Tiwari, S., Ambasta, R.K., Kumar, P., Artificial intelligence to deep learning: machine intelligence approach for drug discovery. *Mol. Divers.*, 25, 3, 1315–60, 2021.
57. Serra, A., Fratello, M., Cattelani, L., Liampa, I., Melagraki, G., Kohonen, P., Nymark, P., Federico, A., Kinaret, P.A., Jagiello, K., Ha, M.K., Transcriptomics in toxicogenomics, part III: Data modelling for risk assessment. *Nanomaterials*,10, 4, 708, 2020 Apr 8.
58. Abd-Alrazaq, A., AlSaad, R., Harfouche, M., Aziz, S., Ahmed, A., Damseh, R., *et al.*, Wearable Artificial Intelligence for Detecting Anxiety: Systematic Review and Meta-Analysis. *J. Med. Internet Res.*, 25, e48754, 2023.
59. Dominguez, L.J., Veronese, N., Vernuccio, L., Catanese, G., Inzerillo, F., Salemi, G., Barbagallo, M., Nutrition, physical activity, and other lifestyle factors in the prevention of cognitive decline and dementia. *Nutrients*, 13, 11, 4080, 2021 Nov 15.
60. Singh, A., Velagala, V.R., Kumar, T., Dutta, R.R., Sontakke, T., The Application of Deep Learning to Electroencephalograms, Magnetic Resonance Imaging, and Implants for the Detection of Epileptic Seizures: A Narrative Review. *Cureus*, 15, 7, e42460, 2023.
61. Jiang, F., Jiang, Y., Zhi, H., Dong, Y., Li, H., Ma, S., *et al.*, Artificial intelligence in healthcare: past, present and future. *Stroke Vasc. Neurol.*, 2, 4, 230–43, 2017.
62. Silberstein, S.D., Holland, S., Freitag, F., Dodick, D.W., Argoff, C., Ashman, E., Evidence-based guideline update: pharmacologic treatment for episodic migraine prevention in adults: report of the Quality Standards Subcommittee of the American Academy of Neurology and the American Headache Society. *Neurology*, 78, 17, 1337–45, 2012.
63. Johnson, K.B., Wei, W.Q., Weeraratne, D., Frisse, M.E., Misulis, K., Rhee, K., *et al.*, Precision Medicine, AI, and the Future of Personalized Health Care. *Clin. Transl. Sci.*, 14, 1, 86–93, 2021.
64. Arvanitakis, Z., Shah, R.C., Bennett, D.A., Diagnosis and Management of Dementia: Review. *Jama*, 322, 16, 1589–99, 2019.
65. Breijyeh, Z. and Karaman R., Comprehensive review on Alzheimer's disease: causes and treatment. *Molecules*, 25, 24, 5789, 2020 Dec 8.
66. Lei, P., Ayton, S., Bush, A.I., The essential elements of Alzheimer's disease. *J. Biol. Chem.*, 296, 100105, 2021.

67. Long, J.M. and Holtzman, D.M., Alzheimer Disease: An Update on Pathobiology and Treatment Strategies. *Cell*, 179, 2, 312–39, 2019.
68. Marucci, G., Buccioni, M., Ben, D.D., Lambertucci, C., Volpini, R., Amenta, F., Efficacy of acetylcholinesterase inhibitors in Alzheimer's disease. *Neuropharmacology*, 190, 108352, 2021.
69. Yang, Z.J., Shao, Q., Jiang, Y., Jurich, C., Ran, X., Juarez, R.J., *et al.*, Mutexa: A Computational Ecosystem for Intelligent Protein Engineering. *J. Chem. Theory Comput.*, 19, 21, 7459–77, 2023.
70. Heneka, M.T., Carson, M.J., El Khoury, J., Landreth, G.E., Brosseron, F., Feinstein, D.L., *et al.*, Neuroinflammation in Alzheimer's disease. *Lancet Neurol.*, 14, 4, 388–405, 2015.
71. Lauretti, E., Dincer, O., Praticò, D., Glycogen synthase kinase-3 signaling in Alzheimer's disease. *Biochim. Biophys. Acta, Mol. Cell. Res.*, 1867, 5, 118664, 2020.
72. Wang, M., Song, W.M., Ming, C., Wang, Q., Zhou, X., Xu, P., *et al.*, Guidelines for bioinformatics of single-cell sequencing data analysis in Alzheimer's disease: review, recommendation, implementation and application. *Mol. Neurodegener.*, 17, 1, 17, 2022.
73. Sarica, A., Cerasa, A., Quattrone, A., Random Forest Algorithm for the Classification of Neuroimaging Data in Alzheimer's Disease: A Systematic Review. *Front. Aging Neurosci.*, 9, 329, 2017.
74. Arega, B., Tilahun, K., Minda, A., Agunie, A., Mengistu, G., Prevalence rate of undiagnosed tuberculosis in the community in Ethiopia from 2001 to 2014: systematic review and meta-analysis. *Arch. Public Health*, 77, 33, 2019.
75. Van Calster, B., Wynants, L., Verbeek, J.F.M., Verbakel, J.Y., Christodoulou, E., Vickers, A.J., *et al.*, Reporting and Interpreting Decision Curve Analysis: A Guide for Investigators. *Eur. Urol.*, 74, 6, 796–804, 2018.
76. Huang, S., Cai, N., Pacheco, P.P., Narrandes, S., Wang, Y., Xu, W., Applications of Support Vector Machine (SVM) Learning in Cancer Genomics. *Cancer Genomics Proteomics*, 15, 1, 41–51, 2018.

13

Computational Attitudes to Drug Discovery in Autism Spectrum Disorder

Himani Nautiyal[1], Shubham Dwivedi[1]*, Silpi Chanda[2] and Raj Kumar Tiwari[3]

[1]*Department of Pharmaceutical Sciences, School of Health Sciences and Technology, UPES, Dehradun, India*
[2]*Department of Pharmacognosy, Amity Institute of Pharmacy, Lucknow, Amity University, Uttar Pradesh, Noida, India*
[3]*Department of Pharmacognosy, Era College of Pharmacy, Era University, Tandon Marg, Sarfaraz Ganj, Lucknow, Uttar Pradesh, India*

Abstract

Autism spectrum disorder (ASD) is a neurodevelopmental ailment characterized by deficits in social interaction and communication along with repetitive behaviors and sensory hypersensitivities. The deficits lead to lifelong infirmities in real-world functioning, thus requiring consistent management of the disease. There is no cure available to treat autism; however, the drugs available possess only symptomatic rescue. Moreover, development of ASD treatment has been difficult due to multifactorial onset, variety of symptoms, limited knowledge on pathophysiology, lack of preclinical models mimicking autism, and unavailability of clinical reports. The accelerating ASD preponderance and insufficient therapies have urged for exploring and redefining the therapeutics against autism. Bioinformatics involving computational, statistical, and mathematical tools has emerged as one of the fancy approaches in exploring the pathophysiology and drug discovery of multifactorial disorders like autism. The approach can retrieve, archive, and analyze biological data along with providing new research directions in preclinical and clinical setup. Even though bioinformatics is in its infancy, it is one of the most rapidly expanding disciplines and an integral part of all biological research, including genomic/proteomics. Prediction and exploration of system biology using computational approaches might help in understanding disease pathophysiology and identifying the targets for drug discovery. This chapter investigates the process involved in drug discovery for ASD using bioinformatic tools. This chapter will give detailed

**Corresponding author*: shubham.dwivedi@ddn.upes.ac.in

Bhupendra Prajapati, Alok Tripathi, Rishabha Malviya and Lucy Mohapatra (eds.) Computational Neuropharmacology: Fundamentals and Clinical Aspects, (375–408) © 2025 Scrivener Publishing LLC

information on data extraction, curation, and analysis from genomics/proteomics/microarray along with high throughput screening in autism.

***Keywords*:** Bioinformatics, autism spectrum disorder, drug discovery, treatment approach

List of Abbreviations

ASD	Autism spectrum disorder
ADOS	Autism Diagnostic Observation Schedule
ANN	Artificial Neural Network
CAMP	Children's Autism Metabolome Project
CNV	Copy number variations
EHRs	electronic health records
ESTs	Expressed Sequence Tags
FDA	The Food and Drug Administration
GWAS	Genome-wide Association Studies
HMM	Hidden Markov Model
INSD	International Nucleotide Sequence Database
KEGG	Kyoto Encyclopedia of Genes and Genomes
LOF	Loss of function
NDDs	neurodevelopmental disorders
ORFs	open reading frames
PDTD	Potential Drug Target Databases
SNPs	Single Nucleotide Polymorphisms
SWISSPROT	Swiss Protein Databank
UniProt	Universal protein sequence database
SRS	Social Responsiveness Scale
SSRIs	Selective serotonin reuptake inhibitors
TrEMBL	Translation of DNA sequences in EMBL
WwPDB	worldwide Protein Data Bank

13.1 Introduction

Bioinformatics is a multidisciplinary field that has emerged through the integration of several disciplines, including biology, mathematics, computer science, and statistics. Its primary objective is to build methodologies for biological data storage, retrieval, and analysis. In 1970, P. Hogeweg, a prominent system biologist, introduced a term "bioinformatics" to describe the application of information technology in the study of biological systems [1]. The server "SWISS-MODEL" was introduced approximately 18 years ago, and was significantly responsible for the growth of bioinformatics due to the interactive automated modeling system [2].

Computational tools are commonly employed in the analysis of genes, the assessment of structural and physiochemical attributes of proteins, the conduct of phylogenetic investigations, and the execution of screening simulations aimed at examining the interactions of biomolecules within a living cell. While *in silico* analyses may not produce information as reliable as experimentation, bioinformatics still assists in making informed decisions on necessity of conducting expensive experiments. For success of any drug into clinics the drug requires to possess specific ADMET properties [3]. Various bioinformatics approaches are available to forecast ADMET properties in compounds to mitigate such defeat, thus enabling scientists for efficient evaluation of a vast array of potential candidates, thereby facilitating the identification of the most promising drug candidates.

The field of psychiatry has experienced a significant acceleration in research activity as a result of recent advancements in multi-omics studies. These investigations have covered a wide range of fields, such as metabolomics, genomics, transcriptomics, proteomics, and epigenomics. The application of multi-omics methods enables the production of a thorough and inclusive analysis of psychiatric disorders, a task that was previously considered implausible. Even though each approach is independent in nature, a multi-omics study that integrate these methodologies may shed light on complicated aetiologies of psychiatric diseases. Prior research has utilized an integrative systems biology methodology to examine ASD. In the domain of psychiatric genomics, there is a common practice of giving priority to the detection of specific genetic variations found exclusively in patients, rather than in healthy individuals, across the entire genome. The Genome-wide Association Studies i.e., GWAS analyze several human genome samples to discover unique Single Nucleotide Polymorphisms (SNPs) and study their association with a disease. The concurrent existence of a wide range of genetic and phenotypic traits in people with ASD makes it difficult to presents multiple biological theories to understand the complex manifestations of ASD. Recent research has identified many potential chromosome's locations and genes including copy number variations (CNV), in psychiatric genetics. These discoveries have facilitated the examination of specific genes and the implementation of comprehensive methodologies in comprehending psychiatric disorders.

In this Chapter, we will cover the bioinformatics approaches and tools which are utilized in the process of genetic analysis to drug discovery of autism spectrum disorder.

13.1.1 Genetic Identification and Analysis of Sequences

Biomolecular characterization of nucleic acids and proteins can be processed using sequence analysis through the examination and interpretation of their functional characteristics. At first, publicly available databases are used to obtain the sequences of the corresponding molecules [4]. Following the process of refinement, if necessary, the specimens undergo analysis using a range of tools facilitating the prediction of their functional, structural, and evolutionary characteristics, also homologous counterparts are identified with a high degree of precision. Tools like Entrez or PubMed enable users to search and retrieve data from various domains [5]. There exists an additional collection of tools that are specifically designed for the purpose of conducting sequence comparison. The genes or protein sequence analysis and their evolution are identified using tools BLAST [6] and ClustalW [7]. Researchers can utilize data visualization tools, including Jalview [8] Genedoc, [9] TreeView [10], and Genes-Graphs [11] to visually represent data (Tables 13.1 and 13.2). The aforementioned tools employ sophisticated mathematic model and inference techniques which statistically analyze sequences using Regression analysis, Clustering dynamic programming, Sequence Mining, Hidden Markov Model (HMM), and Artificial Neural Network (ANN) [12].

Various analytical tools are available for the identification of promoter, terminator, or untranslated locations implicated in the regulation of gene expression, the recognition of open reading frames (ORFs), introns, exons, and transit peptides. Thus, suggesting sequence analysis in the field of bioinformatics.

The majority of predictions are dependent on expressed sequence tags (ESTs) and complementary DNA (cDNA) utilization. Although, cDNA/ESTs data availabilities are frequently limited, thereby posing significant challenges in the identification of novel genes.

13.1.2 Phylogenetic Analysis

Phylogenetic analysis encompasses methodologies employed to restructure the evolution-based relationships within a cluster of interconnected molecules and organisms. These analyses serve the purpose of predicting specific characteristics of a molecule whose functions are yet to be determined, tracing the movement of genes, and ascertaining genetic-relatedness. This information has the potential to be visually represented through the utilization of a genealogical tree diagram. The fundamental principle of phylogeny involves the classification of living organisms based

Table 13.1 Databases of primary sequence.

S. n.	Tool	Description	References
1	BLAST	DNA/Protein sequence search	[13]
2	HMMER	Search tool for homologous protein sequences	[14]
3	Clustal Omega	Search tool for multiple sequence alignments	[15]
4	Sequerome	Sequence profiling tool	[16]
5	ProtParam	Prediction tools for Physical and chemical properties of proteins	[17]
6	JIGSAW	Prediction tool for splicing sites and to identify genes,	[18]
7	novoSNP	Identification tool for single nucleotide variance in the DNA sequence.	[19]
8	ORF Finder	Identify Open Reading Frame (ORF)	[20]
9	PPP	Prediction tool for Prokaryotic promoter	[21]
10	Virtual Foorprint	Whole prokaryotic genome analysis tool with promoter regions with varied regulator pattern	[22]
11	WebGeSTer	Predicts gene transcription end location	[23]
12	Genscan	Tool for genomic sequences exon/ intron sites	[24]
13	Softberry Tools	Annotation tool for animal, plant, and bacterial genomes	[25]

Table 13.2 Tool for phylogenetic analysis.

S. n.	Phylogenetic analysis tools	Description	References
1	MEGA (Molecular Evolutionary Genetic analysis)	Phylogenetic tress generator for study evolutionary closeness	[27]
2	MOLPHY	Tool for Molecular phylogenetic analysis work on the basis of maximum likelihood	[28]
3	PAML	Maximum likelihood based phylogenetic analysis	[29]
4	PHYLIP	Phylogenetic analysis package	[30]
5	JStree	Open-source library for presenting and editing phylogenetic tress	[31]
6	TreeView	Phylogenetic trees viewer, with the provision of changing view	[32]
7	Jalview	Alignment editor used to refine the alignment	[33]

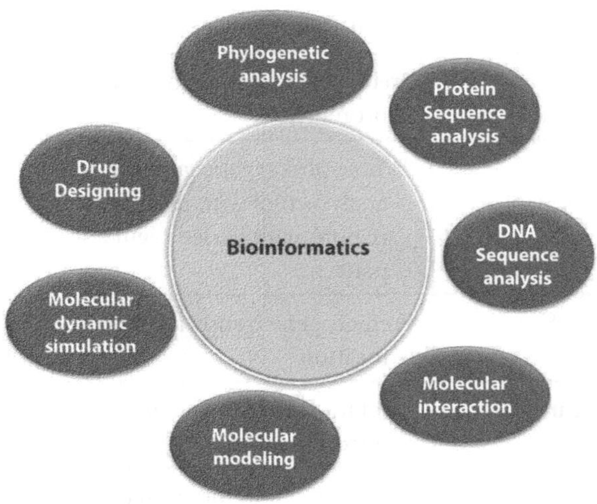

Figure 13.1 Application of bioinformatics on biological sciences.

on their level of similarity. Organisms that exhibit a higher degree of similarity are positioned in closer proximity on a phylogenetic tree. The utilization of phylogenetic comparative analysis is prevalent in order to address the issues related to statistics dependence among species [26] (Figure 13.1).

The construction of a phylogenetic tree using phylogenetic tools are frequently employed to evaluate diverse evolutionary hypotheses and have become essential in the field of functional genomics, particularly in cases where the functions of a gene remain unknown.

13.1.3 Nucleotide Sequence Databases

The primary sequence analysis is performed using the stored information, which also contributes to the future use. Depending on the information stored in, databases they are classified as primary, secondary, or composite databases. SWISS-PROT [34], UniProt [35], and GenBank [36], EMBL [37], DDBJ [38] and the Protein Databank PDB [39] are examples of primary databases. The Secondary databases include SCOP [40], CATH [41], PROSITE [42] eMOTIF [43]. Secondary databases are derived from primary databases conserved secondary motifs. Thus, primary databases are archived and secondary databases are modified.

A composite database contains primary source data. The International Nucleotide Sequence Database (INSD) is a composite database of nucleic acid sequences from EMBL, GenBank, and DDBJ (Table 13.3).

Table 13.3 Databases for nucleotide sequences.

S. n.	Tools	Description	References
1	DNA Data Bank of Japan	One of the biggest nucleotide sequences and member of International Nucleotide Sequence Databases (INSD)	[44]
2	European Nucleotide Archive	Captures and display nucleotide sequencing-based experimental workflows	[45]
3	GenBank	One of the largest International Nucleotide Sequence Databases (INSD)	[36]
4	Rfam	Sequence alignments of RNA families	[46]

13.1.4 Genome Sequence Databases

Genome sequences of nearly 250,000 species are stored in the NCBI GenBank [47]. Large scale genome sequencing programs and individual laboratories submit sequence data to GenBank The data from GenBank can be accessed through the NCBI's integrated retrieval system, Entrez access the GenBank data, and the PubMed searches the literature [5]. Each sequence contains information about organism, literature, bibliography, promoters, untranslated regions, terminators, introns, exons, repeat regions, and translations.

13.1.5 Protein Sequence Databases

The SWISSPROT (Swiss Protein) Databank , TrEMBL (translation of DNA sequences in EMBL) [50], UniProt (Universal Protein Resource) [51], PIR (Protein Information Resource) (Table 13.4), and wwPDB (worldwide Protein DataBank) [52] are most widely used protein sequence databases. SWISS-PROT entries provide experimental and computational data linked to GenBank, EMBL, DDBJ, PDB, domain, posttranslational modification, and species-specific protein databases. SWISS-PROT focus on model organisms and humans, however, TrEMBL contains protein data from all organisms. PIR database provide information on protein sequence alignment, peptide mass calculation, and interactive text search for protein molecules [49]. One of the largest free protein sequence databases is UniProt which is a fusion of SWISS-PROT,PIR and TrEMBL [53].

The worldwide Protein Data Bank (wwPDB) stores every 3D protein structure for scientific research. The databank now contains over 83,000 experimentally generated structures. The PDB also constantly develops tools for the users to provide better access to the data stores every 3D protein structure for scientific research. The databank has around 83,000 experimental constructions. The PDB regularly creates tools for users to improve data access [52] (Table 13.5).

Table 13.4 Database of genome sequencing.

S. n.	Database	Description	Reference
1	Ensembl	Database for eukaryotes annotated genomes	[48]
2	PIR	Integrated public resource to support genomic and proteomic research	[49]

Table 13.5 Database for protein sequence.

S. n.	Tools	Description	References
1	Uniport	largest collection of protein sequences	[35]
2	Protein Data Bank	Provide information on experimentally discovered structure of nucleic acids, protein, other complex assemblies can be found in major databases for protein	[39, 54]
3	Prosite	Information on the active sites, conserved domains, and protein families	[55, 56]
4	SWISS PROT	Protein sequences that have been manually annotated are located in component of UniProt Knowledgebase	[57]
5	InterPro	Explains the families of proteins, conserved domains, and active site	[58]
6	Proteomics Identifications Database	Public source, containing supporting evidence for functional characterization and post-translation modification of proteins and peptides.	[59]

13.1.6 Predicting Protein Structure and Function

A requirement for a protein to operate in a biological system is that it should be active and fold into the proper structure. Gaining insight into a particular protein's function requires knowledge of its 3D structure. In order for a protein to operate in a biological system, it must first fold correctly into the desired structure [60]. The 3D structure of protein is determined using NMR, X-ray crystallography [61]. Therefore, it is lacking information about many proteins tertiary structures on a genome-scale level. As an alternative, several bioinformatics techniques can be used to estimate the 3D structure of a protein [46].

13.1.7 Molecular Interactions

Protein-protein interaction are crucial to cellular signalling, transportation, homeostasis, metabolism, and biochemical processes [62]. Bioinformatics can predict protein-protein interaction without X-ray crystallography or NMR spectroscopy, which are expensive and time-consuming. Because a complex cannot be represented by one structure, crystal structure coordinates often misrepresent interactions. Thus, the 3D structure of a molecules does not reveal all interactions. Thus, computational methods that reliably anticipate protein-protein interactions are essential. However, such research provides crucial information that helps scientists identify a pathway to manipulate to produce cell changes (Table 13.6).

13.1.8 Drug Designing

Various scientific domains, such as clinical sciences, chemistry, and pharmacology, have been widely involved in drug discovery before the advent bioinformatics. These traditional techniques were costly and slow, also the high commercial demand to identify new treatment quickly and safely, the bioinformatic techniques are gaining interest as an alternative drug design strategy (Table 13.7). To find new medicinal compounds, computer-aided drug design (CADD) was created [75], which includes drug target identification, validation, lead identification, and lead optimization [76].

13.1.9 Molecular Dynamic Simulations

The time-dependent molecular interaction results in biological activities, Thus making it important to understand these interactions with drug and target protein. Molecular Dynamics Simulations (MDS) is a tool utilized to investigate time-dependent behavior of a molecules with target site. In biological system, the MDS provide data on fluctuations, dynamic processes, such as ion transport, subtle and significant conformational changes in protein, nucleic acids and their complexes. MDS may help in preliminary prediction of molecules without *in vitro* or *in vivo* experiments and may also assist in XRD and NMR structural determination [84] (Table 13.8).

Table 13.6 Databases for molecular interactions.

S. n.	Tools	Description	References
1	SMART	Simple modular Architecture Retrieval tool	[63]
2	Autodock	Predicts protein-ligand interaction	[64]
3	HADDOCK	Modeling and interaction of bio-molecular complex protein-protein, Protein-DNA	[65]
4	BIND	Database that provides access to molecular interaction and biomolecules	[66]
5	MOE	Integrated set of modeling, visualization, and drug discovery tools	[67]
6	STRING	Database for predicted protein interaction	[68]
7	MIMO	Biological route comparison tool using graph matching	[69]
8	InAct	Open-source database system that offers tools for analyzing data on molecular interaction	[70]
9	Graemlin	Finding of conserved network topologies outside of protein complexes and metabolic pathways by multiple network alignment with its functional evolution model, which also permits generalization of the alignment scoring	[71]
10	PathBLAST	Analyses any organism's protein-protein interaction network and extracts all interaction pathways that match the query	[72]
11	CFinder	Enabled for detecting and visualizing dense overlapping nodes in networks and quantitatively describing evolution	[73]
12	MCODE	Network clustering, cluster discovery parameters, interactive cluster boundary, content exploration	[74]

Table 13.7 Drug target databases.

S. n.	Database	Description	References
1	Potential Drug Target Databases (PDTD)	A comprehensive drug target database with 1207 entries that is available online and accessible worldwide	[77]
2	Drug bank	Exclusive resource that interconnects drug- target data	[78]
3	Therapeutic target database	Assessment of diseases, disease-relate pathways, and drugs that are directed at each of the known and investigated therapeutic protein and DNA target	[79]
4	TDR target database	It is a tool as well as a database, involved in identification and the order of priority of important genes from neglected disease pathogens using straightforward queries	[80]
5	MATADOR	Manually annotated targets and drug online resources are implicated in protein-chemical interactions, and each interaction is connected to PubMed or OMIM entries that were used to determine the interaction	[81]
6	DrugPort	Drug molecules structural data is available on protein data bank	[82]
7	ChemMBL	Collection of drug-like bioactive molecules, along with their 2D structure.	[83]

Table 13.8 Tools for molecular dynamics simulation.

S. n.	Molecular dynamics simulation tool	Description	Reference
1	Ascalaph	Utilizing a molecular modeling tool, using quantum mechanical calculation for the development of models, molecular mechanics and dynamics simulation of hydrocarbons and DNA proteins in either the gaseous or condensed state	[85]
2	Discovery Studio	The features of small molecular simulations, pharmacophore modeling, protein-protein docking, and antibody modeling are all included in comprehensive modeling and simulation program aimed at improving the drug development process	[86]
3	Amber	A group of application that make it easier for users to run molecular dynamics simulations, with a focus on biomolecules	[86]

13.2 Clinical, Genetic, and Molecular Heterogeneity in Autism Spectrum Disorder

Autism Spectrum Disorder (ASD) exhibits heterogeneity across multiple levels, particularly when considering its clinical manifestations. The co-occurrence of ASD with different neuropsychiatric disorders, such as attention defect hyperactivity disorder (ADHD), anxiety, and depression. It has been reported that the autistic people have up to 70% comorbidity rate with other disorders and exert a significant influence on the overall well-being, disease burden, and the rate of death [87]. This co-occurrence leads to the additional neurological symptoms which range from migraine to congenital abnormalities, further corresponding to the heterogenous pattern of ASD [88].

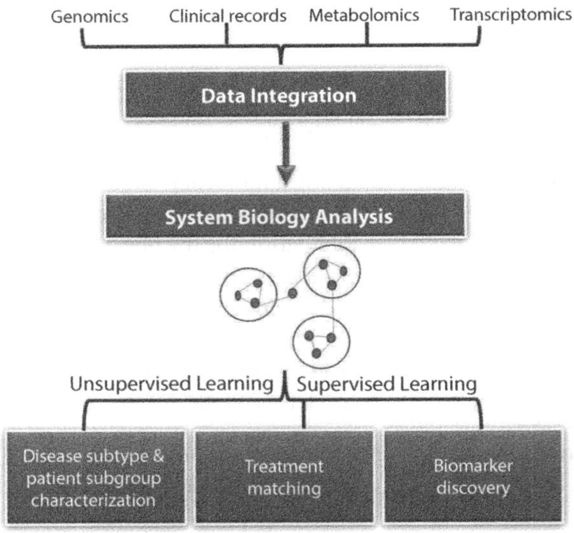

Figure 13.2 Process of drug discovery using computational and system biology approaches.

13.2.1 Genetics of Autism Spectrum Disorder

The scientific studies report the involvement of more than 1000 genes, and respective pathways in complex pathophysiology of ASD [89]. These genes are seldom affected in population producing specific monogenic symptoms associated to ASD. Additionally, it contains genes which are responsible for its polygenic forms, where the common genetic variants play a determining role in this condition [90].

The gene discoveries associated with ASD are mostly the loss of function (LOF) mutations that is highly constrained [91]. ASD-related genetic information is available on https://www.sfari.com. However, the contribution of individual genes and variants linked to an increased likelihood of developing ASD is often minimal, representing only a small fraction of the overall cases [92] (Figure 13.2).

13.2.2 Metabolomics

The research to understand pathophysiology of ASD has significantly enhanced the utilization of metabolomics, which involves the analysis and characterization of small-molecule metabolites [93]. The interplay between environmental and genetic factors in ASD is defined by metabolome which act as a valuable indicator due to its capacities of capturing the intricate interactions among the environment, diet, gut microbiomes, and genome elements [94]. Consequently, it offers valuable insights that can help in bridging the gap between genotype and phenotype (Figure 13.2).

The utilization of potential ASD diagnostic biomarkers, specifically metabolites, holds significant potential in facilitating the early detection of this disorder. This early identification can subsequently facilitate the prompt start of early interventions, which can ameliorate the behavioral symptoms observed in children diagnosed with ASD [95].

Recently, the focus has been shifted to examine the differential metabolites using various specimen, such as feces, urine, and blood in individuals with neurodevelopmental disorders (NDDs) or individuals diagnosed with autism spectrum disorder (ASD) and typically developed individuals (TD) [96].

13.2.3 Transcriptomics

In recent years, much attention has been devoted to conducting transcriptomics profiling on ASD patients, which aims to identify dysregulated genes in brain tissues and peripheral system. Recent research has been primarily concerned with comprehending the modifications in gene expression that contribute to the abnormal functioning of neuronal cells. Additionally, investigations have been conducted to assess changes in gene expression patterns in tissues including peripheral blood to diagnose ASD and developing the effective treatment strategies (Figure 13.2).

Several large cohort studies have been conducted to identify a shared characteristic among autistic patients by comparing them with control groups. Various dysregulated molecular pathways have been reported in ASD, such as pathways associated with development of central nervous system and its function, immune system pathways and pathways associated with ATP synthesis and oxidative phosphorylation [97–99]. Two recent investigations reported [100, 101], a significant shared signal between ASD and other neurological disorders, including schizophrenia and depression. This signal indicates a connection between alterations in transcriptomics and a shared genetic architecture within the disorders of CNS. The findings of this study demonstrate the capacity to identify aberrant gene expression patterns in autistic individuals, and also demonstrate the similarities in dysregulation of gene expression observed in other neurological and immune-related disorders.

13.3 The Necessity of Drug Discovery

The complex nature of ASD at the genetic, molecular and clinical levels presents a substantial obstacle in the realm of drug discovery. A complicated interplay of genetic and environmental factors is responsible for challenges in the drugs discoveries for autism spectrum disorder (ASD). These changes can result in dysregulation and imbalances of various neurotransmitters within nervous system, including 5HT, histamine, dopamine, Ach, GABA, and glutamate [102]. In dysregulated mesolimbic circuit and prefrontal cortex, dopamine serves as a crucial neurotransmitter exerts an influence on behavior, expression, and impaired emotional regulation. Additionally, dysregulation in many neurotransmitters may interfere with glutaminergic signalling leading to diminished expression of GABAergic genes, resulting in emotional imbalance.

The phenomenon of unpaired modulation observed in several brain circuits in ASD patients. While it is true that approved treatments can be effective symptomatically, interventions aimed at alleviating dysregulated pathways and comorbidities can be beneficial. Moreover, the issue of off-target effects and side effects is a matter of concern.

It is noteworthy that risperidone and aripiprazole are the sole pharmaceutical agents that have obtained official approval from regulatory authorities. The Food and Drug Administration (FDA) provides guidelines and regulations pertaining to the treatment of ASD patients. The prescribed medications are provided to alleviate irritation. Selective serotonin reuptake inhibitors (SSRIs), and antianxiety drugs are frequently recommended. However, it is important to note that none of these medications have obtained approval from the FDA in ASD treatment, However, they are used in clinics for the specific symptoms of ASD.

Recent research indicates that a limited proportion of patients will have a positive response to a specific treatment, and the probability of such a response occurring is not guaranteed [103]. Therefore, there is a urge of transition from a generalized strategy to a more individualized method in medicine, wherein the discernment of specific characteristics and factors of each patient is prioritized.

The identification of various biological factors underpinning ASD will play a crucial role in pinpointing molecular targets associated with the pathophysiology of distinct subgroups of patients. This knowledge has the potential to facilitate the development of targeted treatments that are based on a common biological foundation. The lack of success in clinical trials for autism spectrum disorder can be attributed to inadequate research design and outcome measures that are generated from a wide range of factors.

13.4 Computational Model for Drug Discovery

The utilization of computational approaches has the potential to enhance the characterization of autistic individuals. The patient's division into subgroups shows a potential approach to address the heterogeneity in autism spectrum disorder and to establish more homogeneous subtypes of the disease. The identification of patient subgroups with shared molecular characteristics is crucial for the advancement of focused and more efficacious therapeutic approaches. In the last ten years, several methodologies have been suggested, encompassing supervised learning techniques for training models to classify individuals with autism spectrum disorder (ASD) into specific categories, as well as unsupervised learning techniques for grouping individuals with ASD into more homogenous subgroups. This phenomenon has resulted in the birth of computational psychiatry, which seeks to enhance the process of diagnosis by utilizing computer analysis of extensive behavioral and biological datasets [104]. One of the primary constraints associated with these methodologies, is dependence of computer models on limited size of samples within datasets. The diagnosis of ASD poses difficulties when relying solely on behavioral and observational assessments. To address this, clinical diagnostic data for ASD is obtained through the utilization of standardized psychological assessment tools, including the Autism Diagnostic Interview-Revised, the Autism Diagnostic Observation Schedule (ADOS), Social Responsiveness Scale (SRS), and Childhood Autism Rating Scale (CARS). Children who are suspected of having autism spectrum disorder (ASD) are typically assessed using a developmental approach, which involves gathering information from many sources, such as parents and clinicians. The clinical evaluation techniques commonly employed for categorizing ASD in machine learning models have also been utilized in the analysis of neuroimaging data.

There is growing research that indicates the possible usefulness of neuroimaging in capturing brain defičts in autistic individuals. Difference in thickness of cortex, surface area, and volume of brain have been observed in patients of ASD with Copy Number Variants (CNV) in specific regions namely 1q44, 1q21.1, 16p11.2, 7q11.23 and 22q11.21 [55, 105]. Additionally, alterations in pattern connectivity have been identified through functional neuroimaging techniques. In a new investigation employing functional magnetic resonance imaging (fMRI) techniques, researchers examined a cohort of infants at the age of 6 months who possessed a heightened familial susceptibility to ASD. The study predicted diagnostic outcomes at the age of 2, with 100 % specificity and 81.8% sensitivity [106].

13.5 Importance of Multiomics and Endophenotyping-Based Methods Toward Precision Medicine

The efficacy of these approaches has been demonstrated in cancer patients, as they have proven to be effective in disease subtyping and facilitating the selection of optimal treatment options [107–109]. Currently, there is a significant emphasis on increasing the practical use of precision medicine through a multitude of projects [110]. These programs have a special objective of integrating data derived from omics, clinical information obtained from electronic health records (EHRs), and comprehensive patient phenotyping [111]. The utilization of endophenotypes demonstrates the capacity to integrate and analyze data in order to find consistent subgroups among patients, hence enhancing the accuracy of diagnosis and enabling tailored treatment approaches for individuals with neurodevelopmental disorders (NDDs).

A whole exome sequencing data from a cohort of more than 200,000 individuals was integrated with comprehensive phenotyping data. The analysis was conducted on the effects of a specific mutation on a dataset comprising over 3,000 annotated traits. This work has contributed a distinct and extensive comprehension to the scientific community regarding the associations between genes and phenotypes [112].

The ongoing endeavors to establish the association between phenotypes and genetic variations in patients have produced promising results within the realm of uncommon diseases [112]. These results have enabled prompt diagnoses and individualized therapy choices for patients. A thorough understanding of omics-based molecular phenotypes has promise in providing a repertoire of biomarkers suitable for the advancement of drug development and clinical trial approaches in the context of neurodegenerative diseases (NDDs). The field of omics has greatly contributed to the accumulation of comprehensive knowledge pertaining to diverse facets of biological systems. This includes understanding their functions, interactions between proteins and genes, expression of gene products, content of metabolites and lipids, and intricate feedback mechanisms that combine these molecules into pathways spanning across different periods and spatial dimensions. A thorough comprehension of the subject matter will enhance the ability to develop and evaluate molecular biomarkers that can be used to assess the effectiveness of different therapeutic strategies. The acquisition of this knowledge may also facilitate the advancement of customized medicine approaches, thereby augmenting the effectiveness of clinical trials conducted in the context of ASD.

The identification of metabolic biomarkers related to the risk of ASD is a key factor in achieving this objective. This particular methodology has the ability to identify therapeutic targets that may be beneficial for patients exhibiting a specific metabotype.

The Children's Autism Metabolome Project (CAMP) revealed that around 16.7% of the individuals exhibited an aberrant metabolic phenotype, which was defined by an imbalance in the metabolism of branch chain amino acids, with specificity and precision values 96.3% and 93.5%, respectively [113]. Additional investigation is necessary to ascertain the durability of these metabotypes in individual children [114]. The examination of metabolomics data has shown considerable promise in the early detection of ASD. Additional investigation in the domain has potential to significantly influence the management and outlook of individuals [115].

Recent studies have also investigated the transcriptome profile of ASD patients with a precision medicine methodology. The technique of dissecting ASD into subgroups and defining subtype-specific omics-based signatures has promise in terms of diagnosing the disorder and tailoring treatment to account for variations in phenotypes.

13.6 Network-Based Approach for Diseases/Drug Modeling

A key component of computational methodologies used to clarify mechanism of action and potential effects of drugs is swiftly emerging as the network approach. Through the utilization of computational models, one may elucidate the impact of pharmaceuticals and diseases on various biological networks, hence enhancing our understanding of the intricate relationship between disease perturbations and drug targets. Additionally, these models enable us to comprehend how drug molecules elicit desirable biological responses or give rise to unfavourable effects. The application of omics technology has been widely employed to acquire the requisite data for investigating the mechanisms underlying the effects of medications and disorders. The network data mining techniques frequently employed for the investigation of pharmacological mechanisms of action. These methodologies have the capacity to facilitate the foundational aspects of the developmental process of drugs, encompassing potential drug target discovery, drug compounds and drug combinations screening for therapeutic applications in various diseases.

13.6.1 Knowledge-Driven Networks

Knowledge-driven networks are formed in consolidated interaction data across various biological entities that are dispersed throughout the scientific literature. The curation process of knowledge-derived networks is a laborious and time-intensive process; nonetheless, it is notably resilient due to its reliance on manual examination of each interaction data inside published literature. In addition, it should be noted that knowledge-based networks may lack specificity in relation to specific biological situations, including diseased condition. Consequently, the effectiveness is less for the network in these specific relationships that undergo dynamic changes across different contexts. Furthermore, it has been noted that the quantity of experimentally confirmed interactions tends to be disproportionately focused on a small number of extensively researched genes or disease-related biological entities. There are already multiple databases accessible for the collection of interaction and/or association events, which can provide valuable insights into the impacts of medications and diseases. However, it is important to note that these databases typically pertain exclusively to a particular sort of contact. Two often utilized interaction databases in the field are Bio GRID [116] and String DB [68]. These databases are employed extensively to investigate the interactions between mode of action of compounds and the dysregulated molecular pathways in various disorders. Both databases have the objective to define the interaction, such as the STRING database incorporates interaction data acquired from five key sources, including automated text mining of scientific literature, knowledgebases containing primary protein-protein interaction (PPI), co-expression studies, high-throughput experiments, and genomic context prediction. Annotation databases, such as the Kyoto Encyclopedia of Genes and Genomes (KEGG) and REACTOME, can also be utilized to construct PPI networks [117].

Drug Bank is a highly utilized website that offers extensive information on authorized pharmaceuticals, as well as drugs that are still in the experimental and investigative stages [118]. In addition to the molecular details of each medicine, this resource also offers information regarding the relationships among pharmaceuticals, drug targets, and disorders. Drug Bank is recognized as a prominent repository of meticulously curated drug interactions, encompassing harmful and beneficial interactions. In addition to Drug Bank, the ADReCs and SIDER databases serve as valuable resources for comprehending drug toxicity mechanisms by compiling data on reported adverse drug reactions associated with commercially available medications [22, 119]. Additional evidence of genes and diseases connection can be obtained from other databases, such as DisGeNET [120], OpenTargets [121], and PharmGKB [122]. This information can be utilized to construct a network that links diseases with their corresponding genes.

13.6.2 Data-Driven Networks

High-throughput experimental data can serve as the foundation for constructing data-driven networks. These networks are designed to identify potential biomolecular interactions within a particular biological state or disease, and can even be used to create profiles of patient-specific networks. For instance, the utilization of microarray and RNA sequencing techniques has been widely employed in gene expression profiling. This approach has been valuable in delineating co-expression networks, which facilitate comprehension of alterations in interaction patterns resulting from medication or chemical intervention [123, 124], as well as those occurring between healthy and diseased states [125]. Nevertheless, a significant concern associated with data-driven networks is the presence of noise, which can potentially hinder their effectiveness. Additionally, the construction of reliable networks typically necessitates a substantial quantity of samples.

13.6.3 Network Construction by Combining Knowledge-Based and Data-Based Study

In addition to using high-throughput experimental data, efforts have been undertaken to build networks using the data in curated databases. In a study, data were utilized to construct PPI networks which were subsequently used to predict cancer cell line susceptibility [126]. This methodology primarily emphasizes the establishment of homogeneous networks. However, alternative strategies have been investigated to construct heterogeneous networks that connect diverse biological entities. These networks are constructed by incorporating both established interactions and interactions derived from high-throughput data.

13.7 Drug Repurposing Candidates for Treatment of ASD Using Bioinformatic Approaches

No drugs are approved to treat autistic core symptoms, and nonspecific symptom medications have substantial adverse effects due to heredity and clinical variability associated with autism. Repurposing medications to treat ASD core symptoms requires identifying candidates molecules affecting the key pathways in ASD. A bioinformatic study examined ASD risk genes in PPI network and identified the ASD network-based analysis. Loperamide, bromocriptine, drospirenone, and progesterone showed normalize gene expression perturbation in presence of drug versus ASD ($p<0.05$). These drugs act on ASD-related biological systems, suggesting these drugs could treat ASD core symptoms and making them potential drug repurposing candidates for clinical translation [127].

Loperamide, a commonly employed antidiarrheal drug, functions as an agonist for the μ-opioid receptor and exhibits no adverse effects on the central nervous system (CNS) when administered within recommended therapeutic dosages and for a limited duration [128]. However, the μ opioid receptors regulate both human and animal social behavior and rewarding effects [129]. It is noteworthy that mice lacking μ-opioid receptors exhibit behavioral and social impairments that resemble those observed in autistic humans. Furthermore, the administration of μ-opioid receptor agonists to rodents has been found to ameliorate deficiencies in social interaction [129]. Consequently, the potential of loperamide to affect social behavior in ASD patients represents a promising option for repurposing as a drug. Additionally, the gastrointestinal issues in ASD may also be improved in ASD [130].

Bromocriptine, a dopamine D2 receptor agonist, has demonstrated notable disparities in gene expression perturbations between drug usage and ASD. This medication is commonly prescribed to manage prolactin-related disorders [131]. Two initial investigations examining the therapeutic impact of bromocriptine in ASD individuals [132].

The dopaminergic hyperactivity hypothesis was proposed in ASD patients [132]. The presence of dopaminergic dysfunction in ASD patients is already extensively documented due to its manifestation within the midbrain dopaminergic system [133].

It is noteworthy that both pramipexole and piribedil, which are dopamine D2 receptor agonists, as well as fluspirilene, an antagonist of the dopamine D2 receptor, exhibited contrasting alterations in gene expression when comparing drug-induced effects observed with autistic individual at a significance level of $p < 0.1$. Furthermore, the dopamine D2 receptor gene interaction was reported with these medications. It is worth mentioning that the dopamine D2 receptor gene had not been previously characterized as a gene associated with ASD. A recent network based study revealed tight associations of D2 receptor with known ASD risk genes and therefore incorporated it into ASD network [127].

Antipsychotic medications, such as Aripiprazole, are frequently employed to mitigate non-core symptoms of ASD. However, it demonstrates both inconclusive divergent alterations in gene expression in drug treated and ASD control. For instance, the medication risperidone and several other medications exhibited contrasting expression of genes in drug vs. ASD, with significance level of $p < 0.1$ includes anticonvulsants drug, such as lamotrigine and topiramate, selective estrogen receptor modulator raloxifene, the anti-cancer therapy vemurafenib, and the anesthetic agent tetracaine. Previous studies have also demonstrated the potential benefits of utilizing anticonvulsant medications in the treatment of ASD [134].

The presence of positive correlations between medication-induced gene expression perturbations and ASD-associated gene expression within the genes present in ASD network suggests that the drug may potentially amplify ASD-related gene expression perturbations, thereby exacerbating ASD symptoms. a statistically significant level ($p < 0.05$) comprised fluoxetine and epinephrine. The set of interventions with p-values less than 0.1 consisted of sumatriptan and metformin. There has been a longstanding suggestion that epinephrine has a role in the development of ASD, and it is possible that levels of adrenaline and norepinephrine in the bloodstream are higher. In a recent article, efficacy of fluoxetine, a selective serotonin reuptake inhibitor (SSRI), in individuals with autism spectrum disorder (ASD) was evaluated and identified that fluoxetine has potential benefit in addressing repetitive behaviors [135].

However, a separate study focusing on autistic patients and examining impact of antidiabetic drug "metformin" on the memory did not demonstrate any advantageous effects. Although there were no significant differences in measures of memory between participants randomly assigned for metformin treatment or placebo, participants receiving metformin exhibited improvement in verbal learning compared to the placebo group [136].

Sumatriptan, a medication frequently employed for the management of migraine attacks, acts as an agonist on the 5HT1B/D receptors.

Previous research has examined the involvement of the 5HT1D receptor in ASD development and has found that autistic people exhibit increased 5HT1D receptor sensitivity. This heightened sensitivity may be associated with the intensity of repetitive behaviors observed in individuals with ASD [137] . Nevertheless, there is a lack of comprehensive research on the effect of sumatriptan on the CNS due to the prevailing belief that triptans are unable to penetrate the blood-brain barrier [138]. it is unclear if the medications that have a positive correlation with the disruption profile of gene expression associated with ASD, exacerbate autistic symptoms. The effectiveness of medicines that exhibit contrasting gene expression alterations with drug help in mitigating ASD remains uncertain. Given the current lack of knowledge regarding the efficacy of the shortlisted pharmaceuticals in addressing the fundamental symptoms of ASD, it is imperative to conduct clinical trials to assess their potential for repurposing as treatments for ASD. The gene expression patterns were derived from experimental data obtained from cancer cell lines. Nevertheless, there is evidence indicating the usefulness of such data in repurposing medications for diseases other than cancer, as demonstrated by the anticonvulsant medicine. Topiramate has been discovered as a possible candidate for repurposing in inflammatory bowel disease treatment. The *in vivo* validation studies were performed to support these findings [139].

13.8 Conclusion and Future Prospective

The application of bioinformatics in the field of ASD has opened up new avenues for understanding the complex pathophysiology of the disorder and identifying potential targets for drug discovery. The integration of computational and statistical tools has allowed for the efficient retrieval, archiving and analysis of biological data, providing valuable insights into the multifactorial nature of ASD.

The use of bioinformatics in genetic analysis, sequence databases and protein sequence databases has proven to be invaluable. These tools have facilitated the identification and analysis of sequences, the construction of phylogenetic trees, and the prediction of functional, structural, and evolutionary characteristics of molecules.

In conclusion, the advent of bioinformatics has revolutionized the approach toward drug discovery in ASD, making it a promising field for future research and development of effective therapeutic strategies. However, further advancements and refinements in these computational tools and techniques are necessary to fully harness their potential in the fight against ASD.

Acknowledgment

The authors are thankful to the School of Health Sciences and Technology, UPES for supporting the successful completion of this chapter.

References

1. Ouzounis, C.A. and Valencia, A., Early bioinformatics: the birth of a discipline–a personal view. (in eng), *Bioinformatics*, 19, 17, 2176–90, Nov 22 2003, doi: 10.1093/bioinformatics/btg309.
2. Peitsch, M.C., ProMod and Swiss-Model: Internet-based tools for automated comparative protein modelling. (in eng), *Biochem. Soc. Trans.*, 24, 1, 274–9, Feb 1996, doi: 10.1042/bst0240274.
3. Morgan, S., Grootendorst, P., Lexchin, J., Cunningham, C., Greyson, D., The cost of drug development: a systematic review. (in eng), *Health Policy*, 100, 1, 4–17, Apr 2011, doi: 10.1016/j.healthpol.2010.12.002.
4. Pattan, V., Kashyap, R., Bansal, V., Candula, N., Koritala, T., Surani, S., Genomics in medicine: A new era in medicine. (in eng), *World J. Methodol.*, 11, 5, 231–242, Sep 20 2021, doi: 10.5662/wjm.v11.i5.231.
5. Sayers, E.W., *et al.*, Database resources of the National Center for Biotechnology Information in 2023. *Nucleic Acids Res.*, 51, D1, D29–D38, 2022, doi: 10.1093/nar/gkac1032.
6. Johnson, M., Zaretskaya, I., Raytselis, Y., Merezhuk, Y., McGinnis, S., Madden, T.L., NCBI BLAST: a better web interface. *Nucleic Acids Res.*, 36, suppl_2, W5–W9, 2008, doi: 10.1093/nar/gkn201.
7. Chowdhury, B. and Garai, G., A review on multiple sequence alignment from the perspective of genetic algorithm. *Genomics*, 109, 5, 419–431, 2017, 2017/10/01/, doi: https://doi.org/10.1016/j.ygeno.2017.06.007.
8. Waterhouse, A.M., Procter, J.B., Martin, D.M.A., Clamp, M., Barton, G.J., Jalview Version 2—a multiple sequence alignment editor and analysis workbench. *Bioinformatics*, 25, 9, 1189–1191, 2009, doi: 10.1093/bioinformatics/btp033.
9. Nicholas, K.B., Nicholas, H.B., Deerfield, D.W., GeneDoc: Analysis and visualization of genetic variation, 1997.
10. Cheng, K.Y., Lau, K., Shek, Y.T., Liu, Z., Ng, E., Evaluation on the performance of tree view factor in a high-density subtropical city: A case study in Hong Kong. *Build. Environ.*, 239, 110431, 2023, 2023/07/01/, doi: https://doi.org/10.1016/j.buildenv.2023.110431.
11. Paten, B., Novak, A.M., Eizenga, J.M., Garrison, E., Genome graphs and the evolution of genome inference. (in eng), *Genome Res.*, 27, 5, 665–676, May 2017, doi: 10.1101/gr.214155.116.
12. Gujral, H., Kushwaha, A.K., Khurana, S., Utilization of Time Series Tools in Life-sciences and Neuroscience. *Neurosci. Insights*, 15, 2633105520963045, 2020, 2020/01/01, doi: 10.1177/2633105520963045.
13. Reiter, T., Clustering the NCBI nr database to reduce database size and enable faster BLAST searches, 2023.
14. Larralde, M. and Zeller, G., PyHMMER: a Python library binding to HMMER for efficient sequence analysis. *Bioinformatics*, 39, 5, btad214, 2023.

15. Bodri, M. and Webb, S.A., Using the embl-ebi clustal omega tool to calculate diversity of heavy chain phage-display libraries. *Georgia J. Sci.*, 80, 2, 12, 2022.
16. Bairagee, D., Singh, N., Jain, N., Sharma, U., Bioinformatics Tools and Software in Clinical Research, in: *Bioinformatics Tools and Big Data Analytics for Patient Care*, pp. 17–41, Chapman and Hall/CRC, New York, 2022.
17. Ahmad, H.I., *et al.*, *In silico* structural, functional, and phylogenetic analysis of cytochrome (CYPD) protein family. *Biomed Res. Int.*, 2021, 1–13, 2021.
18. Kang, D., Jiang, R., Blanas, S., Jigsaw: A data storage and query processing engine for irregular table partitioning, in: *Proceedings of the 2021 International Conference on Management of Data*, pp. 898–911, 2021.
19. Mousavi, M., *et al.*, *De novo* SNP discovery and genetic linkage mapping in poplar using restriction site associated DNA and whole-genome sequencing technologies. *BMC Genomics*, 17, 1–12, 2016.
20. Arshad, I., *et al.*, Bioinformatics approaches in upgrading microalgal oil for advanced biofuel production through hybrid ORF protein construction. *Biomass Convers. Biorefin.*, 14, 1–21, 2023.
21. Zhang, Y.-C., Luo, W.-Z., Shan, M., Pan, D.-W., Mu, W.-J., Systematic analysis of PPP research in construction journals: from 2009 to 2019. *Eng. Constr. Archit. Manage.*, 27, 10, 3309–3339, 2020.
22. Huang, L.H., *et al.*, ADReCS-Target: target profiles for aiding drug safety research and application. (in eng), *Nucleic Acids Res.*, 46, D1, D911–d917, Jan 4, 2018, doi: 10.1093/nar/gkx899.
23. Gupta, S., INTERPIN: identifying INtrinsic transcription TERminators, hairPINs in bacteria, 2023.
24. Keshavarzi, M., Afolabi, O., Caldas, L., Yang, A.Y., Zakhor, A., Genscan: A generative method for populating parametric 3d scan datasets, arXiv preprint arXiv:2012.03998, 2020.
25. Xu, J.-C., *et al.*, ALDH2 genes are negatively correlated with natural deastringency in Chinese PCNA persimmon (Diospyros kaki Thunb.). *Tree Genet. Genomes*, 13, 1–9, 2017.
26. Freckleton, R.P., Harvey, P.H., Pagel, M., Phylogenetic analysis and comparative data: a test and review of evidence. (in eng), *Am. Nat.*, 160, 6, 712–26, Dec 2002, doi: 10.1086/343873.
27. Tamura, K., Peterson, D., Peterson, N., Stecher, G., Nei, M., Kumar, S., MEGA5: molecular evolutionary genetics analysis using maximum likelihood, evolutionary distance, and maximum parsimony methods. (in eng), *Mol. Biol. Evol.*, 28, 10, 2731–9, Oct 2011, doi: 10.1093/molbev/msr121.
28. Adachi, J. and Hasegawa, M., *MOLPHY version 2.3: programs for molecular phylogenetics based on maximum likelihood*, Institute of Statistical Mathematics, Tokyo, Japan 1996.
29. Cornetta, J., Molecular Evolution of Human Menopause-Associated Genes, Adelphi University, 2020.

30. Dillon-Martin, M., Studying Evolutionary Patterns in Cancer Risk Genes with Computational Tools to Create Sequence Alignments, 2021.
31. Fan, Z., A novel web portal for large biological ontology, University of Delaware, 2020.
32. Endicott, J., *et al.*, TreeView: A small satellite supporting precision forestry for nature-based solutions in a changing climate, 2022.
33. Menard, A., *et al.*, Alignment of Biological Sequences with Jalview. *Methods Mol. Biol.*, 2231, 203–224, 2021.
34. Saar, K.L., *et al.*, Machine learning models for predicting protein condensate formation from sequence determinants and embeddings, bioRxiv, p. 2020.10. 26.354753, 2020.
35. Soudy, M., *et al.*, UniprotR: Retrieving and visualizing protein sequence and functional information from Universal Protein Resource (UniProt knowledgebase). *J. Proteomics*, 213, 103613, 2020.
36. Sayers, E.W., *et al.*, GenBank. *Nucleic Acids Res.*, 50, D1, D161–D164, 2022.
37. Morales, J., *et al.*, A joint NCBI and EMBL-EBI transcript set for clinical genomics and research. *Nature*, 604, 7905, 310–315, 2022.
38. Okido, T., Kodama, Y., Mashima, J., Kosuge, T., Fujisawa, T., Ogasawara, O., DNA Data Bank of Japan (DDBJ) update report 2021. *Nucleic Acids Res.*, 50, D1, D102–D105, 2022.
39. Burley, S.K., *et al.*, Protein data bank: a comprehensive review of 3D structure holdings and worldwide utilization by researchers, educators, and students. *Biomolecules*, 12, 10, 1425, 2022.
40. Chandonia, J.-M., Guan, L., Lin, S., Yu, C., Fox, N.K., Brenner, S.E., SCOPe: improvements to the structural classification of proteins–extended database to facilitate variant interpretation and machine learning. *Nucleic Acids Res.*, 50, D1, D553–D559, 2022.
41. Knudsen, M. and Wiuf, C., The CATH database. *Hum. Genomics*, 4, 3, 1–6, 2010.
42. Liu, Z.-P., Predicting lncRNA-protein interactions by machine learning methods: a review. *Curr. Bioinf.*, 15, 8, 831–840, 2020.
43. Rana, K., Ahammad, K., Salam, M., Bioinformatics: scope and challenges in aquaculture research of Bangladesh-a review. *Int. J. Agric. Res. Innovation Technol. (IJARIT)*, 10, 2, 137–145, 2020.
44. Tanizawa, Y., *et al.*, DNA Data Bank of Japan (DDBJ) update report 2022. *Nucleic Acids Res.*, 51, D1, D101–D105, 2022, doi: 10.1093/nar/gkac1083.
45. Yuan, D., *et al.*, The European Nucleotide Archive in 2023. *Nucleic Acids Res.*, 52, D1, D92–D97, 2023, doi: 10.1093/nar/gkad1067.
46. Timonina, D., Sharapova, Y., Švedas, V., Suplatov, D., Bioinformatic analysis of subfamily-specific regions in 3D-structures of homologs to study functional diversity and conformational plasticity in protein superfamilies. *Comput. Struct. Biotechnol. J.*, 19, 1302–1311, 2021, 2021/01/01/, doi: https://doi.org/10.1016/j.csbj.2021.02.005.

47. Benson, D.A., Karsch-Mizrachi, I., Lipman, D.J., Ostell, J., Sayers, E.W., GenBank. (in eng), *Nucleic Acids Res.*, 38, Database issue, D46–51, Jan 2010, doi: 10.1093/nar/gkp1024.
48. Birney, E., *et al.*, An overview of Ensembl. (in eng), *Genome Res.*, 14, 5, 925–8, May 2004, doi: 10.1101/gr.1860604.
49. Barker, W.C., *et al.*, The protein information resource (PIR). (in eng), *Nucleic Acids Res.*, 28, 1, 41–4, Jan 1 2000, doi: 10.1093/nar/28.1.41.
50. Sarker, B., Ritchie, D.W., Aridhi, S., GrAPFI: predicting enzymatic function of proteins from domain similarity graphs. *BMC Bioinf.*, 21, 1, 1–15, 2020.
51. Consortium, T.U., UniProt: the universal protein knowledgebase in 2021. *Nucleic Acids Res.*, 49, D1, D480–D489, 2020, doi: 10.1093/nar/gkaa1100.
52. Berman, H., Henrick, K., Nakamura, H., Markley, J.L., The worldwide Protein Data Bank (wwPDB): ensuring a single, uniform archive of PDB data. (in eng), *Nucleic Acids Res.*, 35, Database issue, D301–3, Jan 2007, doi: 10.1093/nar/gkl971.
53. The Universal Protein Resource (UniProt) in 2010. (in eng), *Nucleic Acids Res.*, 38, Database issue, D142–8, Jan 2010, doi: 10.1093/nar/gkp846.
54. Markosian, C., Di Costanzo, L., Sekharan, M., Shao, C., Burley, S.K., Zardecki, C., Analysis of impact metrics for the Protein Data Bank. *Sci. Data*, 5, 1, 180212, 2018, 2018/10/16, doi: 10.1038/sdata.2018.212.
55. Maillard, A.M., *et al.*, The 16p11.2 locus modulates brain structures common to autism, schizophrenia and obesity. (in eng), *Mol. Psychiatry*, 20, 1, 140–7, Feb 2015, doi: 10.1038/mp.2014.145.
56. Sigrist, C.J., *et al.*, PROSITE, a protein domain database for functional characterization and annotation. (in eng), *Nucleic Acids Res.*, 38, Database issue, D161–6, Jan 2010, doi: 10.1093/nar/gkp885.
57. Bairoch, A. and Apweiler, R., The SWISS-PROT protein sequence database and its supplement TrEMBL in 2000. (in eng), *Nucleic Acids Res.*, 28, 1, 45–8, Jan 1 2000, doi: 10.1093/nar/28.1.45.
58. Hunter, S., *et al.*, InterPro: the integrative protein signature database. (in eng), *Nucleic Acids Res.*, 37, Database issue, D211–5, Jan 2009, doi: 10.1093/nar/gkn785.
59. Vizcaíno, J.A., *et al.*, A guide to the Proteomics Identifications Database proteomics data repository. (in eng), *Proteomics*, 9, 18, 4276–83, Sep 2009, doi: 10.1002/pmic.200900402.
60. Torrisi, M., Pollastri, G., Le, Q., Deep learning methods in protein structure prediction. *Comput. Struct. Biotechnol. J.*, 18, 1301–1310, 2020.
61. Smyth, M.S. and Martin, J.H., x ray crystallography. (in eng), *Mol. Pathol.*, 53, 1, 8–14, Feb 2000, doi: 10.1136/mp.53.1.8.
62. Lee, M.J. and Yaffe, M.B., Protein Regulation in Signal Transduction. (in eng), *Cold Spring Harbor Perspect. Biol.*, 8, 6, a005918, Jun 1 2016, doi: 10.1101/cshperspect.a005918.

63. Letunic, I., Khedkar, S., Bork, P., SMART: recent updates, new developments and status in 2020. *Nucleic Acids Res.*, 49, D1, D458–D460, 2020, doi: 10.1093/nar/gkaa937.
64. Goodsell, D.S., Sanner, M.F., Olson, A.J., Forli, S., The AutoDock suite at 30. *Protein Sci.*, 30, 1, 31–43, 2021.
65. Koukos, P.I., *et al.*, An overview of data-driven HADDOCK strategies in CAPRI rounds 38-45. *Proteins Struct. Funct. Bioinf.*, 88, 8, 1029–1036, 2020.
66. Wang, M., Cang, Z., Wei, G.-W., A topology-based network tree for the prediction of protein–protein binding affinity changes following mutation. *Nat. Mach. Intell.*, 2, 2, 116–123, 2020.
67. Vilar, S., Cozza, G., Moro, S., Medicinal chemistry and the molecular operating environment (MOE): application of QSAR and molecular docking to drug discovery. *Curr. Top. Med. Chem.*, 8, 18, 1555–1572, 2008.
68. Szklarczyk, D., *et al.*, The STRING database in 2021: customizable protein-protein networks, and functional characterization of user-uploaded gene/measurement sets. (in eng), *Nucleic Acids Res.*, 49, D1, D605–d612, Jan 8 2021, doi: 10.1093/nar/gkaa1074.
69. Di Lena, P., Wu, G., Martelli, P.L., Casadio, R., Nardini, C., MIMO: an efficient tool for molecular interaction maps overlap. *BMC Bioinf.*, 14, 1–11, 2013.
70. Del Toro, N., *et al.*, The IntAct database: efficient access to fine-grained molecular interaction data. *Nucleic Acids Res.*, 50, D1, D648–D653, 2022.
71. Queiroz, F.C., Vargas, A.M.P., Oliveira, M.G.A., Comarela, G.V., Silveira, S.A., ppiGReMLIN: a graph mining based detection of conserved structural arrangements in protein-protein interfaces. *BMC Bioinf.*, 21, 1, 143, 2020, 2020/04/15, doi: 10.1186/s12859-020-3474-1.
72. Koumakis, L., Mizzi, C., Potamias, G., Bioinformatics tools for data analysis, in: *Molecular Diagnostics*, pp. 339–351, Elsevier, London, United Kingdom, 2017.
73. Zheng, S., *et al.*, Network pharmacological screening of herbal monomers that regulate apoptosis-associated genes in acute pancreatitis. *Pancreas*, 46, 1, 89–96, 2017.
74. Ma, C., *et al.*, Network Pharmacology and Bioinformatics Approach Reveals the Therapeutic Mechanism of Action of Baicalein in Hepatocellular Carcinoma. (in eng), *Evid. Based Complement. Alternat. Med.*, 2019, 7518374, 2019, doi: 10.1155/2019/7518374.
75. Sabe, V.T., *et al.*, Current trends in computer aided drug design and a highlight of drugs discovered via computational techniques: A review. (in eng), *Eur. J. Med. Chem.*, 224, 113705, Nov 15 2021, doi: 10.1016/j.ejmech.2021.113705.
76. Hughes, J.P., Rees, S., Kalindjian, S.B., Philpott, K.L., Principles of early drug discovery. (in eng), *Br. J. Pharmacol.*, 162, 6, 1239–49, Mar 2011, doi: 10.1111/j.1476-5381.2010.01127.x.
77. Yadav, M. and Eswari, J.S., Modern Paradigm Towards Potential Target Identification for Antiviral (SARS-nCoV-2) and Anticancer Lipopeptides: A

Pharmacophore-Based Approach. (in eng), *Avicenna J. Med. Biotechnol.*, 14, 1, 70–78, Jan-Mar 2022, doi: 10.18502/ajmb.v14i1.8172.

78. Knox, C., *et al.*, Drugbank 6.0: the drugbank knowledgebase for 2024. *Nucleic Acids Res.*, 52, D1, D1265–D1275, 2024.
79. Wang, Y., *et al.*, Therapeutic target database 2020: enriched resource for facilitating research and early development of targeted therapeutics. *Nucleic Acids Res.*, 48, D1, D1031–D1041, 2020.
80. Magariños, M.P., *et al.*, TDR Targets: a chemogenomics resource for neglected diseases. (in eng), *Nucleic Acids Res.*, 40, Database issue, D1118–27, Jan 2012, doi: 10.1093/nar/gkr1053.
81. Günther, S., *et al.*, SuperTarget and Matador: resources for exploring drug-target relationships. (in eng), *Nucleic Acids Res.*, 36, Database issue, D919–22, Jan 2008, doi: 10.1093/nar/gkm862.
82. de Beer, T.A., Berka, K., Thornton, J.M., Laskowski, R.A., PDBsum additions. (in eng), *Nucleic Acids Res.*, 42, Database issue, D292–6, Jan 2014, doi: 10.1093/nar/gkt940.
83. Gaulton, A., *et al.*, ChEMBL: a large-scale bioactivity database for drug discovery. (in eng), *Nucleic Acids Res.*, 40, Database issue, D1100–7, Jan 2012, doi: 10.1093/nar/gkr777.
84. Tafrishi, H., Sadeghzadeh, S., Ahmadi, R., Molecular dynamics simulations of phase change materials for thermal energy storage: a review. *RSC Adv.*, 12, 23, 14776–14807, 2022.
85. Jebali, A. and Outeiro, T., The adsorption of alpha-synuclein and inhibition of fibrillization by nanocellulose and nanocellulose coated with gold atoms, 2023.
86. Love, O., Pacheco Lima, M.C., Clark, C., Cornillie, S., Roalstad, S., Cheatham Iii, T.E., Evaluating the accuracy of the AMBER protein force fields in modeling dihydrofolate reductase structures: misbalance in the conformational arrangements of the flexible loop domains. (in eng), *J. Biomol. Struct. Dyn.*, 41, 13, 5946–5960, Aug-Sep 2023, doi: 10.1080/07391102.2022.2098823.
87. Lai, M.C., *et al.*, Prevalence of co-occurring mental health diagnoses in the autism population: a systematic review and meta-analysis. (in eng), *Lancet Psychiatry*, 6, 10, 819–829, Oct 2019, doi: 10.1016/s2215-0366(19)30289-5.
88. Croen, L.A., *et al.*, The health status of adults on the autism spectrum. (in eng), *Autism*, 19, 7, 814–23, Oct 2015, doi: 10.1177/1362361315577517.
89. He, X., *et al.*, Integrated model of *de novo* and inherited genetic variants yields greater power to identify risk genes. (in eng), *PLoS Genet.*, 9, 8, e1003671, 2013, doi: 10.1371/journal.pgen.1003671.
90. Weiner, D.J., *et al.*, Polygenic transmission disequilibrium confirms that common and rare variation act additively to create risk for autism spectrum disorders. (in eng), *Nat. Genet.*, 49, 7, 978–985, Jul 2017, doi: 10.1038/ng.3863.

91. Sanders, S.J., *et al.*, Insights into Autism Spectrum Disorder Genomic Architecture and Biology from 71 Risk Loci. (in eng), *Neuron*, 87, 6, 1215–1233, Sep 23 2015, doi: 10.1016/j.neuron.2015.09.016.
92. de la Torre-Ubieta, L., Won, H., Stein, J.L., Geschwind, D.H., Advancing the understanding of autism disease mechanisms through genetics. (in eng), *Nat. Med.*, 22, 4, 345–61, Apr 2016, doi: 10.1038/nm.4071.
93. Sharon, G., *et al.*, Human Gut Microbiota from Autism Spectrum Disorder Promote Behavioral Symptoms in Mice. (in eng), *Cell*, 177, 6, 1600–1618. e17, May 30 2019, doi: 10.1016/j.cell.2019.05.004.
94. Orozco, J.S., Hertz-Picciotto, I., Abbeduto, L., Slupsky, C.M., Metabolomics analysis of children with autism, idiopathic-developmental delays, and Down syndrome. (in eng), *Transl. Psychiatry*, 9, 1, 243, Oct 3 2019, doi: 10.1038/s41398-019-0578-3.
95. Warren, Z., McPheeters, M.L., Sathe, N., Foss-Feig, J.H., Glasser, A., Veenstra-Vanderweele, J., A systematic review of early intensive intervention for autism spectrum disorders. (in eng), *Pediatrics*, 127, 5, e1303–11, May 2011, doi: 10.1542/peds.2011-0426.
96. Shen, L., Liu, X., Zhang, H., Lin, J., Feng, C., Iqbal, J., Biomarkers in autism spectrum disorders: Current progress. (in eng), *Clin. Chim. Acta*, 502, 41–54, Mar 2020, doi: 10.1016/j.cca.2019.12.009.
97. Chow, M.L., *et al.*, Age-dependent brain gene expression and copy number anomalies in autism suggest distinct pathological processes at young versus mature ages. (in eng), *PLoS Genet.*, 8, 3, e1002592, 2012, doi: 10.1371/journal.pgen.1002592.
98. Liu, X., *et al.*, Disruption of an Evolutionarily Novel Synaptic Expression Pattern in Autism. (in eng), *PLoS Biol.*, 14, 9, e1002558, Sep 2016, doi: 10.1371/journal.pbio.1002558.
99. Ramaswami, G., *et al.*, Integrative genomics identifies a convergent molecular subtype that links epigenomic with transcriptomic differences in autism. (in eng), *Nat. Commun.*, 11, 1, 4873, Sep 25 2020, doi: 10.1038/s41467-020-18526-1.
100. Gandal, M.J., *et al.*, Shared molecular neuropathology across major psychiatric disorders parallels polygenic overlap. (in eng), *Science*, 359, 6376, 693–697, Feb 9 2018, doi: 10.1126/science.aad6469.
101. Guan, J., Cai, J.J., Ji, G., Sham, P.C., Commonality in dysregulated expression of gene sets in cortical brains of individuals with autism, schizophrenia, and bipolar disorder. (in eng), *Transl. Psychiatry*, 9, 1, 152, May 24 2019, doi: 10.1038/s41398-019-0488-4.
102. Eissa, N., Al-Houqani, M., Sadeq, A., Ojha, S.K., Sasse, A., Sadek, B., Current Enlightenment About Etiology and Pharmacological Treatment of Autism Spectrum Disorder. (in eng), *Front. Neurosci.*, 12, 304, eaag2882, 2018, doi: 10.3389/fnins.2018.00304.

103. Beversdorf, D.Q., Phenotyping, Etiological Factors, and Biomarkers: Toward Precision Medicine in Autism Spectrum Disorders. (in eng), *J. Dev. Behav. Pediatr.*, 37, 8, 659–73, Oct 2016, doi: 10.1097/dbp.0000000000000351.
104. Jacob, S., Wolff, J.J., Steinbach, M.S., Doyle, C.B., Kumar, V., Elison, J.T., Neurodevelopmental heterogeneity and computational approaches for understanding autism. (in eng), *Transl. Psychiatry*, 9, 1, 63, Feb 4 2019, doi: 10.1038/s41398-019-0390-0.
105. Sønderby, I.E., *et al.*, Dose response of the 16p11.2 distal copy number variant on intracranial volume and basal ganglia. (in eng), *Mol. Psychiatry*, 25, 3, 584–602, Mar 2020, doi: 10.1038/s41380-018-0118-1.
106. Emerson, R.W., *et al.*, Functional neuroimaging of high-risk 6-month-old infants predicts a diagnosis of autism at 24 months of age. (in eng), *Sci. Transl. Med.*, 9, 393, eaag2882, Jun 7 2017, doi: 10.1126/scitranslmed.aag2882.
107. Filipp, F.V., Precision medicine driven by cancer systems biology. (in eng), *Cancer Metastasis Rev.*, 36, 1, 91–108, Mar 2017, doi: 10.1007/s10555-017-9662-4.
108. Sammut, S.-J., *et al.*, Multi-omic machine learning predictor of breast cancer therapy response. *Nature*, 601, 7894, 623–629, 2022, 2022/01/01, doi: 10.1038/s41586-021-04278-5.
109. Parsons, J. and Francavilla, C., 'Omics Approaches to Explore the Breast Cancer Landscape. (in eng), *Front. Cell Dev. Biol.*, 7, 395, 2019, doi: 10.3389/fcell.2019.00395.
110. Guney, E. and Athie, A., A needle for Alzheimer's in a haystack of claims data. (in eng), *Nat. Aging*, 1, 12, 1083–1085, Dec 2021, doi: 10.1038/s43587-021-00139-y.
111. Gottesman, I.I. and Gould, T.D., The endophenotype concept in psychiatry: etymology and strategic intentions. (in eng), *Am. J. Psychiatry*, 160, 4, 636–45, Apr 2003, doi: 10.1176/appi.ajp.160.4.636.
112. Korcsmáros, T., Szalay, M.S., Böde, C., Kovács, I.A., Csermely, P., How to design multi-target drugs. (in eng), *Expert Opin. Drug Discov.*, 2, 6, 799–808, Jun 2007, doi: 10.1517/17460441.2.6.799.
113. Smith, A.M., *et al.*, Amino Acid Dysregulation Metabotypes: Potential Biomarkers for Diagnosis and Individualized Treatment for Subtypes of Autism Spectrum Disorder. (in eng), *Biol. Psychiatry*, 85, 4, 345–354, Feb 15 2019, doi: 10.1016/j.biopsych.2018.08.016.
114. Smith, A.M., *et al.*, A Metabolomics Approach to Screening for Autism Risk in the Children's Autism Metabolome Project. (in eng), *Autism Res.*, 13, 8, 1270–1285, Aug 2020, doi: 10.1002/aur.2330.
115. Bhat, S., Acharya, U.R., Adeli, H., Bairy, G.M., Adeli, A., Autism: cause factors, early diagnosis and therapies. (in eng), *Rev. Neurosci.*, 25, 6, 841–50, 2014, doi: 10.1515/revneuro-2014-0056.
116. Oughtred, R., *et al.*, The BioGRID database: A comprehensive biomedical resource of curated protein, genetic, and chemical interactions. (in eng), *Protein Sci.*, 30, 1, 187–200, Jan 2021, doi: 10.1002/pro.3978.

117. Kanehisa, M., Sato, Y., Kawashima, M., KEGG mapping tools for uncovering hidden features in biological data. (in eng), *Protein Sci.*, 31, 1, 47–53, Jan 2022, doi: 10.1002/pro.4172.
118. Wishart, D.S., *et al.*, DrugBank 5.0: a major update to the DrugBank database for 2018. (in eng), *Nucleic Acids Res.*, 46, D1, D1074–d1082, Jan 4 2018, doi: 10.1093/nar/gkx1037.
119. Kuhn, M., Letunic, I., Jensen, L.J., Bork, P., The SIDER database of drugs and side effects. (in eng), *Nucleic Acids Res.*, 44, D1, D1075–9, Jan 4 2016, doi: 10.1093/nar/gkv1075.
120. Piñero, J., *et al.*, DisGeNET: a discovery platform for the dynamical exploration of human diseases and their genes. (in eng), *Database (Oxford)*, 2015, bav028, 2015, doi: 10.1093/database/bav028.
121. Ochoa, D., *et al.*, Open Targets Platform: supporting systematic drug-target identification and prioritisation. (in eng), *Nucleic Acids Res.*, 49, D1, D1302–d1310, Jan 8 2021, doi: 10.1093/nar/gkaa1027.
122. Whirl-Carrillo, M., *et al.*, An Evidence-Based Framework for Evaluating Pharmacogenomics Knowledge for Personalized Medicine. (in eng), *Clin. Pharmacol. Ther.*, 110, 3, 563–572, Sep 2021, doi: 10.1002/cpt.2350.
123. Subramanian, A., *et al.*, A Next Generation Connectivity Map: L1000 Platform and the First 1,000,000 Profiles. (in eng), *Cell*, 171, 6, 1437–1452. e17, Nov 30 2017, doi: 10.1016/j.cell.2017.10.049.
124. Igarashi, Y., *et al.*, Open TG-GATEs: a large-scale toxicogenomics database. (in eng), *Nucleic Acids Res.*, 43, Database issue, D921–7, Jan 2015, doi: 10.1093/nar/gku955.
125. Hutter, C. and Zenklusen, J.C., The Cancer Genome Atlas: Creating Lasting Value beyond Its Data. (in eng), *Cell*, 173, 2, 283–285, Apr 5 2018, doi: 10.1016/j.cell.2018.03.042.
126. Pouryahya, M., *et al.*, Pan-Cancer Prediction of Cell-Line Drug Sensitivity Using Network-Based Methods. (in eng), *Int. J. Mol. Sci.*, 23, 3, 1074, Jan 19 2022, doi: 10.3390/ijms23031074.
127. Koch, E. and Demontis, D., Drug repurposing candidates to treat core symptoms in autism spectrum disorder. (in eng), *Front. Pharmacol.*, 13, 995439, 2022, doi: 10.3389/fphar.2022.995439.
128. Malinky, C.A., Lindsley, C.W., Han, C., DARK Classics in Chemical Neuroscience: Loperamide. (in eng), *ACS Chem. Neurosci.*, 12, 16, 2964–2973, Aug 18 2021, doi: 10.1021/acschemneuro.1c00382.
129. Pellissier, L.P., Gandía, J., Laboute, T., Becker, J.A.J., Le Merrer, J., μ opioid receptor, social behaviour and autism spectrum disorder: reward matters. (in eng), *Br. J. Pharmacol.*, 175, 14, 2750–2769, Jul 2018, doi: 10.1111/bph.13808.
130. Madra, M., Ringel, R., Margolis, K.G., Gastrointestinal Issues and Autism Spectrum Disorder. (in eng), *Child Adolesc. Psychiatr. Clin. N. Am.*, 29, 3, 501–513, Jul 2020, doi: 10.1016/j.chc.2020.02.005.

131. Kvernmo, T., Härtter, S., Burger, E., A review of the receptor-binding and pharmacokinetic properties of dopamine agonists. (in eng), *Clin. Ther.*, 28, 8, 1065–1078, Aug 2006, doi: 10.1016/j.clinthera.2006.08.004.
132. Dollfus, S., *et al.*, Catecholamines in autistic disorder: effects of amisulpride and bromocriptine in a controlled crossover study. (in eng), *J. Child Adolesc. Psychopharmacol.*, 3, 3, 145–56, Fall 1993, doi: 10.1089/cap.1993.3.145.
133. Pavăl, D. and Miclutia, I., The Dopamine Hypothesis of Autism Spectrum Disorder Revisited: Current Status and Future Prospects. *Dev. Neurosci.*, 43, 1–11, 05/19 2021, doi: 10.1159/000515751.
134. Coleman, D.M., Adams, J.B., Anderson, A.L., Frye, R.E., Rating of the Effectiveness of 26 Psychiatric and Seizure Medications for Autism Spectrum Disorder: Results of a National Survey. (in eng), *J. Child Adolesc. Psychopharmacol.*, 29, 2, 107–123, Mar 2019, doi: 10.1089/cap.2018.0121.
135. Launay, J.M., *et al.*, Catecholamines metabolism in infantile autism: a controlled study of 22 autistic children. (in eng), *J. Autism Dev. Disord.*, 17, 3, 333–47, Sep 1987, doi: 10.1007/bf01487064.
136. Aman, M.G., *et al.*, Effects of Metformin on Spatial and Verbal Memory in Children with ASD and Overweight Associated with Atypical Antipsychotic Use. (in eng), *J. Child Adolesc. Psychopharmacol.*, 28, 4, 266–273, May 2018, doi: 10.1089/cap.2017.0072.
137. Novotny, S., *et al.*, Increased growth hormone response to sumatriptan challenge in adult autistic disorders. (in eng), *Psychiatry Res.*, 94, 2, 173–7, May 15 2000, doi: 10.1016/s0165-1781(00)00134-7.
138. Tfelt-Hansen, P. and Hougaard, A., Sumatriptan: a review of its pharmacokinetics, pharmacodynamics and efficacy in the acute treatment of migraine. (in eng), *Expert Opin. Drug Metab. Toxicol.*, 9, 1, 91–103, Jan 2013, doi: 10.1517/17425255.2013.744394.
139. Dudley, J.T., *et al.*, Computational repositioning of the anticonvulsant topiramate for inflammatory bowel disease. (in eng), *Sci. Transl. Med.*, 3, 96, 96ra76, Aug 17 2011, doi: 10.1126/scitranslmed.3002648.

14

Computational Approaches to Drug Discovery in Depression

Kalpesh Ramdas Patil[1]*, Aman B. Upaganlawar[2], Akhil A. Nagar[3] and Kuldeep U. Bansod[4]

[1]Department of Pharmacology, R. C. Patel Institute of Pharmaceutical Education and Research, Shirpur, Dist. Dhule, Maharashtra, India
[2]Department of Pharmacology, SNJBs SSDJ College of Pharmacy, Neminagar, Chandwad, Dist. Nashik, Maharashtra, India
[3]Department of Pharmaceutical Chemistry, R. C. Patel Institute of Pharmaceutical Education and Research, Shirpur, Dist. Dhule, Maharashtra, India
[4]Government College of Pharmacy, Karad, Dist. Satara, Maharashtra, India

Abstract

Depression is stated as the most prevalent form of mental disorder around the globe by the World Health Organization (WHO). The etiopathogenesis of depression is complex and the disease manifestations vary with the involvement of multiple receptors. Several hypotheses and signaling pathways are described for depression. Computational attitudes are becoming a key component in the process of identification and development of new druggable molecules. Computational attitudes are cost-effective solutions for the discovery of novel leads targeting a variety of human ailments including depression. These computational approaches assist the drug discovery process from identifying leads to discovering clinical biomarkers. Computational attitudes like machine learning, artificial intelligence, docking simulations, and network pharmacology are vital resources for obtaining realistic insights and promoting decision-making in the drug discovery process. Applications of computational approaches range from compound screening, and *de novo* drug design, to the forecast of drug response and drug interactions. The present chapter focuses on various computational approaches for drug discovery in depressive disorders.

**Corresponding author*: kalpeshpharma20@gmail.com; ORCID: 0000-0003-2397-4687

Bhupendra Prajapati, Alok Tripathi, Rishabha Malviya and Lucy Mohapatra (eds.) Computational Neuropharmacology: Fundamentals and Clinical Aspects, (409–436) © 2025 Scrivener Publishing LLC

***Keywords*:** Depression, computational approaches, drug discovery, major depressive disorder, docking, network pharmacology

List of Abbreviations

5-HT	5-hydroxytryptamine
ACTH	Adrenocorticotropic hormone
AMPA	α-amino-3-hydroxy-5- methyl-4-isoxazolepropionic acid
APT-2	Acyl protein thioesterase 2
BDNF	Brain-derived neurotrophic factor
CADD	Computer-aided drug design
CoA	Acetyl-coenzyme A
cPLA2	Cytosolic phospholipase A2
CRF	Corticotropin-releasing factor
CRH	Corticotropin-releasing hormone
DMDD	Disruptive mood dysregulation disorder
EPA	Eicosapentaenoic acid
ERK	Extracellular-signal-regulated kinase
GABA	Gamma-aminobutyric acid
HPA	Hypothalamic pituitary adrenal
IL	Interleukin
INF-γ	Interferon-gamma
L-BDD	Ligand-based drug design
MAO	Monoamine oxidase
MAOIs	Monoamine oxidase inhibitors
MDA	Malondialdehyde
MDD	Major depressive disorder
mTOR	Mammalian target of rapamycin
NF-κB	Nuclear factor kappa B
NMDA	N-methyl-D-aspartate
Nrf2	Nuclear factor erythroid 2–related factor 2
PDD	Persistent depressive disorder
PGE2	Prostaglandin E2
PINE	Psycho-immune-neuroendocrine
PMDD	Premenstrual dysphoric disorder
PMS	Premenstrual syndrome
PUFA	Polyunsaturated fatty acids
ROS	Reactive oxygen species
SAD	Seasonal affective disorder
S-BDD	Structure-based drug design

SNRIs	Serotonin and norepinephrine reuptake inhibitors
SSRIs	Selective serotonin reuptake inhibitors
TAARs	Trace amine-associated receptors
TDO	Tryptophan 2, 3-dioxygenase
TGFβ	Transforming growth factor-β
TNF-α	Tumor necrosis factor alpha
Trp	Tryptophan
TyrB	Tyrosine kinase receptor B
WHO	World Health Organization

14.1 Introduction

Mental illnesses like depression or major depressive disorder (MDD) affect both our physical and mental health. Among all psychiatric disorders, depression is the prevailing and foremost reason for the global disease burden [1, 2]. There is a high rate of morbidity and mortality associated with MDD [3]. Worldwide, depression is predicted as the primary cause of morbidity by 2030 (WHO, Geneva, Switzerland). To better understand depression and monitor treatment outcomes, it is imperative to identify biomarkers [2]. Suicidal thoughts, feelings of guilt, or worthlessness, sleeping and eating disorders, decreased quality of life, low self-esteem, and negative attitudes are among the most common symptoms [1, 2, 4, 5]. Depression remains unclear in terms of its etiology, though it is generally believed to be the result of a multifaceted disorder caused by social, psychological, and biological factors. As a result, genetics, neurobiology, and neuroimaging cannot all independently account for its pathogenesis [6]. Biological markers altered in depression are being studied because of their biological reinforcements [2].

14.2 Types of Depressive Disorders

Depression is a prevalent mental illness that occurs in various forms and intensities with diverse etiopathogenesis. Depressive disorders are categorized into several types [7]. The MDD recognized as clinical depression represents the most common however the most severe form of depression. The symptoms of MDD usually last for at least two weeks and are characterized by persistent sadness, hopelessness, and inactivity. There are numerous possible symptoms, including changes in appetite and sleep patterns, fatigue, trouble concentrating, and even thoughts of death [8]. MDD is influenced by environmental, genetic, biological, and psychosocial factors. Neurotransmitter defects have been linked with MDD and specific antidepressants have been supported to be effective in treating the disorder [9].

Seasonal affective disorder (SAD) or seasonal depression is a type of MDD characterized by recurring major depressions around certain seasons [7]. A repeated incidence of depression in the winter is SAD, which mostly has a remission in the spring or summer [10]. Serotonin dysregulation and melatonin overproduction are correlated with seasonal affective disorder [11, 12]. The depression that arises during pregnancy is termed prenatal depression, while postpartum depression starts when the baby is born. The risk of postpartum depression rises with prenatal depression [13]. Higher levels of cortisol and norepinephrine and lower prenatal maternal dopamine levels could mediate postpartum depression [14].

Atypical depression represents MDD with atypical features in which temporary improvement in the mood appears following positive outcomes. Persistent depressive disorder (PDD) is established as dysthymia or chronic major depression [15]. PDD is often a more disabling and more common form of depression than periodic major depression. Neurotransmitters like serotonin, catecholamines, and their receptors have a significant role in PDD [16, 17]. Disruptive mood dysregulation disorder (DMDD) causes children and youngsters to remain irritable, and angry and repeatedly outburst their moods for long periods [18]. Premenstrual syndrome (PMS) represents a range of, physical, and behavioral symptoms that emerge before and subside following the menstrual cycle. Premenstrual dysphoric disorder (PMDD), is the severe representation of PMS distinguished as somatic and psychological symptoms alongside functional impairments [19, 20].

14.3 Hypotheses and Pathways of Depression

14.3.1 Monoamine Hypothesis

About a half-century ago, the original major hypothesis concerning depression was established. This hypothesis proposed that depression is triggered by a functional deficit of monoamine neurotransmitters including serotonin, noradrenaline, and dopamine, whilst mania is caused by a functional abundance of monoamines [21]. This hypothesis relies on symptomatic improvement in depression with monoamine oxidase inhibitors and tricyclic antidepressants, which act by boosting serotonin and noradrenaline action [22]. The monoamine hypothesis has been an established concept that contributed to the discovery of antidepressant drugs that target these neurotransmitters [23].

This monoamine hypothesis is, however, questioned. Neither a decline in serotonin or dopamine synthesis nor a substantial depletion of monoamines is satisfactory for producing depression in healthy adults. This hypothesis cannot justify the complexity of depression, the gap in responding to drug therapy, and the incapability of antidepressants to treat depression. In addition, no neurotransmitter has been recognized as being significant to depression pathophysiology [21, 24]. The most used antidepressants exhibit many adversities including orthostatic hypotension and sexual malfunction in addition to anticholinergic effects. Furthermore, monoamine oxidase (MAO) breaks down tyramine, so MAO inhibitors interact with other drugs and foods containing tyramine. Presently, MAO inhibitors are only used while all other medications have been unsuccessful. Monoamine-alternate hypotheses are instantly needed to know mood disorders better and develop additional effective treatments [25, 26].

14.3.2 Hypothalamic Pituitary Adrenal Axis

The hypothalamic pituitary adrenal (HPA) axis represents a specific neuroendocrine pathway activated by various stressors. Depression is largely mediated by the HPA axis. MDD results in cognitive dysfunction and miserable mood due to overactivity of the HPA axis [27, 28]. The HPA axis is composed of the hypothalamus, pituitary, and adrenal glands that regulate glucocorticoids [29]. In the HPA axis, the hypothalamus releases corticotropin-releasing factor (CRF) via CRF neurons, which accelerates the anterior pituitary to release adrenocorticotropic hormone (ACTH). Consequently, cortisol is formed by the adrenal cortex. This axis is interrupted by prolonged stress, causing elevated cortisol levels, and disturbed physiology [30, 31]. With consecutive episodes of depression, the hippocampus shows lowered Brain-derived neurotrophic factor (BDNF) expression, hampered neurogenesis in the dental gyrus, and steady volume loss [32]. Depressed individuals may benefit from pharmacotherapy that stabilizes HPA axis activity [28].

14.3.3 Neurotrophic Factor Hypothesis

Depression is characterized by dysfunctions of neuroplasticity or remodeling [33]. The BDNF may contribute to neuronal atrophy in the hippocampus, which is associated with depression [34]. Following the neurotrophic hypothesis, MDD is primarily a result of impaired neurotrophic function. The BDNF, is crucial for the endurance, development, and preservation of neurons that are concerned with emotion and cognition [35]. The Ras/ERK pathway is implicated in depression and BDNF stimulation increases synaptic spine density. In depression, one possible mechanism may be BDNF/mTOR signaling, but other pathways may also be involved [35–38].

14.3.4 Oxidative Stress

The oxidative stress is linked to depression, along with inflammation and immune mechanisms [39–41]. In oxidative stress, inequities occur in the release of reactive oxygen species (ROS) and the body's capacity to repair the resultant damage [39]. Although, neuronal development necessitates moderate ROS levels, increased ROS expression has been correlated to MDD. Advanced ROS levels and compromised antioxidant mechanisms occur due to oxidative stress in MDD. Neuroplasticity, inflammation, and mitochondrial functions are all negatively influenced by this inequality [40]. Free radical and non-radical molecules, like hydrogen peroxide, can damage cells and organs through oxidative damage caused by ROS and reactive nitrogen species (RNS) [42]. Oxidative stress and neuroinflammation are promoted in depressed individuals via extreme ROS and inflammatory biomarkers that result in neuroinflammation and cell death. As an indicator of oxidative stress-induced damage, malondialdehyde (MDA) levels are elevated, while Nrf2 prevents ROS from being generated and inhibits NF-κB [43]. Depression is also linked with decreased levels of antioxidants like vitamins, selenium, and zinc. The oxidative stress is associated with high levels of brain lipid peroxidation, nitric oxide, and cyclooxygenase-2 activity [41].

14.3.5 Cytokines

Depressive disorders are influenced by cytokines [44]. Neurogenesis, neuronal integrity, and synaptic remodeling are supported by cytokines [45]. Inflammatory cytokines are vital to brain function, and an abnormal level of these markers can negatively impact the brain [21]. Studies have shown that TNF-α has an essential role in depression. TNF-α causes monoamine reuptake, HPA hyperactivity, and cognitive decline, potentially disrupting melatonin production and the circadian rhythm. In MDD patients, circulatory levels of IL-6, TNF-α, IL-10, IL-13, IL-18, and IL-12 are upregulated, whereas levels of interferon-gamma (INF-γ) are downregulated [32, 46]. Studies have linked postpartum depression to elevated plasma inflammatory markers, especially IL-1β, soluble IL-1 receptor antagonist, and IL-6 [47]. The severity and progression of depression are influenced by stress which increases the production of proinflammatory cytokines. Major depression has been linked to immune mediator dysregulation, such as elevated proinflammatory cytokines like IL-1β and IL-6 and a decline in anti-inflammatory cytokines like TGFβ [48]. Cytokines, like TNF-α, IL-1β, IFN-α, and IL-6, affect the synthesis of monoamine neurotransmitters, which affects the production of serotonin, noradrenaline, and dopamine. Pro-inflammatory cytokines interrupt neurotrophins, especially BDNF by affecting the phosphorylation of tyrosine kinase receptor B (TyrB) [49].

14.4 Receptors in Depression

14.4.1 Serotonin Receptor

The 5-hydroxytryptamine (5-HT) or serotonin is the major neurotransmitter implicated in a variety of neurobehavioral disorders including depression. Insults like chronic stress or brain injury initiate the 5-HT neurons to restore and reinnervate brain regions which causes 5-HT system interruption and leads to depression and anxiety [50]. 5-HT receptors are present throughout the brain including regions implicated in depression and anxiety. 5-HT1A receptors are found in the cortical, limbic, dorsal, and median raphe nuclei. A total of seven 5-HT receptor families are subdivided into fourteen subtypes [51, 52]. The 5-HT1A receptors link to the Gi/Go pathways, preventing adenylyl cyclase to reduce cAMP levels and triggering G-protein inward rectifying potassium (GIRK) channels [52].

14.4.2 Dopamine Receptor

Several brain parts like the ventral subiculum regulate the complex dopaminergic activity. The dopaminergic system is vital in depression pathology. The deficient dopaminergic system is implicated in depression and dopamine receptor ligands demonstrated efficacy in depression [53]. The dopamine receptors are D1-like receptors (D1-LR) and D2-like receptors (D2-LR). The D1-LR are D1 and D5 receptors, whereas D2, D3, and D4 receptors are D2-LR. Antidepressant treatment is dependent on D2-LR, especially D2 and D3 receptors which are mainly found in the hippocampus, cerebral cortex, striatum, and amygdala and inhibit adenylate cyclase through Gαi/o proteins [54].

14.4.3 Glutamate Receptor

The glutaminergic system has a role in various neurological disorders and its targeting for treating various mood disorders is promising [55]. Depression pathophysiology shows the involvement of glutamatergic neurotransmission [56]. Glutamate receptors are of two types, namely ionotropic glutamate receptors (iGR) and metabotropic glutamate receptors (mGluRs) that are linked to G-protein. The *N*-methyl-D-aspartate (NMDA), α-amino-3-hydroxy-5-methyl-4-isoxazolepropionic acid (AMPA) and kainite receptors belong to ionotropic glutamate receptors. Whereas the mGluRs are further subdivided into mGlu1 and mGlu5 (subgroup-I), mGlu2 and mGlu3 (subgroup-II) and mGlu4 and mGlu6–8 (subgroup-III). The metabotropic receptors of subgroup-I are linked to Gq/G11 and causes activation of phospholipase Cβ. The subgroup-II and subgroup-III receptors are linked with Gi/o proteins and lead to inhibition of adenylate cyclase [57].

14.4.4 Trace Amine-Associated Receptors

The trace amines (TAs) present in the CNS like tyramine, tryptamine, β-phenylethylamine and octopamine act as neurotransmitters. Dysregulation of such trace amines is implicated in the development of depression [58]. Among various types of trace amine-associated receptors, TAAR1 is widely distributed in the CNS, and it acts through Gs proteins and increases the intracellular level of cAMP. Another subtype, TAAR5 is mainly found in the amygdala, substantia nigra, hippocampus, thalamus, and hypothalamus [59]. The pathophysiology of major depression is supposed to be affected by the disruption of TA activity [60].

14.4.5 Cholinergic Receptor

The cholinergic receptors are classified as muscarinic receptors and nicotinic receptors. Muscarinic receptors are GPCRs, whereas nicotinic receptors are ligand-gated ion channels. Muscarinic receptors like M1, M3, and M5 act via G protein to activate the phospholipase C. Whereas Muscarinic receptors M2 and M4 coupled with Gi and Go proteins to reduce cAMP formation. Antidepressant effects of scopolamine are intervened by the M1 receptor blockade and show involvement of the M2 receptor [61, 62]. Earlier studies noted a decline in M2 and M4 cholinergic receptor density in MDD patients [63].

14.4.6 GABA Receptor

The gamma-aminobutyric acid (GABA) is the main inhibitory brain neurotransmitter. The GABAergic receptors are either ionotropic receptors (GABA-A) or metabotropic receptors (GABA-B). The GABA-B receptor is coupled with Gi and Go proteins and affects cAMP production. These metabotropic GABA-B receptors are linked with depression pathology [64, 65].

14.4.7 Cannabinoid Receptor

Depression is substantially affected by the endogenous cannabinoid system, which is comprised of cannabinoid receptor types 1 and 2 (CB1R and CB2R) [66]. The cannabinoid receptors are present in CNS and are linked with Gi/o [67]. Upon activation, these receptors caused adenylyl cyclase inhibition and a decline in cAMP production. The endocannabinoid system (ECS) related to CB1 receptors relates to emotional regulation and the development of depression. The administration of anandamide degradation inhibitors and CB1 receptor agonists results in identical behavioral and biochemical effects to antidepressants [68].

14.5 Computational Approaches to Depression

Drug design is being reformed by computational approaches and information science, which allow the development of chemically stable compounds while speeding up the process. These computational tools, using chemoinformatics, reduce the discovery cycle, making drug discovery more cost-effective and efficient [69]. Data on 3D structures, ligand characteristics, therapeutic targets, and virtual libraries of drug-like compounds are altering computer-aided drug discovery [70]. Drug discovery relies deeply on virtual screening and computational approaches, that can be used to model structures, predict binding sites and cavities, dock molecules, visualize and analyze interactions, map pharmacophores, design new ligands, optimize leads, simulate molecular dynamics, and predict pharmacokinetic, and adverse effects. In comparison with conventional drug discovery methods, computer-aided drug design (CADD) is more competent, less expensive, and faster [71]. Drug design using computer-aided tools includes structure-based drug design (S-BDD) and ligand-based drug design (L-BDD). The S-BDD needs protein structures, while L-BDD can design compounds that interact with protein structures using ligands and ligand activities. S-BDD includes docking, fragment-based drug discovery, *de novo* design, and pharmacophore modeling. The L-BDD approaches involve quantitative structure-affinity relationship and pharmacophore modeling based on ligand properties. Various design strategies can be selected depending on the receptor structure and its interaction with the ligand [72, 73].

Over the last decade, several researchers reported studies focusing on network pharmacology and docking approaches using various biological targets in depression. Based on the articles published in the recent three years (from 2021 to 2023), we represented the major docking targets and their respective PDB IDs in Figure 14.1.

Up until 1999, monoamine oxidase inhibitors (MAOIs), selective serotonin reuptake inhibitors (SSRIs), and serotonin and norepinephrine reuptake inhibitors (SNRIs) were the predominant pharmaceutical treatments for depression. The hydroxy-tryptamine (5-HT) is linked to

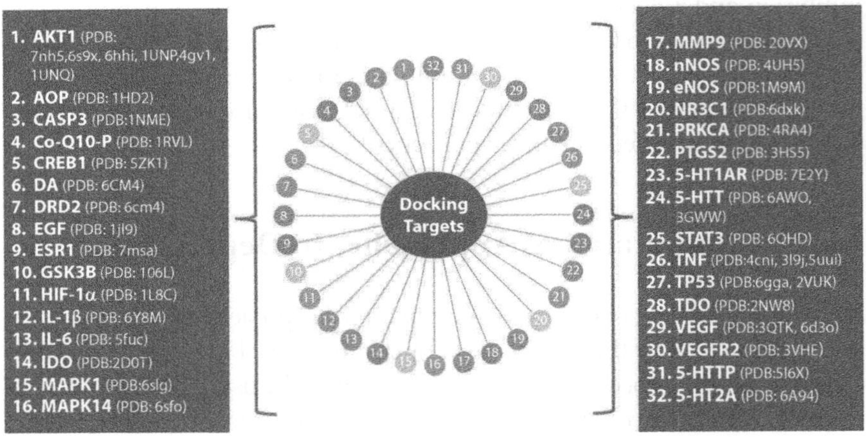

Figure 14.1 Docking targets for anti-depressant drug candidates. AKT1: AKT Serine/Threonine Kinase 1; AOP: Anti-oxidant promoter; CASP3: Caspase-3; Co-Q10-P: Co-Q10 promoter; CREB1: Cyclic AMP-responsive element-binding protein 1; DA: Dopamine; DRD2: D2 Dopamine Receptor; EGF: Epidermal Growth Factor; ESR1: Estrogen receptor alpha; GSK3B: Glycogen synthase kinase 3 beta; HIF-1α: Hypoxia-inducible factor 1-alpha; IL-1β: Interleukin 1 beta; IL-6: Interleukin 6; IDO: Indoleamine 2,3-dioxygenase; MAPK1: Mitogen-activated protein kinase 1; MAPK14: Mitogen-activated protein kinase 14; MMP9: Matrix metalloproteinase 9; nNOS: Neuronal nitric oxide synthase; eNOS: Endothelial nitric oxide synthase; NR3C1: Glucocorticoid receptor gene; PRKCA: Protein Kinase C Alpha; PTGS2: Prostaglandin-endoperoxide synthase (cyclooxygenase-2); 5-HT1AR: Serotonin (5-HT1A) receptor; 5-HTT: Serotonin (5-HT) transporter; STAT3: Signal transducer and activator of transcription 3; TNF: Tumor necrosis factor; TP53: Tumor protein 53 gene; TDO: Tryptophan 2,3-dioxygenase; VEGF: Vascular endothelial growth factor; VEGFR2: Vascular endothelial growth factor receptor 2; 5-HTTP: 5-HT transporter protein; 5-HT2A: 5-HT2A receptor.

biological processes. Numerous investigations discovered a correlation between serotonin-related metrics and mood within the normal range. Greater blood serotonin levels were linked to improved mood, while lower platelet serotonin-2 receptor activity was linked to worse mood. The emotional and cognitive domains have been linked to the serotonergic system. Specifically, a variety of illnesses, including anxiety, eating disorders, schizophrenia, and depression, are linked to 5-HT2A (5-hydroxytryptamine receptor 2A). Since 5-HT2A is a GPCR and a known target for various conditions, it is significant to the pharmaceutical industry. The serotonergic system, which includes 5-HT and its associated receptors, is the extensive neurotransmitter-mediated GPCR. It has been linked to the genesis of numerous mental illnesses including depression [74].

Kanagarajadurai *et al.* investigated 5HT2A, examining a collection of conserved residues. The study was conducted using human 5HT2A with PDB IDs 2RH1 and 1U19, natural ligands with antagonists (ketanserin and haloperidol), and synthetic agonists (lysergic acid diethylamide and 4-iodo-2,5-dimethoxyphenylisopropylamine). Leu22, Phe340, Val 235, Trp151, Phe339, Val336, Val156, Gly238, and Tyr370 were determined to have the most hydrophobic interactions, whereas ASN363 showed the most hydrophilic interactions (Table 14.1). The investigation was first conducted by comparing the sequences of the 5-HT2A receptors in rats and humans. Ala82, Val150, and Ser242 are three residues that are related to one another in the loop area, according to the 91% similarity results. Moreover, the modification of Ser242 of rats by Ala in human 5-HT2A was reported to enhance receptor affinity toward tryptamine derivatives.

An α-amino acid that is necessary to produce proteins is tryptophan. Tryptophan is a polar molecule with a non-polar aromatic beta carbon substituent consisting of an α-amino group, an α-carboxylic acid group,

Table 14.1 Hydrophobic and hydrophilic interactions site binding of 5-HT.

Target	PDB-ID	Antagonist	Agonist	Hydrophilic interaction	Hydrophobic interaction
5-HT2A	2RH1	Haloperidol	4-iodo-2,5-dimethoxy-phenylisopropylamine (DOI)	ASN 363	Tyr370
	1U19	Ketanserin	lysergic acid diethylamide (LSD)		Phe340
					Phe339
					Phe340

and a side chain indole. Additionally, the neurotransmitter serotonin, which is transcribed by the codon UGG, is a precursor to tryptophan. A study by Purnawati *et al.* (Table 14.2), discovered that low mood and even depression might result from an imbalanced nutritional (tryptophan) consumption. Pineapple and banana are known to be strong sources of tryptophan and to interact with the 5-HTTLPR, which is the area where the serotonin transporter gene is expressed. There may be two versions of this gene: the Long (L) and the Short (S). While the S-variant is linked to a reduction in 5-HTT concentration, the L-variant oversees reuptake. The study over, PDB ID-6AWO, was utilized to investigate sertraline, peonidin from bananas, catechins from both bananas and pine trees, dopamine, norepinephrine, L-tryptophan, and serotonin. Tyr-95, Tyr-175, Asp-98, and Phen-341 interacted with catechin, banana peonidin, and L-tryptophan, as the fundamental amino acids [75].

The heme-containing enzyme tryptophan 2, 3-dioxygenase (TDO), which is present in the liver of mammals, oversees catabolizing tryptophan (Trp). Trp is a necessary amino acid that TDO breaks down into

Table 14.2 Antidepressant potency of tryptophan present in banana and pineapple extract.

Target	PDB	Banana metabolites	Pineapple metabolites	H-bond
5HTTLPR	6AWO	Catechin	Catechin	Asp98, Ala169 and Phe335
		L-tryptophan	L-tryptophan	Asp98, Tyr175
		Banana peonidin	Epicatechin	Tyr95, Ser439
TDO	2NW8	Paroxetine, Fluoxetine (Antidepressants)	Sertraline, Hyperforin, Citalopram (Antidepressants)	Arg117, Tyr113 Thr254, Phe51, Tyr24, Tyr27, His55, Leu28.

N-formylkynurenine. One important factor in the structurally based drug design process is the protein-ligand interaction. Dawood *et al.* used *in silico* docking studies to demonstrate how well-known antidepressants bind to the TDO enzyme. The set of antidepressant drugs, Paroxetine, Sertraline, Hyperforin, Citalopram, and Fluoxetine were docked over crystal structures of TDO (PDB ID, 2NW8) recrystallized with L-Trp substrate and heme protein. The maximum binding was with the residues Arg117, Tyr113 and Thr 254, Phe 51, Tyr 24, Tyr 27, His 55, Leu 28 [76].

The 41 amino acid peptide hormone corticotropin-releasing hormone (CRH), sometimes referred to as corticotropin-releasing factor (CRF), functions as a neurotransmitter that is implicated in the stress response. The CRH binds to the CRH receptor (CRFR), a G protein-coupled receptor. The family has two receptors, Type-1 and Type-2, which are each encoded by a different gene (CRHR1 and CRHR2, respectively). One of the two components of the fight-or-flight response to stress is the HPA axis, which is activated by corticotropin-releasing factor receptors (CRFRs). Major depressive disorder and Alzheimer's disease have both been linked to elevated CRH levels. Chronic CRHR1 activation brought on by stress in early life causes anxiety, learning disabilities, and memory problems in adulthood.

The CRF1 antagonists that are currently on the market include Pexacerfont, Antalarmin, CP-316311, and CP-154,526. The pathophysiology of several illnesses, including addiction, post-traumatic stress disorder, anxiety, depression, and stress, is influenced by corticotropin-releasing factor-1, or CRFR. The docking study was conducted over synthetic

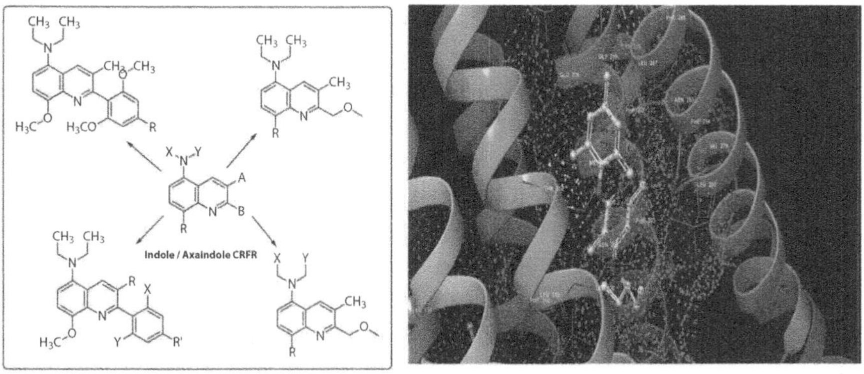

Figure 14.2 Indole and Azaindole aminoquinolines derivatives interacting sites in the 2D structure of PDB-4K5Y.

derivatives of 8-substituted-2-aryl-5-alkyl-aminoquinolines using the human CRFIR crystal structure PDB ID-4K5Y (Figure 14.2).

On analyzing the docked complexes, it was concluded that the residues HIS214, THR215, ARG227, ARG1008, LYS1060, and ASP1061 are important in forming a hydrogen bond with the inhibitors (the indole/azaindole CRFR inhibitor scaffold) [77]. Belhassan *et al.* studied triazolo-benzodiazepine analogs as antidepressants [78]. Since 1957, the well-known class of heterocycles known as diazepines has grown in significance. The psychotropic properties of chlordiazepoxide are already investigated. Analogs of triazolo-benzodiazepines are an important structural motif in several medications with anticancer, sedative, and muscle relaxant properties. Alprazolam, adinazolam and estazolam are drugs based on triazolo-benzodiazepine scaffold that are widely used as anxiolytic and sedative agents. Molecular docking studies on the Drosophila melanogaster dopamine transporter (dDAT) showed that dopamine is reabsorbed from the synaptic cleft using this transmembrane protein. Since DAT inhibitors increase dopamine at the synaptic cleft, they are used as adjuvant therapy for Parkinson's disease (PD) and to treat depressive disorders. Neurotransmitter transporter (DAT) was docked to a novel class of triazolo-benzodiazepines. In a screening process, novel triazolo-benzodiazepine scaffolds were compared to nortriptyline. Nortriptyline participates in alkyl and Pi-alkyl interactions with Ala479, Pi-sigma interactions with Val120, and Pi-Pi T-shaped interactions with Tyr124. Nortriptyline's N-methylpropanamine chain enables it to form carbon hydrogen bonds with Asp46, Ser421, and Phe319, but it can only form conventional hydrogen bonds and Pi-donor hydrogen bonds with Phe43 (Figure 14.3) because it has a nitrogen atom (N) in this position. The most active is M-11 (PDB ID: 4M48). The triazolobenzodiazepine derivatives M1, M2, M3, M8, M10, and M11 displayed less toxicity against daphnia compared to the standard molecule (nortriptyline) during *in vitro* toxicity characteristics prediction [78].

Pratyush *et al.* performed a docking study of 3, 5-diphenyl- pyrazoline on monoamine oxidase A receptor. Common symptoms of depression include minimal energy, loss of interest, guilt emotions, trouble sleeping or eating, and difficulty concentrating. If these enzymes are blocked, monoamine oxidase (MAO) is a powerful site for treating depression by improving patients' moods. We learned more about MAO and its two isoforms, MAO A and MAO B. While MAO B inhibition benefited Parkinson's patients, MAO-A inhibition benefited depression patients. In the study of molecular interactions over MAO receptor (PDB- 2Z5X), The amino acids that the ligand was interacting with were TRP193, TYR410, ILE423, ARG424, PRO275, SER442, LEU298, TRP196, ALA409, ALA438, PHE411, THR437, and GLU436.

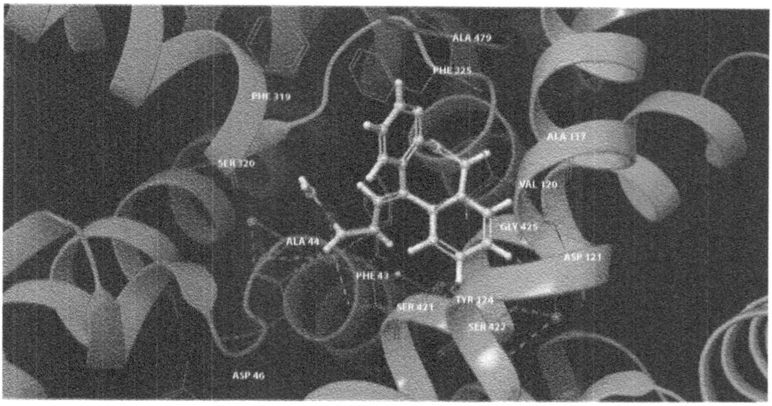

Figure 14.3 The 2D co-crystallize ligand of benzodiazepine with its interacting sites over PDB-4M48.

The molecule selected as ligand possesses phenyl rings at 3 and 5 positions. In their cyclic form, pyrazolines seem like the structure of hydrazine. The pyrazoline molecule's aromatization might increase the selectivity [79].

In the similar study over allosteric site [80], describe that allostery is important for controlling the biological functions of molecules. The use of allosteric modulators will increase the selectivity and lower toxicity.

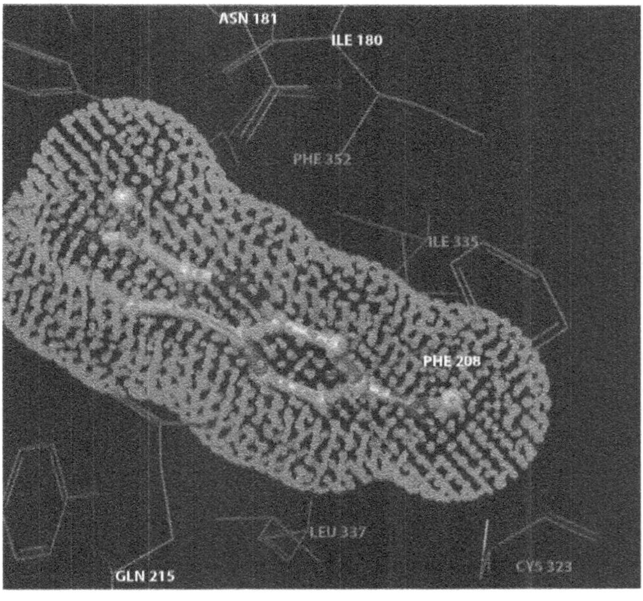

Figure 14.4 The 2D interactions of co-crystalize ligand harmine over human MAO-A.

Table 14.3 Traditional Chinese medicines investigated for depression using computational studies.

Category	Sub-component	Main active components	PDB	Targets	Reference
Yu Zheng	Chaihu Longgu Oyster Decoction	Quercetin, 7-methoxy-2-methyl isoflavone, wogonin, luteolin	4GV1, 4CNI, 4Y6O, 3QKL, 3l9J, 1UNQ, 5UUI	AKT1, IL-6, TNF.	[81]
Genus-Asclepiadaceae	*Dregea sinensis* Hemsl.	Steroidal glycosides	1E8Y, 4GV1, 4JSP, 2Z5Y, 4CRT, 2AZ5, 3ALQ, 2ERJ	PI3K, AKT, mTOR, MAO-A and B TNF-R2, IL2R and TNF-α	[82]
Sinisan Formula	*Bupleurum chinense, Paeoniae Radix Alba, Aurantii Fructus Immaturus* and Licorice	Isosinensetin, formononetin, nobiletin, stigmasterol, betasitosterol, and sitosterol	1O6L, 3HS5, 1M9M, 4RA4, 5I6X, 1L2I, 2OVX, 2VUK, 1NME.	Antidepressant effect via serotoninergic, dopaminergic and GABAergic synapse pathway	[83]
Epicedium	Icariin	luteolin, quercetin, kaempferol	5FUC, 6D3O, 6S9Z, 1Jl9,	IL-1B, IL, LPS, VEGFA,	[84]
Cyperi Rhizoma and *Chuanxiong Rhizoma* herb	*Cyperi Rhizoma* and *Chuanxiong Rhizoma* herb	Quercetin, kaempferol luteolin, stigmasterol, isorhamnetin, beta-sitosterol	6HHI, 1ALU, 6GGA, 6CM4, 6SlG, 6DXK	AKT1, IL-6, TP53, MAPK1, DRD2, NR3C1	[85]

(Continued)

Table 14.3 Traditional Chinese medicines investigated for depression using computational studies. (*Continued*)

Category	Sub-component	Main active components	PDB	Targets	Reference
Curcumae Radix	*Curcuma longa* (Zingiberaceae)	Oxycurcumenol, Beta-sitosterol, sitosterol, zedoalactone E	5HG8, 5F19, 5J2X, 2XRW 3CBP	EGFR, PTGS2, HSP90AA1, MAPK8, TYK2 and ESR1	[86]
Polyrhachis vicina Rogers		Tetradecanoanoic acid, hexadecenoic acid, 14-methyl-pentadecanoic acid, octadecadienoic acid	-	CREB and BDNF an key target in cAMP signaling pathway	[87]

Initially these modulators were used for boosting efficacy and for reduction of catastrophic effect. This time, ligand as isocaboxazid, phelezine, selegiline, and tranylcypromine in chemical forms were tested with natural allosteric MAO inhibitors as homaline and tanimoto over co-crystalize derivative in human monoamine oxidase-A. Phe 208 and Phe 352 with Ile 180 and Ile 335 are found to be important sites of interactions (Figure 14.4). This confirms the binding affinity of natural and chemical derivatives over MAO and its role in depression.

Instead of some common receptors, there are certain Traditional Chinese medicines which are studied over subcategories of a few common receptors (Table 14.3).

14.6 Network Pharmacology of Depression

Formerly depression was thought to be caused by neurotransmitter fluctuations in the brain, particularly serotonin and noradrenaline. The neurotransmitter hypothesis also confirms the involvement of these neurotransmitters in depression. Recently, network pharmacology evolved as a branch of drug research to find out the underlying mechanism of disease and the actions of drugs in diseases. Researchers have recently established the biological pathways that underlie depression, and these pathways have progressively been seen as biological networks [88]. Stapelberg *et al.* 2019, projected the psycho-immune-neuroendocrine (PINE) depression network model. Accordingly, depression pathophysiology and antidepressant medication use are systematically described by the central nervous system (CNS), immune system, endocrine, and autonomic networks [89]. The systems biology and network approaches are verified by developing fields of high throughput, and data-rich biology, that also include genomics, proteomics, and metabolomics.

14.6.1 The Psycho-Immune Neuroendocrine (PINE) Network Model

The CNS, immune system, and endocrine molecular networks are components of the PINE network model, and the three networks are interrelated. There are several nodes and edges in each of the three molecular networks. Small molecules having a variety of features, like as genes, proteins, or metabolites, might serve as the network's nodes. With the help of PINE network model, Stepelberg and co-authors identified plasma metabolic biomarkers and targeted interactions in the pathogenesis of depression [89, 90]. The metabolic biomarker–enzyme interactions network analysis suggested the role of enzymes in affecting the metabolite concentration. Most of these enzymes belong to the hydrolases, transferases, and acyltransferases family. A variety of metabolisms including energy, amino acid, gut microbe, glycerophospholipid, and sphingolipid are all controlled by these enzymes. These enzymes are crucial for the control of metabolic pathways,

Table 14.4 Important enzymes in depression.

Sr. no.	Enzyme	Function	Role in depression	Reference
01	Cytosolic phospholipase A2 (cPLA2)	Polyunsaturated fatty acids (PUFA) metabolism. Release of arachidonic acid and eicosanoids synthesis.	Dysfunction of neurotransmitter systems and the HPA axis.	[96–98]
02	Phospholipase A2	Polyunsaturated fatty acids (PUFA) metabolism. PGE 2 synthesis	Associated with somatic symptoms of depression.	[91, 99]
03	Cyclooxygenase 2	Polyunsaturated fatty acids (PUFA) metabolism, Production of prostaglandins from arachidonic acid.	Neuroinflammation, hyperfunction of the HPA axis, mitochondrial, dysfunction and hippocampal neuronal damage.	[91, 96, 100]
04	Acyl-protein thioesterase 1 and 2	Depalmitoylation of proteins.	Acyl-protein thioesterase levels increased in mild stress and depression.	[95, 99]

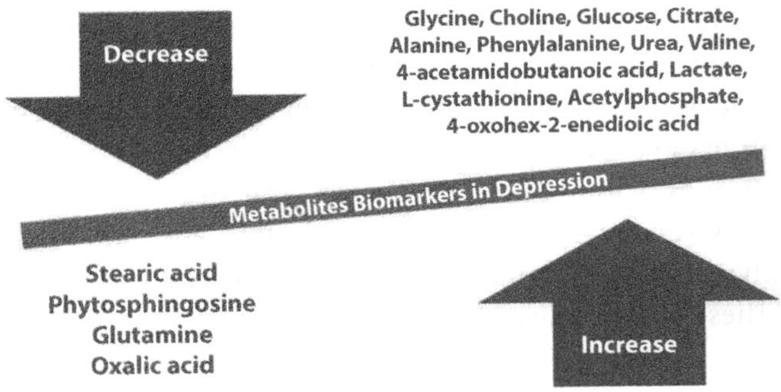

Figure 14.5 Major metabolite biomarkers in depression.

which has an impact on the molecular basis of depression. Enzymes and metabolite biomarkers are listed in Table 14.4 and Figure 14.5, respectively.

According to recent literature, cytosolic phospholipase A2 (cPLA2) is crucial for polyunsaturated fatty acids (PUFA) metabolism and prostaglandin E2 (PGE2) generation, which is the mechanism of mood stabilizers in preclinical research. Additionally, depression patients' peripheral blood cells have a substantial increase in the mRNA expression of the COX-2 gene, which is crucial in the pathophysiology of depression. It is crucial to research the role of phospholipase in depression since the phospholipase A2 family of metabolite-related enzymes is connected to several metabolites. Different phospholipase A2 subtypes may be linked to somatic depressive symptoms, according to prior study [91]. The danger of depression brought on by IFN- α is thought to be increased by genetic variation of the phospholipase A2 gene [91]. Phospholipase A2 inhibitors may also have therapeutic benefits for conditions like depression that are aggravated by inflammation [92]. Eicosapentaenoic acid (EPA) may also up-regulate the expression of the cytosolic phospholipase A2 gene and have an antidepressant effect in a clinical setting, demonstrating EPA's superior antidepressant impact [93]. The identical G allele of the PLA2 Ban I polymorphism was also discovered by another study team as a risk factor for depression in Korean populations [94]. In chronic moderate stress (CMS), the expression of acyl protein thioesterase 2 (APT-2) and oleamide is elevated. This serves as a guide for how to target these proteins in the therapy of depression [95]. Studies to establish the role of acetylhydrolase in depression are a fresh area for study.

Stearic acid, phytosphingosine, glutamine, and phospholipids were up-regulated whereas most metabolites quantities were down-regulated

(Figure 14.5). By oxidizing -fatty acids, acetyl-coenzyme A (CoA) is produced. CoA may then be converted to ATP in the TCA cycle, which can then be used to create ketone bodies that can be stored in the liver and kidney [101]. According to studies, depressed individuals have considerably higher plasma concentrations of stearic acid, which may prevent fatty acid transport and disrupt the TCA cycle [102, 103]. These findings are in line with earlier study findings. Moreover, oxalic acid and stearic acid both are increased indicating strong clinical diagnostic value. Stearic acid might therefore be used to diagnose depression. These suggest that the biological basis of depression involves the metabolic indicators in the metabolic biomarker–enzyme interactions network.

Additionally, most metabolic indicators are long-chain fatty acids and phytosphingosine-related, suggesting dysfunction of lipid metabolism as a key factor in depression. Sphingolipid and glycerophospholipid metabolism are also impacted by these metabolic indicators. Sphingosine has a role in several biological processes, including cell-cell communication, cell division, proliferation, and apoptosis. High levels of hemolytic phosphatidylcholine also lead to oxidative stress, which is a major contributor to depression development [104].

14.7 Conclusion

Depression pathophysiology is multifactorial which shows an involvement of multiple mechanisms, and receptors. Several hypotheses are proposed to support the etiopathogenesis of depression. Computational approaches including network pharmacology and docking studies are key tools to explore the mechanisms of various anti-depressant drugs. These computational tools will further guide the discovery and development of potential anti-depressant therapeutics.

References

1. Wang, S., Lu, T., Sun, J., Huang, L., Li, R., Wang, T., *et al.*, Long-term trends in the incidence of depressive disorders in China, the United States, India and globally: A comparative study from 1990 to 2019. *Front. Psychol.*, 13, 1066706, 2022.
2. Nobis, A., Zalewski, D., Waszkiewicz, N., Peripheral Markers of Depression. *J. Clin. Med.*, 9, 12, 3793, 2020.

3. Norkeviciene, A., Gocentiene, R., Sestokaite, A., Sabaliauskaite, R., Dabkeviciene, D., Jarmalaite, S., *et al.*, A Systematic Review of Candidate Genes for Major Depression. *Medicina (Kaunas)*, 58, 2, 285, 2022.
4. Fava, M. and Kendler, K.S., Major depressive disorder. *N*, 28, 2, 335–41, 2000.
5. Nestler, E.J., Barrot, M., DiLeone, R.J., Eisch, A.J., Gold, S.J., Monteggia, L.M., Neurobiology of depression. *Neuron*, 34, 1, 13–25, 2002.
6. Li, Z., Ruan, M., Chen, J., Fang, Y., Major depressive disorder: advances in neuroscience research and translational applications. *NB*, 37, 863–80, 2021.
7. American Psychiatric Association D and Association AP, *Diagnostic and statistical manual of mental disorders: DSM-5*, American psychiatric association, Washington, DC, 2013.
8. Vahia, V.N., Diagnostic and statistical manual of mental disorders 5: A quick glance. *Indian J. Psychiatry*, 55, 3, 220–3, 2013.
9. Bains, N. and Abdijadid S., *Major Depressive Disorder,* in: StatPearls. Treasure Island (FL): StatPearls Publishing, USA, 2024.
10. Magnusson, A. and Boivin, D., Seasonal affective disorder: an overview. *Chronobiol. Int.*, 20, 2, 189–207, 2003.
11. Mc Mahon, B., Andersen, S.B., Madsen, M.K., Hjordt, L.V., Hageman, I., Dam, H., *et al.*, Seasonal difference in brain serotonin transporter binding predicts symptom severity in patients with seasonal affective disorder. *Brain*, 139, Pt 5, 1605–14, 2016.
12. Lewy, A.J., Lefler, B.J., Emens, J.S., Bauer, V.K., The circadian basis of winter depression. *Proc. Natl. Acad. Sci. U.S.A.*, 103, 19, 7414–9, 2006.
13. Field, T., Prenatal depression effects on early development: a review. *Infant Behav. Dev.*, 34, 1, 1–14, 2011.
14. Patel, M., Bailey, R.K., Jabeen, S., Ali, S., Barker, N.C., Osiezagha, K., Postpartum depression: a review. *J. Health Care Poor Underserved*, 23, 2, 534–42, 2012.
15. Cristancho, M.A. and Thase, M.E., Persistent depressive disorders: dysthymia and chronic major depressive disorder. *The Encyclopedia of Clinical Psychology*, 1–2. 2015. https://doi.org/10.1002/9781118625392.wbecp188
16. Duman, R.S., Heninger, G.R., Nestler, E.J., A molecular and cellular theory of depression. *Arch. Gen. Psychiatry*, 54, 7, 597–606, 1997.
17. Patel, R.K. and Rose, G.M., Persistent Depressive Disorder, in StatPearls [Internet]. Treasure Island (FL): StatPearls Publishing; USA, 2024.
18. Health UDo and Services H, *Disruptive mood dysregulation disorder: The basics,* National Institute of Mental Health, 2022, Retrieved May.
19. Freeman, E.W., Premenstrual syndrome and premenstrual dysphoric disorder: definitions and diagnosis. *Psychoneuroendocrinology*, 28 Suppl 3, 25–37, 2003.
20. di Scalea, T.L. and Pearlstein, T., Premenstrual dysphoric disorder. *PCNA*, 40, 2, 201–16, 2017.
21. Filatova, E.V., Shadrina, M.I., Slominsky, P.A., Major Depression: One Brain, One Disease, One Set of Intertwined Processes. *Cells*, 10, 6, 1283, 2021.

22. Kaltenboeck, A. and Harmer, C., The neuroscience of depressive disorders: A brief review of the past and some considerations about the future. *BNA*, 2, 2398212818799269, 2018.
23. Jiang, Y., Zou, D., Li, Y., Gu, S., Dong, J., Ma, X., *et al.*, Monoamine Neurotransmitters Control Basic Emotions and Affect Major Depressive Disorders. *Pharmaceuticals (Basel)*, 15, 10, 1203, 2022.
24. Boku, S., Nakagawa, S., Toda, H., Hishimoto, A., Neural basis of major depressive disorder: Beyond monoamine hypothesis. *Psychiatry Clin. Neurosci.*, 72, 1, 3–12, 2018.
25. Karabin, T., Biala, G., Kruk-Slomka, M., The monoamine theory of depression as a target to effective pharmacotherapy. *CIPMS*, 36, 2, 108–13, 2023.
26. Sabella, D., Antidepressant Medications. *Am. J. Nurs.*, 118, 9, 52–9, 2018.
27. De Kloet, E.R., Joëls, M., Holsboer, F., Stress and the brain: from adaptation to disease. *NRN*, 6, 6, 463–75, 2005.
28. Mikulska, J., Juszczyk, G., Gawrońska-Grzywacz, M., Herbet, M., HPA axis in the pathomechanism of depression and schizophrenia: new therapeutic strategies based on its participation. *BS*, 11, 10, 1298, 2021.
29. Varghese, F.P. and Brown, E.S., The hypothalamic-pituitary-adrenal axis in major depressive disorder: a brief primer for primary care physicians. *PCCJCP*, 3, 4, 151, 2001.
30. Keller, P.A., McCluskey, A., Morgan, J., O'connor, S.M.J., The role of the HPA axis in psychiatric disorders and CRF antagonists as potential treatments. *AP*, 339, 7, 346–55, 2006.
31. Rein, T., Ambree, O., Fries, G.R., Rappeneau, V., Schmidt, U., Touma, C., The hypothalamic-pituitary-adrenal axis in depression: molecular regulation, pathophysiological role, and translational implications, in: *Neurobiology of Depression*, pp. 89–96, Academic Press, USA, Elsevier, 2019.
32. Gałecki, P. and Talarowska, M., Inflammatory theory of depression. *PP*, 52, 3, 437–47, 2018.
33. Kiss, J.P., Theory of active antidepressants: a nonsynaptic approach to the treatment of depression. *NI*, 52, 1-2, 34–9, 2008.
34. Yu, H. and Chen, Z.-Y., The role of BDNF in depression on the basis of its location in the neural circuitry. *APS*, 32, 1, 3–11, 2011.
35. Liu, Q., He, H., Yang, J., Feng, X., Zhao, F., Lyu, J., Changes in the global burden of depression from 1990 to 2017: Findings from the Global Burden of Disease study. *J. Psychiatr. Res.*, 126, 134–40, 2020.
36. Sun, Z., Jia, L., Shi, D., He, Y., Ren, Y., Yang, J., *et al.*, Deep brain stimulation improved depressive-like behaviors and hippocampal synapse deficits by activating the BDNF/mTOR signaling pathway. *Behav. Brain Res.*, 419, 113709, 2022.
37. Fries, G.R., Saldana, V.A., Finnstein, J., Rein, T., Molecular pathways of major depressive disorder converge on the synapse. *MP*, 28, 1, 284–97, 2023.

38. Li, X., Teng, T., Yan, W., Fan, L., Liu, X., Clarke, G., *et al.*, AKT and MAPK signaling pathways in hippocampus reveals the pathogenesis of depression in four stress-induced models. *Transl. Psychiatry*, 13, 1, 200, 2023.
39. Bajpai, A., Verma, A.K., Srivastava, M., Srivastava, R., Oxidative stress and major depression. *J. Clin. Diagn. Res.*, 8, 12, CC04–7, 2014.
40. Ait Tayeb, A.E.K., Poinsignon, V., Chappell, K., Bouligand, J., Becquemont, L., Verstuyft, C., Major Depressive Disorder and Oxidative Stress: A Review of Peripheral and Genetic Biomarkers According to Clinical Characteristics and Disease Stages. *Antioxidants (Basel)*, 12, 4, 942, 2023.
41. Correia, A.S., Cardoso, A., Vale, N., Oxidative Stress in Depression: The Link with the Stress Response, Neuroinflammation, Serotonin, Neurogenesis and Synaptic Plasticity. *Antioxidants (Basel)*, 12, 2, 470, 2023.
42. Vavakova, M., Durackova, Z., Trebaticka, J., Markers of Oxidative Stress and Neuroprogression in Depression Disorder. *Oxid. Med. Cell Longev.*, 2015, 898393, 2015.
43. Hamed, R.A., Elmalt, H.A., Salama, A.A., Abozaid, S.Y., Ahmed, A.S., Biomarkers of oxidative stress in major depressive disorder. *OAMJMS*, 8, B, 501–6, 2020.
44. Himmerich, H., Patsalos, O., Lichtblau, N., Ibrahim, M.A., Dalton, B., Cytokine research in depression: principles, challenges, and open questions. *FP*, 10, 30, 2019.
45. Yirmiya, R. and Goshen, I., Immune modulation of learning, memory, neural plasticity and neurogenesis. *Brain Behav. Immun.*, 25, 2, 181–213, 2011.
46. Islam, M.R., Sohan, M., Daria, S., Masud, A.A., Ahmed, M.U., Roy, A., *et al.*, Evaluation of inflammatory cytokines in drug-naive major depressive disorder: A systematic review and meta-analysis. *Int. J. Immunopathol. Pharmacol. (IJIP)*, 37, 03946320231198828, 2023.
47. Miller, E.S., Sakowicz, A., Roy, A., Yang, A., Sullivan, J.T., Plasma and cerebrospinal fluid inflammatory cytokines in perinatal depression. *American Journal Of Obstetrics and Gynecology (AJOG)*, 220, 3,271.e1–e10, 2019.
48. Haimovici, F., Anderson, J.L., Bates, G.W., Racowsky, C., Ginsburg, E.S., Simovici, D., *et al.*, Stress, anxiety, and depression of both partners in infertile couples are associated with cytokine levels and adverse IVF outcome. *Am. J. Reprod. Immunol.*, 79, 4, e12832, 2018.
49. Dionisie, V., Filip, G.A., Manea, M.C., Manea, M., Riga, S., The anti-inflammatory role of SSRI and SNRI in the treatment of depression: a review of human and rodent research studies. *I*, 29, 75–90, 2021.
50. Vahid-Ansari, F. and Albert, P.R., Rewiring of the serotonin system in major depression. *FP*, 12, 802581, 2021.
51. Hoyer, D., Hannon, J.P., Martin, G.R., Molecular, pharmacological and functional diversity of 5-HT receptors. *PB Behav.*, 71, 4, 533–54, 2002.
52. Albert, P.R. and Vahid-Ansari, F., The 5-HT1A receptor: signaling to behavior. *B*, 161, 34–45, 2019.

53. Dunlop, B.W. and Nemeroff, C.B., The role of dopamine in the pathophysiology of depression. *AGP*, 64, 3, 327–37, 2007.
54. Zhao, F., Cheng, Z., Piao, J., Cui, R., Li, B., Dopamine Receptors: Is It Possible to Become a Therapeutic Target for Depression? *FP*, 13, 947785, 2022.
55. Mathews, D.C., Henter, I.D., Zarate, C.A., Targeting the glutamatergic system to treat major depressive disorder: rationale and progress to date. *Drugs*, 72, 10, 1313–33, 2012.
56. Moriguchi, S., Takamiya, A., Noda, Y., Horita, N., Wada, M., Tsugawa, S., *et al.*, Glutamatergic neurometabolite levels in major depressive disorder: a systematic review and meta-analysis of proton magnetic resonance spectroscopy studies. *Mol. Psychiatry*, 24, 7, 952–64, 2019.
57. Niciu, M.J., Ionescu, D.F., Richards, E.M., Zarate, C.A., Jr., Glutamate and its receptors in the pathophysiology and treatment of major depressive disorder. *J. Neural. Transm. (Vienna)*, 121, 8, 907–24, 2014.
58. Wolf, M.E. and Mosnaim, A.D., Phenylethylamine in neuropsychiatric disorders. *Gen. Pharmacol.*, 14, 4, 385–90, 1983.
59. Zeng, Z., Fan, P., Rand, E., Kyaw, H., Su, K., Madike, V., *et al.*, Cloning of a putative human neurotransmitter receptor expressed in skeletal muscle and brain. *Biochem. Biophys. Res. Commun.*, 242, 3, 575–8, 1998.
60. Mantas, I., Millan, M.J., Di Cara, B., Groenink, L., Veiga, S., Cistarelli, L., *et al.*, Trace Amine-Associated Receptor 1 Contributes to Diverse Functional Actions of O-Phenyl-Iodotyramine in Mice but Not to the Effects of Monoamine-Based Antidepressants. *Int. J. Mol. Sci.*, 22, 16, 8907, 2021.
61. Navarria, A., Wohleb, E.S., Voleti, B., Ota, K.T., Dutheil, S., Lepack, A.E., *et al.*, Rapid antidepressant actions of scopolamine: Role of medial prefrontal cortex and M1-subtype muscarinic acetylcholine receptors. *Neurobiol. Dis.*, 82, 254–61, 2015.
62. Liu, S., Shi, D., Sun, Z., He, Y., Yang, J., Wang, G., M2-AChR mediates rapid antidepressant effects of scopolamine through activating the mTORC1-BDNF signaling pathway in the medial prefrontal cortex. *FP*, 12, 601985, 2021.
63. Dulawa, S.C. and Janowsky, D.S., Cholinergic regulation of mood: from basic and clinical studies to emerging therapeutics. *Mol. Psychiatry*, 24, 5, 694–709, 2019.
64. Jarończyk, M. and Walory, J., Novel molecular targets of antidepressants. *M*, 27, 2, 533, 2022.
65. Pehrson, A.L. and Sanchez, C., Altered gamma-aminobutyric acid neurotransmission in major depressive disorder: a critical review of the supporting evidence and the influence of serotonergic antidepressants. *Drug Des. Dev. Ther.*, 9, 603–24, 2015.
66. Bambico, F.R., Katz, N., Debonnel, G., Gobbi, G., Cannabinoids elicit antidepressant-like behavior and activate serotonergic neurons through the medial prefrontal cortex. *JN*, 27, 43, 11700–11, 2007.

67. Howlett, A.C., Qualy, J.M., Khachatrian, L.L., Involvement of Gi in the inhibition of adenylate cyclase by cannabimimetic drugs. *Mol. Pharmacol.*, 29, 3, 307–13, 1986.
68. Zhou, D., Li, Y., Tian, T., Quan, W., Wang, L., Shao, Q., *et al.*, Role of the endocannabinoid system in the formation and development of depression. *Pharmazie*, 72, 8, 435–9, 2017.
69. Baldi, A., Computational approaches for drug design and discovery: An overview. *SRP*, 1, 1, 99, 2010.
70. Sadybekov, A.V. and Katritch, V., Computational approaches streamlining drug discovery. *Nature*, 616, 7958, 673–85, 2023.
71. Singh, D.B. and Pathak, R.K., Computational approaches in drug designing and their applications, in: *Experimental Protocols in Biotechnology*, pp. 95–117, 2020.
72. Tiwari, A. and Singh, S., Computational approaches in drug designing, *Bioinformatics, Methods and Applications*, pp. 207–217, Academic Press USA, Elsevier, 2022.
73. Hung, C.L. and Chen, C.C., Computational approaches for drug discovery. *Drug Dev. Res.*, 75, 6, 412–8, 2014.
74. Kanagarajadurai, K., Malini, M., Bhattacharya, A., Panicker, M.M., Sowdhamini, R., Molecular modeling and docking studies of human 5-hydroxytryptamine 2A (5-HT2A) receptor for the identification of hotspots for ligand binding. *Mol. Biosyst.*, 5, 12, 1877–88, 2009.
75. Purnawati, S., Wrasiati, L.P., Lesmana, C.B.J., Megantara, S., Lesmana, R., A study of molecular docking of l-tryptophan ligand as a compound in pineapples and bananas binding with the human serotonin transporter (SERT). *BMJ*, 11, 3, 1243–9, 2022.
76. Dawood, S., Zarina, S., Bano, S., Docking studies of antidepressants against single crystal structure of tryptophan 2, 3-dioxygenase using Molegro Virtual Docker software. *Pak. J. Pharm. Sci.*, 27, 5 Spec no, 1529–39, 2014.
77. Babu, S., Docking Study of Corticotropin-Releasing Factor-1 Receptor with Its Antagonists. *CNS*, 11, 1, 19–24, 2018.
78. Belhassan, A., Zaki, H., Benlyas, M., Lakhlifi, T., Bouachrine, M., Study of novel triazolo-benzodiazepine analogues as antidepressants targeting by molecular docking and ADMET properties prediction. *Heliyon*, 5, 9, e02446, 1-7, 2019.
79. Pratyush, K. and Asnani, A., Docking of 3, 5-diphenyl-pyrazoline with monoamine oxidase A receptor and *In-Silico* structural property calculation. *JDD Ther.*, 9, 3-s, 43–5, 2019.
80. Srivastava, P. and Jain, C.K., *In-Silico* Investigation of Plant-Derived Natural Allosteric Compounds Towards Enhanced Drug-Protein Interaction of MOA Protein Complex in Depression Based on Molecular Docking and Molecular Dynamic Simulation Approaches. *CTB Pharm.*, 16, 4, 529–39, 2022.
81. Li, C., Li, B., Liu, H., Qu, L., Wang, H., Mechanism of Chaihu Longgu oyster adjusted decoction for the treatment of depression based on network

pharmacology and molecular docking technology. *Ann. Transl. Med.*, 11, 4, 172, 2023.

82. Liu, X., Shi, Y., Deng, Y., Dai, R., Using Molecular Docking Analysis to Discovery Dregea sinensis Hemsl. Potential Mechanism of Anticancer, Antidepression, and Immunoregulation. *Pharmacogn. Mag.*, 13, 51, 358–62, 2017.
83. Wang, H., Liu, J., He, J., Huang, D., Xi, Y., Xiao, T., *et al.*, Potential mechanisms underlying the therapeutic roles of sinisan formula in depression: Based on network pharmacology and molecular docking study. *Front. Psychiatry*, 13, 1063489, 2022.
84. Dong, Y., Tao, B., Xue, X., Feng, C., Ren, Y., Ma, H., *et al.*, Molecular mechanism of Epicedium treatment for depression based on network pharmacology and molecular docking technology. *BMC Complement. Med. Ther.*, 21, 1, 222, 2021.
85. Shi, Y., Chen, M., Zhao, Z., Pan, J., Huang, S., Network Pharmacology and Molecular Docking Analyses of Mechanisms Underlying Effects of the Cyperi Rhizoma-Chuanxiong Rhizoma Herb Pair on Depression. *Evid. Based Complement. Alternat. Med.*, 2021, 5704578, 2021.
86. Wang, X., Lin, Q., Shen, M., Lin, H., Feng, J., Peng, L., *et al.*, Identification of the ingredients and mechanisms of Curcumae Radix for depression based on network pharmacology and molecular docking, *Nat. Prod. Commun.*, 16, 5, 1934578X211016643, 2021.
87. He, J., Han, D., Jia, C., Xie, J., Zhu, F., Wei, J., *et al.*, Integrating Network Pharmacology, Molecular Docking and Pharmacological Evaluation for Exploring the Polyrhachis vicina Rogers in Ameliorating Depression. *Drug Des. Dev. Ther.*, 17, 717–35, 2023.
88. Headrick, J.P., Peart, J.N., Budiono, B.P., Shum, D.H., Neumann, D.L., Stapelberg, N., *et al.*, The heartbreak of depression: 'Psycho-cardiac' coupling in myocardial infarction. *JMCC*, 106, 14–28, 2017.
89. Stapelberg, N., Neumann, D., Shum, D., Headrick, J., Health, pre-disease and critical transition to disease in the psycho-immune-neuroendocrine network: Are there distinct states in the progression from health to major depressive disorder? *P. Behav.*, 198, 108–19, 2019.
90. Stapelberg, N., Neumann, D., Shum, D., McConnell, H., Hamilton-Craig, I., From physiome to pathome: a systems biology model of major depressive disorder and the psycho-immune-neuroendocrine network. *CPR*, 11, 1, 32–62, 2015.
91. Su, K.-P., Huang, S.-Y., Peng, C.-Y., Lai, H.-C., Huang, C.-L., Chen, Y.-C., *et al.*, Phospholipase A2 and cyclooxygenase 2 genes influence the risk of interferon- α-induced depression by regulating polyunsaturated fatty acids levels. *Biol. psychiatry*, 67, 6, 550–57, 2010.
92. Magrioti, V. and Kokotos, G., Phospholipase A2 inhibitors as potential therapeutic agents for the treatment of inflammatory diseases. *EOTP*, 20, 1, 1–18, 2010.

93. Su, K.-P., Yang, H.-T., Chang, J.P.-C., Shih, Y.-H., Guu, T.-W., Kumaran, S.S., *et al.*, Eicosapentaenoic and docosahexaenoic acids have different effects on peripheral phospholipase A2 gene expressions in acute depressed patients. *Prog. Neuropsychopharmacol. Biol. Psychiatry*, 80, 227–33, 2018.
94. Pae, C.-U., Yu, H.-S., Kim, J.-J., Lee, C.-U., Lee, S.-J., Lee, K.-U., *et al.*, Ban I Polymorphism of the Cytosolic Phospholipase A2 Gene and Mood Disorders in the Korean Population. *Neuropsychobiology*, 49, 4, 185–88, 2004.
95. Ge, L., Zhu, M.M., Yang, J.Y., Wang, F., Zhang, R., Zhang, J.H., *et al.*, Differential proteomic analysis of the anti-depressive effects of oleamide in a rat chronic mild stress model of depression. *Pharmacol. Biochem. Behav.*, 131, 77–86, 2015.
96. Chang, J.P., Guu, T.W., Chen, Y.C., Galecki, P., Walczewska, A., Su, K.P., BanI polymorphism of cytosolic phospholipase A2 gene and somatic symptoms in medication-free acute depressed patients. *Prostaglandins Leukot. Essent. Fatty Acids*, 136, 111–5, 2018.
97. Ng, C.Y., Kannan, S., Chen, Y.J., Tan, F.C.K., Ong, W.Y., Go, M.L., *et al.*, A new generation of arachidonic acid analogues as potential neurological agent targeting cytosolic phospholipase A2. *Sci. Rep.*, 7, 1, 13683, 2017
98. Song, C., Zhang, X.Y., Manku, M., Increased phospholipase A2 activity and inflammatory response but decreased nerve growth factor expression in the olfactory bulbectomized rat model of depression: effects of chronic ethyl-eicosapentaenoate treatment. *J. Neurosci.*, 29, 1, 14–22, 2009.
99. Gao, Y., Xu, T., Zhao, Y.X., Ling-Hu, T., Liu, S.B., Tian, J.S., *et al.*, A Novel Network Pharmacology Strategy to Decode Metabolic Biomarkers and Targets Interactions for Depression. *Front. Psychiatry*, 11, 667, 2020.
100. He, Y., Han, Y., Liao, X., Zou, M., Wang, Y., Biology of cyclooxygenase-2: An application in depression therapeutics. *Front. Psychiatry*, 13, 1037588, 2022.
101. Lin, L., Huang, Z., Gao, Y., Chen, Y., Hang, W., Xing, J., *et al.*, LC-MS-based serum metabolic profiling for genitourinary cancer classification and cancer type-specific biomarker discovery. *Proteomics*, 12, 14, 2238–46, 2012.
102. Gao, X., Zheng, X., Li, Z., Zhou, Y., Sun, H., Zhang, L., *et al.*, Metabonomic study on chronic unpredictable mild stress and intervention effects of Xiaoyaosan in rats using gas chromatography coupled with mass spectrometry. *J. Ethnopharmacol.*, 137, 1, 690–9, 2011.
103. Tian, J.S., Peng, G.J., Gao, X.X., Zhou, Y.Z., Xing, J., Qin, X.M., *et al.*, Dynamic analysis of the endogenous metabolites in depressed patients treated with TCM formula Xiaoyaosan using urinary (1)H NMR-based metabolomics. *J. Ethnopharmacol.*, 158 Pt A, 1–10, 2014.
104. Nunes, S.O., Vargas, H.O., Prado, E., Barbosa, D.S., de Melo, L.P., Moylan, S., *et al.*, The shared role of oxidative stress and inflammation in major depressive disorder and nicotine dependence. *Neurosci. Biobehav. Rev.*, 37, 8, 1336–45, 2013.

15

Computational Attitudes to Drug Discovery in Anxiety

Meenakshi Attri[1], Asha Raghav[2], Piyush Vatsha[3], Mohit Agrawal[1]*, Manmohan Singhal[4], Hema Chaudhary[1], Nalini Kanta Sahoo[5] and Bhupendra Prajapati[6,7]

[1]*School of Medical & Allied Sciences, K. R. Mangalam University, Gurugram, Haryana, India*
[2]*Department of Pharmaceutics, School of Health Sciences, Sushant University, Gurugram, Haryana, India*
[3]*Department of Pharmacology, Faculty of Pharmacy, M.S. Ramaiah University of Applied Sciences, Bengaluru, Karnataka, India*
[4]*Faculty of Pharmacy, SoPPHI, DIT University, Dehradun, Uttrakhand, India*
[5]*MIT College of Pharmacy, MIT Campus, Ram Ganga Vihar, Phase-II, Moradabad, Uttar Pradesh, India*
[6]*Shree S.K. Patel College of Pharmaceutical Education & Research, Faculty of Pharmacy, Ganpat University, Gujarat, India*
[7]*Department of Industrial Pharmacy, Faculty of Pharmacy, Silpakorn University, Nakhon Pathom, Thailand*

Abstract

In order for organisms to avoid injury and to survive, defensive activities are induced by the adaptive states of fear and anxiety. however, they can be harmful in excess and result in poor mental health. The development of effective drugs for anxiety is challenging due to the complex and multifactorial nature of the disorder. The existing *in vivo* models frequently use artificial systems that inadequately mimic clinical presentation, such as directed evolution, mutations, and fear induction. Whereas there is growing interest in the field of computational modeling of behavior because it offers a mechanical framework for comprehending the cognitive and neurological underpinnings of fear and anxiety. In recent years, computational approaches have emerged as valuable tools in drug discovery for anxiety. The review offers an overview of computational attitudes, techniques,

**Corresponding author*: mohitagrawalmohu@gmail.com; ORCID: 0000-0003-0200-7882

Bhupendra Prajapati, Alok Tripathi, Rishabha Malviya and Lucy Mohapatra (eds.) Computational Neuropharmacology: Fundamentals and Clinical Aspects, (437–482)

and advancements in the field of drug discovery for anxiety. Molecular modelling techniques, including molecular docking, molecular dynamics simulations, and quantum mechanical calculations, are used to study the interactions between drug molecules and their target receptors or enzymes. Virtual screening methods, such as structure-based virtual screening and ligand-based virtual screening, are employed to identify potential drug candidates from large chemical databases. Network analysis is employed to study the complex interactions in anxiety-related pathways. The identification of promising therapeutic candidates has been expedited by these computational methods and also aided in the optimization of drug properties, and provided insights into the molecular mechanisms underlying anxiety. The review also discusses challenges and future perspectives in computational drug discovery for anxiety, including the need for improved accuracy and availability of data, incorporation of multi-omics approaches, and the importance of experimental validation. Overall, computational approaches have become invaluable in drug discovery for anxiety, paving the way for the development of novel and effective therapies for this debilitating mental health condition.

Keywords: Computational modelling, anxiety, model-based planning, drug discovery, virtual screening method, fear, uncertainty

List of Abbreviations

VS	Virtual screening
LBDD	Ligand based drug design
SBDD	Structure based drug design
3D-QSAR	Quantitative Structure Activity Relationship
HTS	High throughput screening
5-HT1A	5-hydroxytryptamine-1-A
CRP1	Corticotropin-releasing factor 1
SERT	Serotonin transporter
CB1	Cannabinoid Receptor 1
ADT	Autodock tools
AI	Artificial Intelligence
HAM-A	Hamilton Anxiety Rating Scale
HADA	Hospital Anxiety & Depression Scale
GADSS	Generalized Anxiety Disorder Severity Scale
FMRI	Functional magnetic resonance imaging
BAI	Back Anxiety Inventory
AUC	Area under the curve
GPCR	G-Protein Couple Receptor

15.1 Introduction

Among the most prevalent disorders encountered in health care facilities are anxiety and depression. The co-occurrence of anxiety and depression with medical problems is a topic of great interest to both academics and clinicians. Depression is a serious mental health issue that has serious negative effects on people's quality of life, lost productivity, and even death. Computational approaches are being used more and more in our modern era to examine psychiatric diseases. Creating models that can be fitted to the conduct of individuals who display subclinical character traits that are known to be risk factors for the formation of mental illnesses is frequently required for this. The conduct of people who exhibit a high level of "trait" anxiety, that is a well-known risk factor for the formation of mood and anxiety disorders, can be used to study anxiety disorders. The most promising candidates to concentrate on experimental work in contemporary medicinal chemistry are molecular designs created using computers. This will help to lower the overall cost of finding and creating a novel medication for the treatment of anxiety [1]. Every stage of the drug development and discovery process benefits from the use of computational techniques, which are significant and fundamental tools. Identifying and evaluating possible novel drug candidates researchers might benefit from a variety of computational approaches. Virtual screening [2–5], (3D-QSAR) 3-D Quantitative Structure-Activity Relationship [6–9], and molecular docking investigations [10–17] are a few of these *in-silico* approaches.

15.2 Computational Approaches for Drug Discovery

15.2.1 Virtual Screening (VS)

VS is a computer-based method to recognize the structures most likely to bind to a therapeutic target, often an enzyme or protein receptor [18, 19].

"Automatically evaluating very large libraries of compounds" is how virtual screening is defined [20]. This remark implies that the primary goal of VS has been to find ways to condense the enormous chemical space of over 1060 potential molecules [21] into a manageable amount that can be synthesised, obtained, and assessed. Theoretically, it may be interesting to search the entire chemical universe, but in more practical VS settings, the emphasis is on building and improving specific combinatorial libraries and expanding libraries of existing compounds using information from vendor offerings or internal compound repositories. Virtual screening has developed into a crucial step in the drug development process as the technique's accuracy

FLOWCHART OF VS [23]

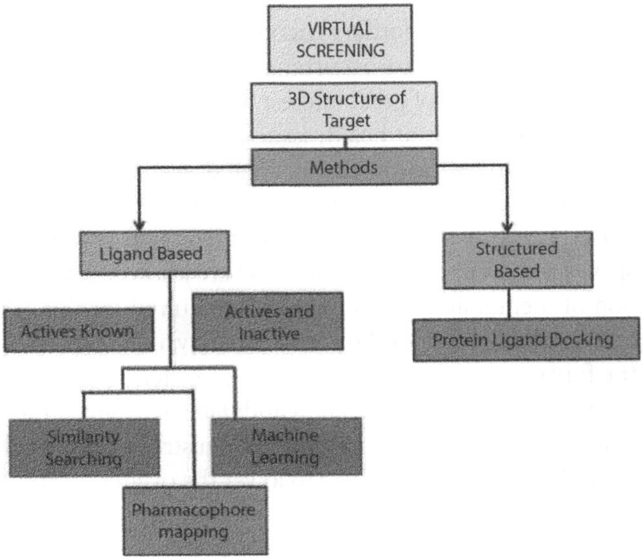

Figure 15.1 Flowchart of virtual screening.

has improved [22]. Compounds from the internal database, those that may be purchased externally, and the next chemical to be synthesised can all be chosen for screening, the virtual screening flowchart separates methods into ligand-based and structure-based approaches, as seen in Figure 15.1 [23].

15.2.1.1 Methods of VS

The two main categories of screening techniques are ligand-based and structure-based [24].

15.2.1.1.1 Ligand-Based Approach [25–33]

Ligand-based methods refer to approaches in drug development and discovery that rely on the characteristics and properties of specific molecules (ligands) to design potential therapeutics. In the context of anxiety diseases, ligand-based methods are used to identify and develop compounds that can modulate the activity of specific targets involved in anxiety pathways.

Here are a few ligand-based methods commonly used in the investigation and development of drugs for anxiety disorder:

- High-throughput screening (HTS): High through output screening perform by screening of large chemical libraries to identify ligands that interact with specific targets relevant to anxiety disorders. This method typically utilizes assays designed to measure the activity or potential for binding of potential ligands against the target of interest.
- Virtual screening VS: Virtual screening involves the application of computational methods to identify potential ligands from large databases or libraries. Through the analysis of molecular structures and properties, virtual screening can prioritize compounds with a higher likelihood of binding to target proteins involved in anxiety pathways.
- QSAR (Quantitative Structure-Activity Relationship Modeling): QSAR models are computational tools that correlate the structural features of ligands with their biological activities. By analyzing the connection between the structure of known ligands and its activity against a target, QSAR models can predict the activity of new compounds, helping researchers identify potential anti-anxiety drug candidates.
- SBDD (Structure-Based Drug Design): In SBDD, ligands that can selectively bind to the target are generated by using three-dimensional structural data of the target protein, which is often collected through methods like NMR spectroscopy or X-ray crystallography.
- Understanding the binding location and interactions in a visual between ligands and the target protein, researchers can design and optimize ligands with improved potency and selectivity.
- Pharmacophore Modeling: Pharmacophore models represent the essential features and spatial arrangement of ligands required for binding to a specific target. By identifying common pharmacophoric elements among known ligands, pharmacophore models can guide the design of novel ligands with similar properties and potential anti-anxiety effects.

These ligand-based methods play a significant part in the development and discovery of new drugs for anxiety disorders by identifying compounds that can modulate specific targets involved in anxiety pathways. They provide valuable tools for rational drug design and optimization, accelerating the process of finding effective treatments for anxiety diseases.

15.2.1.1.2 Structured-Based Methods

Structure-based virtual screening involves using computational methods to identify potential ligands for a target of interest by analyzing their interactions with the target's three-dimensional structure [34–40]. In the context of anxiety disorders, this approach can aid in the discovery of compounds that modulate specific targets involved in anxiety pathways. Here are the steps typically involved:

- Target Selection: Identify and select the relevant protein target associated with anxiety disorders. This could be a receptor, enzyme, transporter, or other protein involved in anxiety pathways.
- Protein structure: Experimental approaches for determining protein structure like Nuclear Magnetic Resonance spectroscopy, X-ray diffraction or cryo-electron microscopy, determine target protein's three-dimensional structure.
- Alternatively, if an experimental structure is unavailable, homology modeling techniques can be employed to generate a predicted structure based on the known structures of related proteins.
- Target Preparation: Prepare the protein structure for virtual screening by removing water molecules, ions, and other non-essential components. Additionally, protonation states and correct bond orders of the protein should be assigned.
- Ligand Library Preparation: Assemble a library of compounds that will be screened virtually. This can include commercially available compounds, natural products, or compounds generated through computational chemistry techniques.
- Molecular Docking: Run molecular docking simulations to determine the potential of binding and modes of the ligands within the target protein's active site. The most advantageous binding orientations and scores for each ligand are predicted by docking algorithms using the protein structure and ligand library.
- Scoring and Ranking: Score and rank the ligands based on their predicted binding affinity or other relevant properties. To assess the quality of ligand-protein interactions and select potential hits various scoring methods are used.

- Hit Validation and Optimization: Experimentally validate the top-ranked virtual hits through *in vivo* or *in vitro* assays to check their activity against the target and their potential for anxiety-related effects. Subsequent medicinal chemistry efforts can be employed to optimize the hits for improved potency, selectivity, and other drug-like properties.

15.3 Ligand-Based Techniques

15.3.1 Modeling Using (Quantitative Structure-Activity Relationship) QSAR

Hansch and Fujita (1964) created the LBDD (ligand-based drug design) technique known as quantitative structure-activity relationship (QSAR) analysis. Since that time and up to the present, QSAR has remained an effective technique to develop mathematical models that uses regression and classification techniques, respectively. To figure out whether the chemical structure and a continuous (Ki, pEC50, etc.) or categorical/binary (active, inactive, dangerous, harmless, etc.) toxicological or biological feature are statistically related. Over the past few decades. The dimensions of molecular descriptors have changed significantly (from 1D to nD), and there are many different approaches to determining the connection between chemical structures and biological features. Quantitative Structure-Activity Relationship modeling was first limited to simple regression methods and

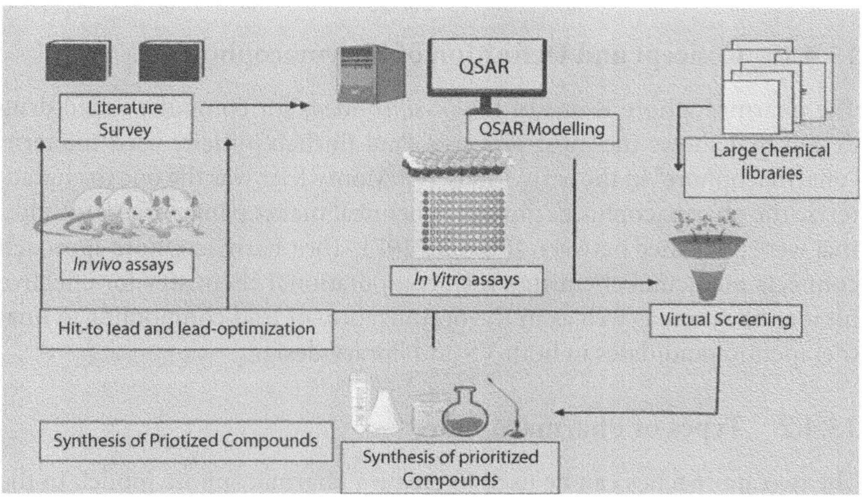

Figure 15.2 Workflow of QSAR.

small series of congeneric compounds. A range of machine learning techniques are now used in the expansion, diversification, and evolution of QSAR modelling to model and perform VS on very large data sets containing hundreds of distinct chemical structures. The QSAR approach is depicted in Figure 15.2, which begins with chemical libraries and moves on to virtual screening, testing, and compound optimisation [41–44].

15.3.2 QSAR Used in Mood and Anxiety Disorder

Schizophrenia and other mood and anxiety disorders have been successfully treated by targeting 5-HT1A (5-hydroxytryptamine 1A) serotonin receptor [45, 46].

However, the 5-hydroxytryptamine 1A receptor-targeting drugs that are now available on the market have serious adverse effects. To develop a new hit compound procedure based on QSAR that targets the 5-HT1A receptor. Initially, machine learning techniques and Dragon descriptors were used to create binary QSAR models. Following thorough validation, four commercial chemical databases were used in consensus for the developed QSAR models. Nine of the fifteen compounds underwent experimental testing, and it was discovered that they were all active at low nanomolar quantities. With an extremely high binding affinity (Ki) of 2.3 nm for the 5-HT1A receptor, one of the validated hits, which is (methanol) [(8)-6-methyl-9, 10-didehydroergolin-8-y] [47].

15.4 Pharmacophore

15.4.1 Concept and Definition of Pharmacophore

The pharmacophore is one of the *in-silico* ideas for computer-aided drug design that shows the most promise. Paul Ehrlich initially used the term "pharmacophore" in the early 1900s, but Monty Kier was the one to characterize the pharmacophore's physical-chemical makeup in a series of studies that were published between 1967 and 1971. The pharmacophore approach connects medicinal chemistry with computational chemistry for effective hit identification as well as in the optimization of lead compounds to final therapeutic candidates in both VS and library design.

15.4.2 Types of Pharmacophore

The two approaches can be used to create a pharmacophore model. In the first approach, known as ligand-based modelling, a collection of active compounds is superimposed in order to identify the common chemical characteristics necessary for their bioactivity. The second technique, structure-based modelling, explores potential locations of interaction between ligands and macromolecular targets.

These scenarios show the use of pharmacophore queries as positive filters to find drugs; however, they may also be employed as negative filters to weed out undesirable effects [48, 49].

- **The structures of neither ligands nor proteins are known.**

Pharmacophore modelling is difficult if the target structure and all of its ligands are unknown. The pharmacophore concept can only be applied by creating a diverse library using a diversity metric based on pharmacophore fingerprints. This will ensure that the library is as diverse as possible and contains a variety of medications with different combinations of discrete pharmacophore features. The trend is to carefully build libraries in order to cover chemical space effectively in any search technique, given the vast quantity of current and future chemicals [50–53].

- **The structures of active ligands are known but there is no known protein structure.**

The receptor's structure (as well as any interactions with ligands) may also be unknown. Situations like this are typical in the drug discovery process. Because it is impossible to transfer the important contributing pharmacophore features onto the molecule when only one active molecule is known, similarity searches (such as using pharmacophore fingerprints to locate analogous molecules) may be the sole alternative [54].

- **Protein and ligand structures are known**

In the third illustration, both the ligands and the receptor protein have the structural data. The main characteristics of a small molecule that enable it to bind to a receptor are typically represented by a pharmacophore model, however this principle can be flipped and a pharmacophore query can be created from the characteristics of a protein active site [55]. These characteristics, which may be mapped onto the ligand's bioactive conformation, explain the fundamental interactions between the protein and its ligands [56].

- **The only known protein structure**

In the fourth situation, the structure of the protein receptor is known, but there are no active ligands. In this instance, analysing the chemical characteristics of the pertinent binding site may help create a potential pharmacophore model. The three-dimensional atomic configurations of protein binding sites can be easily converted into queries using a variety of computer techniques. The pharmacophore query can be made using the interaction maps of the *de novo* drug discovery tool LUDI. The interaction maps of the *de novo* drug discovery tool LUDI can be used to make the pharmacophore query. The interaction maps of the de novo drug discovery tool LUDI [57] can be used to do a pharmacophore inquiry. The knowledge-based approach HS-Pharm prioritizes the most exciting interacting atoms and uses machine learning techniques to construct an interaction map within the binding site [58]. The features of the pharmacophore are then created from the interaction map. The GRID programme [59] is another technique for examining the pocket and identifying crucial interactions. Using molecular interaction fields, the binding site's most favourable atomic probe sites can be identified and converted into pharmacophore properties [60].

- **Anxiety pharmacophore model**

Since the 1981 discovery of the stress-secreted peptide, also known as the adrenocorticotropin-releasing hypothalamic peptide, corticotropin-releasing factor 1 (CRF 1) receptor antagonists have been search for after. Despite inconsistent research on the effectiveness of CRF 1 receptor antagonists as antidepressants, they may provide a potential pharmacotherapy for anxiety and addiction [61]. CRF 1 and CRF 2 are two distinct receptor subtypes that have been found. The central and peripheral neurological systems each include many of these G-protein-coupled receptors [62]. Clinical data supports the idea that depression, anxiety, and stress-associated diseases may have a pathophysiology related to overproduction of CRF1, and it suggests that CRF 1 receptor antagonists may be helpful in treating these conditions [63].

Antibodies to the CRF 1 receptor may be beneficial in the treatment of diseases like anxiety and depression that are brought on by high levels of CRF 1. In an effort to find structurally diverse CRF 1 receptor antagonists, pyrazinone-based compounds were created, and it was shown that the branching point on the carbon atom immediately linked to the pyrazinone ring increased the CRF 1 receptor binding affinity [64]. Recognizingcc additional chemicals that bind to the target biological target is essential for ligand-based medication design. These molecules can be utilized to create a pharmacophore that defines the minimum structural criteria for a molecule to bind to a target [65]. To put it another way, a biological target is defined by what binds to it, and this model may then be used to create new molecular entities that interact with the target [66].

15.5 Structure-Based Methods for Screening

15.5.1 Protein–Ligand Docking Techniques Based on Structure

Structure-based protein-ligand docking methods have been applied to various protein targets associated with anxiety disorders. Here are a few examples of protein targets and relevant references where structure-based docking has been used in anxiety research [67]:

- GABA-A Receptor: GABA-A receptors are key targets for anxiolytic drugs. Docking studies have been conducted to identify ligands that interact with GABA-A receptor subunits and modulate receptor activity [68].
- Serotonin Transporter (SERT): SERT plays a major role in the regulation of serotonin levels and its dysfunction is associated with anxiety disorders. Docking studies have been employed to identify ligands that interact with SERT and modulate serotonin reuptake [69].
- Corticotropin-Releasing Factor Receptor 1 (CRF1): CRF1 receptors play a role in stress and anxiety regulation. Structure-based docking has been used to identify ligands that interact with CRF1 receptors and modulate their activity [70].
- Cannabinoid Receptors (CB1, CB2): The endocannabinoid and Cannabinoid receptors system are both implicated in the modulation of anxiety are potential targets for anxiolytic drugs. Docking studies have been conducted to identify ligands that interact with cannabinoid receptors [71].

15.5.2 Molecular Docking

Concept and definition of docking

The study of molecular docking focuses on how to fit two or more molecules together, such as a medication and an enzyme or protein [72]. A molecular modelling method called docking is used to forecast how an enzyme will interact with tiny ligands. The ability of a protein (or enzyme) and nucleic acid to interact with microscopic molecules to form a supramolecular complex, which can either promote or impede its biological activity, has a substantial influence on its dynamics. Molecular docking can explain the behavior of tiny compounds in target protein binding sites.

The technique seeks to foresee ligand affinities and the proper positioning of ligands within a protein's binding pocket. Docking can be categorized into the following categories:

- Docking of small molecule of protein (ligand)
- Protein-nucleic acids Docking
- Docking of proteins

15.5.3 Autodocking

It is a molecular docking tool. Active SZRT Compounds Automatically Dock with Target Protein. The main tactics used, such as Ligand-Based Virtual Screening (LBVS) and Structure-Based Virtual Screening (SBVS), are described in Figure 15.3. Following pharmacophore searches and molecular docking, compounds are chosen for additional assessment based on factors like charge, volume, and clog P. By selecting the most promising compounds from enormous compound databases, this all-encompassing method expedites the drug discovery process [72, 73]. For flexible docking, the molecular docking programme AutoDock 4.2.6 was used. The MAOB protein code was PDB 2V60, and the PDB database provided the structure of MAOB. MAOB is a key protein target, according to analysis of KEGG pathways. The target protein's main ligand conformation was extracted using PyMOL 2.7 software and saved as a PDB format file.en. The water molecules were then taken out using AutoDock tools (ADT). Using AutoDock and PyMOL 2.7, the results of docking chemicals and proteins

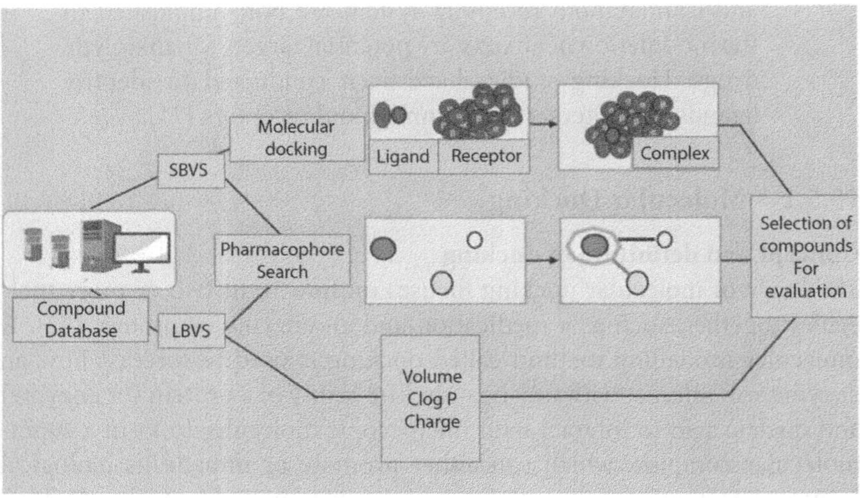

Figure 15.3 Strategies of drug design through molecular docking.

were investigated. MAOB underwent hydrogenation and charge treatment, and the results were recorded in a PDBQT format file. ADT was used to process Mol2 format ligands before they were saved in PDBQT format.

Active SZRT and MAOB Compound Docking Results. Using AutoDock and PyMOL 2.7, ten active substances, including stigmasterol, glypallichalcone, coumaroyltyramine, quercetin, and HMO, were docked with the primary target MAOB. The minimum energy of binding with MAOB is 9.6 kcal/mol for coumaroyltyramine. MAOB's LEU-164 and PHE-168 amino acid residues bonded to coumaroyltyramine via hydrogen bonding. In MAOB, With the amino acid residues TYR-435 and TYR-326, HMO formed hydrogen bonds. Similar to glypallichalcone, quercetin, isorhamnetin, and MAOB exhibit low binding enthalpies.

In this study, molecular docking techniques and network pharmacology depend on the GeneCards record, KEGG record and TCMSP record, were used to evaluate the probable structural foundations and anxiolytic mechanism of SZRT. By examining the identify pathways and targets, it was shown that 22 anxiety targets correspond to 57 potential active SZRT components. These active targets and ingredients may further support anxiolytic mechanism of SZRT. According to analyses of the PPI network and C-T network, the main active components of SZRT, which include kaempferol, sitosterol, 7-methoxy-2-methyl isoflavone and quercetin, may interact with targets like MAOB, IL1B, MAOA, TNF, NR3CI and HTR3A, Affecting immunoregulation pathways and regulation of neurotransmitter to generate anxiety-reducing effects.

15.6 AI

Recent advances in wearable and artificial intelligence (AI) technology for clinical and healthcare use have been made quickly and continue to offer many benefits for individualized diagnosis and treatment of psychiatric conditions, such as anxiety and depression [74–76]. Electrocardiogram electrodes are one example of an electronic device that may be worn close to the body. Other examples include smartwatches, eyewear, wristbands, and implanted smart patches. Wearable technology also includes electronic textiles, such as smart clothing. The aim of wearable technology is to provide a constant flow of medical data for diagnosing and treating illnesses. This involves ongoing monitoring of physiological metrics like blood pressure, body temperature, blood oxygen levels, respiratory rate, physical activity, as well as skin, heart, and brain activity. Real-time data collected through wearable devices can be leveraged to diagnose and monitor individuals experiencing symptoms of anxiety and depression by analyzing various parameters.

However, a growing demand for automatic, effective, and real-time techniques to analyze the massive amounts of data acquired by wearable sensors has arisen as a result of the radically accelerated pace of wearables' development and acceptance, as well as a dearth of qualified carers. This is what inspired the development of "Wearable AI" technology, which incorporates AI techniques into wearable gadgets. Intelligent electronic gadgets that are built to be worn on the user's body and have intelligent operations are referred to as wearable AI. Health data monitoring and analysis are typical tasks for wearable technology. Wearable technology, however, introduces major breakthroughs in the diagnosis and treatment of anxiety and depression when combined with AI. It may enable better individualized treatment for patients with anxiety and depression and help create preventive strategies for populations at risk of anxiety and depression [77]. It also has the potential to offer an early and accurate diagnosis of anxiety and depression.

15.7 Machine Learning Algorithms for Anxiety Disorder Detection and Prediction

Following are some examples of how machine learning algorithms is useful in the anxiety field disorders:

- Identification or diagnosis of an anxiety problem
- Prediction of future anxiety disorder risk
- Prediction of how well medical treatment will work to treat anxiety disorder

The existence or absence of a specific anxiety disorder, treatment or risk and response levels can all be categorized using machine learning algorithms. The classification procedure is explained in Figure 15.4. Data can be gathered from a variety of sources, such as demographic information, health records, medical histories, various measurement devices, etc. The data can be used to extract a large number of features, and significant features can be chosen using a suitable feature selection technique. The classifier's training and testing data sets make up these features. The final phase involves choosing and fine-tuning a good classifier based on performance measures from the training data set.

Details of the figure blocks will be discussed in the following subsections:

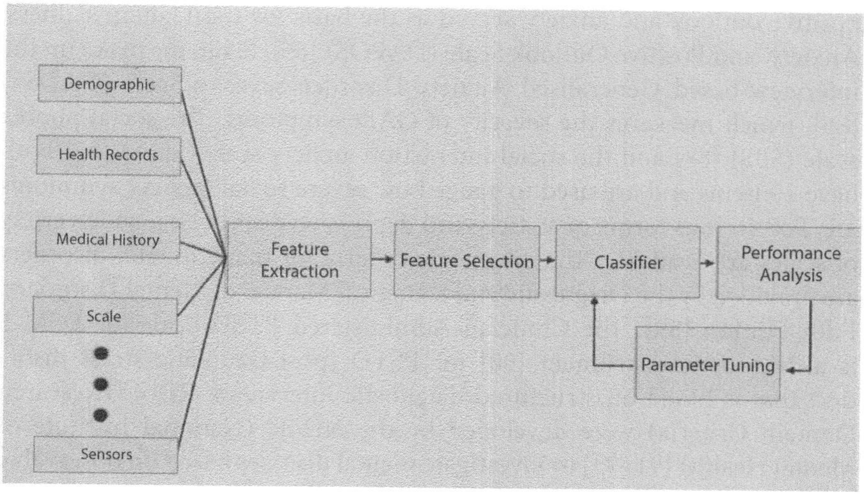

Figure 15.4 Workflow of machine learning.

- **Data collection**

The machine learning algorithms can be trained using many sorts of data, including demographic information, medical records, treatment histories, and anxiety assessment scales. Clinical and healthcare practitioners employ a wide variety of scales. Numerous techniques, including electrocardiograms, electroencephalograms, and FMRI (functional magnetic resonance imaging), are frequently utilized in the clinical environment to diagnose anxiety disorders. Researchers currently use sensor data from wearable and non-wearable devices without patient engagement in addition to this type of data. The aforementioned tests and raw sensor data from non-wearable and wearable sensors need feature extraction and pre-processing algorithms. Recently, DNA methylation profiles have also become widely used to distinguish between healthy control participants and individuals with mental illnesses [78]. One of the four DNA bases, cytosine, is methylated when a methyl group is covalently attached to it. Methylation has been linked to a number of psychological disorders, for example PTSD, MDD, and a propensity towards suicide [79–81].

A questionnaire named the HAM-A (Hamilton Anxiety Rating Scale) [82] is used by healthcare professionals to determine how severe the anxiety is. Seven items make up HADA (Hospital Anxiety and Depression Scale) [83] that many studies use to identify anxiety and depression. The 21-item Beck Anxiety Inventory (BAI) [84] measures three different levels of anxiety: low, medium, and high. A inquiry for varied levels of depression, positive outlook and anxiety served as the basis for the 11-item sadness, Anxiety, and Positive Outlook Scale (DAPOS) [85]. Six items make up the interview-based Generalised Anxiety Disorder Severity Scale (GADSS) [86], which measures the severity of GAD symptoms. The social phobia scale (SBS) [88] and the social interaction anxiety scale (SIAS) [87] both have 20 items and are used to gauge how severe social anxiety symptoms are. For various forms of anxiety and depressive disorders, a wide variety of scales are available. The diagnostic criteria for many mental disorders are provided by the Diagnostic and Statistical Manual of Mental Disorders, Fifth Edition [89]. The Clinician-Administered PTSD Scale for DSM-5 is a diagnostic instrument [90] for PSTD (post-traumatic stress disorder) that is based on structured diagnostic interviews. RDoC (Research Domain Criteria) were developed by the NIMH (National Institute of Mental Health) [91–93] to investigate mental diseases based on observable behavioural characteristics and neurobiological measurements. The RDoC paradigm emphasises the significance of symptom modelling to determine the causal genesis of mental illness, and exact symptom grouping may improve the diagnosis of psychiatric disorders [94].

- **Selection and extraction of features**

Features can be retrieved from data after it has been collected. The unprocessed data collected from multiple sources is either considered to be characteristics or features need to be retrieved from the data. It could be necessary to pre-process or normalize these traits. Depending on the classification algorithm being used, normalization may be necessary. Certain classification algorithms don't need to be normalized. In order to limit the feature set's dimension and the computational work involved in gathering test data, feature extraction, and classification, it can sometimes be necessary to choose the key features before moving on to the classification process. Filter methods, wrapper methods, and embedding methods are the three main categories of feature selection methods. There might be additional ways to choose features. You can find helpful surveys of these techniques in [95–97].

- **Classifier selection and optimization**

The choice of a suitable classifier for a classification task must follow a strict guideline. To improve classification accuracy, researchers experiment with various based on their own judgement or literature assessments, classify data. Discriminative or generative classifiers are two different types of classifiers. Many different fields also successfully employ recurrent neural networks, convolutional neural networks, and other deep learning architectures [98].

15.8 A Review of the Literature on Machine Learning Approaches for Anxiety-Related Disorders

The patient interview needs a minimum of 20 to 30 minutes to complete in order to look into and diagnose the patient with an anxiety problem. Additionally, an inquiry must be completed by the patient with a thorough grasp of its contents in order for a diagnosis based on a particular scale to be made. Early detection of an anxiety disorder's hazards may enable early intervention and make treatment simpler.

In this study, we examined research on machine learning algorithms used for various forms of anxiety disorders over the preceding 10 years and classified the publications into the many types of anxiety disorders. More than 1000 people with a history of serious depression supplied baseline reports, which Kessler *et al.* [99] used. Machine learning algorithms were used to predict the findings assessed 10 to 12 years following the baseline data. The machine learning approach outperformed traditional logistic regression models in the prediction models, with an area under the curve (AUC) of 0.63 for chronicity and 0.71 to 0.76 for all outcomes.

Features derived from multi-modal magnetic resonance imaging are combined with demographic data and measures of cognitive ability [100]. To estimate depression and treatment response, variety models of machine learning, such as optimized ADTree, logistic regression, and (SVM) support vector machines are applied. The highest classification accuracy was achieved by ADTree, with 87% for diagnosis and 89% for treatment response. A contextual data mining approach was utilized by Panagiotakopoulos *et al.* [101] to help therapists manage anxiety diseases. Only 10 subjects were utilized to forecast how well the algorithm for prior-to rule extraction would perform and to demonstrate to therapists the effectiveness of the suggested aided algorithm.

- **GAD (Generalized anxiety related disorder)**

The most typical form of anxiety disorder is GAD. There are numerous GAD screening and treatment techniques used in clinical practices [102]. A neural network-based model for the identifying of GAD was created by Sribala [103] using the general characteristics and qualities linked to the (DSM) Diagnostic and Statistical Manual-IV standard inquiry. [104]. Results without sensitivity analysis were 90.32% [103], while those with it were 96.43%. Data from the BDI (Beck Depression Inventory), which is most popular psychometric tests for assessing the seriousness of depression, were gathered by Hussain *et al.* [105, 106]. The GAD data set was created using survey findings from 182 respondents, including 70 men and 112 women. A 99.3% accuracy rate, 100 random forest trees, and balanced data (equal numbers of male and female respondents) were used to provide the best results [104].

- **Social psychological disorders**

Social media has a significant influence on a person's life since it allows users to express their feelings and emotions anonymously. Social media is a valuable resource for data scientists who are researching psychological aspects of human behaviour. In their lives and Twitter posts, Reece *et al.* connected anxiety disorders to childhood tragedies. With 63 participants, 234.000 posts who experienced anxiety, and 111 Healthy peoples, the Random Forest algorithm was used in the study. The study gathered (n = 279.951) predictive data from people tweets that measure the effect, context and language style, and used this information in models with supervised learning algorithms. With 89% accuracy, the Random Forest Model identified anxiety as a clinical symptom [107].

- **Suicidal tendency**

Suicidal tendencies are not classified as anxiety disorders by DSM-5. Thoughts and impulses towards suicide are more common in persons with anxiety disorders [108–110]. A review of machine learning methods for identifying or forecasting suicidal inclinations is done in this subsection. Whether successful or not, suicide attempts are a serious problem for mental health everywhere. Suicide claims the lives of 800,000 individuals every year [111]. Predicting non-fatal suicide attempts is challenging. The prevention of suicides and suicidal attempts depends heavily on effective intervention. Suicidal attempts cost governments, families, and individuals a lot of money. Therefore, it is crucial to foresee the suicidal tendency in those with mental diseases. Although there are several techniques available to estimate the possibility of suicide, people may be hesitant to use tools that rely heavily on questionnaires [112–114].

15.9 Molecular Dynamic Simulation

As everyone is aware, the word "simulation" literally means the imitation of an upcoming event. Analysing the physical mobility and chemical interactions of atoms and molecules after briefly allowing them to interact, a technique known as molecular dynamic simulation uses computer methods to understand movement in molecules of biology. The, fluctuations, structure and dynamics, thermodynamics of biological molecules, conformational changes, and their complexes are all covered in great detail. There is little doubt that comprehension of these intricate biomolecular processes is essential for the creation of novel medications [115].

Emil Fischer's original "lock-and-key" ligand binding mechanism, which assumed a motionless, fixed receptor could house a small molecule without any conformational rearrangements, has been replaced by new binding models that account for both conformational changes and the random motions of ligands and receptors [116–120]. This supports the claim made by Richard Feynman. According to this Nobel Prize winner in physics in 1965, all things are formed of atoms, and all living things can be explained in terms of the shaking and vibrating of atoms [121].

- **Aim and Concept**

In order to provide the most accurate information about the atoms based on the physics atomic relationships are governed by algorithms, molecular dynamic simulation attempts to anticipate how atoms behave in biological systems and they move in a manner that depends on time [122]. By doing this, we hope to gain a better understanding of how molecules behave under various conditions and how their traits relate to their structural makeup. It functions as an essential suffix to the laboratory experiments, saving the scientists' time, money, and effort. This bridges a significant gap between the conventional experimental scientists and the most recent technological developments in the scientific world today.

In addition to educating students about simulations that are challenging or impossible to carry out in the lab, it is intended to decrease the number of assumptions and adaptations that conventional scientists make. It's vital to remember that a molecular model may not be entirely accurate, but it should be able to represent the most important chemistry and physics concepts as well as follow the relevant mathematical laws and have the right biological characteristics, and that should be enough.

To examine the dynamic behavior of molecules using computer simulation in order to comprehend some aspects of the biological universe complexity is a difficult endeavour. It necessitates complex calculations that account for the approach's temporal component, highly developed models that can accurately reflect physical forces that can replicate the physics and thermodynamics laws, as well as the biological environment. The technology has become more dependable as a result of the development of tools for energy calculations, molecular modelling, algorithms that reproduce the chemical properties of real systems, procedures, docking scoring etc.

Structure is submerged in a "bath" of tens of thousands of water molecules to give the simulation realism [123].

- **Molecular dynamic simulation used in Anxiety**

More than 40% of drugs with a license target GPCR, the largest class of cell surface receptors with exceptional therapeutic potential. Through the use of G protein independent signaling pathways, G protein dependent signaling pathways, and other complex regulatory mechanisms, cells are able to transform extracellular inputs into intracellular responses. Seven transmembrane helices (7TM), alternate interior cellular and exterior cellular loops (ICLs and ECLs), an intracellular C-terminal and an extracellular N-terminal, make up the conserved structural element that joins them. Based on sequence of functional and homology similarity, five different classes of GPCRs have been identified: Class A, Class B, Class C, Class D, and Class E. Because there is so much structural data available, molecular comprehension of activation mechanisms and functions has greatly improved. Class A is the largest GPCR class that is best investigated. Class B, on the other hand, is rarely investigated because there aren't many viable structural arrangements. Extracellular domains (ECDs) of this class's members are substantial and cysteine-rich, and they are crucial for activation. This family of 15 discovered receptors has been linked to anxiety, stress, and other related neuro-disorders through a number of illnesses.

Adrenocorticotropic hormone (ACTH), a crucial regulator of the stress response, is controlled by the CRF1R (corticotropin-releasing factor receptor type 1), a representative member of class B that is mostly found in the CNS (central nervous system). CRF1R is a helpful therapeutic taegets of stress-related neuro-disorders for example anxiety, inflammation, depression and other conditions.

The atomic-level and high-resolution natural mobility of proteins and other biomolecules can be studied via molecular dynamics modelling. The CRF1R transmembrane domain (TMD) activation mechanism, which necessitates large conformational changes of the TM helices, was discovered using MD modelling. We created a model of the full-length CRF1R utilising the crystal structures of the transmembrane domain (TMD) (PDB ID: 4KBY), and N-terminal ECD (PDB ID: 3EHU) as uncertainty existed over the completely CRF1R's stimulation mechanism. Using molecular dynamics simulation, we were able to show that the CRF1R undergoes significant conformational changes as a result of the peptide agonist urocortin 1. These alterations include the dissociation of salt bridges, migration of transmembrane helix 6 (TM6), and networks of inter-helical/regional H-bonds. On the other hand, the transmembrane domain (TMD) is kept in its initial inactive conformation when CRF1R binds to the small molecule antagonist CP-376395.

In this study, a SBVS of the 17 million molecule ZINC "Druglike" library was carried out using unique structures produced by the previous MD simulations. We identified the binding energies among the best 30 drugs that interacted with the CRF1R using MM-GBSA and MD modelling. Potential CRF1R antagonists have higher results for the MM-GBSA free energy test of binding and superior physicochemical qualities than the antagonist CP-376395. The transmembrane helices that bridge the CRF1R's allosteric location are strongly impacted by these substances. The interactions could keep the receptor in an inactive state [124].

15.9.1 Network Analysis

A network is a visual representation of a project's operations. To accomplish a project's end goal, a series of events and actions must take place [125].

In a large project, several tasks are carried out concurrently, and some tasks can only be started after others have been completed. Finding out information about the project and then coming up with a new, more efficient, and quicker manner to complete the task are the major tasks for a thorough analysis of the product. Thus, a detailed analysis of the project is carried out using a suitable diagram that highlights the various activities and their relative places within it. It is also useful to understand how the delay in any activity may impact the project as a whole in terms of both time and money. A project network consists of a number of nodes, which are commonly represented as small circles or rectangles, and a number of arcs, which are symbolized by arrows and connect various nodes. In order to help with project planning, scheduling, and control, a series of techniques known as network analysis are used to display information about the time and resources required for the project. The sequences, interdependencies, interrelationships, and important activity of the many project activities are among the information that is typically represented by a network.

Goal of Network Analysis

1) Reduce production lag, disruptions, and disputes
2) Reduction of Project Cost Overall Project time and cost trade-offs are made, and the overall project duration is kept to a minimum.
3) Reduction of Wasted Resources

- **Network analysis used in anxiety**

The ability to comprehend and describe how various anxiety disorder symptoms, or anxious behavior in general, interact and reinforce one another, may be another benefit of network analysis.

Network analysis, as described in reference [126], offers a means of representing and understanding interactions within intricate systems, such as biological cells or social networks. Its application in scrutinizing psychiatric symptoms represents a recent development akin to other computational methodologies, as highlighted in references [127–129]. Initially inspired by the theory of psychological measurement (reference [130]) and diagnostic frameworks (reference [131], which approached mental illnesses (reference [131]) or personality traits (reference [132]) by conceiving them as emanating from a singular latent factor, network analysis has emerged as a novel tool in this domain. This work contends that unlike many physical illnesses, where each symptom is caused independently by a single hidden variable, psychiatric illnesses are wholly made up of distinctive sets of symptoms that interact causally with one another ([129, 131]; see also Ref. [133]). Comorbidity patterns [127], conversions among sickness and health [134], and susceptibility to mental illness [135] have all been explained using this concept. In such studies, network analysis typically begins by examining correlations between self-reported symptoms that point to possible relationships between them, such as the tendency of subgroups of symptoms to group together within a disorder [136] or the likelihood that some stressful life events will cause more symptoms of depression than others [137].

Self-reported and experimental evaluations of anxiety have been gathered from individuals diagnosed with social anxiety disorder within the realm of anxiety disorders [138], making it easier to look for relationships between these measures. Network analytic techniques are successfully coupled with other experimental procedures in the analysis of attentional components, which have a well-established but complex association with anxiety symptoms [139]. They can be combined with the accurate quantification of neural and behavioural phenotypes found employing alternative computational techniques outlined in this section [140] to unveil the manner in which interconnected neural processes interact to produce and maintain psychopathological behaviours, including clinically relevant anxiety over-expression.

15.9.2 Bioinformatics

By combining biological data with computer methods to find prospective drug targets, create new medications, and foresee how they will interact with biological molecules, bioinformatics plays a critical role in drug discovery [141]. The general technique can be modified to examine anxiety-related targets and pathways, even though there may not be specialized studies on bioinformatics in anxiety drug discovery.

Here are several ways that bioinformatics can help with the search for anxiety medications:

- Target Identification: Biological resources, such as genomic and proteomic data, can be analyzed using bioinformatics to find prospective therapeutic targets involved in pathways connected to anxiety. Researchers can identify important molecules that are involved in anxiety disorders by looking at gene expression profiles or protein-protein interactions.
- Pathway analysis: Complex biological pathways and brain circuits are involved in anxiety disorders. These pathways can be clarified and the important nodes that might be the focus of medication intervention might be found using bioinformatics methods.
  - Drug repurposing: Based on interactions with known anxiety-related targets, bioinformatics can be used to analyze current drug databases and uncover medications that may have potential therapeutic benefits on anxiety.
- Virtual Screening: Molecular docking and virtual screening are two computational bioinformatics techniques that can be used to anticipate the interactions between prospective medication candidates and target molecules involved in anxiety disorders.
- Pharmacogenomics: Bioinformatics can be used to explore the genetic variants that affect how well people with anxiety disorders respond to medications. Using this information, medication treatments can be tailored to a person's genetic profile.
- Bioinformatics can be used to forecast probable side effects and evaluate drug safety prior to moving therapeutic candidates to preclinical and clinical trials. It's important to bear in mind that the journey of uncovering novel medications is complex and multi-dimensional, requiring cooperation between experts from different fields like biology, pharmacology, chemistry, and computer sciences. As bioinformatics is increasingly incorporated into drug development pipelines, it is predictable that it will show a significant part in the forthcoming of anxiety drug discovery.

15.9.3 Recent Advancementon Computation Approaches in Anxiety [142–147]

The comprehension, diagnosis, and treatment of anxiety disorders have been greatly improved by recent developments in computational approaches. To find new therapeutic targets and drug candidates, these methods combine technologies such as molecular docking, virtual screening, machine learning, and artificial intelligence (AI). Table 15.1 provides an overview of current developments in computational methods, This table summarizes key methods such as molecular docking, pharmacophore modeling, virtual screening, machine learning, and molecular dynamics simulations, highlighting their contributions to identifying potential therapeutic agents targeting anxiety-related receptors and pathways [77, 141–148].

15.10 Future Prospective

There are, however, both opportunities and difficulties for the future. In the future, computational methods could be improved to comprehend fundamental causes and cures for diseases linked to fear and anxiety.

The followings are the approaches which could be developed in future for better results

- **Virtual Reality Exposure Therapy (VRET) with Biofeedback:** Virtual reality environments combined with biofeedback technologies could enhance exposure therapy for anxiety disorders. The computational aspect would involve creating dynamic and personalized virtual scenarios and using physiological feedback to adjust the virtual experience based on the individual's responses.
- **Mobile Apps for Monitoring and Intervention:** Mobile applications could be developed to monitor individuals' anxiety levels, triggers, and behaviors in real-time. These apps might implement machine learning algorithms to analyze data and deliver personalized coping strategies or therapeutic interventions on the go [150].
- **Personalized Digital Therapeutics and Chatbots:** AI-powered chatbots and digital therapeutics could offer personalized support and cognitive-behavioral interventions for individuals with anxiety. These systems might adapt their responses and strategies based on an individual's unique needs and progress [151].

Table 15.1 Recent advancement on computational approaches.

Author	Tittle	Computational technique	Outcomes	Ref.
Alaa Abd-Alrazaq *et al.*, 2023	Wearable Artificial Intelligence for Anxiety and Depression: Scoping Review	Artificial Intelligence	69 (5.74%) of the 1203 identified studies were included in this study. Wearable AI was employed for depression in almost two-thirds of the research, and anxiety in the remaining studies. The most common use of wearable AI was for depression and anxiety diagnosis; none of the research, however, employed it for treatment. Most research focused on people between the ages of 18 and 65. Actiwatch AW4 (Cambridge Neurotechnology Ltd.) was the most often utilised wearable technology in the research. The majority of wearable gadgets in the research were wrist-worn. Physical activity data were the most often employed type of data for model construction, followed by heart rate and sleep data. Depresjon was the open source data set that was used the most. Random Forest was the most often utilised algorithm, followed by support vector machine.	[77]

(*Continued*)

Table 15.1 Recent advancement on computational approaches. (*Continued*)

Author	Tittle	Computational technique	Outcomes	Ref.
Jennifer Wright-Berryman1* *et al.*, 2023	Virtually Screening Aults for Depression, Anxiety, and Suicide Risk using Machine Learning and language from an Open-ended interview	Virtual screening techniques	Jennifer conducted an open-ended interview to assess suicide risk in adults, depression, and anxiety. To efficiently screen large populations, machine learning was employed to analyze language patterns and identify potential indicators of mental health conditions. Natural language processing techniques were used to extract relevant features from the interview data. The machine learning model was trained on a diverse dataset with labeled mental health outcomes. By analyzing patterns in language, the model learned to recognize subtle cues and linguistic markers associated with depression, anxiety, and suicide risk. The system's performance was validated against clinical assessments to ensure accuracy and reliability. This approach enables scalable and cost-effective screening, providing valuable insights into individuals' mental well-being and aiding in the early identification of those at risk. This technology has the potential to revolutionize mental health care by facilitating timely interventions and support for those in need.	[142]

(*Continued*)

Table 15.1 Recent advancement on computational approaches. (*Continued*)

Author	Tittle	Computational technique	Outcomes	Ref.
Gamal A. Mohamed *et al.*, 2023	Structure-Based Virtual Screening and Molecular Dynamics Simulation Assessments of Depsidones as Possible Selective Cannabinoid Receptor Type 2 Agonists	Structure based VS technique, Molecular Dynamic Simulatuion	Molecular dynamics simulations and Structure-based virtual screening were utilized to explore the potential of depsidones as selective agonists for the cannabinoid receptor type 2 (CB2). The virtual screening utilized computer algorithms to efficiently screen a large database of depsidone compounds and identify those with the most favorable binding affinity for the CB2 receptor's active site. Subsequently, molecular dynamics simulations were employed to study the dynamic behavior and stability of the identified depsidone-CB2 receptor complexes over time, providing insights into their interactions at the atomic level. This integrated approach aimed to uncover novel depsidones with high selectivity and specificity towards CB2, which employ a vital role in modulating immune response and inflammation. The findings from this study hold significant promise for the design and development of new therapeutics targeting CB2 for various immune-related disorders and inflammatory conditions.	[143]

(*Continued*)

Table 15.1 Recent advancement on computational approaches. (*Continued*)

Author	Tittle	Computational technique	Outcomes	Ref.
Faiq Amin *et al.*, 2022	Interactions of Apigenin and Safranal with the 5HT1A and 5HT2A Receptors and Behavioral Effects in Depression and Anxiety: A Molecular Docking, Lipid-Mediated Molecular Dynamics, and *In Vivo* Analysis	AutoDock Vina software, Protein-lipid molecular dynamics simulations, Molecular dynamic simulation	Apigenin and Safranal were studied for their interactions with 5HT1A and 5HT2A receptors using molecular docking and lipid-mediated molecular dynamics simulations. The research aimed to investigate their binding affinity and stability within the receptor's active sites. *In vivo* analysis was conducted to evaluate their behavioral effects in depression and anxiety models. The study identified potential ligands with strong interactions, shedding light on their therapeutic potential as modulators of serotonin receptors. These outcomes provide valuable understanding into the molecular mechanisms underlying their pharmacological activities, offering a promising direction for the production of new treatments for depression and anxiety disease.	[144]

(*Continued*)

Table 15.1 Recent advancement on computational approaches. (*Continued*)

Author	Tittle	Computational technique	Outcomes	Ref.
Neyder Contreras *et al.*, 2022	*In silico* study of dimethyltryptamine analogues against 5-HT1B receptor: Molecular docking, dynamic simulations and ADMET prediction	Molecular Docking, Molecular dynamic simulation, ADMET Prdiction	An *in-silico* study was conducted to investigate the interactions of dimethyltryptamine (DMT) analogues with the 5-HT1B receptor. Molecular docking was used to identify the binding modes of these analogues to the receptor's active site. Subsequently, dynamic simulations were performed to analyze the stability and dynamic behavior of the ligand-receptor complexes. The study aimed to identify potential analogues that exhibit strong binding affinity and favorable interactions with the 5-HT1B receptor. These computational insights offering essential insights for the systematic design and enhancement of novel compounds with therapeutic potential for various neurological and psychiatric disorders related to the serotonin receptor system.	[145]

(*Continued*)

Table 15.1 Recent advancement on computational approaches. (*Continued*)

Author	Tittle	Computational technique	Outcomes	Ref.
Shushanik Sh Dashyan *et al.*, 2022	Evaluation of Neurotropic Activity and Molecular Docking Study of New Derivatives of pyrano[4″,3″:4′,5′] pyrido[3′,2′:4,5] thieno[3,2-d] pyrimidines on the Basis of pyrano[3,4-c] pyridines	Molecular Docking technique	Eleven novel tri- and tetracyclic heterocyclic systems were produced as a result of multistep reactions. Pentylenetetrazole (PTZ) seizures are prevented by the investigated substances, however they also have certain psychoactive effects. Nine out of the eleven compounds under study demonstrated strong anticonvulsant action through antagonism with pentylenetetrazole, according to the biological experiments. The chemicals have a minimal level of toxicity and do not cause muscular relaxation at the tested levels. The investigation into psychotropic activity utilized the "open field" and "elevated plus maze" (EPM) models, revealing that the chosen medications exhibit anxiolytic effects and promote activity in these models. The results indicate the anxiolytic properties of derivatives belonging to tricyclic thieno[2,3-b]pyridines and tetracyclic pyridothieno[3,2-d]pyrimidin-8-ones. Compounds 3b-f and 4e have a very strong display of this activity. On the "forced swimming" (FS) paradigm, the tested compounds, particularly compounds 3e and 3f, demonstrate these effects that are comparable to those of diazepam and prolong the latent time of first immobilisation.	[146]

(*Continued*)

Table 15.1 Recent advancement on computational approaches. (*Continued*)

Author	Tittle	Computational technique	Outcomes	Ref.
Kariyappa N. Ankali *et al.*, 2021	Synthesis and Molecular Docking of novel 1,3-Thiazole Derived 1,2,3-Triazoles and *In vivo* Biological Evaluation for their Anti-anxiety and Anti-inflammatory Activity	Molecular Docking technique	1,2,3-triazoles generated from novel 1,3-thiazoles were synthesized and subjected to molecular docking studies to evaluate their potential as anti-inflammatory and anti-anxiety agents. *In vivo* biological evaluations were conducted to assess their pharmacological activity. Promising candidates were identified with potent activity against anxiety and inflammation, providing potential leads for further drug development. This research demonstrates the importance of thiazole-triazole hybrids in drug discovery, showcasing their potential as therapeutic agents for anxiety disorders and inflammatory conditions. These findings offer valuable insights into the design and development of novel pharmaceuticals for improved mental health and inflammatory disease management.	[147]

(*Continued*)

Table 15.1 Recent advancement on computational approaches. (*Continued*)

Author	**Tittle**	**Computational technique**	**Outcomes**	**Ref.**
Wei Bai *et al.*, 2021	Network analysis of anxiety and depressive symptoms among nursing students during the COVID-19 pandemic	Network Analysis	The predominant indicators sustaining the network structure of depression and anxiety among Chinese nursing students throughout the pandemic included persistent anxiety, irritability, challenges in relaxation, and feelings of low mood. For this demographic, implementing comprehensive, timely interventions targeting both core and connecting symptoms could effectively mitigate the concurrent manifestations of anxiety and depression.	[148]

(*Continued*)

Table 15.1 Recent advancement on computational approaches. (*Continued*)

Author	Tittle	Computational technique	Outcomes	Ref.
Paramjit Kaur *et al.*, 2012	Pharmacophore Modelling and 3D-QSAR Studies on N3-Phenylpyrazinones as Corticotropin-Releasing Factor 1 Receptor Antagonists	Pharmacophore modeling based Virtual screening	In the study conducted by Paramjit Kaur, pharmacophore modeling and Quantitative Structure-Activity Relationship in Three Dimensions analyses were performed on N3-phenylpyrazinones to investigate their potential as corticotropin-releasing factor 1 (CRF1) receptor antagonists. Pharmacophore modeling involved identifying essential structural features required for interaction with the CRF1 receptor, aiding in the rational design of new compounds with improved activity. Additionally, 3D-QSAR studies provided valuable understanding the quantitative correlation among molecular structures of N3-phenylpyrazinones and their biological activity against the CRF1 receptor. By combining these computational approaches, the research aimed to identify key structural characteristics essential for potent CRF1 receptor antagonism. This information is crucial for guiding future drug development efforts, enabling the discovery of novel and more effective CRF1 receptor antagonists for potential therapeutic applications in stress-related disorders and other conditions where CRF1 modulation plays a significant role.	[149]

- **Multi-modal Data Fusion:** Integrating data from various sources, such as genetics, neuroimaging, behavioral assessments, and environmental factors, into comprehensive models could provide a more holistic understanding of anxiety disorders and their underlying mechanisms [152].

These technologies should be developed and implemented ethically, with appropriate considerations for privacy, consent, and potential biases in the data used to train them. Additionally, the effectiveness and safety of these approaches need to be rigorously tested through well-designed clinical trials and research studies [153].

15.11 Conclusion

The processes used in computational drug design are mentioned in this chapter. Anxiety computational techniques have become important resources for comprehending and treating this difficult mental health problem. Computational models provide distinctive insights into the underlying mechanisms of anxiety disorders by utilizing technological breakthroughs in data analytics, artificial intelligence, and modelling, assisting in precise diagnosis, individualized therapy, and therapeutic interventions.

In the discipline of computational drug design, drug molecules are designed and improved to have particular therapeutic qualities. With CADD, researchers can quickly find possible drug candidates that can be put through additional testing in a lab, saving time and resources. CADD has become a crucial component of the drug development and discovery process. The computational drug design method entails a number of processes. First, the biological system or target protein that the medication is intended to interact with is determined. Various techniques, including as molecular modelling, structural biology, and bioinformatics, can be used to accomplish this. The next stage is to develop or recognize prospective medication candidates after the target has been identified. Virtual screening, molecular dynamics simulations and molecular docking are frequently used to accomplish this. Virtual screening is the process of searching through huge databases of chemical compounds to search molecules that might linked with the target protein. The dynamic behaviour of the drug-target complex is predicted using Molecular simulation, Molecular docking, however focuses on predicting the affinity and orientation of prospective therapeutic compounds with respect to the target protein. Promising drug candidates are first identified, and then additional optimization is carried out utilising a range of computational techniques, including as ligand-based design, fragment based design, and structure-based design. The target protein's known structure is used in structure-based design to produce substances that will make interaction with particular protein domains. A known therapeutic molecule's chemical structure may be improved as part of ligand-based design in order to boost its activity and selectivity. In fragment-based design, smaller molecular fragments are combined to form larger therapeutic molecules.

In modern drug discovery and development, computational drug design plays an essential role, facilitating the rapid identification and enhancement of potential drug candidates. Although computational methods' accuracy still has some limitations, the discipline is always developing and getting better, which results in the discovery of entirely novel, innovative drugs.

In conclusion, computational approaches to anxiety constitute a promising new area for study and application in the field of mental health. These methods have the potential to revolutionize the field as technology advances, offering more individualized and effective interventions that will ultimately improve the lives of people with anxiety disorders. In order to fully utilize computational techniques to battle anxiety and advance mental well-being in our society, continued cooperation between researchers, physicians, and technologists will be essential.

References

1. Vicidomini, C., Fontanella, F., D'Alessandro, T., Roviello, G.N., A Survey on Computational Methods in Drug Discovery for Neurodegenerative Diseases. *Biomolecules*, 14, 10, 1330, 2024 Oct 19. doi: 10.3390/biom14101330.
2. Brogi, S., Kladi, M., Vagias, C., Papazafiri, P., Roussis, V., Tafi, A., Pharmacophore modeling for qualitative prediction of antiestrogenic activity. *J. Chem. Inf. Model.*, 49, 2489–2497, 2009.
3. Brogi, S., Papazafiri, P., Roussis, V., Tafi, A., 3D-QSAR using pharmacophore-based alignment and virtual screening for discovery of novel MCF-7 cell line inhibitors. *Eur. J. Med. Chem.*, 67, 344–351, 2013.
4. Zaccagnini, L., Brogi, S., Brindisi, M., Gemma, S., Chemi, G., Legname, G., Campiani, G., Butini, S., Identification of novel fluorescent probes preventing PrP(Sc) replication in prion diseases. *Eur. J. Med. Chem.*, 127, 859–873, 2017.
5. Vallone, A., D'Alessandro, S., Brogi, S., Brindisi, M., Chemi, G., Alfano, G., Lamponi, S., Lee, S.G., Jez, J.M., Koolen, K.J.M., *et al.*, Antimalarial agents against both sexual and asexual parasites stages: structure-activity relationships and biological studies of the Malaria Box compound 1-[5-(4-bromo-

2-chlorophenyl)furan-2-yl]-N-[(piperidin-4-yl)methyl]methanamine (MMV019918) and analogues. *Eur. J. Med. Chem.*, 150, 698–718, 2018.
6. Brogi, S., Brindisi, M., Joshi, B.P., Sanna Coccone, S., Parapini, S., Basilico, N., Novellino, E., Campiani, G., Gemma, S., Butini, S., Exploring clotrimazole-based pharmacophore: 3D-QSAR studies and synthesis of novel antiplasmodial agents. *Bioorg. Med. Chem. Lett.*, 25, 5412–5418, 2015.
7. Brogi, S., Corelli, F., Di Marzo, V., Ligresti, A., Mugnaini, C., Pasquini, S., Tafi, A., Three-dimensional quantitative structure-selectivity relationships analysis guided rational design of a highly selective ligand for the cannabinoid receptor 2. *Eur. J. Med. Chem.*, 46, 547–555, 2011.
8. Chemi, G., Gemma, S., Campiani, G., Brogi, S., Butini, S., Brindisi, M., Computational Tool for Fast *in silico* Evaluation of hERG K(+) Channel Affinity. *Front. Chem.*, 5, 7, 2017.
9. Pasquini, S., Mugnaini, C., Ligresti, A., Tafi, A., Brogi, S., Falciani, C., Pedani, V., Pesco, N., Guida, F., Luongo, L., *et al.*, Design, synthesis, and pharmacological characterization of indol-3-ylacetamides, indol-3-yloxoacetamides, and indol-3-ylcarboxamides: potent and selective CB2 cannabinoid receptor inverse agonists. *J. Med. Chem.*, 55, 5391–5402, 2012.
10. Gasser, A., Brogi, S., Urayama, K., Nishi, T., Kurose, H., Tafi, A., Ribeiro, N., Desaubry, L., Nebigil, C.G., Discovery and cardioprotective effects of the first non-Peptide agonists of the G protein-coupled prokineticin receptor-1. *PLoS One*, 10, e0121027, 2015.
11. Cappelli, A., Manini, M., Valenti, S., Castriconi, F., Giuliani, G., Anzini, M., Brogi, S., Butini, S., Gemma, S., Campiani, G., *et al.*, Synthesis and structure-activity relationship studies in serotonin 5-HT(1A) receptor agonists based on fused pyrrolidone scaffolds. *Eur. J. Med. Chem.*, 63, 85–94, 2013.
12. Brindisi, M., Brogi, S., Relitti, N., Vallone, A., Butini, S., Gemma, S., Novellino, E., Colotti, G., Angiulli, G., Di Chiaro, F., *et al.*, Structure-based discovery of the first non-covalent inhibitors of Leishmania major tryparedoxin peroxidase by high throughput docking. *Sci. Rep.*, 5, 9705, 2015.
13. Brogi, S., Fiorillo, A., Chemi, G., Butini, S., Lalle, M., Ilari, A., Gemma, S., Campiani, G., Structural characterization of Giardia duodenalis thioredoxin reductase (gTrxR) and computational analysis of its interaction with NBDHEX. *Eur. J. Med. Chem.*, 135, 479–490, 2017.
14. Brogi, S., Giovani, S., Brindisi, M., Gemma, S., Novellino, E., Campiani, G., Blackman, M.J., Butini, S., *In silico* study of subtilisin-like protease 1 (SUB1) from different Plasmodium species in complex with peptidyl-difluorostatones and characterization of potent pan-SUB1 inhibitors. *J. Mol. Graphics Modell.*, 64, 121–130, 2016.
15. Brindisi, M., Brogi, S., Giovani, S., Gemma, S., Lamponi, S., De Luca, F., Novellino, E., Campiani, G., Docquier, J.D., Butini, S., Targeting clinically-relevant metallo-beta-lactamases: from high-throughput docking to broad-spectrum inhibitors. *J. Enzyme Inhib. Med. Chem.*, 31, 98–109, 2016.

16. Brogi, S., Ramunno, A., Savi, L., Chemi, G., Alfano, G., Pecorelli, A., Pambianchi, E., Galatello, P., Compagnoni, G., Focher, F., *et al.*, First dual AK/GSK-3beta inhibitors endowed with antioxidant properties as multifunctional, potential neuroprotective agents. *Eur. J. Med. Chem.*, 138, 438–457, 2017.
17. Brogi, S., Computational approaches for drug discovery. *Molecules*, 24, 17, 3061, 2019.
18. Rester, U., From virtuality to reality - Virtual screening in lead discovery and lead optimization: a medicinal chemistry perspective. *Curr. Opin. Drug Discov. Dev.*, 11, 4, 559–68, July 2008, PMID 18600572.
19. Rollinger, J.M., Stuppner, H., Langer, T., Virtual screening for the discovery of bioactive natural products. Natural Compounds as Drugs Volume I. Progress in Drug Research Fortschritte der Arzneimittelforschung. Progrès des Recherches Pharmaceutiques. *Prog. Drug Res.*, 65, 211, 213–49, 2008.
20. Walters, W.P., Stahl, M.T., Murcko, M.A., Virtual screening – an overview. *Drug Discov. Today*, 3, 4, 160–178, 1998, doi:10.1016/S1359-6446(97)01163-X.
21. Bohacek, R.S., McMartin, C., Guida, W.C., The art and practice of structure-based drug design: a molecular modeling perspective. *Med. Res. Rev.*, ***16***, 1, 3–50, 1996, doi:10.1002/(SICI)1098-1128(199601)16:1<3::AID-MED1>3.0.CO;2-6. PMID 8788213.
22. McGregor, M.J., Luo, Z., Jiang, X., Chapter 3: Virtual screening in drug discovery, in: *Drug Discovery Research. New Frontiers in the Post-Genomic Era*, Z. Huang (Ed.), p. 63–88, Wiley-VCH, Weinheim, Germany, June 11, 2007, ISBN 978-0-471-67200-5.
23. Gillet, V., Ligand-Based and Structure-Based Virtual Screening (PDF), The University of Sheffield, 2013.
24. McInnes, C., Virtual screening strategies in drug discovery. *Curr. Opin. Chem. Biol.*, ***11***, 5, 494–502, October 2007.
25. Santana, K., do Nascimento, L.D., Lima e Lima, A., Damasceno, V., Nahum, C., Braga, R.C., Lameira, J., Applications of Virtual Screening in Bioprospecting: Facts, Shifts, and Perspectives to Explore the Chemo-Structural Diversity of Natural Products. *Front. Chem.*, **9**, 662688, 2021-04-29, Bibcode:2021FrCh..9.155S. doi:10.3389/fchem.2021.662688. ISSN 2296-2646. PMC 8117418. PMID 33996755.
26. Sun, H., Pharmacophore-based virtual screening. *Curr. Med. Chem.*, ***15***, 10, 1018–24, 2008, doi:10.2174/092986708784049630. PMID 18393859.
27. Willet, P., Barnard, J.M., Downs, G.M., Chemical similarity searching. *J. Chem. Inf. Comput. Sci.*, **38**, 6, 983–996, 1998, CiteSeerX 10.1.1.453.1788. doi:10.1021/ci9800211.
28. Rush, T.S., Grant, J.A., Mosyak, L., Nicholls, A., A shape-based 3-D scaffold hopping method and its application to a bacterial protein-protein interaction. *J. Med. Chem.*, **48**, 5, 1489–95, March 2005, CiteSeerX 10.1.1.455.4728. doi:10.1021/jm040163o. PMID 15743191.
29. Ballester, P.J., Westwood, I., Laurieri, N., Sim, E., Richards, W.G., Prospective virtual screening with Ultrafast Shape Recognition: the identification of novel

inhibitors of arylamine N-acetyltransferases. *J. R. Soc. Interface*, **7**, 43, 335–42, February 2010, doi:10.1098/rsif.2009.0170. PMC 2842611. PMID 19586957.

30. Kumar, A. and Zhang, K.Y., Advances in the Development of Shape Similarity Methods and Their Application in Drug Discovery. *Front. Chem.*, **6**, 315, 2018, Bibcode:2018FrCh..6.315K. doi:10.3389/fchem.2018.00315. PMC 6068280. PMID 30090808.
31. Li, H., Leung, K.S., Wong, M.H., Ballester, P.J., USR-VS: a web server for large-scale prospective virtual screening using ultrafast shape recognition techniques. *Nucleic Acids Res.*, **44**, W1, W436–41, July 2016, doi:10.1093/nar/gkw320. PMC 4987897. PMID 27106057.
32. Sperandio, O., Petitjean, M., Tuffery, P., wwLigCSRre: a 3D ligand-based server for hit identification and optimization. *Nucleic Acids Res.*, **37**, Web Server issue, W504–9, July 2009, doi:10.1093/nar/gkp324. PMC 2703967. PMID 19429687.
33. Kirchmair, J., Distinto, S., Markt, P., Schuster, D., Spitzer, G.M., Liedl, K.R., Wolber, G., How To Optimize Shape-Based Virtual Screening: Choosing the Right Query and Including Chemical Information. *J. Chem. Inf. Model.*, **49**, 678–692, 2009, doi:10.1021/ci8004226.
34. Toledo Warshaviak, D., Golan, G., Borrelli, K.W., Zhu, K., Kalid, O., Structure-based virtual screening approach for discovery of covalently bound ligands. *J. Chem. Inf. Model.*, **54**, 7, 1941–50, July 2014, doi:10.1021/ci500175r. PMID 24932913.
35. Maia, E.H., Assis, L.C., de Oliveira, T.A., da Silva, A.M., Taranto, A.G., Structure-Based Virtual Screening: From Classical to Artificial Intelligence. *Front. Chem.*, **8**, 343, 2020-04-28, Bibcode:2020FrCh..8.343M. doi:10.3389/fchem.2020.00343. PMC 7200080. PMID 32411671.
36. Kroemer, R.T., Structure-based drug design: docking and scoring. *Curr. Protein Pept. Sci.*, **8**, 4, 312–28, August 2007, CiteSeerX 10.1.1.225.959. doi:10.2174/138920307781369382. PMID 17696866.
37. Cavasotto, C.N. and Orry, A.J., Ligand docking and structure-based virtual screening in drug discovery. *Curr. Top. Med. Chem.*, **7**, 10, 1006–14, 2007, doi:10.2174/156802607780906753. PMID 17508934.
38. Kooistra, A.J., Vischer, H.F., McNaught-Flores, D., Leurs, R., de Esch, I.J., de Graaf, C., Function-specific virtual screening for GPCR ligands using a combined scoring method. *Sci. Rep.*, **6**, 28288, June 2016, Bibcode:2016NatSR..628288K. doi:10.1038/srep28288. PMC 4919634. PMID 27339552.
39. Irwin, J.J., Shoichet, B.K., Mysinger, M.M., Huang, N., Colizzi, F., Wassam, P., Cao, Y., Automated docking screens: a feasibility study. *J. Med. Chem.*, **52**, 18, 5712–20, September 2009, doi:10.1021/jm9006966. PMC 2745826. PMID 19719084.
40. Li, H., Leung, K.S., Ballester, P.J., Wong, M.H., istar: a web platform for large-scale protein-ligand docking. *PLoS One*, **9**, 1, e85678, 2014-01-24, Bibcode:2014PLoSO..985678L. doi:10.1371/journal.pone.0085678. PMC 3901662.

41. Cherkasov, A., Muratov, E.N., Fourches, D., Varnek, A., Baskin, I.I., Cronin, M., *et al.*, QSAR modeling: where have you been? Where are you going to? *J. Med. Chem.*, 57, 4977–5010, 2014, doi: 10.1021/jm4004285.
42. Ekins, S., Lage de Siqueira-Neto, J., McCall, L.-I., Sarker, M., Yadav, M., Ponder, E.L., *et al.*, Machine learning models and pathway genome data base for Trypanosoma cruzi drug discovery. *PLoS Negl. Trop. Dis.*, 9, e0003878, 2015, doi: 10.1371/journal.pntd.0003878.
43. Goh, G.B., Hodas, N.O., Vishnu, A., Deep learning for computational chemistry. *J. Comput. Chem.*, 38, 1291–1307, 2017, doi: 10.1002/jcc.24764.
44. Mitchell, J.B.O., Machine learning methods in chemoinformatics. *Wiley Interdiscip. Rev. Comput. Mol. Sci.*, 4, 468–481, 2014, doi: 10.1002/wcms.1183.
45. Lacivita, E., Di Pilato, P., De Giorgio, P., Colabufo, N.A., Berardi, F., Perrone, R., *et al.*, The therapeutic potential of 5-HT1A receptors: a patent review. *Expert Opin. Ther. Pat.*, 22, 887–902, 2012, doi: 10.1517/13543776.2012.703654.
46. Nichols, D.E. and Nichols, C.D., Serotonin receptors. *Chem. Rev.*, 108, 1614–1641, 2008, doi: 10.1021/cr078224o.
47. Luo, M., Wang, X.S., Roth, B.L., Golbraikh, A., Tropsha, A., Application of quantitative structure-activity relationship models of 5-HT1A receptor binding to virtual screening identifies novel and potent 5-HT1A ligands. *J. Chem. Inf. Model.*, 54, 634–647, 2014, doi: 10.1021/ci400460q.
48. Liu, X., Zhu, F., Ma, X.H., *et al.*, Predicting targeted polypharmacology for drug repositioning and multi-target drug discovery. *Curr. Med. Chem.*, 20, 13, 1646–1661, 2013.
49. Thai, K.M., Ngo, T.D., Tran, T.D., Le, M.T., Pharmacophore modeling for antitargets. *Curr. Top. Med. Chem.*, 13, 9, 1002–1014, 2013.
50. McGregor, M.J. and Muskal, S.M., Pharmacophore fingerprinting. 1. Application to QSAR and focused library design. *J. Chem. Inf. Comput. Sci.*, 39, 3, 569–574, 1999.
51. McGregor, M.J. and Muskal, S.M., Pharmacophore fingerprinting. 2. Application to primary library design. *J. Chem. Inf. Comput. Sci.*, 40, 1, 117–125, 2000.
52. Luu, T.T., Malcolm, N., Nadassy, K., Pharmacophore modeling methods in focused library selection – applications in the context of a new classification scheme. *Comb. Chem. High Throughput Screening*, 14, 6, 488–499, 2011.
53. Jose, R.A., Voet, A., Broos, K., *et al.*, An integrated fragment based screening approach for the discovery of small molecule modulators of the VWF-GPIbalpha interaction. *Chem. Commun. (Camb)*, 48, 92, 11349–11351, 2012.
54. Voet, A.R., Kumar, A., Berenger, F., Zhang, K.Y., Combining *in silico* and in cerebro approaches for virtual screening and pose prediction in SAMPL4. *J. Comput.-Aided Mol. Des.*, 28, 4, 363–373, 2014.
55. Sanders, M.P.A., McGuire, R., Roumen, L., *et al.*, From the protein's perspective: the benefits and challenges of protein structure-based pharmacophore modeling. *Med. Chem. Commun.*, 3, 28–38, 2012.

56. Wolber, G. and Langer, T., LigandScout: 3-D pharmacophores derived from protein-bound ligands and their use as virtual screening filters. *J. Chem. Inf. Model.*, 45, 1, 160–169, 2005.
57. Böhm, H.J., The computer program LUDI: a new method for the *de novo* design of enzyme inhibitors. *J. Comput.-Aided Mol. Des.*, 6, 1, 61–78, 1992.
58. Barillari, C., Marcou, G., Rognan, D., Hot-spots-guided receptor-based pharmacophores (HS-Pharm): a knowledge-based approach to identify ligand-anchoring atoms in protein cavities and prioritize structure-based pharmacophores. *J. Chem. Inf. Model.*, 48, 7, 1396–1410, 2008.
59. Goodford, P.J., A computational procedure for determining energetically favorable binding sites on biologically important macromolecules. *J. Med. Chem.*, 28, 7, 849–857, 1985.
60. Tintori, C., Corradi, V., Magnani, M., Manetti, F., Botta, M., Targets looking for drugs: a multistep computational protocol for the development of structure-based pharmacophores and their applications for hit discovery. *J. Chem. Inf. Model.*, 48, 11, 2166–2179, 2008.
61. Zorrilla, E.P. and Koob, G.F., Progress in corticotropin-releasing factor-1 antagonist development. *Drug Discov. Today*, 15, 9-10, 371–383, 2010.
62. Hartz, R.A., Ahuja, V.T., Schmitz, W.D., *et al.*, Synthesis and structure-activity relationships of N3-pyridylpyrazinones as corticotropin-releasing factor-1 (CRF1) receptor antagonists. *Bioorg. Med. Chem. Lett.*, 20, 6, 1890–1894, 2010.
63. Han, X., Pin, S.S., Burris, K., *et al.*, Synthesis and structureactivity relationship of imidazo[1,2-a] benzimidazoles as corticotropin-releasing factor 1 receptor antagonists. *Bioorg. Med. Chem. Lett.*, 15, 18, 4029–4032, 2005.
64. Hartz, R.A., Ahuja, V.T., Arvanitis, A.G., *et al.*, Synthesis, structure-activity relationships, and *in vivo* evaluation of N3- phenylpyrazinones as novel corticotropin-releasing factor-1 (CRF1) receptor antagonists. *J. Med. Chem.*, 52, 14, 4173–4191, 2009.
65. Guner, O.F., *Pharmacophore Perception, Development and Use in Drug Design*, International University Line, San Diego, Calif, USA, 2000.
66. Bao, L., Wang, S., Song, D., Wang, J., Cao, X., Ke, S., Synthesis and Bio-Evaluation of Natural Butenolides-Acrylate Conjugates. *Molecules*, 24, 7, 1304, 2019 Apr 3. doi: 10.3390/molecules24071304.
67. Pradeepkiran, J.A. and Reddy, P.H., Structure Based Design and Molecular Docking Studies for Phosphorylated Tau Inhibitors in Alzheimer's Disease. *Cells*, **8**, 3, 260, March 2019, doi:10.3390/cells8030260. PMC 6468864. PMID 30893872.
68. Jacob, T.C., Moss, S.J., Jurd, R., GABA(A) receptor trafficking and its role in the dynamic modulation of neuronal inhibition. *Nat. Rev. Neurosci.*, 9, 5, 331–343, 2008, doi: 10.1038/nrn2370.
69. Rudnick, G. and Wall, S.C., The molecular mechanism of "ecstasy" [3,4-methylenedioxymethamphetamine (MDMA)]: serotonin transporters are targets

for MDMA-induced serotonin release. *Proc. Natl. Acad. Sci. U.S.A.*, 89, 5, 1817–1821, 1992, doi: 10.1073/pnas.89.5.1817.
70. Bale, T.L. and Vale, W.W., CRF and CRF receptors: role in stress responsivity and other behaviors. *Annu. Rev. Pharmacol. Toxicol.*, 44, 525–557, 2004, doi: 10.1146/annurev.pharmtox.44.101802.121410.
71. Howlett, A.C., Barth, F., Bonner, T.I., *et al.*, International Union of Pharmacology. XXVII. Classification of cannabinoid receptors. *Pharmacol. Rev.*, 54, 2, 161–202, 2002, doi: 10.1124/pr.54.2.161.
72. Roy, K., Kar, S., Das, R.N., *Understanding the basics of QSAR for applications in pharmaceutical sciences and risk assessment*, Academic press, Cambridge, Massachusetts, 2015.
73. Xu, X., Gao, B., Li, X., Lei, S., Study on the Antianxiety Mechanism of Suanzaoren Decoction Based on Network Pharmacology and Molecular Docking. *Evidence-Based Complementary Altern. Med.*, 2021, 1–28, 2021 Jul 31.
74. Welch, V., Wy, T.J., Ligezka, A., Hassett, L.C., Croarkin, P.E., Athreya, A.P., Romanowicz, M., Use of mobile and wearable artificial intelligence in child and adolescent psychiatry: scoping review. *J. Med. Internet Res.*, 24, 3, e33560, 2022 Mar 14, doi: 10.2196/33560. https://www.jmir.org/2022/3/e33560/v24i3e33560 [PMC free article] [PubMed] [CrossRef] [Google Scholar].
75. Elgendi, M. and Menon, C., Assessing anxiety disorders using wearable devices: challenges and future directions. *Brain Sci.*, 9, 3, 50, 2019 Mar 01, doi: 10.3390/brainsci9030050. https://www.mdpi.com/resolver?pii=brainsci9030050 .brainsci9030050 [PMC free article] [PubMed] [CrossRef] [Google Scholar].
76. Sequeira, L., Perrotta, S., LaGrassa, J., Merikangas, K., Kreindler, D., Kundur, D., Courtney, D., Szatmari, P., Battaglia, M., Strauss, J., Mobile and wearable technology for monitoring depressive symptoms in children and adolescents: a scoping review. *J. Affect. Disord.*, 265, 314–24, 2020 Mar 15, doi: 10.1016/j.jad.2019.11.156.S0165-0327(19)31030-4.
77. Abd-Alrazaq, A., AlSaad, R., Aziz, S., Ahmed, A., Denecke, K., Househ, M., Farooq, F., Sheikh, J., Wearable Artificial Intelligence for Anxiety and Depression: Scoping Review. *J. Med. Internet Res.*, 25, e42672, 2023 Jan 19, doi: 10.2196/42672. Erratum in: J Med Internet Res. 2023 Feb 7;25:e46233. PMID: 36656625; PMCID: PMC9896355.
78. Bartlett, C. and Bichindaritz, I., Machine learning based classification of mental disorders from methylation data, in: *Innovation in Medicine and Healthcare Systems, and Multimedia*, pp. 355–365, Springer, Berlin, 2019.
79. Sinha, R. and Jastreboff, A.M., Stress as a common risk factor for obesity and addiction. *Biol. Psychiatry*, 73, 827–835, 2013.
80. Tyrka, A., Parade, S., Welch, E., *et al.*, Methylation of the leukocyte glucocorticoid receptor gene promoter in adults: associations with early adversity and depressive, anxiety and substance-use disorders. *Transl. Psychiatry*, 6, e848, 2016.

81. Klengel, T., Pape, J., Binder, E.B., *et al.*, The role of DNA methylation in stress-related psychiatric disorders. *Neuropharmacology*, 80, 115–132, 2014.
82. Hamilton, M., The assessment of anxiety states by rating. *Br. J. Med. Psychol.*, 32, 50–55, 1959.
83. Zigmond, A.S. and Snaith, R.P., The hospital anxiety and depression scale. *Acta Psychiatr. Scand.*, 67, 361–370, 1983.
84. Beck, A.T., Epstein, N., Brown, G., *et al.*, An inventory for measuring clinical anxiety: Psychometric properties. *J. Consult. Clin. Psychol.*, 56, 893, 1988.
85. Pincus, T., de C Williams, A.C., Vogel, S., *et al.*, The development and testing of the depression, anxiety, and positive outlook scale (DAPOS). *Pain*, 109, 181–188, 2004.
86. Shear, K., Belnap, B.H., Mazumdar, S., *et al.*, Generalized anxiety disorder severity scale (GADSS): A preliminary validation study. *Depress. Anxiety*, 23, 77–82, 2006.
87. Rodebaugh, T.L., Heimberg, R.G., Brown, P.J., *et al.*, More reasons to be straightforward: Findings and norms for two scales relevant to social anxiety. *J. Anxiety Disord.*, 25, 623–630, 2011.
88. Mattick, R.P. and Clarke, J.C., Development and validation of measures of social phobia scrutiny fear and social interaction anxiety. *Behav. Res. Ther.*, 36, 455–470, 1998.
89. AP Association, *Diagnostic and statistical manual of mental disorders (DSM-5)*, American Psychiatric Pub, Washington, 2013.
90. Weathers, F.W., Bovin, M.J., Lee, D.J., *et al.*, The clinician-administered ptsd scale for DSM-5 (CAPS- 5): Development and initial psychometric evaluation in military veterans. *Psychol. Assess.*, 30, 383, 2018.
91. Cuthbert, B.N., The NIMH research domain criteria project: Toward an integrated neuroscience of mental disorders, in: *Genomics, circuits, and pathways in clinical neuropsychiatry*, T. Lehner, B.L. Miller, M.W. State (Eds.), pp. 397–409, Elsevier, Netherland, 2016.
92. Kozak, M.J. and Cuthbert, B.N., The NIMH research domain criteria initiative: Back- ground, issues, and pragmatics. *Psychophysiology*, 53, 286–297, 2016.
93. Simmons, J. and Quinn, K., The NIMH research domain criteria (RDoC) project: Implications for genetics research. *Mamm. Genome*, 25, 23–31, 2014.
94. Fellowes, S., Symptom modelling can be influenced by psychiatric categories: Choices for research domain criteria (RDoC). *Theor. Med. Bioeth.*, 38, 279–294, 2017.
95. Chandrashekar, G. and Sahin, F., A survey on feature selection methods. *Comput. Electr. Eng.*, 40, 16–28, 2014.
96. Bolón-Canedo, V., Sánchez-Maroño, N., Alonso-Betanzos, A., A review of feature selection methods on synthetic data. *Knowl. Inf. Syst.*, 34, 483–519, 2013.
97. Sheikhpour, R., Sarram, M.A., Gharaghani, S., *et al.*, A survey on semi-supervised feature selection methods. *Pattern Recognit.*, 64, 141–158, 2017.
98. Rokach, L., *Ensemble Learning: Pattern classification using ensemble methods*, World Scientific Publishing Co Pvt Ltd, Singapore, 2019.

99. Kessler, R.C., van Loo, H.M., Wardenaar, K.J., *et al.*, Testing a machine-learning algorithm to predict the persistence and severity of major depressive disorder from baseline self-reports. *Mol. Psychiatry*, 21, 1366–1371, 2016.
100. Dwyer, D.B., Falkai, P., Koutsouleris, N., Machine learning approaches for clinical psychology and psychiatry. *Annu. Rev. Clin. Psychol.*, 14, 91–118, 2018.
101. Panagiotakopoulos, T.C., Lyras, D.P., Livaditis, M., *et al.*, A contextual data mining approach toward assisting the treatment of anxiety disorders. *IEEE Trans. Inf. Technol. Biomed.*, 14, 567–581, 2010.
102. Robichaud, M., Koerner, N., Dugas, M.J., *Cognitive behavioral treatment for generalized anxiety disorder: From science to practice*, Routledge, NY, 2019.
103. Sribala, M. and Sribala, M., An approach of artificial neural networks for prediction of generalized anxiety disorder. *Int. J. Res. Comput. Appl. Rob.*, 3, 118–124, 2015.
104. Spitzer, R.L., Gibbon, M.E., Skodol, A.E., *et al.*, DSM-IV casebook: A learning companion to the diagnostic and statistical manual of mental disorders, 4th edn, American Psychiatric Association, Washington, 1994.
105. Husain, W., Xin, L.K., Jothi, N., *et al.*, Predicting generalized anxiety disorder among women using random forest approach, in: *2016 3rd International Conference on Computer and Information Sciences (ICCOINS)*, Malaysia, pp. 37–42, IEEE, 2016.
106. Richter, P., Werner, J., Heerlein, A., *et al.*, On the validity of the beck depression inventory. *Psychopathology*, 31, 160–168, 1998.
107. Reece, A.G., Reagan, A.J., Lix, K.L., *et al.*, Forecasting the onset and course of mental illness with twitter data. *Sci. Rep.*, 7, 1–11, 2017.
108. Kanwar, A., Malik, S., Prokop, L.J., *et al.*, The association between anxiety disorders and suicidal behaviors: A systematic review and meta-analysis. *Depress. Anxiety*, 30, 917–929, 2013.
109. De La Vega, D., Giner, L., Courtet, P., Suicidality in subjects with anxiety or obsessive-compulsive and related disorders: recent advances. *Curr. Psychiatry Rep.*, 20, 26, 2018.
110. Stanley, I.H., Boffa, J.W., Rogers, M.L., *et al.*, Anxiety sensitivity and suicidal ideation/suicide risk: A meta-analysis. *J. Consult. Clin. Psychol.*, 86, 946, 2018.
111. *Suicide*, World Health Organization, 2019.
112. Crawford, J.R. and Henry, J.D., The depression anxiety stress scales (DASS): Normative data and latent structure in a large non-clinical sample. *Br. J. Clin. Psychol.*, 42, 111–131, 2003.
113. Taiminen, T., Huttunen, J., Heila, H., *et al.*, The schizophrenia suicide risk scale (SSRS): Development and initial validation. *Schizophr. Res.*, 47, 199–213, 2001.
114. Kliem, S., Lohmann, A., Mößle, T., *et al.*, German beck scale for suicide ideation (BSS): Psychometric properties from a representative population survey. *BMC Psychiatry*, 17, 389, 2017.
115. Durrant, J.D. and McCammon, J.A., Molecular dynamics simulations and drug discovery. *BMC Biol.*, 9, 1, 1–9, 2011.

116. Ma, B., Kumar, S., Tsai, C.-J., Nussinov, R., Folding funnels and binding mechanisms. *Protein Eng.*, 12, 9, 713–720, 1999.
117. Ma, B., Shatsky, M., Wolfson, H.J., Nussinov, R., Multiple diverse ligands binding at a single protein site: a matter of pre-existing populations. *Protein Sci.*, 11, 2, 184–197, 2002.
118. Fischer, E., Einfluss der Configuration auf die Wirkung der Enzyme. *Ber. Dtsch. Chem. Ges.*, 27, 3, 2985–2993, 1894.
119. Teague, S.J., Implications of protein flexibility for drug discovery. *Nat. Rev. Drug Discov.*, 2, 7, 527–541 22, 2003.
120. Tsai, C.J., Kumar, S., Ma, B., Nussinov, R., Folding funnels, binding funnels, and protein function. *Protein Sci.*, 8, 6, 1181–1190, 1999.
121. Feynman, R.P., Leighton, R.B., Sands, M., *The Feynman lectures on physics, Vol. I: The new millennium edition: mainly mechanics, radiation, and heat*, vol. 1, Basic books, New York, 2011.
122. Hollingsworth, S.A. and Dror, R.O., Molecular dynamics simulation for all. *Neuron*, 99, 6, 1129–1143, 2018.
123. Ishida, H. and Agag, T., *Handbook of benzoxazine resins*, Elsevier, Netherlands, 2011.
124. Uba, A.I. and Wu, C., Identification of potential antagonists of CRF1R for possible treatment of stress and anxiety neuro-disorders using structure-based virtual screening and molecular dynamics simulation. *Comput. Biol. Chem.*, 100, 107743, 2022.
125. https://www.toppers4u.com/2021/01/network-analysis-objectives-advantages.html
126. Newman, M.E.J., *Networks: An Introduction*, Oxford University Press, Oxford, 2010, [Google Scholar].
127. Cramer, A.O.J., Waldorp, L.J., van der Maas, H.L.J., Borsboom, D., Comorbidity: a network perspective. *Behav. Brain Sci.*, 33, 2–3, 137–50, 2010, 10.1017/S0140525X09991567 [PubMed] [CrossRef] [Google Scholar].
128. Borsboom, D., Cramer, A.O.J., Schmittmann, V.D., Epskamp, S., Waldorp, L.J., The small world of psychopathology. *PLoS One*, 6, 11, e27407, 2011, 10.1371/journal.pone.0027407 [PMC free article] [PubMed] [CrossRef] [Google Scholar].
129. Borsboom, D. and Cramer, A.O.J., Network analysis: an integrative approach to the structure of psychopathology. *Annu. Rev. Clin. Psychol.*, 9, 91–121, 2013, 10.1146/annurev-clinpsy-050212-185608 [PubMed] [CrossRef] [Google Scholar].
130. Borsboom, D., Mellenbergh, G.J., van Heerden, J., The theoretical status of latent variables. *Psychol. Rev.*, 110, 203–19, 2003, 10.1037/0033-295X.110.2.203 [PubMed] [CrossRef] [Google Scholar].
131. Cramer, A.O.J., van der Sluis, S., Noordhof, A., Wichers, M., Geschwind, N., Aggen, S.H., *et al.*, Dimensions of normal personality as networks in search of equilibrium: you can't like parties if you don't like people. *Eur. J. Pers.*, 26, 4, 414–31, 2012, 10.1002/per.1866.

132. Borsboom, D., Psychometric perspectives on diagnostic systems. *J. Clin. Psychol.*, 64, 1089–108, 2008, 10.1002/jclp.20503 [PubMed] [CrossRef] [Google Scholar].
133. Kendler, K.S., Zachar, P., Craver, C., What kinds of things are psychiatric disorders? *Psychol. Med.*, 41, 1143–50, 2011, 10.1017/S0033291710001844 [PubMed] [CrossRef] [Google Scholar].
134. van de Leemput, I.A., Wichers, M., Cramer, A.O.J., Borsboom, D., Tuerlinckx, F., Kuppens, P., *et al.*, Critical slowing down as early warning for the onset and termination of depression. *Proc. Natl. Acad. Sci. U.S.A.*, 111, 1, 87–92, 2014, 10.1073/pnas.1312114110 [PMC free article] [PubMed] [CrossRef] [Google Scholar].
135. Bringmann, L.F., Vissers, N., Wichers, M., Geschwind, N., Kuppens, P., Peeters, F., *et al.*, A network approach to psychopathology: new insights into clinical longitudinal data. *PLoS One*, 8, 4, e60188, 2013, 10.1371/journal.pone.0060188 [PMC free article] [PubMed] [CrossRef] [Google Scholar].
136. Bringmann, L.F., Lemmens, L.H.J.M., Huibers, M.J.H., Borsboom, D., Tuerlinckx, F., Revealing the dynamic network structure of the Beck Depression Inventory-II. *Psychol. Med.*, 45, 4, 747–57, 2015, 10.1017/S0033291714001809 [PubMed] [CrossRef] [Google Scholar].
137. Cramer, A.O.J., Borsboom, D., Aggen, S.H., Kendler, K.S., The pathoplasticity of dysphoric episodes: differential impact of stressful life events on the pattern of depressive symptom inter-correlations. *Psychol. Med.*, 42, 5, 957–65, 2012, 10.1017/S003329171100211X [PMC free article] [PubMed] [CrossRef] [Google Scholar].
138. Heeren, A. and McNally, R.J., An integrative network approach to social anxiety disorder: the complex dynamic interplay among attentional bias for threat, attentional control, and symptoms. *J. Anxiety Disord.*, 42, 95–104, 2016, 10.1016/j.janxdis.2016.06.009 [PubMed] [CrossRef] [Google Scholar].
139. Heeren, A., De Raedt, R., Koster, E.H.W., Philippot, P., The (neuro)cognitive mechanisms behind attention bias modification in anxiety: proposals based on theoretical accounts of attentional bias. *Front. Hum. Neurosci.*, 7, 119, 2013, 10.3389/fnhum.2013.00119 [PMC free article] [PubMed] [CrossRef] [Google Scholar].
140. Raymond, J.G., Steele, J.D., Seriès, P., Modeling Trait Anxiety: From Computational Processes to Personality. *Front. Psychiatry*, 8, 1, 2017 Jan 23, doi: 10.3389/fpsyt.2017.00001. PMID: 28167920; PMCID: PMC5253387.
141. Katara, P., Role of bioinformatics and pharmacogenomics in drug discovery and development process. *Netw. Model. Anal. Health Inf. Bioinf.*, 2, 225–230, 2013.
142. Wright-Berryman, J., Cohen, J., Haq, A., Black, D.P., Pease, J.L., Virtually screening adults for depression, anxiety, and suicide risk using machine learning and language from an open-ended interview. *Front. Psychiatry*, 14, 1143175, 2023.
143. Mohamed, G.A., Omar, A.M., AlKharboush, D.F., Fallatah, M.A., Sindi, I.A., El-Agamy, D.S., Ibrahim, S.R., Structure-Based Virtual Screening and

Molecular Dynamics Simulation Assessments of Depsidones as Possible Selective Cannabinoid Receptor Type 2 Agonists. *Molecules*, 28, 4, 1761, 2023.

144. Amin, F., Ibrahim, M.A., Rizwan-ul-Hasan, S., Khaliq, S., Gabr, G.A., Muhammad, Abidi, S.H., Interactions of apigenin and safranal with the 5HT1A and 5HT2A receptors and behavioral effects in depression and anxiety: a molecular docking, lipid-mediated molecular dynamics, and *in vivo* analysis. *Molecules*, 27, 24, 8658, 2022.
145. Contreras, N., Alviz-Amador, A., Manzur-Villalobos, I., *In silico* study of dimethyltryptamine analogues against 5-HT1B receptor: Molecular docking, dynamic simulations and ADMET prediction. *J. Herbmed Pharmacol.*, 11, 2, 204–212, 2022.
146. Dashyan, S.S., Babaev, E.V., Paronikyan, E.G., Ayvazyan, A.G., Paronikyan, R.G., Hunanyan, L.S., Evaluation of Neurotropic Activity and Molecular Docking Study of New Derivatives of pyrano [4 ″, 3 ″: 4′, 5′] pyrido [3′, 2′: 4, 5] thieno [3, 2-d] pyrimidines on the Basis of pyrano [3, 4-c] pyridines. *Molecules*, 27, 11, 3380, 2022.
147. Ankali, K.N., Rangaswamy, J., Shalavadi, M., Naik, N., naik Krishnamurthy, G., Synthesis and molecular docking of novel 1, 3-thiazole derived 1, 2, 3-triazoles and *in vivo* biological evaluation for their anti anxiety and anti inflammatory activity. *J. Mol. Struct., 1236*, 130357, 2021.
148. Bai, W., Xi, H.T., Zhu, Q., Ji, M., Zhang, H., Yang, B.X., Cai, H., Liu, R., Zhao, Y.J., Chen, L., Ge, Z.M., Wang, Z., Han, L., Chen, P., Liu, S., Cheung, T., Tang, Y.L., Jackson, T., An, F., Xiang, Y.T., Network analysis of anxiety and depressive symptoms among nursing students during the COVID-19 pandemic. *J. Affect. Disord.*, 294, 753–760, 2021 Nov 1, doi: 10.1016/j.jad.2021.07.072. Epub 2021 Jul 22. Erratum in: J Affect Disord. 2023 May 1;328:355. PMID: 34375200; PMCID: PMC8433813.
149. Kaur, P., Sharma, V., Kumar, V., Pharmacophore Modelling and 3D-QSAR Studies on N3-Phenylpyrazinones as Corticotropin-Releasing Factor 1 Receptor Antagonists. *Int. J. Med. Chem., 2012*, 2012.
150. Kothgassner, O.D., Goreis, A., Bauda, I., Ziegenaus, A., Glenk, L.M., Felnhofer, A., Virtual reality biofeedback interventions for treating anxiety: A systematic review, meta-analysis and future perspective. *Wien. Klin. Wochenschr.*, 1–11, 2022.
151. Balaskas, A., Schueller, S.M., Cox, A.L., Doherty, G., The functionality of mobile apps for anxiety: systematic search and analysis of engagement and tailoring features. *JMIR mHealth uHealth*, 9, 10, e26712, 2021.
152. Haque, M.R. and Rubya, S., An Overview of Chatbot-Based Mobile Mental Health Apps: Insights From App Description and User Reviews. *JMIR mHealth uHealth*, 11, 1, e44838.w, 2023.
153. Xie, W., Wang, C., Lin, Z., Luo, X., Chen, W., Xu, M., Cheng, M., Multimodal fusion diagnosis of depression and anxiety based on CNN-LSTM model. *Comput. Med. Imaging Graphics*, 102, 102128, 2022.

Index

Printed and bound by CPI Group (UK) Ltd, Croydon, CR0 4YY
17/09/2026
14964261-0008